## HARE BRAIN, TORTOISE MIND

Guy Claxton received his D.Phil. in Psychology from the University of Oxford and has developed, over the ensuing twenty years, an international reputation for his work on learning and the mind. His twelve previous books include *Wholly Human* (1981), *Educating the Inquiring Mind* (1991) and *Noises from the Darkroom: The Science and Mystery of the Mind* (1994). He has contributed to journals and magazines ranging from the *Times Educational Supplement* and *New Scientist* to *Arena*. Guy Claxton is presently Visiting Professor of Psychology and Education at the University of Bristol.

# *Hare Brain*
# Tortoise Mind

## Why Intelligence Increases
## When You Think Less

*Guy Claxton*

FOURTH ESTATE · *London*

For Jo

This paperback edition published in 1998

First published in Great Britain in 1997 by
Fourth Estate Limited
6 Salem Road
London W2 4BU

3 5 7 9 10 8 6 4 2

A catalogue record for this book is available from the British Library

ISBN 1-85702-709-4

Text design by Richard Kelly
Text figures drawn by Carole Vincer
Typeset by Rowland Phototypesetting Ltd,
Bury St Edmunds, Suffolk
Printed in Great Britain by
Clays Ltd, St Ives, plc

Everything is gestation and bringing forth. To let each impression and each germ of a feeling come to completion wholly in itself, in the dark, in the inexpressible, the unconscious, beyond the reach of one's own intelligence, and await with deep humility and patience the birth-hour of a new clarity: that alone is living the artist's life. Being an artist means not reckoning and counting, but ripening like the tree which does not force its sap, and stands confident in the storms of spring without the fear that after them may come no summer. It does come. But it comes only to the patient, who are there as though eternity lay before them, so unconcernedly still and wide.

*Rainer Maria Rilke*

# Contents

*Acknowledgements*                                                viii

*List of Figures*                                                  xi

 1  The Speed of Thought                                            1

 2  Basic Intelligence: Learning by Osmosis                        15

 3  Premature Articulation: How Thinking Gets in
    the Way of Learning                                            28

 4  Knowing More than We Think: Intuition and Creativity           48

 5  Having an Idea: the Gentle Art of Mental Gestation             68

 6  Thinking Too Much? Reason and Intuition
    as Antagonists and Allies                                      85

 7  Perception without Consciousness                              100

 8  Self-Consciousness                                            116

 9  The Brains behind the Operation                               133

10  The Point of Consciousness                                    148

11  Paying Attention                                              164

12  The Rudiments of Wisdom                                       188

13  The Undermind Society: Putting the Tortoise to Work           201

*Notes*                                                          227

*Index*                                                          249

# Acknowledgements

There are many people who have supported, encouraged and guided me throughout the long gestation of this book, and to whom thanks are due. They include Stephen Batchelor (for duck dinners), Mark Brown (for his enthusiasm), Merophie Carr, Polly Carr, Isabelle Gall, Rod Jenkinson (who wanted it to be *The Sin of Certainty*), Kikan Massara, Helen and Colin Moore (for their room), and my mother, Ruby Claxton (for her love and for not interrupting). Among those who offered me scholarly advice and generously shared their time and knowledge were Peter Abbs, Maurice Ash, Brian Bates, Susan Blackmore, Alan Bleakley, Laurinda Brown, Fritjof Capra, Martin Conway, Peter Fenwick, Brian Goodwin, Susan Greenfield, Valerie Hall, Jane Henry, Tony Marcel, Richard Morris, Brian Nicholson, Dick Passingham, Mark Price, Robin Skynner, John Teasdale, Francisco Varela, Max Velmans and Larry Weiskrantz. Special thanks to Margaret Carr for wonderful conversations, and to her and her husband Malcolm for their friendship, and the use, yet again, of their muse-filled beach-house in Raglan, New Zealand. Michelle Macdonald, Steven Smith and Christopher Titmuss helped me to practise the art of thinking slowly. And Christopher Potter and Emma Rhind-Tutt of Fourth Estate believed in the book enough to chivvy me into improving it. Emma's love of language and her persistent refusal to let me get away with sloppiness of mind or prose helped enormously to shape the book for the better. Such inaccuracies and infelicities as remain are, of course, down to me.

I am grateful to the following authors and publishers for permission to reproduce quotations and illustrations.

Academic Press Inc., and Professor Patricia Bowers (executor of the estate of the late Professor Kenneth S. Bowers), for two panels from figure 2, p83 in 'Intuition in the context of discovery', by K.S. Bowers, G. Regehr, C. Balthazard and K. Parker, reprinted with kind permission from *Cognitive Psychology*, vol 22, pp72–110, © 1990 Academic Press.

American Psychological Association and Professor J. Schooler for illustrations from the appendix (p182) to J.W. Schooler, S. Ohlsson and K. Brooks (1993) 'Thoughts beyond words: when language overshadows insight', *Journal of Experimental Psychology: General*, vol 122, pp166–183; © 1993 by the American Psychological Association. Reprinted with permission.

The British Psychological Society for extracts from the report 'Fostering Innovation: A Psychological Perspective' by M.A. West, C. Fletcher and J. Toplis, March 1994.

Cambridge University Press for extracts from A.E. Housman, *The Name and Nature of Poetry*.

Elsevier Science Ltd for the extract from 'How does cognitive therapy prevent depressive relapse and why should attentional control (mindfulness) training help?' by John Teasdale, Zindel Segal and Mark Williams, reprinted from *Behavior Research and Therapy*, 1995, vol 33, pp25–39. © 1995, with kind permission from Elsevier Science Ltd, The Boulevard, Langford Lane, Kidlington, OX5 1GB, UK.

Faber and Faber Ltd. for the extract from Ted Hughes, *Poetry in the Making*, published by Faber and Faber Ltd., London 1967. © Ted Hughes 1967, reprinted with kind permission.

HarperCollins Publishers for extracts from *Discourse on Thinking*, translated by John M. Anderson and E. Hans Freund by Martin Heidegger. Copyright © 1959 by Verlag Gunther Neske. Copyright © in the English Translation by Harper & Row, Publishers Inc. Reprinted by permission of HarperCollins Publishers, Inc. Excerpts from *Women's Ways of Knowing* by Mary Field Belenky et al. Copyright © 1986 by Basic Books, Inc. Reprinted by permission of BasicBooks, a division of HarperCollins Publishers, Inc.

The MIT Press for the extract from P.S. Churchland. *Neurophilosophy*, MIT Press, Cambridge, MA. © 1986 Patricia S. Churchland.

Oxford University Press for the extract from Arthur Reber, *Implicit Learning and Tacit Knowledge: An Essay on the Cognitive Unconscious*, OUP, Oxford, 1993.

Princeton University Press for the extract from D.T. Suzuki, *Zen and Japanese Culture*, Princeton University Press, 1959; and for the extract from J. Hademard, *The Psychology of Invention in the Mathematical Field*, Princeton University Press, 1945.

Random House Inc. for the excerpt from F. Scott Fitzgerald, *Tender is the Night*, published by The Bodley Head; and for the excerpts from Tom Peters, *The Pursuit of Wow!*, published by Vintage Books.

Scientific American Inc., New York, for permission to reproduce the schematic representation of a neuron first published in 'The chemistry of the brain' by Leslie L. Iversen, in *Scientific American*, September, 1979, and later in *The Brain: A Scientific American Book*.

Simon and Schuster Inc. for the extract from Richard Selzer, *Mortal Lessons*, © 1974, 1975, 1976.

John Wiley and Sons Ltd. for the extract from Nicholas Humphrey, in discussion of Chapter by John Kihlstrom, in CIBA Symposium 174, *Experimental and Theoretical Studies of Consciousness*, Wiley, Chichester, 1993.

# *Figures*

1. Sample grid of numbers used in the
   Lewicki experiments     24

2. The Rubik cube     29

3. The mutilated chessboard     36

4. The polar planimeter     40

5. Bowers' degraded images     64

6. Insight problems     89

7. What does the doctor reply?     111

8. A stylised neuron     136

9. A simple neural net for distinguishing rocks
   from mines     142

10. A map of neural pathways     152

11. Brainscape and wordscape     154

12. Illusory shapes and contours, after Kanizsa (1979)     181

13. The solutions to the insight problems in Figure 6     236

## CHAPTER 1

# The Speed of Thought

Turtle buries its thoughts, like its eggs, in the sand, and allows the sun to hatch the little ones. Look at the old fable of the tortoise and the hare, and decide for yourself whether or not you would like to align with Turtle.

*Native American Medicine Cards*

There is an old Polish saying, 'Sleep faster; we need the pillows', which reminds us that there are some activities which just will not be rushed. They take the time that they take. If you are late for a meeting, you can hurry. If the roast potatoes are slow to brown, you can turn up the oven. But if you try to speed up the baking of meringues, they burn. If you are impatient with the mayonnaise and add the oil too quickly, it curdles. If you start tugging with frustration on a tangled fishing line, the knot just becomes tighter.

The mind, too, works at different speeds. Some of its functions are performed at lightning speed; others take seconds, minutes, hours, days or even years to complete their course. Some can be speeded up – we can become quicker at solving crossword puzzles or doing mental arithmetic. But others cannot be rushed, and if they are, then they will break down, like the mayonnaise, or get tangled up, like the fishing line. 'Think fast; we need the results' may sometimes be as absurd a notion, or at least as counterproductive, as the attempt to cram a night's rest into half the time. We learn, think and know in a variety of different ways, and these modes of the mind operate at different speeds, and are good for different mental jobs. 'He who hesitates is lost', says one proverb. 'Look before you leap', says another. And both are true.

Roughly speaking, the mind possesses three different processing speeds. The first is faster than thought. Some situations demand an unselfconscious, instantaneous reaction. When my motor-bike skidded on a wet manhole cover in London some years ago, my brain and my body immediately choreographed for me an intricate and effective set of movements that enabled me to keep my seat – and it was only after the action was all over that my conscious mind and my emotions started to catch up. Neither a concert pianist nor an Olympic fencer has time to figure out what to do next. There is a kind of 'intelligence' that works more rapidly than thinking. This mode of fast, physical intelligence could be called our 'wits'. (The five senses were originally known as 'the five wits'.)

Then there is thought itself: the sort of intelligence which does involve figuring matters out, weighing up the pros and cons, constructing arguments and solving problems. A mechanic working out why an engine will not fire, a family arguing over the brochures about where to go for next summer's holiday, a scientist trying to interpret an intriguing experimental result, a student wrestling with an examination question: all are employing a way of knowing that relies on reason and logic, on deliberate conscious thinking. We often call this kind of intelligence 'intellect' – though to make the idea more precise, I shall call it *d-mode*, where the 'd' stands for 'deliberation'. Someone who is good at solving these sorts of problems we call 'bright' or 'clever'.

But below this, there is another mental register that proceeds more slowly still. It is often less purposeful and clear-cut, more playful, leisurely or dreamy. In this mode we are ruminating or mulling things over; being contemplative or meditative. We may be pondering a problem, rather than earnestly trying to solve it, or just idly watching the world go by. What is going on in the mind may be quite fragmentary. What we are thinking may not make sense. We may even not be aware of much at all. As the English yokel is reported to have said: 'sometimes I sits and thinks, but mostly I just sits'. Perched on a seaside rock, lost in the sound and the motion of the surf, or hovering just on the brink of sleep or waking, we are in a different mental mode from the one we find ourselves in as we plan a meal or dictate a letter. These leisurely, apparently aimless, ways of knowing and experiencing are just as 'intelligent' as the other, faster ones. Allowing the mind time to meander is not a luxury that can safely be cut back as life or work gets more demanding. On the contrary, thinking slowly is a vital part of the cognitive armamentarium. We need the tortoise mind just as much as we need the hare brain.

Some kinds of everyday predicament are better, more effectively approached with a slow mind. Some mysteries can *only* be penetrated with a relaxed, unquesting mental attitude. Some kinds of understanding simply refuse to come when they are called. As the *Tao Te Ching* puts it:

> Truth waits for eyes unclouded by longing.
> Those who are bound by desire see only the outward
> container.

Recent scientific evidence shows convincingly that the more patient, less deliberate modes of mind are particularly suited to making sense of situations that are intricate, shadowy or ill defined. Deliberate thinking, d-mode, works well when the problem it is facing is easily conceptualised. When we are trying to decide where to spend our holidays, it may well be perfectly obvious what the parameters are: how much we can afford, when we can get away, what kinds of things we enjoy doing, and so on. But when we are not sure what needs to be taken into account, or even which questions to pose – or when the issue is too subtle to be captured by the familiar categories of conscious thought – we need recourse to the tortoise mind. If the problem is not whether to go to Turkey or Greece, but how best to manage a difficult group of people at work, or whether to give up being a manager completely and retrain as a teacher, we may be better advised to sit quietly and ponder than to search frantically for explanations and solutions. This third type of intelligence is associated with what we call creativity, or even 'wisdom'.

Poets have always known the limitations of conscious, deliberate thinking, and have sought to cultivate these slower, mistier ways of knowing. Philosophers from Spinoza and Leibniz to Martin Heidegger and Suzanne Langer have written about the realms of mind that lie beyond and beneath the conscious intellect. Psychotherapists know that 'the unconscious' is not just a source of personal difficulties; a revised *relationship* with one's unconscious is also part of the 'cure'. And the sages and mystics of all religious traditions attest to the spontaneous transformation of experience that occurs when one embraces the 'impersonal mystery' at the core of mental life – whether this mystery be the 'godhead' of Meister Eckhart or the 'Unborn' of Zen master Bankei. Even scientists themselves, or at least the most creative of them, admit that their genius comes to them from layers of mind over which they have little or no control (and they may even feel somehow fraudulent for taking personal credit for insights that simply 'occurred to them').[1]

It is only recently, however, that scientists have started to explore the slower, less deliberate ways of knowing directly. The newly formed hybrid discipline of 'cognitive science', an alliance of neuroscience, philosophy, artificial intelligence and experimental psychology, is revealing that the unconscious realms of the human mind will successfully accomplish a number of unusual, interesting and important tasks *if they are given the time*. They will learn patterns of a degree of subtlety which normal consciousness cannot even see; make sense out of situations that are too complex to analyse; and get to the bottom of certain difficult issues much more successfully than the questing intellect. They will detect and respond to meanings, in poetry and art, as well as in relationships, that cannot be clearly articulated.

One of my main aims in writing this book is to bring this fascinating research to a wider audience, for it offers a profound and salutary challenge to our everyday view of our own minds and how they work. These empirical demonstrations are more than interesting: they are important. For my argument is not just that the slow ways of knowing exist, and are useful. It is that our culture has come to ignore and undervalue them, to treat them as marginal or merely recreational, and in so doing has foreclosed on areas of our psychological resources that we need. Just like the computer, the Western mind has come to adopt as its 'default mode' just one of its possible modes of knowing: d-mode. (The 'd' can stand for 'default' as well as 'deliberation'.)

The individuals and societies of the West have rather lost touch with the value of contemplation. Only active thinking is regarded as productive. Sitting gazing absently at your office wall or out of the classroom window is not of value. Yet many of those whom our society admires as icons of creativity and wisdom have spent much of their time doing nothing. Einstein, it is said, would frequently be found in his office at Princeton staring into space. The Dalai Lama spends hours each day in meditation. Even that paragon of penetrating insight, Sherlock Holmes, is described by his creator as entering a meditative state 'with a dreamy vacant expression in his eyes'.

There are a number of reasons why slow knowing has fallen into disuse. Partly it is due to our changing conception of, and attitude towards, time. In pre-seventeenth-century Europe a leisurely approach to thinking was much more common, and in other cultures it still is. A tribal meeting at a Maori *marae* can last for days, until everyone has had time to assimilate the issues, to have their say, and to form a consensus. However, the idea that time is plentiful is in many parts of the world now seen as laughably old-fashioned and self-indulgent.

Swedish anthropologist Helena Norberg-Hodge has documented

the way in which the introduction of Western culture has radically altered the pace of life in the traditional society of Ladakh, for example.[2] Until ten years ago, a Ladakhi wedding lasted a fortnight. But their lifestyle rapidly altered following the introduction of some simple 'labour-saving' changes: tools, such as the Rotovator, to make ploughing quicker and easier; and some new crops and livestock, such as dairy cows. Compared to the traditional yak, cows yield more milk than a family needs, creating a surplus which can be turned into cheese and sold to bring in some extra cash. While there is no harm in making life a little easier, in encouraging families to accumulate a little 'wealth', unfortunately this apparently benign 'aid package' also gave the Ladakhis a new view of time – as something in short supply. Instead of the Rotovators and the cows generating *more* leisure, they have in fact reduced it. People are now busier than they were: busy creating wealth – and 'saving time'. Today a Ladakhi wedding lasts less than a day, just like an English one. Within the Western mindset, time becomes a commodity, and one inevitable consequence is the urge to 'think faster': to solve problems and make decisions quickly.

Partly the decline of slow thinking is to do with the rise of what the American social critic Neil Postman has called 'technopoly' – the widespread view that every ill is a problem which has a potential solution; solutions are provided by technological advances, which are generated by clear, purposeful, disciplined thinking; and the faster problems are solved, the better. Thus, as the Ladakhis have recently joined us in believing, time is an adversary over which technology can triumph. For Postman, technopoly is based on

> the beliefs that the primary, if not the only goal of human labour and thought is efficiency; that technical calculation is in all respects superior to human judgment; that in fact human judgment cannot be trusted, because it is plagued by laxity, ambiguity, and unnecessary complexity; that subjectivity is an obstacle to clear thinking; that what cannot be measured either does not exist or is of no value; and that the affairs of citizens are best guided and conducted by 'experts.'[3]

In such a culture, time spent exploring the question is only justified to the extent that it clearly leads towards a solution to the problem. To spend time dwelling on the question to see if it may lead to a *deeper* question seems inefficient, self-indulgent or perverse.

In contemporary 'Western' society (which now effectively covers the globe), we seem to have generated an inner, psychological culture of speed, pressure and the need for control – mirroring the outer culture of efficiency and productivity – in which access to the slower modes of mind has been lost. People are in a hurry to know, to have answers, to plan and solve. We urgently want explanations: Theories of Everything, from marital mishaps to the origin of the universe. We want more data, more ideas; we want them faster; and we want them, with just a little thought, to tell us clearly what to do.

We find ourselves in a culture which has lost sight (not least in its education system) of some fundamental distinctions, like those between being wise, being clever, having your 'wits' about you, and being merely well informed. We have been inadvertently trapped in a single mode of mind that is characterised by information-gathering, intellect and impatience, one that requires you to be explicit, articulate, purposeful and to 'show your working'. We are thus committed (and restricted) to those ways of knowing that can function in such a high-speed mental climate: predominantly those that use language (or other symbol systems) as a medium and deliberation as a method. As a culture we are, in consequence, very good at solving analytic and technological problems. The trouble is that we tend, increasingly, to treat all human predicaments as if they were of this type, including those for which such mental tools are inappropriate. We meet with cleverness, focus and deliberation those challenges that can only properly be handled with patience, intuition and relaxation.

To tap into the leisurely ways of knowing, one must dare to wait. Knowing emerges from, and is a response to, not-knowing. Learning – the process of coming to know – emerges from uncertainty. Ambivalently, learning seeks to reduce uncertainty, by transmuting the strange into the familiar, but it also needs to tolerate uncertainty, as the seedbed in which ideas germinate and responses form. If either one of these two aspects of learning predominates, then the balance of the mind is disturbed. If the passive acceptance of not-knowing overwhelms the active search for meaning and control, then one may fall into fatalism and dependency. While if the need for certainty becomes intemperate, undermining the ability to tolerate confusion, then one may develop a vulnerability to demagoguery and dogma, liable to cling to opinions and beliefs that may not fit the bill, but which do assuage the anxiety.

Perhaps the most fundamental cause of the decline of slow knowing, though, is that as a culture we have lost our sense of the *unconscious intelligence* to which these more patient modes of mind give access, a loss for which René Descartes conventionally takes the blame. If the busy conscious mind is to allow itself to wait, mute, for something to come, presumably from a source beyond its ken and its control, it has, minimally, to acknowledge the existence of such a source. Modern Western culture has so neglected the intelligent unconscious – the *undermind*, I shall sometimes call it – that we no longer know that we have it, do not remember what it is for, and so cannot find it when we need it.[4] We do not think of the unconscious as a valuable resource, but (if we think of it at all) as a wild and unruly 'thing' that threatens our reason and control, and lives in the dangerous Freudian dungeon of the mind.[5] Instead, we give exclusive credence to conscious, deliberate, purposeful thinking – d-mode. Broader than strict logic or scientific reasoning, though it includes these, d-mode has a number of different facets.

*D-mode is much more interested in finding answers and solutions than in examining the questions.* Being the primary instrument of technopoly, and as such centrally concerned with problem-solving, d-mode treats any unwanted or inconvenient condition in life as if it were a 'fault' in need of fixing; as if one's loss of libido or turnover were technical malfunctions which one ought – either by oneself, or with the aid of an 'expert', such as a counsellor or a market analyst – to be able to put right.

*D-mode treats perception as unproblematic.* It assumes that the way it sees the situation is the way it is. The diagnosis is taken for granted. The idea that the fault may be in the way the situation is perceived or 'framed', or that things might look different 'on closer inspection', does not come naturally to d-mode.

*D-mode sees conscious, articulate understanding as the essential basis for action, and thought as the essential problem-solving tool.* The activity in d-mode is predominantly that of gaining a mental grasp, or figuring out. This may involve the impeccable rationality of the proto-typical scientist, with her equations and flow charts and technical terms. Or it may involve the more common-or-garden kinds of thinking: weighing up the pros and cons of a decision; talking things through with a friend; jotting down thoughts or making lists on the back of an envelope; trying out arguments over dinner, discussing family arrangements, making a sales pitch. Though this latter kind of thinking may not match up to the exacting standards of the professional philosopher or mathematician, and is often full of unnoticed holes, nevertheless it is, in its form and intent, 'quasi' or 'proto'-rational.

*D-mode values explanation over observation,* and is more concerned about 'why' than 'what'. Sometimes figuring out is designed to get directly to the point of action. But commonly, either as a means or an end in itself, what it seeks is understanding or explanation. The need to have a mental grasp, to be able to offer, to oneself if not to others, an acceptable account of things, is an integral part of d-mode. Right from playschool, adults will be asking children: 'What are you trying to do?', or 'That's interesting; why did you do that?' And children quickly get the idea that they ought to know what they are up to, what they are trying to achieve; and to be able to give an account of themselves, their actions and their motives, to other people. They come to assume, with their parents and teachers, that it is normal to be intentional, and proper to have explanations to offer. As ever, there is no problem with this *per se*; it is a very useful ability. But when this purposeful, justificatory, 'always-show-your-working' attitude becomes part of the dominant default mode of the mind, it then tends to suppress other ways of knowing, and makes one sceptical of any activity whose 'point' you cannot immediately, consciously see.

*D-mode likes explanations and plans that are 'reasonable' and justifiable, rather than intuitive.* The demand that ideas always come with supporting arguments and explanations may lead one to reject out of hand thoughts that are in fact extremely fruitful, but which arrive without any indication of their pedigree or antecedents. The productive intuition can be overlooked in favour of the well-argued case. And if explanation comes to be seen as a necessary intermediary between a problem and a plan of action – if one does not feel qualified to act without a conscious rationale – then again one might miss out on some short cuts and bright ideas. Doubt, in the sense of a lack of conscious comprehension, becomes stultifying rather than facilitating; a trap rather than a springboard.

*D-mode seeks and prefers clarity, and neither likes nor values confusion.* Because of its concern with justification, d-mode likes to move along a well-lit path from problem to solution, preserving, as it goes, as much of a mental grasp as it can. It prefers learning that hops from stepping-stone to stepping-stone, without getting its feet wet, like a mathematical proof, or a well-argued report that progresses smoothly from a problem, to a clear analysis, to a plausible solution, to an action plan. And while some learning may proceed in this point-by-point fashion, much does not. Often learning emerges in a more gradual, holistic way, only after a period of casting around for a vague sense of direction, like a pack of hounds that has lost the scent. An artist composing a still life, a client in psychotherapy, even a scientist on the verge of a breakthrough: none of these (as we shall see) would be functioning optimally in d-mode. To undertake this kind of slow learning, one needs to be able to feel comfortable being 'at sea' for a while.

*D-mode operates with a sense of urgency and impatience.* It is accompanied by a subtle – or sometimes gross – sense of not having enough time; of wanting things to be sorted out soon; of getting irritable when the fix is not quick enough. Fuelled by this sense of urgency, we find ourselves living, increasingly, in the fast lane. And the technology – be it planes or Powerbooks, microwaves or modems – tracks this need, but also channels and exacerbates it. If you have to wait for the TV news, or tomorrow's newspapers, to hear about the rumours on Wall Street, or a small earthquake in Peru, you're not a serious player. Our intolerance of dissatisfaction, or even of a delay in information, comes to dictate the kind of mind-mode with which we meet *any* kind of adversity.

*D-mode is purposeful and effortful rather than playful.* With problem-solving and impatience comes a feeling of mental strain, of pushing for answers that would not arrive by themselves, or certainly not quickly enough. In d-mode there is always this sense, vague or acute, of being under time pressure, and of being intentional, purposeful, questing: of needing to have an answer to a pre-existing question, whether it concerns a fault in the production line or the meaning of life. Once this busy activity becomes all we know how to do, the default mode, then we are going to miss any fruits of *relaxed cognition.*

*D-mode is precise*; it tends to work with propositions made up of clearly defined symbols, preferably the hyper-precise languages of mathematics and science, where every term seems to be transparent and complete. A model of the national economy which can be represented as a sophisticated computer program, in which everything that counts can be given a measure – and in which therefore everything which has no measure has no place and no value – is taken more seriously than one which may subsume a richer view of human nature, but which is less explicit and precise. The history of scientific psychology – a d-mode enterprise if ever there was one – is full of precise theories about how memory works, for example, which make quantitative predictions about arcane laboratory tasks, but which simply ignore almost everything that people find interesting about their own powers of retention. When I was working on memory for my doctorate, I stopped telling people at parties because they would inevitably start to ask me all kinds of fascinating

questions to which my detailed knowledge was embarrassingly irrelevant. (Happily things in memory research have improved somewhat in the last twenty-five years.)

*D-mode relies on language that appears to be literal and explicit*, and
tends to be suspicious of what it sees as the slippery, evocative world
of metaphor and imagery. If something can be understood, it can
be understood clearly and unambiguously, says the intellect. An
intimation of understanding that does not quite reveal itself, that
remains shrouded or indistinct, is, to d-mode, only an impoverished
kind of understanding; one that should either be forced to explain
itself more fully, or treated with disdain. Poetry does not capture
anything that cannot ultimately be better, more clearly rendered in
prose, and rhetoric is a poor cousin of reasoned explanation.

*D-mode works with concepts and generalizations*, and likes to apply
'rules' and 'principles' where possible. D-mode favours abstraction
over particularity. It works with what is generic or prototypical. It
talks about 'the workforce', 'the rational consumer', 'the typical
teacher', 'the environment', 'holidays', 'feelings'. Even individuals
are treated as generalizations, collections of traits and dispositions.
'John Major' and 'Cher' are as much abstractions as 'the national
debt' or 'the state of Welsh rugby'. The idea that a kind of truth
could be derived from a close, sustained but unthinking attention
to a single object is foreign to d-mode.

Language necessarily imposes a certain speed, a particular time
frame, on cognition, so *d-mode must operate at the rates at which
language can be received, produced and processed*. If you speed speech
up it soon becomes unintelligible. If you slow it down beyond a
certain point it loses its meaning. (Old-fashioned vinyl '45s', when
played at either 33 or 78 revolutions per minute, demonstrate this
phenomenon nicely.) Those modes of mind that work very slowly
(or, for that matter, very fast) cannot, therefore, operate with the
familiar tools of words and sentences. They need different contents,
different elements – or perhaps no conscious elements at all. And
without the familiar ticker tape of words rolling across the screen
of consciousness, there may come a disconcerting feeling of having
lost predictability and control. Thus *d-mode maintains a sense of
thinking as being controlled and deliberate*, rather than spontaneous
or wilful.

*D-mode works well when tackling problems which can be treated as
an assemblage of nameable parts*. It is in the nature of language to
segment and analyse. The world seen through language is one that
is perforated, capable of being gently pulled apart into concepts

that seem, for the most part, self-evidently 'real' or 'natural', and which can be analysed in terms of the relationships between these concepts. Much of traditional science works so well precisely because the world of which it treats is this kind of world. But when the mind turns its attention to situations that are ecological or 'systemic', too intricate to be decomposed in this way without serious misrepresentation, the limitations of d-mode's linguistic, analytical approach are quickly reached. Any situation that is organic rather than mechanical is likely to be of this kind. The new 'sciences' of chaos and complexity are in part a response to the realisation that d-mode is *in principle* unequal to the task of explaining systems as complicated as the weather, or the behaviour of animals in the natural world. Along with the rise of these new sciences must come a re-evaluation of the slower ways of knowing; of intuition as an essential complement to reason.

The fact that language can handle only so much complexity is easy to demonstrate. Take the sentence

The ecologist hated the accountant.

This is trivially easy to understand. Now take

The accountant the ecologist hated abused the waiter.

This is still perfectly manageable. Add another (quite grammatical) embedded clause

The waiter the accountant the ecologist hated abused loved the archbishop.

Understanding begins to get slightly tenuous. And when we get to

The archbishop the waiter the accountant the ecologist hated abused loved joined the conspiracy

you have to work quite hard. You begin to need some kind of cognitive prosthesis, like a diagram, if you are to overcome the limitations of memory and understanding that are being revealed. Without the build-up, it would take some very deliberate unpacking to figure out who it was who abused whom. D-mode stretched to its limit becomes cumbersome and inept.

Here are two other examples of perfectly grammatical language that are, in practice, virtually incomprehensible.

We cannot prove the statement which is arrived at by substituting for the variable in the statement form 'We cannot prove the statement which is arrived at by substituting for the variable in the statement form Y the name of the statement form in question', the name of the statement form in question.

And:

Both is preferable to neither; but naturally both both and neither is preferable to neither both nor neither; but naturally both both both and neither and neither both nor neither is preferable to neither both both and neither nor neither both nor neither; but – naturally – both both both both and neither and neither both nor neither and neither both both and neither nor neither both nor neither is preferable to neither both both both and neither and neither both nor neither nor neither both both and neither nor neither both nor neither.

Unless we have spent years getting used to statements like this, d-mode simply has to give up. A professor of logic might be able to make her way through these abstract jungles, but the fact that d-mode admits of levels of expertise should not blind us to its inherent limitations. Even language and logic can rapidly get out of control if we let them. And it is therefore an open question whether there are kinds and degrees of complexity which might be handled better in a different way.

If we see d-mode as the only form of intelligence, we must suppose, when it fails, that we are not 'bright' enough, or did not think 'hard' enough, or have not got enough 'data'. The lesson we learn from such failures is that we must develop better models, collect more data, and ponder more carefully. What we do not learn is that we may have been thinking *in the wrong way*. While this epistemological stance remains invisible and unchallenged, therefore, the search for better answers to personal, social, political and environmental predicaments has to be conducted by the light of conscious thought. Our efforts are like those of the man who was searching for his car keys under the streetlight – though he has lost them elsewhere – because that was the only place he could *see*. Thus scientists, researchers, intellectuals and those who program computers with complicated formulae in order to try to predict economic trends remain the people on whom we tend to pin our hopes in the face of difficulties and uncertainties. They are the ones who, by general acclaim, have the best, most explicit models; who have the most information; and who are the most skilled thinkers. We trust them. Where else could we look for guidance?

The 'slow ways of knowing' are, in general, those that lack any or all of the characteristics of d-mode. They spend time on uncovering what may lie behind a particular question. They do not rush into conceptualisation, but are content to explore more fully the situation itself before deciding what to make of it. They like to stay close to the particular. They are tolerant of information that is faint, fleeting, ephemeral, marginal or ambiguous; they like to dwell on details which do not 'fit' or immediately make sense. They are relaxed, leisurely and playful; willing to explore without knowing what they are looking for. They see ignorance and confusion as the ground from which understanding may spring. They use the rich, allusive media of imagination, myth and dream. They are receptive rather than proactive. They are happy to relinquish the sense of control over the directions that the mind spontaneously takes. And they are prepared to take seriously ideas that come 'out of the blue', without any ready-made train of rational thought to justify them. These are the modes of mind that the following chapters will explore, in order both to reveal their nature and their value and also to uncover ways in which they might be rehabilitated.

In order to rehabilitate the slow ways of knowing, we need to adopt a different view of the mind as a whole: one which embraces sources of knowledge that are less articulate, less conscious and less predictable. The undermind is the key resource on which slow knowing draws, so we need new metaphors and images for the relationship between conscious and unconscious which escape from the polarisation to which both Descartes and Freud, from their different sides, subscribed. Only in the light of new models of the mind will we see the possibility and the point of more patient, receptive ways of knowing, and be able to cultivate – and tolerate – the conditions which they require.

The crucial step in this recovery is not the acquisition of a new psychological technology (brainstorming, visualisation, mnemonics and so on), but a revised understanding of the human mind, and a willingness to move into, and to enjoy, the life of the mind as it is lived in the shadowlands rather than under the bright lights of consciousness. Clever mental techniques – devices that 'tap' the resources of the 'right hemisphere', as if it were a barrel of beer – miss the point if they leave in place the same questing, restless attitude of mind. In many courses on 'creative management' or 'experiential learning', it is a case of *plus ça change, plus c'est la même chose*. Instead of calling a meeting to 'discuss' the problem, you call one to 'brainstorm' it, or to get people to draw it with crayons. But the pressure for results, the underlying impatience, is still there. The key to the undermind is not an overlay of technique but radical reconceptualisation. When the mind slows and relaxes, other ways of knowing automatically reappear. If and when this shift of mental mood takes place, *then* some different strategies of thought may indeed be helpful, but, without it, they are useless. (This, incidentally, explains why the enthusiasm for each new, much-hyped mental technology has such a disappointingly short half-life.)[6]

Another step in the recovery of the slower ways of knowing is to recognise that these forms of cognition are not the exclusive province of special groups of people – poets, mystics or sages – nor do they appear only on special occasions. They have sometimes been talked about in rather mystifying ways, as the work of 'the muse', or as signifying great gifts, or special states of grace. Such talk makes slow knowing look rather awesome and arcane. One feels intimidated, as if such mental modes were beyond the reach of ordinary mortals, or had little to do with the mundane realities of modern life. This is a false and unhelpful impression. A 'poetic way of knowing' is not the special prerogative of those who string words together in certain ways. It is accessible, and of value, to anyone. And though it cannot be trained, taught or engineered, it can be cultivated by anyone.

So *Hare Brain, Tortoise Mind* is about why it is sometimes a good idea to pull off the Information Super-Highway into the Information Super Lay-By; to stop chasing after more data and better solutions and to rest for a while. It is about why it is sometimes more intelligent to be less busy; why there are mental places one can gain access to by loafing which are inaccessible to earnest, purposeful cognition. And it is about the reasons why these natural endowments of the human mind have tended to become neglected in twentieth-century Euro-American culture, and why, in this culture, they are sorely missed.

## CHAPTER 2

# Basic Intelligence:
# Learning by Osmosis

It is a profoundly erroneous truism, repeated by all
copybooks and by eminent people when they are
making speeches, that we should cultivate the habit of
thinking of what we are doing. The precise opposite is
the case. Civilization advances by extending the
number of important operations which we can perform *without* thinking about them. Operations of
thought are like cavalry charges in battle – they are
strictly limited in number, they require fresh horses,
and must only be made at decisive moments.

*A. N. Whitehead*

It is February: summer in New Zealand. I am closeted with my
laptop in a beach-house on the west coast of the North Island
(overlooking what surfers tell me is the best left-hand break in the
Southern Hemisphere), and there are a lot of flies. I find them,
especially the big brown ones, very distracting, so, despite my
Buddhist leanings, I swat them. There are also a number of spiders,
long-legged and small-bodied, which I rather like. This morning I
dropped a freshly swatted fly into the web of one of the spiders. I
then proceeded to watch, rapt, for twenty minutes as the spider
manoeuvred the fly from where I had placed it to its own dining
area, a distance of some twelve centimetres. First it spun a coat
round the fly to hold it secure. Then it delicately cut away the
strands of web that were supporting it, until it was left dangling
only by a few threads. Holding on to the web above with two legs,
and clasping the fly with the others, it pulled it towards its destination
by about half a centimetre, and secured it with another thread. It
released the strands that were now holding the fly back, allowing it
to swing a little towards the goal, and spun some more ties to hold

it in its new position. Taking up the diagonal position, it hauled its prize sideways again, secured it, and then cast off the ropes that were restraining it. And so it went, until lunch was finally in the right place.

The equivalent task for me, I computed, would have been something like single-handedly transporting a blue whale a distance of 120 metres across a bottomless abyss, equipped only with some very strong elastic, grappling hooks, and a sharp knife. This perilous feat of engineering would have taken a great deal of thought and calculation to counter the constant risk that one false move, such as cutting the wrong string at the wrong moment, would send the whale, and very possibly me as well, plummeting into the void. The spider, whose whole body was two millimetres long, with a minute brain, didn't make a mistake. I was impressed. I did not feel obliged to credit the spider with consciousness; but I had to marvel at its intelligence.

There is a resurgence of interest in the concept of 'intelligence' these days, prompted, in large measure, by a growing dissatisfaction with the assumption that d-mode is the be-all and end-all of human cognition. Harvard psychologist Howard Gardner has suggested that there are 'multiple intelligences', of which he claims to have identified eight and a half, and which resemble quite closely the subjects of the traditional school curriculum.[1] Daniel Goleman argues for a rapprochement between reason and feeling with his notion of 'emotional intelligence'.[2] But to understand more broadly how the different facets of intelligence fit together, we have to find an approach that does not presuppose the primacy of the intellect.

At its most basic, intelligence is what enables an organism to pursue its goals and interests as successfully as possible in the whole intricate predicament in which it finds itself. My spider had been designed by evolution to perform, within its own world, the most challenging of tasks in an efficient and sophisticated way. These miracles of intelligent adaptation are commonplace in the animal kingdom, and many of them have been documented rather more systematically than my spider. If a rat eats a meal that consists of a mixture of a familiar and an unfamiliar food, and subsequently becomes sick, it will in future avoid the new food but not the familiar one.[3] That, I think, is intelligent.

Much of human intelligence, too, has little to do with d-mode. A baby is being intelligent when it smiles hopefully at its mother, or turns its head away from a looming object. A teenager is being intelligent when learning to get along in school by blending into the background, or deploying a disarming humour. A poet is being intelligent when considering a variety of candidates for the *mot juste*. And though a mathematician is also being intelligent as she tries to work out the solution to a complex problem, her finely honed intellectual ability is just one variety of intelligence, and a rather peculiar and arcane one at that. Intelligence may be associated with words, logical argument, explicit trains of thought or articulate explanation, but it may equally well not be. Fundamentally, intelligence helps animals, including human beings, to survive.

The most basic of these strategies, common to all levels of life from amoebas to archbishops, is the bred-in-the-bone tendency to approach and maintain conditions which favour survival, and to avoid or escape from conditions which are aversive. The former conditions we call 'needs'; the latter 'threats'. Evolution has equipped every animal with a repertoire, large or small, of ways to minimise the risk of damage and enhance its wellbeing. The spider weaves its web, manoeuvres its prey, and freezes when the air moves in a disturbing way. The digger wasp, *sphex ichneumoneus*, which cleverly buries a paralysed cricket alongside her eggs for the new-born grubs to feed on, always leaves the cricket outside her burrow while she goes in to make sure all is well, before dragging it in.[4] Even potential threats are allowed for in such reflex behaviour.

But the genetically given reactions to threats of 'fight, flight, freeze and check', though helpful, are by no means infallible. A spider may go still, despite being dangerously highlighted against the background of a white bathtub. If an interfering ethologist steals the cricket every time *sphex* makes her subterranean safety check, she cannot adapt. She never realises that in this new world it may make more sense to drag the cricket in with her first time. A baby shows distress even though the looming object is actually a rapidly inflated balloon, and not a projectile. Reflexes, though intelligent, may be turned on their heads by unprecedented events: those for which evolution has not had time to prepare you. Such reflexes provide a vital starter kit of survival intelligence, but if the ability to build on these wired-in, entry-level responses is lacking, an animal remains highly vulnerable to change.

So the next stage in the evolution of intelligence is *learning*. Gathering knowledge and developing expertise are survival strategies. In unfamiliar situations, animals are at risk. They are unable to predict and control what is going on. Potential sources of succour may go unrecognised. Actual threats may not be perceived until it is too late. Uncertainty may always conceal danger. The ability to reduce uncertainty, to convert strangeness into known-ness, therefore offers a powerful evolutionary advantage. All the different ways of learning and knowing which human beings possess, however sophisticated, spring ultimately from this biological imperative. Crudely, we might say that 'knowing' is a state in which useful patterns in the world have been registered, and can be used to guide future action. 'Learning' is the activity whereby these patterns are detected. And, at this level, 'intelligence' refers to the resources that make learning, and therefore knowing, possible.

This ability to detect, register and make use of the patterns of relationships that happen to characterise your particular environment is widespread in the animal world. Take the gobiid fish, for example. It has been shown that certain of these fish can find their way from one rock-pool to another, at low tide, by jumping accurately across the exposed rocks. Jumping in this way is a high-risk exercise; if they get it wrong the fish are stranded or injured. In fact, they do it without error. Studies of these fish have ruled out the possibility that they are using some sensory clues such as reflections or smells. If they are placed in an unfamiliar pool, they will not jump. The only possible explanation for this remarkable ability is that, during high tide, the fish swim over and around the crevices and hollows in the sea floor, forming a detailed map of the area which is stored in their memories and used as the basis on which to compute their jumps when they are 'imprisoned' in the low-tide pools.[5]

In the same way, the baby soon comes to know not just the difference between a ball, a balloon and a face, but between her mother's and her father's faces, and tunes her responses accordingly. Her brain is malleable: it is formed not just by the experience of her ancestors, but is also moulded – like that of the gobiid fish, but enormously more so – by the idiosyncrasies of her own experience. A brain is plastic: it transmutes ignorance into competence, and is extraordinarily adept at doing so. Categories and concepts are distilled from particular encounters so that, by a process of spontaneous analogy, 'what I do next' can be informed by records of 'what happened before'. Past mistakes can be avoided and new mistakes made, until, with luck, an effective way of dealing with 'this type of thing' – a big dog, a puncture, an angry face, a new teacher – emerges and confidence is restored. Coming to know the world in this way, to register its patterns and to develop and coordinate skilful responses, is what a sophisticated nervous system – what I shall call a 'brain-mind' – does. It is built to tune itself to certain wavebands of information, and to coordinate these with its own expanding range of capabilities.

After plasticity, the next great development in the evolution of knowing is *curiosity*. Instead of learning simply by reacting to uncertainty, animals became proactive – inquisitive, adventurous, playful. When no more urgent need is occupying your attention, it pays to extend your knowledge, and hence your competence and your security, by going out and actively exploring. So useful is this that evolution has installed it in many species as one of their basic drives. Rats who are allowed to become thoroughly familiar with a maze will quickly explore a new section that is added to it, even though they are being consistently and adequately fed elsewhere. Monkeys kept in a box will repeatedly push open a heavy door to see what is going on outside, and will spend hours fiddling with mechanical puzzles even though they receive no reward for doing so. Human beings who have volunteered to take part in a 'sensory deprivation' experiment, in which all they have to do, to earn forty dollars a day, is to stay in a room with no stimulation, rapidly become desperate for something – anything – to feed their minds, and will repeatedly press a button to hear a voice reading out-of-date stock market quotations.[6]

Being receptive, attentive and experimental, seeking to expand competence and reduce uncertainty, are the design functions of a plastic and enquiring brain-mind. No added encouragement or discipline – no conscious intention, no effort, no deliberation, no articulation – is needed to fulfil this brilliant function. The original design specification of learning does not include the production of conscious rationales. Knowing, at root, is implicit, practical, intuitive. The brain discovers patterns and tunes responses, it is programmed by experience, but this programming is recorded in millions of minute functional changes in the neurons, and manifested in the way the whole organism behaves.

Given the evolutionary priority of this unconscious intelligence – the primacy of *know-how* over *knowledge* – what would we expect the main differences between unconscious and conscious ways of knowing to be?[7] First, we might expect the unconscious to be more robust and resilient, more resistant to disruption, than our conscious abilities. This is exactly what neurological studies of brain damage reveal. When memory, perception or the control of action are degraded, it is the conscious aspects that tend to be lost first, while abilities that are managed unconsciously are spared.[8]

If unconscious abilities are more primitive, more a function of evolution than of culture, we might also expect them to vary less from person to person than do their conscious deliberations. In particular, we should not expect intuitive know-how to show much correlation with measures of 'conscious intelligence' such as IQ – and it doesn't. People's ability to pick up the skills that their everyday lives require – their 'practical intelligence', as Harvard psychologist Robert Sternberg calls it – is independent of their intellectual or linguistic facility. Brazilian street children are able to do the mental arithmetic that their businesses require – quite complex sums, by school standards – without error, despite having very low mathematical ability as measured by tests. People who work as handicappers at American racecourses are able to make calculations, based on a highly intricate model involving as many as seven different variables, yet their ability to do so is completely unrelated to their IQ score.[9]

The minds of children, being immature, must rely more on unconscious than conscious operations. Babies learn to recognise the important people in their lives, and to take an increasingly sophisticated part in the rituals of family life – bathtimes, mealtimes, bedtimes and so on – long before they are able to comment or reflect on what they are doing. They learn to walk by a vast amount of trial and error, out of which is gradually distilled thousands of inarticulate correspondences between the muscles of shoulders, torso and legs, and the sensations of vision, touch and balance. They learn to speak by picking up the language of their culture without any explicit knowledge of grammar. They develop styles of relating without recourse to any instruction book. And as they get older they will learn to ride bicycles, play violins, kick balls, take part in meetings, prepare meals, shop, catch planes and make love, for the most part without being able to explain how it is that they do what they do, or how they learnt it.

The greater part of the useful understanding we acquire throughout life is not explicit knowledge, but implicit know-how. Our fundamental priority is not to be able to talk about what we are doing, but to do it – competently, effortlessly, and largely unconsciously and unreflectingly. And the corresponding need for the kind of learning that delivers know-how – which I shall call *learning by osmosis* – is not one that we outgrow. The brain-mind's ability to detect subtle regularities in experience, and to use them as a guide to the development and deployment of effective action, is our biological birthright. The evolution of more sophisticated strategies complements this basic capability; it does not supersede it. Although the presence of unconscious intelligence is much more obvious in animals and small children, not being overlaid by their conscious, articulate intellect, it is a mistake to suppose that we grow out of it as we get older.

Yet this mistake is made, and it is partly the fault of the renowned Swiss developmental psychologist (or, as he preferred, 'genetic epistemologist') Jean Piaget. Piaget called this ability to master the intuitive craft of living 'sensorimotor intelligence', and claimed that it was of pre-eminent importance during the first two years of life, but was subsequently overtaken and transformed by other more powerful, abstract and increasingly intellectual ways of knowing. In his tremendously influential 'stage theory' of development, Piaget implicitly accepted the cultural assumption that d-mode was the highest form of intelligence, and through the impact that his approach has had on several generations of educators, he inadvertently made sure that schools, even primary schools and kindergartens, saw their job as weaning children off their reliance on their senses and their intuition, and encouraging them to become deliberators and explainers as fast as possible.

The ability to distil out of our everyday experience useful maps and models of the world around us is very down-to-earth; so mundane that it is, in many ways, the unsung hero of the cognitive repertoire. We do this so continuously, so automatically and so unconsciously that it is very easy to overlook just how valuable, and how 'intelligent,' this ability is. It represents the 'poor bloody infantry' of the mind: much less glamorous than the flamboyant cavalry charges of conscious thought. Yet we ignore or disparage this constant honing or sharpening of our 'wits' (in the practical sensory sense of 'wits' that I used in the previous chapter) at our peril, for it turns out that there are things we can learn through this gradual, tacit process which d-mode cannot master; and also that d-mode, if used over-enthusiastically, can actively interfere with this way of knowing. The conscious human intellect stands on the shoulders of learning by osmosis. D-mode is an evolutionary and cultural parvenu, and we cannot properly reassess either its nature or its limitations without looking at its evolutionary underpinnings.

We continue throughout our lives to make use of this unsung ability to pick up patterns and tune our actions accordingly, without being able to say what we have learnt, or even, very often, that we have learnt anything at all. When you start to listen to the works of a particular composer, your mind begins to detect all kinds of characteristics of instrumentation, harmony, rhythm and so on, which enable you to say, on hearing a new piece, 'Isn't that

Bruckner?' Yet how you can tell, unless you are a music scholar, you may be quite unable to say. People who read a lot of whodunnits become, often unconsciously, so familiar with the genre that they know, without thinking, that the murderer is going to be some incidental character in chapter two. When we take a new job, we may consciously collect as much low-down on colleagues-to-be, and the ethos of the workplace, as we can; yet during the first few days and weeks we are also learning a tremendous amount quite automatically: how people greet each other in the morning; how to look busy when we aren't; what kinds of jokes are 'funny' and what are 'crude' or 'sexist'; and so on. As people gain promotions, form stable relationships, have children and are faced with bereavement, so the usefulness of this ability to soak up know-how through their pores does not diminish.

Recent research by psychologists in both Britain and America has reaffirmed the importance of this implicit learning, and shown how it gradually develops over time. Take a professional problem such as learning to regulate the flow of traffic in a city by adjusting the number of buses and the provision and cost of private parking; or to manage a school budget; or to control a complicated industrial process such as the output of a factory or a power station. Situations like this have been studied by Dianne Berry and the late Donald Broadbent at the University of Oxford.[10] Consider the factory production problem. It can be simulated as a 'computer game' in which the levels of various factors, such as the size of the workforce, or of financial incentives, are displayed on the screen, along with the level of the output, and the 'player's' job is to stabilise the output by manipulating the input variables. The effect of each of the variables is actually determined by a reasonably complex equation which the players are not explicitly told.

Players, in their role as 'trainee managers', come, over a period of time, to be able to make adjustments to the input variables that do in fact bring production to the required level – but they are not able to say what they are doing, or explain why it works. When asked to justify a particular 'move', all they may be able to say is that they 'had a hunch', or 'it felt right'. They may even, having made a perfectly good move, say that they thought they were guessing. When the task is quite a difficult one, and people's performance is monitored over several days, their practical know-how and their explicit knowledge – what they can say about their own performance – develop at startlingly different rates. The ability to *do* the job develops relatively quickly, and in some cases quite abruptly; but the ability to articulate that knowledge emerges, if at all, much more slowly.

Broadbent and Berry's laboratory results are by no means unfamiliar in everyday life. Sportspeople and musicians develop high levels of expertise which they are often hard put to analyse or explain. Teachers come to be able to make on-the-spot decisions about how to present a topic or manage a classroom situation, yet may be quite unable to justify their actions to an inquisitive student. In the introduction to a fascinating account of their work on 'principles of problem formation and problem resolution', American psychotherapist Paul Watzlawick and his colleagues describe how the book came about. Working together over several years, they developed some powerful new ways of, as they put it, 'intervening in human problem situations' so as to break through apparent impasses and bring about welcome change. However, as more and more people became interested in their methods through demonstrations and training courses, they became increasingly embarrassed to realise that they had no way of explaining how or why their methods were so successful. 'Only gradually were we ourselves able to conceptualise our approach,' they write – the approach which, at another level, they understood inside out.[11]

Other aspects of this 'implicit learning' have been investigated experimentally by Pawel Lewicki and his colleagues at the University of Tulsa in the United States.[12] Though some of their long series of experiments are rather stylised, they are very illuminating. Like the British researchers, they explored kinds of learning in which people can get better at doing a particular job by picking up subtle patterns embedded in hundreds of examples, but the experimental designs are rather different. In one of these designs, the people taking part in the experiment sit in front of a computer screen which is divided into quarters. Every so often a random-looking array of digits appears on the screen, covering all the four quadrants, and the subjects' job is to detect a particular predetermined number – 6, say – and to push one of four buttons in front of them to indicate which of the four quadrants the 6 is in. The computer automatically records how long it took them to spot the target, and whether their choice of button was right or wrong. There is a brief pause, and then another (different) display appears, and they have to find the 6 again; and so on, for a large number of such 'trials'. The trials are grouped into blocks of seven, with a short break between each block.

As you might expect, the first thing the computer shows is that people get faster at detecting the target as they get more used to,

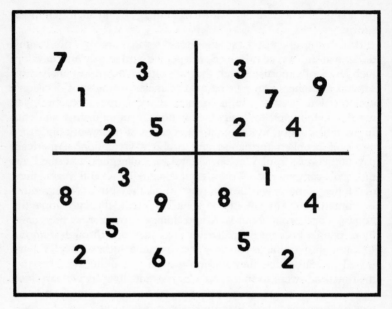

*Figure 1. 'In which quadrant is the 6?' – sample grid of numbers used in the Lewicki experiments*

and practised at, the task. But now comes the twist. Although it looks to the subjects as if the position of the 6 varies randomly from trial to trial, in fact there is a subtle pattern. Specifically, if you take the positions of the target on the first, third, fourth and sixth trials in a block, you could theoretically predict in which quadrant it is going to appear on the seventh. For example, if the 6 had been in the upper left on trial 1, lower right on trial 3, upper right on trial 4, and lower left on trial 6, then it will appear in the lower right on trial 7. Note that you have to register the positions on *each* of trials 1, 3, 4 and 6, in *each* block of seven. Nothing less than this will give you any useful information at all. Subjects, of course, are not told about this faint pattern. The question is: do they none the less pick it up and make use of it? If they do, this will be shown by the fact that their response times to the seventh target become faster than to the other six. (The general effect of practice and familiarity would obviously not be able to account for this differential effect.)

Sure enough, over a long series of blocks the response to the seventh target becomes progressively quicker than the responses to the other six. Clearly people are picking up the pattern and making use of it. However, when Lewicki showed them the results, all the subjects were surprised at the 'seventh trial' effect – they themselves had not noticed that they were getting selectively faster – and they had no conscious idea what the information might have been that they were, apparently, using. If they were given some more trials, and asked to make a *conscious* prediction for trial 7, they could do no better than chance: 1 in 4.

Lewicki tried very hard to induce in his subjects some conscious awareness of the situation. At the end of several of his studies he told subjects that there *was* a pattern which they had been using, and offered them unlimited time to study all the stimulus arrays, and a sizeable financial reward if they could come up with a suggestion that was close to the actual pattern which he had been using. Nobody was able to say what the pattern was. Next, he ran a group of subjects who were actually his colleagues on the faculty of the University of Tulsa Psychology Department. They should have been able to work out what was going on, if anyone could; they all knew what his research was about. But even they could not consciously detect the pattern. In fact, when they were shown the data proving that they were responding differently to the last trial in each block, some of them confidently accused Lewicki of using subliminal messages to speed them up or slow them down. Now he came to mention it, they said, they had definitely seen something 'fishy' about the displays. Yet there had been nothing fishy at all; merely a pattern that was perfectly visible, if only the conscious mind could have seen it.

The evidence from these studies is clear: we are able unconsciously to detect, learn and use intricate patterns of information which deliberate conscious scrutiny cannot even *see*, under favourable conditions, let alone register and recall. The complexity of Lewicki's patterns (like my impossible-to-understand sentences in the last chapter) was just too great for d-mode to deal with. But when the hare of conscious comprehension ran out of ideas, tortoise mind just kept going. Simply by attending and responding to the situation, without thinking about it, people are able to extract complex patterns of useful information. Of course there are limits to the powers of observation and detection even of the unconscious brain-mind. There must be a great deal of potentially valuable information in the world that is too faint or subtle even for the undermind to detect. But we might, *en passant*, wonder at the wisdom of a society which ignores these unconscious powers, or treats them as ephemera; and of an education system that persists in privileging just one form of conscious, intellectual intelligence over all others.

Mention of education should remind us that even intellectual understanding itself often benefits from this gradual, soaking-it-up-through-the-pores approach. Really 'getting your brain round' a topic seems to depend at least as much on the slower processes of 'mulling over' and 'cogitating' as it does on being mentally busy. Yet many educators seem to be under the impression that people can (and should) master a body of knowledge entirely through d-mode, via intentional study and 'hard work'. One of the 'fathers' of research on unconscious learning, Arthur Reber of Brooklyn College in New York, described in a recent overview of the field how it was that he first came to be interested in it.

> I was drawn to the problem of implicit learning simply because that has always been, for me, the most natural way to get a grip on a complex problem. I just never felt comfortable with the overt, sequential struggles that characterised so much of standard learning ... As a result of this stance I was not a particularly good 'standard' student ... I found that what seemed for me to be the most satisfactory of 'learnings' were those that took place through what we used to call 'osmosis', that is, one simply steeped oneself in the material, often in an uncontrolled fashion, and *allowed understanding to emerge magically over time*. The kind of knowledge that seemed to result was often *not easily articulated*; and most interesting, *the process itself seemed to occur in the absence of the effort* to learn what was in fact learned.[13] (Emphasis added)

The studies by Broadbent and Berry, Lewicki and others have made it very clear what learning by osmosis is, what its value is, and the conditions it needs to operate. It extracts significant patterns, contingencies and relationships that are distributed across a diversity of situations in both time and space. It works through a relaxed yet precise non-verbal attention to the details of these situations, and to the actual effect of one's interventions, without any explicit commentary of justification or judgement, and without deliberately hunting for a conscious, articulate mental grasp. Learning by osmosis echoes the insight of the Japanese proverb: 'Don't learn it; get used to it'. It operates in complicated situations which cannot be clearly analysed or defined, and where the goal is to achieve a measure of practical mastery rather than to pursue explanation. And it takes time, as it gradually extracts the patterns that are latent within a whole diversity of superficially different experiences. This form of basic intelligence, inherited from our animal forebears, remains both active and valuable throughout life – if it is unimpeded. It is the first, and the most fundamental, of the slow ways of knowing. Unfortunately, it is all too easy for it to become neglected and overshadowed by d-mode.

## CHAPTER 3

# Premature Articulation: How Thinking Gets in the Way of Learning

Our simplest act, our most familiar gesture, could not be performed, the least of our powers might become an obstacle to us, if we had to bring it before the mind and know it thoroughly in order to exercise it. Achilles cannot win over the tortoise if he meditates on space and time.

*Paul Valéry*

About ten years ago, when I was involved in helping people learn to become teachers, I remember sitting at the back of a school science laboratory observing one of my students taking a lesson on photosynthesis. The class of twelve-year-olds had been set a little practical to do, and the student teacher was walking around the lab responding to the pupils' queries. All was going well. Sitting in front of me was a pair of girls working together who had got 'stuck'. They were chatting quietly while one of them kept her hand in the air, waiting patiently for the teacher to notice them and come across to help. The girl who had her hand up was also playing with the fashionable puzzle of the time: the Rubik cube. (This was composed of smaller cubes – each large face having nine such – cunningly engineered in such a way that the faces could be rotated with respect to each other. The mini-cubes were of different colours, and the idea was to manipulate the cube as a whole in such a way that each face of the big cube ended up composed of mini-cubes all of the same colour.)

Having only one free hand, the girl was holding the cube in the other, and turning the faces with her teeth – all the while keeping up her conversation with her friend. She seemed to be giving only

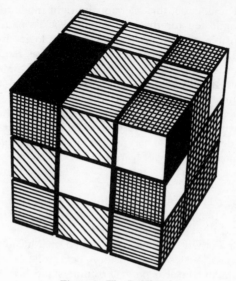

*Figure 2. The Rubik cube*

the most minimal attention to the manipulation of the cube. Yet, as I watched, I could see that she was making some kind of progress, and every so often stopped to reverse the last few moves and take a different tack. I went over to her and asked her to tell me what she was doing with the cube. She looked startled, both because she thought I might be ticking her off in the indirect way that teachers sometimes adopt, but also because she hadn't realised what she had been doing. It was as if she was surprised to find the Rubik cube in her hand. She looked at me to see if I was 'cross', and on reassuring herself that I was genuinely interested, explained, I think to the best of her ability, what she had been doing. 'Nothing,' she said. 'Just messing about.'

Adults, like myself, were prone to become rather frustrated – even infuriated – with 'the stupid cube', and to feel embarrassed and inadequate in the face of the apparent ease with which children – even not very 'bright' children – seemed able to master it.[1] We could not *understand* how to do it, and after toying with it for a while, we would give it back to its small owner, as if it were too trivial to be worth bothering with, and find something with which to repair the small dent to our self-esteem. The trouble was that we grown-ups went immediately into d-mode, trying to *figure it out*, and, in the case of the Rubik cube, this was not the right mode to be in. It is just too complicated for that. As with Lewicki's patterns, or the incomprehensible sentences, the powers of logic and memory needed were beyond our normal range. What was required, if one was to master the cube, was a gradual build-up of the ability to *see* various recurrent patterns, and to adjust one's moves accordingly: to sharpen our wits through the non-intellectual process of observing and experimenting that we have just discussed. And this is just the kind of 'knowing' that my twelve-year-old scientist's 'messing about' was good at delivering. She had not yet lost the knack of this casual, apparently incidental, way of learning; nor did she seem to mind if she could not articulate its results. I, a long-time d-mode addict, had, and did.

What is the relationship between implicit know-how – the practical intelligence that enables us to function well in the world – on the one hand, and the explicit, articulated understanding that d-mode delivers on the other? It is widely assumed, in education and elsewhere, that conscious comprehension – being able to articulate and explain – is of universal benefit. To understand how and why to do something ought to help us to do it. But does it? In the case of the adults' response to the Rubik cube, it seems as if there is an acquired *need* to understand which may actually block the use of our non-intellectual ways of knowing. We have forgotten them, or do not 'believe' in them any more. There is now good evidence that this suspicion is well founded.

The 'stupid cube' effect appeared in Broadbent and Berry's studies. Not only does people's intuitive ability to control the factory output develop much faster than their ability to explain what they were doing; their *confidence* in their ability tends to follow their explicit knowledge, rather than their know-how. Unless they are able to explain what they are doing, they tend to underestimate quite severely how well they are doing it. People feel as if they are merely guessing, even when they are in fact doing well, and, if they had felt free to, many of the subjects would probably have withdrawn from the game, for fear of looking foolish. It is only because they would have felt even more foolish dropping out that they persevered with the task, despite their lack of confidence – and actually gave their unconscious learning a chance to reveal itself. The subjects have learnt to put their faith in d-mode as the indicator of how much they know, and therefore to distrust, at least initially, perfectly effective knowledge that has not (yet) crystallised into a conscious explanation.

You would at least imagine that there would be a positive link between the two kinds of knowledge, implicit and explicit: that people's sense of having a conscious handle on what they are up to should correlate with how well they are in fact doing. After all, we expect airline pilots and medical students to take written examinations, as well as practical ones, so we must assume that the verbal tests of knowledge and understanding are assessing something relevant. Unfortunately, this does not always seem to be the case. In several investigations of the Broadbent and Berry type, people's ability to articulate the rules which they think are underlying their decisions turns out to be *negatively* related to their actual competence.[2] People who are better at controlling the situation are actually worse at talking about what they are doing. And conversely, in some situations it appears that the more you think you know what you are doing, the less well you are in fact doing. You can either be a pundit or you can be a practitioner, it appears; not always can you become both by the same means.

The situations where this dislocation between expertise and explanation appears most strongly are those that are novel, complicated and to some extent counterintuitive; where the relevant patterns you need to discover are different from what 'common sense' – the 'reasonable assumptions' on which d-mode rests – might predict.[3] In situations where a small number of factors interact in a predictable fashion, and where these interactions are in line with what seems 'plausible' or 'obvious', then d-mode does the job, and trying to figure out what is going on can successfully short-circuit the more protracted business of 'messing about'. But where these conditions are not met, then d-mode gets in the way. It is not the right tool for the job, and if d-mode is persistently misused, the job cannot be successfully completed. Trying to force the situation to fit your expectations, even when they are demonstrably wrong, allows you to continue to operate in d-mode – but prevents you from solving the problem.

For example, consider a classic experiment performed by Peter Wason of the University of London. Undergraduate students were shown the three numbers 2, 4 and 6 and told that these conformed to a rule that Wason had in mind.[4] The students' job was to generate other trios of numbers – in response to which Wason would say whether they did or did not conform to the rule – until they thought they knew what the rule was, at which point they should announce it. Typically the conversation would go like this.

Student: 3, 5, 7?
Wason: Yes (that meets the rule).
S: 10, 12, 14?
W: Yes.
S: 97, 99, 101?
W: Yes.
S: OK, the rule is obviously n, n+2, n+4.
W: No it isn't.
S: (very disconcerted) Oh. But it must be!

The problem is that the students thought the rule was obvious from the start, and were making up numbers only with the intention of confirming what they thought they already knew. If their assumption had been correct, their way of tackling the problem would have been logical and economical. But when what is plausible is not what is actually there, those operating in this manner are in for a nasty shock. In fact, Wason's rule is much more general: it is 'any three ascending numbers'. So '2, 4, 183' would have been a much more informative combination to try – even though, to someone who thinks they know the rule, it looks 'silly'.

When d-mode is disconcerted like this, it often responds by trying all the harder. Instead of flipping into a more playful or lateral mode, in which silly suggestions may reveal some interesting information, people start to devise more and more baroque solutions. 'Ah ha,' they may think, 'maybe the rule is the middle number has to be halfway between the first and the third. So let's try 2, 5, 8 and 10, 15, 20.' When Wason agrees that these too conform to the rule, they heave a sigh of relief – only to be flummoxed once again when they announce the rule and are told it is incorrect. Or, even more ingeniously, they may cling to the original hypothesis – which they have clearly been told is not the solution – by rephrasing it. So they might say, 'OK, it's not n, n+2, n+4, but perhaps it is take one number, add four to it to make the third number, and then add the first and third together and divide by two to get the middle number' – which is, of course, the same thing. Having articulated a misleading account, people then proceed to use this faulty map to guide their further interactions with the task, rather than relying on the ability of trial and error, 'messing about', to deliver the knowledge they need. Attention gets diverted from *watching how the system actually behaves* to *trying to figure out what is going on, and using these putative explanations as the basis for action.*

What happens when you introduce into the Broadbent and Berry task some instruction, in the form of potentially helpful hints and suggestions? Does this give learners a head start, or does it handicap them? Again, conventional educational 'wisdom' would strongly back the former, and once more the research shows that things are not so straightforward. Conscious information is not always an asset, especially when it is given early on in the learning process, or when it serves to direct attention to features of the situation which may be true, but which are not strictly relevant to the way it behaves, or which interact in unexpected ways with other features.

For example, if, in the factory task, I give you a hint that the *workers' age* is worth paying attention to, this information may send you off on a mental wild-goose chase if it eventually turns out that what matters (in this hypothetical factory) is doing the job not too fast and not too slow – and that work-rate is related to age, so that people in their thirties and forties are to be preferred to those in either their twenties (who are too quick) or their fifties (who are too slow). If this correlation is something that would never have occurred to you, then my suggestion has flipped you into what is a doomed attempt to try to understand how age is relevant, and, by the same token, diverted you away from just seeing what happens. People tend to assume, quite reasonably, that the information they have been given ought to be useful, so they keep trying to use it, even when that is not the best thing to do. And in doing so, they may effectively starve the unconscious brain-mind of the rich perceptual data on which its efficacy depends. The time when some instruction *may* be of practical benefit, it turns out, is later on, after the learner has had time to build up a solid body of first-hand experience to which the explicit information can be related.

The fact that giving instruction and advice, in the context of developing practical expertise, is a delicate business, is well known (or should be) to sports coaches, music teachers, and trainers of management or other vocational skills. Most coaches and trainers understand very well that the major learning vehicles, in their lines of work, are observation and practice, and that hints, tips and explanations need to be introduced into learners' minds slowly and appropriately. Whatever is offered needs to be capable of being bound by learners into their gradually developing practical mastery. It must be tested against existing experience and incorporated into it, and this takes time. Coaching is, to draw on my earlier analogy, like making mayonnaise: you need to add advice, like oil, very sparingly. If you add too much, too quickly – if you are in a hurry – then the mind curdles, conceptual knowledge separates out from working knowledge, and you will be on the way to producing (or becoming) a pundit rather than a practitioner.

The corollary of these results is that, when people find themselves in situations where learning by osmosis is what is called for, then they ought to learn better if they have given up trying to make conscious sense of it. If you have abandoned d-mode, it cannot get in the way. A recent experiment by Mark Coulson of the University of Middlesex suggests that this may well be so.[5] He employed two variants of the 'factory' task, in one of which the relationship between the subjects' responses and the system's behaviour was fairly 'logical', and in the other of which it was not. In this second 'illogical' version, the system was programmed to respond in a way that depended on what the subject's response had been one or two trials *previously*, rather than on the current trial – a relationship that does not make a great deal of intuitive sense. (This is somewhat analogous to the party game in which one person has to try to discover, by asking Yes/No questions, the nature of a 'dream' that everyone else has apparently agreed upon. Unbeknownst to the victim, the others respond Yes or No to her questions purely on the basis of whether her question ends in a vowel or a consonant. The fun comes from the fact that this rule starts to generate some fairly bizarre information about the 'dream', and that the more the victim tries to make rational sense of this information, the stranger the 'dream' becomes, and the less likely she is to discover the 'trick'.)

Similar studies have shown that the logical task is amenable to the d-mode approach, while the illogical task is not. As with the dream game, the correlations between question and answer in the illogical version are so obscure that the attempt to follow sensible lines of thought and construct reasonable hypotheses about what is going on is unlikely to uncover them. The only effective strategy is to try to observe what is happening with as few preconceptions as possible. Thus subjects should do better on the illogical task if they have somehow been persuaded to give up d-mode before they start. Conversely, if they have abandoned d-mode they should do worse on the logical version.

Subjects in Coulson's study took part in either the 'logical' or the 'illogical' version of the task. Their job, as before, was to learn, over a series of trials, to control the factory process. However, in each version, half of the subjects had been given some advance 'training' in which the behaviour of the computer was *completely random* – an experience which, Coulson reasoned, would have weakened their faith in d-mode, as no amount of clever thinking could reveal patterns where there were none to be found. While overall subjects took longer to learn the illogical than the logical version, the group who had had the prior 'random' experience learnt the illogical task faster than those who had not. Subjects who had had the random experience, on the other hand, learnt the logical version more slowly than those who had not. Coulson argued that the preliminary experience of grappling with the random version of the task induces a state of confusion, so that, when the main task comes along, subjects have dropped d-mode in favour of a learning by osmosis approach. If the main task is actually the illogical one, this puts you at an advantage. Your learning-by-osmosis is unimpeded by the intellect. But if the main task is one which is amenable to being figured out, then you are disadvantaged if you have abandoned d-mode. Whether to back the hare or the tortoise depends crucially on the nature of the situation. If it is complex, unfamiliar, or behaves unexpectedly, tortoise mind is the better bet. If it is a nice logical puzzle, try the hare brain first.

There are indeed many cases in which d-mode is the right tool, and in which the hare clearly comes out the winner. Imagine that you have a regular chessboard, an $8 \times 8$ chequered square, and you cut out two diagonally opposite corner squares (leaving 62 squares, see Figure 3). You make up 31 domino-shaped bits of cardboard, each of which neatly covers two squares on the board. You give me the mutilated board and the oblong pieces of card and ask me if I can exactly cover the 62 squares using the 31 bits of board, without cutting, bending or overlapping them. What do I do?

My first thought may be that of course I can do it – 31 dominoes, each covering two squares; $31 \times 2 = 62$; QED. Your quizzical look, however, strongly suggests that it is not quite that simple. So what I then do is start laying out the dominoes on the board . . . but every time I try it, I always seem to be left with an odd square on the opposite side of the board from the available piece of cardboard. As I am deep down convinced that it *is* possible, I keep shuffling the dominoes around hopefully; but finally have to confess that I do not seem to be able to find a solution. A large amount of time, and some emotional energy, are consumed.

You then invite me to think about the *colours* of the squares . . . especially the ones that have been cut out. I, having implicitly decided that the colours are irrelevant to the problem, and therefore given them no thought, wonder what you are talking about. Then I realise that the two opposite corner squares must be the *same* colour, either both black or both white. If you have taken two white

*Figure 3. The mutilated chessboard*

squares away, that means that there are 30 white ones and 32 black ones left – an unequal number. But each domino has to cover two adjacent squares, i.e. one black one and one white one. So for the puzzle to be soluble there has to be not just an even number of squares, but an equal number of blacks and whites. Obviously – now I come to think about it – it can't be done. (Imagine a 2 × 2 board consisting of four squares; take away diagonally opposite squares and, by analogy, the answer is plain.) Some straightforward deliberation could have saved me time and trouble.

Pawel Lewicki's group at the University of Tulsa have investigated a slightly different aspect of the relationship between know-how and knowledge: whether they automatically change together, or whether learning that affects one can leave the other unchanged. The researchers focused on one particular set of patterns that we have all been developing since babyhood, and in which we might be considered to be quite expert: those that associate how people look with how they are likely to react – their faces, most obviously, with their moods and personalities. Even if much of this knowledge is implicit, we should have developed some conscious self-knowledge about the interpersonal rules of thumb that we tend to use. Spectacle wearers are likely to be studious, for example. People who don't make eye contact are shy or shifty. People whose eyes have large pupils are more warm and friendly than those with small pupils. People whose heads loll about in an alarming way are probably not Cambridge professors. Everyone has their idiosyncratic set of diagnostic features. We think we can recognise 'sad eyes'. 'mean mouths' or 'business-like moustaches'.

Lewicki first elicited from his subjects as many of these personal associations as they could give. He then asked them to look at a long succession of photographs of unfamiliar people, and to try to predict what their personalities were like. After each picture, they were given 'feedback' about how good their predictions were. Unbeknownst to the subjects, Lewicki had again 'stacked the deck', by determining the character which he attributed to each photographed person on the basis of some subtle combination of facial features. As with the experiments of his which were discussed in Chapter 2, subjects gradually got significantly better at making the predictions, even though they had absolutely no conscious knowledge of any connection between the facial features and the supposed personalities.

However, there is a new twist. Lewicki had rigged his character attributions so that, for each subject, some of the connections between face and personality were the exact opposite of the ones which they had told him they relied on in everyday life. So in order to learn the patterns in the experiment, they were required to go against their normal assumptions. What effect did this have, Lewicki asked, either on the speed with which the subjects learnt the experimental pattern, or on the strength of their pre-existing rules of thumb? Should the mismatch not slow down the new learning, and/or cause some shifting in what is known consciously? You would think so – if you make the commonsense assumption that people's self-knowledge is an accurate reflection of the way they go about things.

In fact, Lewicki found that the subjects' pre-existing conscious beliefs a) had no effect on the speed or efficiency with which the contrary associations were learnt through experience; and b) were themselves unaffected by the unconscious learning that had taken place. The undermind is acquiring knowledge of which consciousness is unaware, and by which it is unchanged, and using it to influence the way people behave. Consequently a schism develops between what people think they know (about themselves), and the information that is unconsciously driving their perceptions and reactions. The views that they *espouse* about themselves, we might say, become at odds with the ones that their behaviour in fact *embodies*.

This small experiment thus furnishes us with a neat illustration of the kind of 'split personality' phenomenon with which we are all familiar, but which it is often convenient to ignore: the existence in the mind of a second centre of operations which is capable of going its own way, untroubled by what conscious 'headquarters' happens to be saying. And consciousness itself can remain unruffled by the discrepancy, by the simple expedient of not noticing that it exists. At the end of a review of his experiments, Pawel Lewicki concludes that 'our non-conscious information processing system appears to be faster and "smarter" overall than our ability to think and identify meanings . . . in a consciously controlled manner. Most of the "real work" [of the mind] is being done at a level to which our consciousness has no access.'[6] This, from the hard-nosed world of cognitive science, is an extraordinary conclusion. Yet it is what these carefully controlled experiments reveal.

The studies we have reviewed so far demonstrate that the urge to be articulate is a mixed blessing when it comes to learning. But there are other areas of life where the same might be said. What about the execution of a skill that is already well learnt, for example? Does it make any difference to one's expertise whether one is able to put what one knows into words, or not? In a paper entitled 'Knowledge, knerves and know-how', R. S. Masters of the University of York has shown that people who can articulate what they are doing may go to pieces under pressure more than those whose skill in entirely intuitive.[7] He studied people who were learning to play golf, focusing particularly on their putting skill. One group of learners was taught how to putt 'explicitly' – they were given a set of very specific instructions which they were asked to follow as carefully as they could. Another group was given no instructions – they simply practised – and they were even asked to occupy their minds with an irrelevant task to prevent them thinking about their putting as they were doing it. After their training, both groups were tested on their putting ability by an imposing 'golfing expert' whom they had not met before; and there were also significant financial rewards and penalties depending on how well they did. Both the 'expert' and the money were designed to take the test stressful.

Masters discovered that the performance of those who had learnt intuitively held up much better than that of those who had been following instructions. His explanation was that the breakdown of performance under pressure – what sportspeople refer to as 'choking', or 'the yips' – was due to the instructed group flipping back into d-mode and trying to remember and follow instructions, rather than just play the shot. People who had learnt intuitively were not able to do this, as they had no explicit knowledge to fall back on. They just had to carry on as normal – and this, it turned out, stood them in better stead. Thinking about what you are doing may introduce a kind of analytic self-consciousness which gets in the way of fluent performance – an effect reminiscent of the famous centipede who was rooted to the spot when asked which leg he moved first.

Know-how is acquired in different ways from verbal knowledge, as we have seen; but it is also 'formatted' differently, and is good for different kinds of purposes. For example, Euclidean geometry – the kind we did at school – is an extremely elegant and powerful tool for describing a family of idealised shapes, those that can be made out of straight lines and mathematical curves drawn on a flat surface, and regular three-dimensional objects such as spheres, cubes and cones. In this arcane universe, all kinds of strange and beautiful properties appear, and precise calculations can be made. The areas of circles and parallelograms, for example, can be computed exactly with the aid of certain formulae. However, if you ask geometry about the area of an *untidy* shape, one that cannot be described by equations, it immediately loses its power and grace. The real, irregular world is too awkward and intractable, and it has to be neatened up, in the way the axioms of geometry demand, before it can be treated. To calculate the area of France, using Euclid, we would have to suppose it to be a badly drawn hexagon, or to superimpose upon it a grid of little squares. Unless we force it into such *a priori* shapes and categories, we cannot get our strong generalisations to work.

Now, in contrast, consider a humble device called the polar planimeter, invented in 1854 by a German mechanic, Jacob Amsler.[8] It consists of two sticks flexibly joined together as in Figure 4. The top end of the 'vertical' stick is fixed to the table. At the left-hand end of the 'horizontal' stick is a wheel that sits on the table which can both rotate and skid (if it is pulled sideways). At the right-hand end is a pointer that also sits on the table. The cunning thing about this simple tool is that, if you trace the outline of any shape with the pointer, the wheel will rotate by an amount that is directly proportional to the area of the shape. All you have to do is calibrate

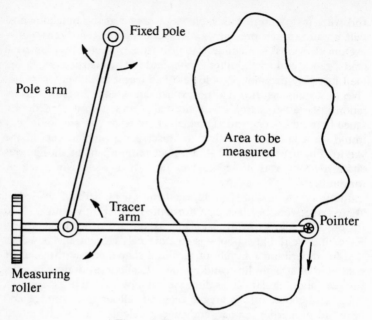

*Figure 4. The polar planimeter*

the wheel by tracing out a shape with a known area – a 5cm square, say – and then you can use your polar planimeter to measure the area of any shape, no matter how odd.

The knowledge which d-mode generates, and on which it relies, is more like Euclidean geometry. It tends to be general, abstract and powerful, and to apply it to particular cases you often have to make the world appear neater than it truly is. The polar planimeter corresponds to 'know-how'. It capitalises intelligently on a curious fact, and uses a 'trick' to solve cheaply and easily what for geometry is embarrassingly difficult. Geometry can do lots of things that the polar planimeter cannot, but for the particular job of measuring irregular areas (within a certain range of sizes) the planimeter is much more accurate and efficient. *How* it does it neither I nor Herr Amsler can tell you – and provided it works, our explanatory shortcomings are of little consequence (though they pose a nice intellectual challenge: the basis for a PhD perhaps). It is said that the painter Giotto could draw perfect freehand circles, and would leave them as calling cards. It is doubtful whether he knew the algebraic formula for a circle, or how to calculate its circumference, and certain that a course in geometry would have not improved his skill, and might well have impeded it.

Our know-how is in general of this ad hoc, smart, opportunistic kind. The know-how regions of our minds are organised less like the Library of Congress than a well-used kitchen: logic continually gives way to convenience. I do not have to make my kitchen so rational that anyone could come in and figure out where the Tabasco sauce is from first principles. If I did have such a logical layout, I would not be so well set up to cook the kinds of things that *I* like to cook, and habitually do. Know-how is, as I say, formatted differently to knowledge in that it grows by osmosis (rather than comprehension); manifests itself in specific domains of expertise (rather than in abstractions); capitalises on serendipity (rather than first principles); and is organised idiosyncratically (rather than systematically). No wonder that the ways of knowing that use and create it are very different in their time characteristics from d-mode.

Western culture's over-reliance on d-mode reflects a lack of appreciation of these vital differences between knowledge and know-how. We tend, as a society, to make what was originally only an academic error: what Pierre Bourdieu refers to as the *scholastic fallacy*. 'This fallacy . . . induces people to think that agents involved in action, in practice, in life, think, know and see as someone who has the leisure to think thinks, knows and sees.'[9] By assuming that knowledge is similar to know-how, we are led to suppose that know-how can – even should – be acquired through knowledge, and that knowledge, once acquired, ought to transform itself automatically into know-how. Managers are sent on a five-day course on 'leadership', and are immediately supposed to come into work the following Monday and start leading. The frustration with, and frequent cynicism about, such short courses, in the business world and elsewhere, is not due to lack of commitment on the part of participants, nor of skill on the part of the trainer. It reflects a deep confusion about the nature of learning and knowing.

The confusion makes us promote 'book learning' and formal education (and training) as the proper medium for acquiring everything. Adults pore over the instruction manual for a new computer, afraid to plug it in until they know how it works and what to do, while their children have already discovered, just by 'messing about', how to make it do the most complicated tricks. Apprentice midwives used to learn their craft by assisting their more experienced mentors at hundreds of births. Now they have to have a degree. There are even those who argue that couples should have to attend a series of seminars on 'parenting skills' before they are allowed to have a baby. The tragedy is that now there may even be some sense to this. If the other ways of knowing *have* been effectively disabled by the belief that intellect is the only mode we have, or the only mode we need, then the belief becomes the reality. D-mode does then provide the only avenue open to us for learning, however limited or inappropriate it may actually be for the job in hand.

A MORI poll in 1996 on learning attitudes revealed that two-thirds of people 'prefer to learn from books', while another 19 per cent prefer CD-ROM and computers. Nobody, apparently, said that they prefer to learn by messing about, by osmosis, or just by watching. Learning has become something that you do in a special place, with special equipment, under the instruction of experts, using your deliberate, conscious intellect. No other possibilities seem to be catered for: a pity, at the very least, if learning by osmosis is, when faced with certain kinds of complexity, a more intelligent option than d-mode.

But learning by osmosis has its own limitations, just as d-mode has. Not only may it be deployed at the wrong time, leading to a protracted process of trial and error which could have been short-circuited by a little logical thought; it often cannot be communicated, or only very crudely, and there are many occasions on which this is a definite handicap. The first time I went ice-skating with my twelve-year-old cousin, I strapped on the skates and stood nervously on the edge of the rink sliding my feet backwards and forwards, convinced of the physical impossibility of what dozens of people were doing around me. Eventually I swallowed my pride and asked Dany to tell me how it was done. 'It's easy,' she said, 'watch,' and she sped off round the ice. When she got back I was beginning to get irritated. 'I know you can *do* it,' said, 'but I want to know *how* to do it.' 'It's easy,' she said, 'watch,' and sped off again.

Her know-how was completely inarticulate, yet there *are* useful tips and explanations that can be given. To be a practitioner, it may be best not to be able to think too much about your skill or your art. But to be a coach is different. A whole new phase of learning may be required if the virtuoso wants to become a teacher, for she may have laboriously to unpick her seamless expertise and turn it into the descriptions and explanations that, judiciously administered, help learning to happen.

The most important limitation of know-how, however, is its relative inflexibility. Practical knowledge that has been learnt without thinking may work smoothly and fluidly within the original domain.

But many psychological studies have shown that when the appearance of a task is changed, even if only a little, while the underlying logic remains exactly the same, know-how often fails to transfer. People who have learnt to control the factory process may function no better than a complete beginner if effectively the same problem is now presented as being about the control of 'traffic flow'. The polar planimeter is useless for determining the volume of three-dimensional objects, while the principles of Euclidean geometry can easily be extended. With know-how, perceiving and doing are wrapped up together in one tightly interwoven package.

From an evolutionary perspective, the 'bundling' of know-how does not matter at all if your world consists of a small number of separate scenarios – looking for food, cleaning your fur, mating, sleeping, avoiding predators and raising your young, for example. If your life consists of such a neat set of discrete jobs, then your main problem, apart from keeping your wits sharpened, is to know which scenario you are in, or which one you need to switch to. To have your know-how organised and integrated under separate headings is economical and efficient. But if your world is more complicated, the scenarios or 'scripts' you take part in become more numerous, and they begin to interweave. The same individuals in your community may take different roles in different scripts. For the male black widow spider, or praying mantis, your mate may suddenly turn into your executioner.

As life gets more intricate, so it becomes a matter of survival to be able to deconstruct situations into familiar parts, and to be able to construct responses to hybrid situations by putting together different facets of different scripts. Parties, for example, may be stressful because they bring together friends and family with whom one has quite different kinds of relationship, and which separately bring out contrasting sides of your personality. If you could only relate to them 'in context', and had only a single stereotyped 'party' script, you would have no way of solving this intricate social problem. But if you have a sense of your friends that is somewhat disembedded from the contexts in which you usually meet, you may be able to integrate all the pieces that originally came from different 'jigsaw puzzles' into a new and hopefully coherent picture.

This carving of scenarios into recombinable 'concepts' is, basically, the ability conferred by language, and by d-mode. When understanding has been 'articulated', it has not only been turned into words. Articulated also means 'jointed; composed of distinct parts which may move independently of each other'. Know-how is not articulated in either sense. It is not capable of being taken apart, reflected upon, or put together again in novel ways when expertise breaks down or conditions change. It can only shift gradually under the influence of learning by osmosis. And because it cannot be discussed, it cannot easily be influenced by what other people may say or bring to it. The risk with fluent know-how is that it will be deployed *mindlessly*, in a way that takes no account of considerations or information that is held in a different sub-compartment of the mind. The ability to see that some aspect of what you have learnt in one situation is of relevance in another which looks different superficially is a highly valuable one, and it has been shown in several experiments that it can be increased by the use of conscious reflection.

In one classic study of what has come to be referred to as 'functional fixedness', people were set a problem which could only be solved by seeing that a familiar object could be used in an unfamiliar way. The task was to tie together two pieces of rope hanging from the ceiling. The problem was that they were too far apart to be grasped simultaneously. In the experimental room there was a variety of everyday objects, including a pair of pliers. The problem could be solved by seeing that the pliers could be used as a pendulum bob: you tie them to the bottom of one of the ropes and set it swinging so that, when you are holding the other rope, the first now comes within your grasp. Left to themselves, a high proportion of people fail to solve the problem. But if the experimenter waits until they have got stuck and then simply says 'Think! Think!', many subjects then spontaneously see the solution.

Without d-mode, without the benefits that concentrated analytical awareness bring, the lower animals are smart, but within limits. The spider, *sphex*, the digger wasp, and even the gobiid fish are good at what they are designed by evolution to do: they meet a range of challenges with conspicuous intelligence, but when the world throws a different *kind* of challenge at them, they are found wanting: inflexible and uncreative. They cannot turn around what they 'know' and recombine it in ways that are both novel and appropriate. They cannot dismember their abilities and perceptions – cannot segment and articulate them – and so cannot *re*-member them to suit an unprecedented present. Their unconscious intelligence is more or less crystallised; they lack the ability to dissolve it and reconstitute it.

The same is true, initially, of children, but they are able to transcend these limitations. As they develop, the range and complexity of the scenarios in which children take part start to expand dramatically. They go to playgroup and on to school, where they meet different kinds of adults with different agendas, and with whom they have very different kinds of relationship from those they have with their parents. They take part in new social groups of various sizes and compositions. They start to meet many different *kinds* of things to be learnt about, and to discover new ways of going about learning them. And as they do so, they face a choice: whether to keep multiplying the number of separate mental scenarios; or whether to start to seek a higher-order level of knowing that enables these different scripts to be integrated, compared and combined. If they take the former option, their mental landscape develops into a patchwork of separate 'modules' of know-how that are unable to share what they know among themselves. If they take the latter they need to develop a new form of learning, one which enables them to *ruminate* over their experience; to bring back, as the cow does, what has been separately ingested, and by chewing it over make it more homogeneous. They would have to be concerned not just to meet new challenges one by one, but to look actively for points of segmentation and integration.

And children do start to develop this ability to ruminate, it turns out, around the age they first go to school. Annette Karmiloff-Smith of the Medical Research Council Cognitive Development Unit in London has demonstrated the beginnings of rumination in the context of what she calls 'learning beyond success'. Across many different kinds of task, she has observed that children will first learn how to 'get it right' – and will then, if they are given the opportunity, continue to 'play' with the situation in ways that actually reduce their apparent control and competence for a while. In language learning, for example, a child will very often learn to say (correctly) 'went', but will then go through a phase of using the 'regular' (but wrong) form 'goed', before finally reverting to 'went'. Or they will learn to balance different-shaped rulers on a fulcrum, and then make mistakes, and then get it right again.

Karmiloff-Smith argues that these dips in performance are symptomatic of exactly the kind of searching for coherence and conceptualisation that I have described. It is as if, when faced with a challenge, children use whatever is at hand to respond to it, like someone after a shipwreck constructing an emergency raft out of all kinds of flotsam in order to keep afloat. But later, when they have a little more leisure, after the storm has passed, they move into a more reflective mode in which they experiment with taking this lash-up to bits again to see what happens, and where it might fruit-fully draw on pre-existing pockets of know-how developed to cope with different situations, to make their know-how as a whole more elegant, integrated and powerful.

Language, and the ways of knowing which it affords, liberates; but it comes with snares of its own. Although it allows us to learn from the experience of others, and to segment and recombine our own knowledge in novel ways, it creates a different kind of rigidity. As Aldous Huxley said: 'Every individual is at once the beneficiary and the victim of the linguistic tradition into which he has been born – the beneficiary inasmuch as language gives access to the accumulated records of other people's experience; the victim insofar as it . . . bedevils his sense of reality, so that he is all too apt to take his concepts for data, his words for actual things.'[10] Know-how is tied to particular domains and purposes, but within those bounds it is detailed, accurate, efficient and flexible. D-mode creates a superordinate stratum of knowledge that transcends particular con-texts, but is, by the same token, more abstract, and liable to become detached from the shifting layers of experience that originally under-pinned it. As the Lewicki experiment showed, once this detachment has taken place, know-how can develop pliably in response to new exigencies of experience while knowledge is left unaltered, cast in stone.

Language is not only the internal code in which knowledge is inscribed; it also relies upon, and enshrines, a public system of categories. A language represents a consensus about how the world is to be segmented, and thus determines heavily how things are categorised, talked about, and even perceived. Much has been writ-ten about the relationship between language and 'reality', but the only point to note here is that we are obliged to articulate our know-how in terms that we ourselves have not chosen, and which may well not be the most congenial or accurate descriptors of our personal experience. As we articulate our experience, so we have to pour what is intrinsically fluid and ill-defined into moulds that are more clear-cut, and not of our own making. The language of d-mode implies a 'reality' that is neater, more solid, more impersonal and more agreed-upon than the one that often confronts us. It is both an approximation – leaving out much of the detail – and a distortion – introducing fictional elements that actually have no referent.

D-mode is like map-reading: with a map we are able to get our bearings, and see how one area relates geographically to another.

But maps must be simpler and more static than the world they represent; and they contain conventions that aid the interpretation of the map, but which are not 'real'. As we climb the mountain, we do not periodically have to step over the contour lines. As we cross from England to Wales, the terrain does not change from pink to blue. It is not the case that we cannot go where there is no track, nor, certainly, that the motorway is always the best route. When the map is good enough, and we understand the status of the conventions, then d-mode works well. When we forget, as Alfred Korzybski[11] insisted, that 'the map is not the territory', or when we need, to resolve a predicament, a finer-grain, more subtle or more holistic image than language provides – it is then that we need recourse to our other, slower ways of knowing. Some predicaments cannot be dealt with effectively with the tools of analysis and reason. And there are some, too, that will not succumb to an increase in expertise, such as learning by osmosis delivers. To deal with such problems, we need access to those slow ways of knowing we have preliminarily called rumination or contemplation; mental modes which deliver, it is claimed, forms of *creativity* and *intuition*.

## CHAPTER 4

# Knowing More than We Think:
## Intuition and Creativity

'Did you make that song up?'
'Well, I sort of made it up,' said Pooh. 'It isn't Brain
. . . but it comes to me sometimes.'
'Ah,' said Rabbit, who never let things come to him,
but always went and fetched them.

*A. A. Milne, The House at Pooh Corner*

In his autobiography the nineteenth-century English philosopher Herbert Spencer recounts a conversation with his friend Mary Ann Evans – the novelist George Eliot. They had been discussing Spencer's recently published book *Social Statics*, and George Eliot suddenly observed that, given the amount of thinking he must have done, his forehead remained remarkably unlined. 'I suppose it is because I am never puzzled,' said Spencer – to which Eliot, understandably, replied that this was the most arrogant remark she had ever heard. Spencer says that he went on to justify his remark by explaining that

my mode of thinking does not involve the concentrated effort which is commonly accompanied by wrinkling of the brows. The conclusions, at which I have from time to time arrived . . . have been arrived at unawares – each as the ultimate outcome of a body of thoughts that slowly grew from a germ . . . Little by little, in unobtrusive ways, without conscious intention or appreciable effort, there would grow up a coherent and organised theory. Habitually the process was one of slow unforced development, often extending over years; and it was, I believe, because the thinking done went on in this gradual, almost spontaneous way, without strain, that there was an absence of those lines of thought which Miss Evans remarked.

In Spencer's opinion 'a solution reached in the way described is more likely to be true than one reached in pursuance of a determined effort to find a solution. *The determined effort causes perversion of thought.* . . An effort to arrive forthwith at some answer to a problem acts as a distorting factor in consciousness and causes error, [whereas] a quiet contemplation of the problem from time to time allows those proclivities of thought which have probably been caused unawares by experience, to make themselves felt, and to guide the mind to the right conclusion.' (Emphasis added)[1]

The ways of knowing with which both Pooh Bear and Herbert Spencer are familiar are different from d-mode in a number of ways. Most obviously, they take time, and therefore they require patience: a relaxed, unhurried, unanxious approach to problems. In this they resemble 'learning by osmosis', but they are not the same. In learning by osmosis, the undermind gradually uncovers patterns that are embedded in, or distributed across, a wide variety of experiences. Know-how is distilled from the residue of hundreds of specific instances and events. But while Spencer's insights into the organisation of society undoubtedly drew on much prior thought and observation, the process that he is referring to is one that goes beyond this unconscious distillation. This process seems to reflect not the acquisition of new information so much as the mind's ability to discover, over time, new patterns or meanings within the information which it already possesses, and to register these consciously as *insight* or *intuition*. Though experience provides the data, the process is not acquisitive but ruminative. Pooh's song that prompted Rabbit's question demonstrates the same process on a smaller scale. He was not announcing the inductive discovery of a new generalisation, but simply producing something which came 'out of the blue'.

Despite the widely-held assumption that d-mode represents the most powerful thinking tool we possess – which makes it the one we call upon, or revert to, in the face of urgent demands for solutions – the truth is that our ideas, and often our best, most ingenious ideas, do not arrive as the result of faultless chains of reasoning. They 'occur to us'. They 'pop into our heads'. They come out of the blue. When we are relaxed we operate very largely by intuition. We don't usually offer a detailed rationale for our restaurant preference: we say 'I feel like Thai'. We happily allow ourselves to be nudged by feelings and impulses that do not come with an explicit justification. Yet when we are put 'on the spot' in a meeting, or are faced with an urgent 'problem' that demands 'solution', we may act as if these promptings were weak, unreliable or negligible. We feel as if intuition will not stand up to scrutiny, and will not bear much weight. There is now a body of research which shows that intuition is more valuable and more trustworthy than we think; and that we disdain it, when we are 'on duty', to our practical detriment.

We need a more accurate understanding of the nature and status of intuition: one which neither under nor overvalues it. Those who disparage intuition are reacting, often unwittingly, against the presumption that intuition constitutes a form of knowledge that is 'higher' than mere reason, or even infallible. The dictionary definitions still carry some of that inflated view, and by doing so they create expectations that are patently false. Chambers' dictionary gives intuition as 'the power of the mind by which it *immediately perceives the truth* of things, without reasoning or analysis'. The Shorter Oxford is more poetic and more presumptuous still: it gives intuition as 'the immediate knowledge ascribed to angelic and spiritual beings, with whom vision and knowledge are identical'.[2] Now while it may be the case that there is a certain quality of intuition, one which may take much careful cultivation to acquire, which does give access to a qualitatively different kind of knowledge, it is self-evident that everyday intuition falls far short of this ideal. Our promptings are notoriously fallible, whether they concern a career move or a life partner, a book that we misjudged by its cover or a new route that the 'nose' confidently said was a short cut, but which only succeeded in getting us lost.

Intuitions can be wrong, but that does not mean they are worthless. Intuitions are properly seen as 'good guesses'; hunches or hypotheses thrown up by the undermind which deserve serious, but not uncritical, attention. They offer an overall 'take' on a situation that manifests not – not yet – as a reasoned analysis, but as an inkling or an image. Behind the scenes, the undermind may have integrated into this tangible prompt a host of different considerations, including analogies to past experience and aspects of the present situation, of which the conscious mind may not have even been aware. And this integration can happen, as the dictionary definitions say, 'immediately', or it may take time – even, as in Spencer's case, up to years. But the result, when it does 'pop up', is always provisional. It is a pudding, served up by the unconscious, whose proof is in the eating: a critical testing which may be the reaction of the audience to an impromptu witticism, à la Pooh, the rigorous checking of logical implications, or the detailed working out of a creative poetic or artistic theme.

Fast intuitions – 'snap judgements' and quick reactions – are vital responses for the human being, just as they are for animals. When the present event is a variation on a familiar theme, it pays to be able to classify it and react in habitual fashion. To spend time pondering on insignificant details is sometimes wasteful, or even dangerous. No need to inspect the number plate of the bus as it bears down upon you. But these reflexes work to our detriment when a new situation looks similar to ones we have experienced in the past, but is actually different. Then the balance of priorities shifts, and it is now the quick, stereotyped response that is the risky one, while more leisurely scrutiny can pay dividends.

The importance of this shift from fast to slow thinking was graphically demonstrated in the laboratory by Abraham and Edith Luchins as long ago as the 1950s. They set people puzzles of the following sort. 'Imagine that you are standing beside a lake, and that you are given three empty jars of different sizes. The first jar holds 17 pints of water; the second holds 37 pints; and the third holds 6 pints. Your job is to see whether, using these three jars, you can measure out exactly 8 pints.' After some thought (which may, to start with, be quite logical), most people are able to end up with 8 pints in the largest jar. Then they are set another problem of the same type, except this time the jars hold respectively 31, 61 and 4 pints, and the target is to get 22 pints. And then another, with jars holding 10, 39 and 4 pints where the target is 21 pints. (You may like to try to solve these puzzles before consulting the notes for the solutions.)[3] You will find that the same strategy will work for all three problems. But now comes the critical shift. You are next given jars of capacity 23, 49 and 3 pints, and asked to make 20 pints. If you have stopped thinking, and are now applying your new-found rule mindlessly, you will solve the problem – but you will not spot that there is now a much simpler solution. The problem looks the same, but this particular one admits of *two* solutions, one of which is more elegant and economical than the other.

We can easily imagine a business company – or any other kind of organisation – falling into the same trap. They may 'think they are thinking' about each problem as it comes along; but if they are unable to think *freshly*, they will keep coming up with the same kinds of answers – even when circumstances have changed and new possibilities are there to be discovered. And one of the strongest forces that prevents the discovery of these new avenues may be the habit of thinking fast: of taking your first intuitive assessment of the situation for granted, and not bothering to stop and check. Milton Rokeach tested this hypothesis, using the Luchins jars, by forcing people to slow down when they were looking at the new problems. If they were allowed to give the 'solution' in their own time, most people immediately applied the rule that had worked previously without question. But when they were prevented from writing down their answer for a minute, some of them pondered the problem in greater detail – and were able to discover the new solution.

Not surprisingly, this benefit only accrued to people who did actually attend to the details of the new problem during the delay. Many people reported that they made up their minds quickly about the answer, and then spent the enforced interval thinking about all kinds of unrelated things – 'making plans for Saturday night's party', 'thinking about letters I had to write', 'counting the holes in the tiles on the ceiling', and so on – and for them, the extra time obviously did nothing to improve their creativity. What was more interesting, however, was the mental activity of the subjects who did find the new solution. They were *not* earnestly figuring out the answers, or making calculations on bits of scrap paper. They were actually musing in a much more general way on what *type* of questions were being asked, and what the experimenter was up to. One said, 'I was wondering what the experiment was trying to prove.' Another said, 'I was thinking what the results would indicate.' It was this kind of 'meta-level' questioning that led to the insight, not the disciplined application of procedures.

Let me illustrate how intuition works with the aid of a slightly more complicated example (one, incidentally, that Wittgenstein was fond of using in his philosophy seminars). Imagine that the Earth has been smoothed over so that it forms a perfect sphere, and that a piece of (non-elastic) string has been tied snugly round the equator. Now suppose that the string is untied, and another 2 metres added to the total length, which is then spaced out so that the gap which has been created between the string and the Earth's surface is the same all the way round. How big is this gap? Could you slide a hair under the string? A coin? A paperback book? Could you crawl under it? Most people's strong intuition is that the gap would be tiny, of the order of a millimetre or two at the most. In fact, it is easy to prove mathematically that it is about 32 centimetres, or just over a foot; so you could indeed crawl under the string. (The proof is in the notes, for those who wish to follow it.)[4] The strange thing, when you work out the geometry, is that the size of the gap turns out to be independent of the size of the original sphere (or circle: the problem is not essentially three-dimensional). So you would get the same-sized gap whether you started with a tennis ball, a circus ring, or the universe. Most people's intuition, on the other hand, insists that the larger the original object, the less 'difference' the 2 metre extension is going to make: in other words, the smaller the gap.

Intuition goes awry here because it is based on the unconscious assumption that this situation is analogous to other, apparently similar, situations where the idea that 'the larger the object, the smaller the change' *does* apply. If we were to change the puzzle slightly, and say: 'Supposing the oceans were neatened up into a huge cylinder, how much would the level rise if we added 20 litres of water?', then the answer is indeed 'not very much'; and we would be right, in this instance, to assume that the larger the original volume, the smaller the difference to its depth. The 20 litres would make much more difference to the depth of a paddling pool. It just turns out that this plausible assumption works for the height of cylinders, but not for the radius of circles. It is a good guess that in one case turns out to be right, and in the other case wrong. Fast intuitions depend on the undermind taking a quick look at the situation and finding an analogy which seems to offer understanding and prediction. These unconscious analogies surface as intuitions. Whether they are right or not depends not on how 'intuitive' they are, but on the appropriateness of the underlying analogy. Often we are absolutely right. But sometimes the undermind is fooled by appearances, and then it leads us off in the wrong direction.

This example also demonstrates how the way of knowing you employ may give different answers to the very same question. D-mode and intuition may well draw on different processes, knowledge and beliefs, and thus may produce conflicting solutions. If you followed the mathematics in the notes, then you might be rationally persuaded that the gap is a foot, while intuitively you persist in believing it to be minute. Below the surface, some assumptions are being made that lead to one answer. Above the surface, so it seems, different premises lead to a different answer. In this case, it turns out that the 'rational' answer is the right one. In other cases (as when intuition told you that there was something suspicious about the well-spoken woman at the door 'collecting for charity', but you persuaded yourself you were being 'silly') it may be intuition that is right and reason that is wrong. It is an empirical issue.[5]

Which mental mode is engaged – and therefore which answer you get – may depend on how you happen to be thinking when the question arises; or on some – possibly quite incidental – feature of the situation. If you catch a physics undergraduate in the bar one night and ask her why, when you throw a ball, it moves through an arc, she is likely (if she can be bothered) to tell you a story about the 'energy' or 'impetus' you give to the ball when you throw it, and how this gets used up overcoming the drag of the air and the force of gravity. When the upwards 'oomph' has been depleted to a certain level, she says, gravity starts to 'win', the ball reaches its zenith and begins to fall. However, if you then remind her that this is a *physics* problem, she may well stop for a moment (as she switches from intuitive mode to physicists' d-mode) and say, 'Of course. Silly me. There isn't any "oomph" you put in to the ball as you throw it. The only forces are gravity and the air resistance.'[6] Her first 'take' is an everyday, intuitive one; her second switches her into a different frame of reference, giving access to a different data-base and different ways of thinking. If the question had been on an examination paper, she would have selected d-mode automatically.

The power of context to flip people into one way of knowing rather than another – and to produce quite different responses to what is logically the same problem – is widespread, and very striking. In a study of ten-year-olds by Ceci and Bronfenbrenner in 1985, for example, the children sat in front of a computer screen in the centre of which one of a variety of geometric shapes would period-ically appear.[7] Their job was to predict (by moving the cursor with a mouse) in which direction, and how far, the shape was about to 'jump'. The shapes were circles, squares and triangles that could be dark- or light-coloured, and large or small. In theory, the children could have predicted the jump on the basis of the shape, because squares always went to the right, circles to the left, and triangles stayed in the middle; dark things went up and light things down; and large things went a short distance and small things a long dis-tance. After 750 trials, the children had learnt virtually nothing.

However, after making a small change to the task, *which had no effect at all on its logical difficulty*, things looked very different. All the experimenters did was replace the three geometrical shapes with animals (birds, bees and butterflies); swap the normal computer cursor for an image of a 'net'; add some sound effects; and tell the children that this was a game in which they had to try to catch the animals as they moved. After less than half as many goes, all the children were placing the net in the right position to 'capture' the animals with near-perfect accuracy. The geometrical shapes told the children that this was a 'school-type task', and so automatically flipped them into d-mode. They tried to figure out the rules, and couldn't. So they made no progress. The other version led them to reinterpret the display as a 'video game' – and this flipped them into an intuitive mode which enabled them to pick up the relevant relationships easily and unconsciously.[8]

Intuitions can also go wrong when they are based on inaccurate judgements about what is relevant and what is not, as we saw earlier with the 'mutilated chessboard'. Here is another example.

> A certain town is served by two hospitals. In the larger hospital about 45 babies are born each day, and in the smaller one, about 15. As you know, about 50% of all babies are boys and 50% girls. The exact percentage of girls however naturally varies from day to day. Some days it may be over the 50%; some days under. As a check on this variation, for a period of one year, both of the hospitals recorded the days on which more than 60% of the babies born were girls. Over the year, which hospital do you think recorded more such days? The large one? The small one? Or about the same?

When psychologists Daniel Kahneman and Amos Tversky asked nearly a hundred people this question, 22 per cent said the larger; 22 per cent said the smaller; and 56 per cent said 'about the same'.[9] Nobody sat down and worked it out with a calculator, so we must suppose that all these were intuitions. But more than three-quarters of them were wrong. (I was one of those who said 'about the same'.) A moment's reflection should be enough, however, to convince you – as it did me – that the correct answer is 'the small one'. The smaller the sample, the easier it is to get a larger percentage skew by chance. (It only takes two 'boys' to turn out to be girls for the small hospital to exceed its 60 per cent point.) A relevant piece of information – the size of the hospital – is actually being tacitly disregarded by half the population when they are generating their intuitive response (even though they are perfectly able to see its relevance when it is pointed out). These 'fast intuitions' are suscep-tible to all kinds of invisible influences, some of which will be appro-priate and beneficial, and others of which will, in a particular instance, be misleading.

If fast intuition is vulnerable when responding to predicaments that look familiar but which are not as they seem, in what circum-stances are the slower ways of knowing of most value? As with learning by osmosis, it turns out that slow intuition is good at uncovering non-obvious relationships between areas of knowledge; at seeing 'the pattern that connects' experiences that are superficially disparate. Intuition proves its worth in any situation that is shadowy, intricate or ill defined – regardless of whether the focus of concern is a mid-life crisis, a knotted-up relationship, an artistic project or a scientific conundrum.

In science, intuition is the faculty that comes up with the metaphor, the image or the idea that binds together and makes sense of experimental results which cumulatively seem to embarrass an existing theory, but which up to that point had lacked any alternative coherence. Both Darwin's account of the mechanism of evolution and Einstein's theories of special and general relativity offered just such explanatory patterns. They took a pile of details and transformed them into a theoretical structure that gave them meaning, and predicted new findings. And these, like many other scientific breakthroughs, came about through a way of knowing that was patient, playful and mysterious, not rational, earnest and explicit. As Einstein himself famously said, of his own creative process:

> The words of the language as they are written or spoken do not seem to play *any role* in my mechanism of thought. The psychical entities which seem to serve as elements of thought are certain signs and more or less clear images which . . . are in my case of visual and some of muscular type. [These elements take part in] a rather vague play . . . in which they can be voluntarily reproduced and combined . . . This combinatory play seems to be *the* essential feature in productive thought, before there is any connection with logical construction in words or other kinds of sign which can be communicated to others . . . In a stage where words intervene at all, they are, in my case, purely auditive, but they interfere [note, 'interfere'] only in a secondary stage. (Emphasis added)

Sometimes, as in the case of Herbert Spencer, one is aware of the pattern of thought gradually forming itself, as a large crystalline structure may slowly appear out of a saturated chemical solution in which a seed crystal has been placed. While at other times, the work proceeds unconsciously until the point at which the binding idea as a whole is delivered into consciousness. Rita Levi-Montalcini, who shared the Nobel Prize for medicine in 1986, for example, said: 'You've been thinking about something without willing to for a long time . . . Then, all of a sudden, the problem is opened to you in a flash, and you suddenly see the answer.' While Sir Neville Mott, physics laureate in 1977, confirms both the suddenness of the insight, and the difficulty of finding the right way of expressing it in d-mode: 'You suddenly see: "It must be like this". That's intuition . . . if you can't convince anybody else. This certainly happened to me in the work for which I got the Nobel prize. It took me years to get my stuff across.'[10]

Intuition may deliver its produce to consciousness in the form of more or less connected and coherent thoughts. But at other times, even for scientists, the undermind speaks in a variety of different voices. For Einstein, as for many creators, the language of intuition drew on visual and other forms of imagery. Kekulé first discovered that the carbon atoms of the benzene molecule linked up into a ring through watching the flames of his fire transform themselves, in his mind's eye, into snakes that turned round and bit their own tails. Sometimes intuition emanates in an almost aesthetic judgement: what Nobel chemistry laureate Paul Berg calls 'taste'. 'There is another aspect I would add to [intuition], and that is, I think, taste. Taste is almost the artistic sense. Certain individuals . . . in some undefinable way, can put together something which has a certain style, or a certain class, to it. A certain rightness to it.'

For others intuition manifests itself as a vague but trustworthy feeling of direction or evaluation – one 'just knows' which of several lines of enquiry to pursue, or which of a range of experimental results to take seriously, and which to ignore. Michael Brown (Nobel medicine laureate, 1985) describes how 'as we did our work, we felt at times that there was almost a hand guiding us. Because we would go from one step to the next, and somehow we would know which was the right way to go. And I really can't tell how we knew that . . .' While Stanley Cohen (Nobel medicine laureate, 1986), in similar vein, commented on the importance of developing a 'nose' for the important result – and of seeing this intuitive response as a valuable guide. 'To me it is a feeling of . . . "Well, I really don't believe this result", or "This is a trivial result" and "This is an important result" and "Let us follow this path". I am not always right, but I do have feelings about what is an important observation and what is probably trivial.' Note that Cohen acknowledges both the value and the fallibility of intuition. It can be wrong, and needs checking; but it none the less acts as source of guidance that is to be heeded and respected.

There are many accounts by creative artists and scientists of the need for patience and receptivity. In science, Konrad Lorenz, who won the Nobel Prize for medicine in 1973, stressed the importance of waiting. 'This apparatus ... which intuits ... plays in a very mysterious manner, because it sort of keeps all known facts afloat, waiting for them to fall in place, like a jigsaw puzzle. And if you press ... if you try to permutate your knowledge, nothing comes of it. You must give a sort of mysterious pressure, and then rest, and suddenly BING, the solution comes.' While mathematician and philosopher George Spencer Brown declares, in his book *Laws of Form*:

> To arrive at the simplest truth, as Newton knew and practised, requires years of contemplation. Not activity. Not reasoning. Not calculating. Not busy behaviour of any kind. Not reading. Not talking. Not making an effort. Not thinking. Simply *bearing in mind* what it is that one needs to know.[11]

It is not, according to Lorenz and Spencer Brown, that one gives up on an intractable problem and drops it completely. The process is subtler than that. You do not try to figure it out, yet you 'give a sort of mysterious pressure'. You do not actively think, but you somehow 'bear the problem in mind'. It is as if you allow the problem to be there, to continue to exist on the edge of consciousness, yet without any purposeful attempt to bring it to a resolution. Nel Noddings, the American philosopher, describes this delicate balance of seeking and receiving in the more mundane context of studying a book.

> The mind remains, or may remain, remarkably active, *but instrumental striving is suspended*. In such modes we do not try to impose order on the situation but rather we let order-that-is-there present itself to us. This is not to say, certainly, that purposes and goals play no role in our submitting ourselves to a receptive state. Clearly they do. We may sit down with our mathematics or literature because we want to achieve something – a grade, a degree, a job – but if we are fortunate and willing, *the goal drops away*, and we are captured by the object itself.[12] (Emphasis added)

The gradual formation and development of an idea over a long time, perhaps from the tiniest of beginnings, and its delivery unwilled into consciousness, is a process that is as well known to artists as it is to scientists and mathematicians. Playwright Jean Cocteau both enthusiastically endorses the need to let the mind lie fallow, and attempts to scotch the idea that 'the muse' which springs from a patient state has anything magical or supernatural about it.

Often the public forms an idea of inspiration that is quite false, almost a religious notion. Alas! I do not believe that inspiration falls from heaven. I think it rather the result of a profound indolence and of our incapacity to put to work certain forces in us. These unknown forces work deep within us, with the aid of the elements of daily life, its scenes and its passions, and, when . . . the work that makes itself in us, and in spite of us, demands to be born, we can believe that this work comes to us from beyond and is offered us by the gods. The artist is more slumberous in order that he shall not work . . . The poet is at the disposal of his night. His role is humble, he must clean house and await its due visitation.

The historian John Livingston Lowe has made a detailed study of the sources and materials on which Coleridge based 'The Ancient Mariner', and has been able to trace in these sources the forgotten antecedents of every word and phrase that appears in the most vivid stanzas.[13] He summarises the processes that must have been occurring, out of sight, in the poet's mind thus:

> Facts which sank at intervals out of conscious recollection drew together beneath the surface through almost chemical affinities of common elements . . . And there in Coleridge's unconscious mind, while his consciousness was busy with the toothache, or Hartley's infant ills, or pleasant strollings with the Wordsworths between Nether Stowey and Alfoxden, or what is dreamt in this or that philosophy – there in the dark moved the phantasms of the fishes and animalculae and serpentine forms of his vicarious voyagings, thrusting out tentacles of association, and interweaving beyond disengagement.[14]

Coleridge himself has described the composition of his other famous epic, 'Kubla Khan'. Feeling slightly 'indisposed', as he puts it, he took some opium, and settled down to continue his reading of a work called 'Purchas's Pilgrimage'. Shortly he dozed off, just as he was reading, 'Here the Khan Kubla commanded a palace to be built, and a stately garden thereunto. And thus ten miles of fertile ground were enclosed with a wall.' Three hours later he awoke 'with the most vivid confidence that he could not have composed less than two to three hundred lines – if that indeed can be called composition in which all the images rose up before him . . . without any sensation or consciousness of effort.' Immediately Coleridge grabbed pen, ink and paper and 'eagerly wrote down the lines that are here preserved'.[15]

American poet Amy Lowell describes how she uses incubation quite consciously as a trustworthy technique. 'An idea will come into my head for no apparent reason; "The Bronze Horses", for instance. I registered the horses as a good subject for a poem; and, having so registered them, I consciously thought no more about the matter. But what I had really done was to drop my subject into the subconscious, much as one drops a letter into the mailbox. Six months later, the words of the poem began to come into my head, the poem – to use my private vocabulary – was "there".'

Incubation is a process that may last for months or years, but its value is not confined to such long periods of gestation. It works over days (as when we 'sleep on it', and find the problem clarified, or even resolved, in the morning), or such short spans as a few minutes. The French mathematician Henri Poincaré, well known for his reflections on his own creative process, concluded:

> Often when one works at a hard question, nothing good is accomplished at the first attack. Then one takes a rest, longer or shorter, and sits down anew to the work. During the first half-hour, as before, nothing is found, and then all of a sudden the decisive idea presents itself to the mind ... The role of this unconscious work in mathematical invention appears to me incontestable, and traces of it would be found in other cases where it is less evident ...

There is now experimental evidence that corroborates these vivid anecdotes, and which helps us to understand how it is that incubation does its work. Steven Smith and colleagues at Texas A&M University have carried out a series of studies in which they were able to demonstrate incubation in the laboratory. Of course, they have not been able to reproduce the full complexity of the real-life creative experiences of an Einstein or a Coleridge. It is of the essence of such experiences that they cannot be directly manipulated or controlled. The undermind will not perform to order. Nevertheless, the results are informative.

The kinds of problems which Smith set his subjects were designed to mimic one of the key features of real-life creative insight: the discovery of a meaningful, but non-obvious, connection between different elements of the situation. So-called 'rebus' problems arrange words and images in such a way that they suggest an every-day phrase. For example:

ME             JUST            YOU

represents spatially the phrase 'just between me and you'. Or

TIMING            TIM  ING

is a visual pun on the expression 'split second timing'.

Subjects were shown a succession of such puzzles, and initially given thirty seconds in which to attempt to solve each one. Some of the puzzles were accompanied by helpful 'clues' (such as 'precise' for the second example above), or unhelpful ones (such as 'beside' for the first one). Those problems that the subjects failed to solve the first time round were re-presented for a second try either immediately, or after a delay of five or fifteen minutes. When they had a second go immediately, subjects showed no improvement over their initial score. But when they were retested after a delay, performance improved by 30 per cent on the puzzles that had been accompanied by the unhelpful clues; and the longer (fifteen-minute) delay produced greater improvement than the shorter (five-minute) one. Significantly, the improvement did not depend on whether subjects had been able to work consciously on the problems during the delay, or had been given an irrelevant task to occupy their attention. So the benefit of incubation in this situation cannot be explained on the basis of having longer to think purposefully.

In another study, Smith elicited the incubation effect using the 'tip-of-the-tongue' (TOT) phenomenon, which occurs when you are trying to recall something – a name, typically – which just won't come to mind, though you have the strong feeling that it is 'on the tip of your tongue'. Using computer graphics, Smith invented pictures of imaginary animals, to which he attached names and a brief description of their supposed habits, habitats and diets. Subjects were given twelve such animals to study briefly, and then were asked to recall their names. As in the previous study, for the names they were unable to remember they were given a second recall test either immediately or after a five-minute delay. On this second test, they were asked to indicate, if they still could not get the name, what the first letter might be, if they thought they would recognise the name if they were shown it, and whether they felt that the name was on the tip of their tongue or not. The delay improved memory by between 17 per cent and 44 per cent And furthermore, even if subjects were unable to recall the whole name, their 'guesses' as to the first letter were more accurate when they said they were in the TOT state.

Smith suggests that in both studies there is a common explanation for the positive effect of incubation: the delay allows time for wrong guesses and blind alleys to be forgotten, so that when you come back to the task, you do so with a more open mind. There is a tendency to get fixated on a particular approach, even when it is patently not working. The delay increases the chances that your mind will stop barking up the wrong tree. 'When the thinker makes a false start, he slides insensibly into a groove and may not be able to escape at the moment. The incubation period allows time for an erroneous set to die out and leave the thinker free to take a fresh look at the problem.'[16]

The idea that delay encourages a release from fixation, that it enables you to shake off unpromising approaches or assumptions that have been blocking progress, is certainly one aspect of incubation, but it cannot be the whole story, for it fails to take into account the active workings of unconscious intelligence. The fact that we can tell with a fair degree of accuracy when we are in the TOT state, whether we would be able to recognise the name if it were shown to us, and even, sometimes, retrieve its initial letter, or some other characteristic such as the number of syllables, suggests that the undermind has an idea what the word is, but for some reason is not yet willing or able to release that hypothesis fully into consciousness.

Yaniv and Meyer have shown directly that this sort of subliminal knowledge does exist. Like Smith, they investigated the TOT effect, this time reading to their subjects definitions of rare words and collecting those that subjects could not recall but felt sure they knew. They then used these words, along with other new words, in a 'lexical decision task', in which strings of letters were flashed on to a computer screen, and subjects had to press one of two keys to indicate, as quickly as they could, whether the string was a real word or not. It has been shown previously that words which have recently been seen, prior to the test, are recognised as being real words faster than other words which are equally familiar but which have not been recently 'activated' in memory. Yaniv and Meyer found that, even though the TOT words had not been *consciously* recalled, they still showed this 'priming' effect, indicating that they had been activated in memory, despite the fact that the 'strength' of the activation had not been great enough to exceed the threshold required for consciousness.[17]

One of the effects of this partial activation is to increase the likelihood that some chance event may provide the extra little 'nudge' that is needed to get the word to tip over the threshold into consciousness – and this provides another way in which incubation can come about. Consciously you may think that you have made no progress towards the solution of the problem, and may even feel that you have given up. But unconsciously some progress might have been made; not enough to satisfy the criteria for consciousness, but enough to leave the 'candidate' somewhat primed. If some random daily occurrence serves to remind you, even if only subliminally, of the same word or concept, that may be sufficient to tip the scales, and you have the kind of sudden, out-of-the-blue experience of insight to which personal accounts of creativity often refer. Many people have had the experience of suddenly remembering a dream during the course of the day, when some trivial stimulus, such as a fragment of overheard conversation, acts as a sufficient trigger for conscious recollection.

In the discussion of 'learning by osmosis', we saw that the undermind may be making progress in picking out a useful pattern of which the conscious mind is unaware. In such situations, we can show that we know more than we think we know. Does the same apply to the kinds of problem-solving that we are looking at in this chapter? Can we demonstrate directly that the undermind is closer to the solution of a problem than we think? And can we learn to be more sensitive to the subtle clues or indications that this is the case? Should we place greater trust, perhaps, in ideas that just pop into our minds, rather than treating them as random noise in the system, to be ignored? Recent studies by Kenneth Bowers and his colleagues at the University of Waterloo in Canada have suggested positive answers to these questions.

Like Smith, Bowers assumes that intuition is closely related to the ability to detect the underlying link or pattern that makes sense of seemingly disparate elements, and he has used both visual and verbal stimuli to explore the ways in which the undermind can home in on such patterns before conscious, deliberate thought has any idea what is going on. Consider the images shown in Figure 5.

One of each pair of pictures, either A or B, shows a highly degraded image of a real object.[18] The other shows the same visual elements arranged in a different configuration. Subjects were shown a series of such pairs, and asked to write down the name of the object depicted in one of the drawings. If they were unable to do so, they were asked to guess which of the two images actually represented the real object, and to indicate their degree of confidence in this 'guess'. The results showed that people's guesses were better

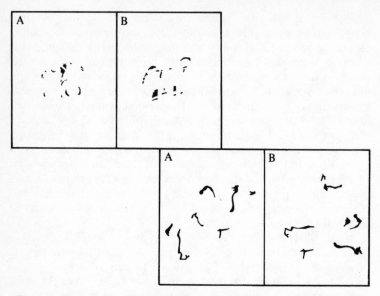

*Figure 5. Bowers' degraded images. One of each pair represents a real object: a camera (top) and a camel (bottom)*

than chance, and that this was so even when they indicated that they had zero confidence in their guess. The possibility that the visual fragments of the real objects were somehow more coherently arranged, and that it was this perceptual clue, rather than any unconscious activity, that was informing the guesses, was discounted by showing the pairs of shapes to naïve subjects and asking them to rate directly which looked the more 'coherent'. There was no difference in this judgement between the real and the rearranged images. Thus it appears, just as with the tip-of-the-tongue state, that the unconscious is able to indicate, in the form of what consciousness judges to be a complete guess, that it has got some way towards detecting the pattern, even when it has not yet been unequivocally identified.

The same finding is obtained with verbal rather than visual stimuli. Below are three pairs of three words. Within each pair, one of the sets of three words has a common (but not very obvious) associate – a single word that is related in some way to each of the three – while the other set of words is not connected in this way.[19]

|   | A | B |
|---|---|---|
| 1 | STICK | PARTY |
|   | LIGHT | ROUND |
|   | BIRTHDAY | MARK |
| 2 | HOUSE | MAGIC |
|   | LION | PLUSH |
|   | BUTTER | FLOOR |
| 3 | WATER | SIXTEEN |
|   | TOBACCO | SPIN |
|   | LINE | TENDER |

As with the pictures, people were asked to try to spot the connection, and if they could not, to indicate which of the two sets was the one that *did* have the (undetected) link. The results were essentially the same as with the visual stimuli: people were able, some of the time, to detect the presence of a pattern that they could not identify, and were able to do so more reliably than their own confidence ratings would suggest.

In an ingenious elaboration of this last study, Bowers devised what he called the 'accumulated clues task'. The problem was similar to the one just described: subjects had to discover the single word that formed the common associate to a number of other words. But now there was a list of fifteen such words, and they were presented in sequence, one at a time, rather than all together.[20]

*Accumulated Clues Test*
1. RED
2. NUT
3. BOWL
4. FRESH
5. PUNCH
6. CUP
7. BASKET
8. JELLY
9. COCKTAIL
10. GUM
11. PIE
12. TREE
13. BAKED
14. SALAD
15. FLY

The first word was presented for ten seconds or so, during which time subjects were obliged to write down at least one association. Then the second cue word was revealed, and another response was required; and so on. When subjects thought that they had found a response that was a viable hunch or hypothesis, they marked it, but continued making further responses until they either changed their mind or were convinced that they had found the target word. Typically, over a number of such tests, people found a viable candidate on about the tenth word, and settled on a firm answer after receiving about twelve of the cue words.

If the unconscious can run ahead of consciousness, then people's 'guesses' might begin to converge on the target word before they realise it. In order to check this, the responses that subjects gave *before* they settled on a plausible hypothesis were presented to a panel of judges, to see if they bore any meaningful relationship to the as yet unidentified target word. Sure enough, they did. If they looked back over people's guesses, independent judges, who knew the solution, could see a pattern of steady approximation to the target; a pattern of which the subjects themselves were unaware. It appears that the ideas that just pop into our heads may have greater validity than we think, and that we therefore deprive ourselves of useful information if we ignore them, or treat them as 'complete guesses'.

Bowers himself notes one important way in which these stylised problems are unrepresentative of problem-solving in real life. In the real world, a major part of the 'problem' is often that one does not know in advance what is relevant and what is not. The predicaments confronted by a business executive, an architect, a research scientist or a teacher are 'messy', in the sense that it is often not at all clear, at the beginning, how to conceptualise the problem, or what aspects of the available information to pay attention to and what to discard. The novice driver or medical student frequently feel overwhelmed with data because they have not yet discovered through experience what matters and what does not; what needs placing in the foreground of awareness and what can recede into the background. Bowers' puzzles, like many of those used by psychologists (and by those who design school curricula), are carefully tidied up before they are presented. The image of the camera is degraded, but there is no 'noise' in it. So in his most recent experiments, Bowers has made his problems more messy, and more lifelike. His tests now include some information that is irrelevant or distracting, as well as information that is relevant and valuable. Similar results seem to be emerging. For example, as they get closer (unbeknownst to them-selves) to a solution, so subjects get better at 'guessing' which of the details of the problem are actually relevant.

We now have empirical evidence for the existence of the mysteri-ous 'guiding hand' that told Nobel laureate Michael Brown which step to take next, and Stanley Cohen which result to 'believe' and which to doubt. There is evidence, in other words, for the undermind, the intelligent unconscious that works quietly below, and in some cases ahead of, conscious apprehension. Both poets and scientists have always suspected as much. If they are observant, as they must be, they know it through their direct experience. When Amy Lowell was asked, 'How are poems made?', she replied: 'I don't know. It makes not the slightest difference that the question as asked refers solely to my own poems, for I know as little of how they are made as I do of anyone else's. What I do know about them is only a millionth part of what there must be to know. *I meet them where they touch consciousness, and that is already a considerable distance along the road of evolution.*' (Emphasis added)

While R. W. Gerard, writing in *The Scientific Monthly* in 1946, foreshadowed, with his acute speculations, much of what cognitive science is finally beginning to demonstrate beyond question.

Much attention has been given to the phenomena of learning: by the slow cumulation of a correct response in the course of experience ['learning by osmosis']; and by the sudden grasp of a solution and abrupt performance of the correct response ['intuition']. They seem very different ... but it is possible, perhaps probable, that they are basically quite similar. In both cases, new functional connections must be established in the brain; and this process may be more gradual and cumulative in the case of 'insight' than it appears. For here, also, much brain work precedes the imaginative flash – the theory of gravi-tation may result only when the metaphorical apple falls on the prepared mind – and only when the process has progressed to some threshold level does it overflow into a conscious insight.

## CHAPTER 5

# Having an Idea: the Gentle Art of Mental Gestation

You cannot go into the womb to form the child; it is
there and makes itself and comes forth whole ... Of
course you have a little more control over your writing
than that; but let it take you and if it seems to take
you off the track don't hold back.

*Gertrude Stein*

There are a number of metaphors that creators use to describe their
process, but none more common than that of gestation. 'Having' a
good idea is akin, they say, to having a baby. It is something that
needs a seed to get started. It needs a 'womb' to grow in that is
safe and nurturing, and which is inaccessible. The progenitor is a
host, providing the conditions for growth, but is not the manufac-
turer. You 'have' a baby, you do not 'make' it – and so with insight
and inspiration. Gestation has its own timetable: psychologically, as
biologically, it is the process *par excellence* that cannot be hurried.
And it cannot be controlled; once the process has been set in motion
it happens by itself, and will, barring any major accident or inter-
vention, carry through to fruition.

It is not just romantics who see the mind this way. Even the
arch-behaviourist B. F. Skinner once gave a lecture at the Poetry
Center in New York which he entitled 'On "Having" a Poem', and
which he started by explaining that his talk had the curious property
of illustrating itself, in that he was at that moment in the throes of
'having' a lecture.[1] And he went on to develop the metaphor in
more detail. 'When we say that a woman "bears" a child, we suggest
little by way of creative achievement. The verb refers to carrying
the foetus to term.' And then, after she has 'given' birth to the child

– as if birth were some kind of property or gift that can be bestowed – we tend to say merely that she has 'had' a baby, where 'had' can seem to mean little more than 'came into possession of'.

What precisely is the nature of the mother's contribution? She does not decide upon the colour of the baby's eyes or skin. She gives it her genes, but are they really 'hers', when she inherited them from *her* parents, and through them from an entirely unwilled lineage? She surely cannot take much personal pride in the hazel eyes and the auburn hair she is handing on. 'A biologist', says Skinner, 'has no difficulty in describing the role of the mother. She is a place, a locus in which a very important biological process takes place. She provides warmth, protection and nourishment, but she does not design the baby who profits from them. The poet is also a locus, a place in which certain genetic and environmental causes come together.' And, as we have seen, what is true of the poet can be equally true of the scientist, the novelist, the sculptor or the product designer.

The analogy reminds us that, although the process of creativity is essentially organic rather than mechanical, nevertheless the nature of the 'incubator' is vital to the germination of the seed. The mother does not engineer her child's intrauterine development, but she influences it enormously through her lifestyle and her sensitivity, her anxieties, appetites and attitudes, her history and her constitution. Who she is, and the physical and emotional environment that she herself inhabits, affects the nature and the quality of the sanctum that she provides for the growing form of life within her. And so it seems to be with intuition: there are conditions which render the mental womb more or less hospitable to the growth and birth of ideas; and differing ways in which, and extents to which, different people are able, wittingly or unwittingly, to provide those conducive conditions. The more clearly we can identify what these conditions are, the more able we shall be to see how they can be fostered.

First, one needs to find the seed – and this process, for the creator, requires curiosity: an openness to what is new or puzzling. One must allow oneself to be impregnated. Unless one is piqued by a detail that obstinately refuses to fit the conventional pattern, or a chance remark that somehow resonates with one's own unexplicated views or feelings, there is nothing for the creative process to work upon. A. E. Housman breaths life into a hackneyed image when he says: 'If I were obliged to name the class of things to which [poetry] belongs, I should call it a secretion; whether a natural secretion, like turpentine in the fir, or a morbid secretion, like the pearl in the oyster. I think that my own case, though I may not deal with the material so cleverly as the oyster does, is the latter.'[2]

For scientists, the stimulus is often an unexplained detail or incongruity. The imaginative seed that finally flowered in the theory of relativity was the teenage Einstein's attempt to imagine what it would be like to ride on a beam of light. While making a routine check through three miles of computer print-out from the radio telescope, a young Cambridge astrophysicist spotted just a few traces that puzzled her. They could easily have been ignored, or written off as noise. But, with a lot of subsequent hard work, this observation finally resulted in the discovery of a completely new type of star. Out of hundreds of tiny fruit flies, one had a misshapen eye. A biologist could not help wondering why. Five years later, his investigations have led him to the discovery of a kind of receptor protein that may well be implicated in the production of cancer cells.[3]

In the commercial world, the competitive edge belongs to the executive or product developer who is capable of sensing the potential in an apparent setback, or taking time to mull over the meaning of a quirk in the market. And the reflective accounts of artists, too, reveal the importance of this sensitivity to poignant trifles. In the preface to his story 'The Spoils of Poynton', Henry James explains how essential such details are. One Christmas Eve he was dining with friends when the lady beside him made, as he puts it, 'one of those allusions that I have always found myself recognising on the spot as "germs" ... Most of the stories straining to shape under my hand have sprung from [such] a ... precious particle. Such is the interesting truth about the stray suggestion, the wandering word, the vague echo, at the touch of which the novelist's imagination winces as at the prick of some sharp point: its virtue is all in its needle-like quality, the power to penetrate as finely as possible.'[4]

It seems that such seeds implant themselves only in those who at an unconscious level are already prepared. Even if the issue is an intellectual rather than an artistic one, its recognition is personal, affective, and even aesthetic (such as Nobel laureate Paul Berg talking about how important, in his work, was the sense of 'taste' for a problem or an approach). Novelist Dorothy Canfield, in the same vein as Henry James, recounts the incident that formed the nucleus of her story 'Flint and Fire'. She had some business with a neighbour, and to get to his house had to walk along a narrow path through dark pines, beside a brook swollen with melted snow. Emerging from the wood, she found the old man sitting silent and

alone in front of his cottage. Having done her business, and keen
not to offend against country protocol, she sat beside him to chat
for a few minutes.

> We talked very little, odds and ends of neighbourhood gossip,
> until the old man, shifting his position, drew a long breath and
> said, 'Seems to me I never heard the brook sound so loud
> as it has this spring.' There came instantly to my mind the
> recollection that his grandfather had drowned himself in that
> brook, and I sat silent, shaken by that thought and by the sound
> of his voice . . . I felt my own heart contract dreadfully with
> helpless sympathy . . . and, I hope this is not as ugly as it
> sounds, I knew at the same instant that I would try to get that
> pang of emotion into a story and make other people feel it.[5]

Stephen Spender said that his experience of inspiration was that of
a 'line or a phrase or a word or sometimes something still vague, a
dim cloud of an idea which I feel must be condensed into a shower
of words'.

So the seed will not germinate unless it makes contact with a
'body of knowledge' of the right kind, in a congenial state. But what
exactly is the 'right' kind? The evidence from studies of conspicuous
innovators suggests that this pre-existing body is most fecund when
it is full of rich experience – but not to the point where it has
become so familiar that it is automated and fixed. One must have
the evidence on which to draw, and one must know enough to be
able to recognise a good idea when it comes. Clearly one cannot
be creative *in vacuo*. But if one is too steeped in the problem, the
danger is that the grooves of thought become so worn that they do
not allow a fresh perception, or a mingling of different currents of
ideas, to occur. Recall the experiments with the water jars, and the
fact that people quite quickly became 'set in their ways'. The more
experience they had had with the complex rule, the less likely they
were to spot the simpler solution when it became available. Studies
of creative individuals generally show an inverted U-shaped relationship between creativity and age. In mathematics and the physical
sciences, for example, the age of peak creativity is between twenty-
five and thirty-five.[6]

To give a more specific example: the *New York Times* carried a
front-page article on 18 February 1993 reporting the discovery of
the first successful technique for eliminating the AIDS virus from
human cells *in vitro*, and also for preventing the infection of healthy
cells. The inventor of this method was a medical graduate student,

Yung Kang Chow, who, precisely because of his relative inexperi-
ence, was able to see through a blocking assumption which
researchers had, up to that point, been unconsciously making. Chow
speculated: 'Perhaps by virtue of being a graduate student and not
having learned much medicine yet, I had more naive insight into the
problem.' Seeing through an existing, invisible assumption, which is
often the key to creativity, requires a mind that is informed but not
deformed; channelled but not rutted.[7]

Intuition, as we have seen, tends to work best in situations that
are complex or unclear, in which the information that is given may
be sketchy or incomplete, and in which progress can only be made
by those who can, in Jerome Bruner's famous phrase, 'go beyond
the information given', and are able to draw upon their own knowledge in order to develop fruitful hunches and hypotheses. Both
novelist and scientist may well need to go out and collect more
'data', but the creative idea comes from bringing into maximum
contact the 'problem specification', the data, and one's own store
of experience and expertise; allowing these to resonate together as
intimately and as flexibly as possible, so that the full range of meaning and possibility of both current data and past experience are
extracted. The good intuitive is the person who is ready, willing
and able to make a lot out of a little.

If you insist on having high-quality information from impeccable
sources before you are willing to form a judgement, you may reduce
the occasions on which you are obviously 'wrong'. You will make
instead 'errors of omission' that are often less visible. By adopting
such a conservative attitude, you may also fail to make use of the
more tentative, holistic responses that are authorised by the unconscious. On the other hand, if you are indiscriminately intuitive, you
are more ready to back hunches on the faintest of whims. The
crucial question concerning intuition, therefore, is how to relate to
conscious and unconscious in such a way that *both* kinds of mistake
are minimised; so that you are open to the promptings of the
undermind, willing to hear and acknowledge them, yet not over-
respectful or lacking in discernment.

Are people differentially willing to make judgements and decisions
on the basis of inadequate (conscious) information? And if so, of
those who are willing, are some better at it than others? Studies by
Malcolm Westcott at Vassar College in America show that the
answer to both these questions is a clear 'yes'. Westcott gave his
undergraduate subjects one example of a relationship that could
hold between two words or two numbers, and their task was to

show that they had discovered the rule or relationship by adding the correct 'partner' to another word or number. So they might be shown '2, 6', and asked to complete the pair '10, ?', or 'mouse, rat', and then 'weekend, ?'. But subjects were also given other sealed clues, which they could ask to be revealed, one by one, before they were ready to give their answer. They were free to look at as many or as few of these other clues as they wished before responding. When the students gave their answers, they were also asked to rate how confident they were that they were in fact correct. Westcott was thus able to take three measures for each problem: whether the solution was right or wrong; how many clues people wanted to see before they gave their response; and how confident they were. He repeated the experiment with several different groups of subjects both in England and America, and with different kinds of problems.

He found that people differ markedly, and consistently, on all three measures of their performance, so much so that it was possible to identify four quite different sub-groups. There were those who typically required very little information before offering their solutions, and who were likely to be correct. These he called the 'successful intuitives'. Then there were those who also took little extra information, but who tended to be wrong – the 'wild guessers'. The third group required a lot of information before being willing to respond, but were generally successful when they did: the 'cautious successes'. And finally there were those who made use of all the information they could lay their hands on, but who still made a lot of mistakes, the 'cautious failures'.

Westcott also gave various personality tests to his subjects, so that he was able to see what the characteristics of the successful intuitives (and of the other groups) were. He found that the good intuiters tend to be rather 'introverted'; they like to keep out of the social limelight, but feel self-sufficient and trust their own judgement. They like to make up their own minds about things, and to resist being controlled by others. They tend to be unconventional, and comfortable in their unconventionality. In social gatherings they are 'composed', but are capable of feeling strongly, and showing their feelings in more intimate or solitary situations. They enjoy taking risks, and are willing to expose themselves to criticism and challenge. They can accept or reject criticism as necessary, and they are willing to change in ways they deem to be appropriate. They describe themselves as 'independent', 'foresighted', 'confident' and 'spontaneous'. '*They explore uncertainties and entertain doubts far more than the other groups do, and they live with these doubts and uncertainties without fear.*'[8]

In contrast, the 'wild guessers' are much more socially orientated. But 'their interactions are characterized by considerable strife, they seem to be quite self-absorbed, and their "affective investments" seem to be directed towards themselves'. These traits often manifest themselves as 'a driven and anxious unconventionality, coupled with strong and rigid opinions, and overlaid with cynicism . . .' They describe themselves as 'alert', 'quick', 'headstrong' and 'cynical'. Westcott comments that these people 'appear to be striving for a grasp of reality which so far eludes them, and they are likely to attempt different modes of attack [on uncertainty] in a somewhat chaotic manner'.

The 'cautious successes' are distinguished by 'a very strong pref-erence for order, certainty and control', and they have a high respect for authority. They are well-socialized, in the sense that their stated interests and values are in the mainstream of their culture, but they do not recognize that they have been influenced in this. Their desire for certainty and order seems to lead them to some social awkward-ness and anxiety in the uncertain world of interpersonal relations. Affect is difficult for them to handle, unless it is very well structured, and they describe themselves as 'cautious', 'kind', 'modest' and 'confident'. The overall picture of this group, according to Westcott, is 'one of conservative, cautious, somewhat repressive people who function well in situations where expectations are well-established and well met': d-mode types, we may assume.

Finally, the 'cautious failures' have a view of the world 'in which everything is risky at best, and they are essentially powerless to influence or control it. There is a broadly generalized passivity, a sensitivity to – and felt inability to deal with – injustice, and a wish for a quiet, certain *status quo*; through all of this they lack self-confidence . . . They are quite conservative, presumably as the best defence against the great uncertainties of life, and they seem to wander through life just managing to keep their heads above water, not making waves. They see themselves as "cautious", "kind" and "modest".'

Perhaps the most significant of all these interesting findings is that the group who are most at ease with uncertainty and doubt, the most able to 'live with it', are the group who are most able to make successful use of the inadequate information they have. They can use their unconscious resources to help them make *good guesses* in uncertain situations, and are willing to do so. This provides some powerful empirical corroboration for the idea that the flight from the experience of uncertainty pushes people into the exclusive use of a cognitive mode which is ill suited to dealing with some puzzling situations. It is also very significant that, having reviewed the relevant studies to date, Westcott concludes that 'intuition is most likely to occur when the information on which an inference is based is excessively complex, apparently absent or limited, or when the time necessary for explicit manipulation of data is not available . . . These are all conditions which remove the thinker from direct application of adult, socially validated logic.'[9] American social scientist Donald Schon has recently argued that it is situations of just this sort that are routinely faced by professionals such as teachers and lawyers.[10] Although there are bodies of helpful precedents and maxims, such people spend much of their time dealing with cases that are sufficiently unique, and sufficiently complex, to prevent the straightforward application of any rule-book. They are off the well-laid-out highways of 'technical rationality', trying to find their way through what Schon refers to as the 'swampy lowlands' of professional practice.

Sometimes this resonating of data and experience – perception and cognition – happens quickly. Westcott's puzzles are sufficiently simplified and stylised to allow intuition to work quite rapidly. Not much experience has to be brought to bear. No remote analogies or metaphors need to be found. No very subtle patterns connecting apparently disparate elements have to be uncovered. Very often, though, when the predicament is more intricate, the undermind needs to be left to its own devices for a while, and then the need for patience – the ability to tolerate uncertainty, to stay with the feeling of not-knowing for a while, to stand aside and let a mental process that can neither be observed nor directed take its course – becomes all important.

Someone who cannot abide uncertainty is therefore unable to provide the womb that creative intuition needs. Milton Rokeach, having, as we saw in Chapter 4, showed that creativity is enhanced when people are forced to slow down, concludes that 'differences between people characterised as rigid, and others characterised as less rigid, may be attributable . . . to personality differences in time availability . . . Time availability [i.e. the willingness to think slowly] makes possible broader cognitions, more abstract thinking . . . and consequently greater flexibility.' And he goes on to offer a plausible speculation as to how these differences may arise. 'Some individuals, because of past experiences with frustrating situations involving

delay of need satisfaction, become generally incapable of tolerating
frustrating situations. To allay anxiety, such individuals learn to
react relatively quickly to new problems . . . The inevitable consequence is behavioural rigidity.'[11] Whether one is or is not a good
intuiter therefore turns out to be a matter of cognitive habits or
dispositions – but these are underpinned by emotional and personal
characteristics that may be quite deep-seated. If one is threatened
by the experience of ignorance, then one cannot dare to wait, and
may, as a result, cling to a mode of cognition – d-mode – that is
purposeful and busy, seeming to offer a sense of direction and
control, which may be the wrong tool for the job in hand.

There is a wealth of evidence to confirm the common impression
that when people feel threatened, pressurised, judged or stressed,
they tend to revert to ways of thinking that are more clear-cut, more
tried and tested and more conventional: in a word, less creative.
Studies with the Luchins' water jars problem have shown that
adherence to the over-complicated solution, when an easier one
becomes available, is increased by stress. In an old study (which
would certainly not be approved these days by an ethics committee), students were told that, on the basis of a previously administered questionnaire, there was evidence that they possessed some
'maladapted personality features', and that their performance on
the water jars problem would clarify the situation. The more threatened the subjects felt, the more tenaciously they clung to the out-
dated solution, and the less likely they were to spot the new
possibility.[12]

Less severe degrees of stress also disrupt performance. Arthur
Combs and Charles Taylor gave people the task of encrypting some
sentences according to the kind of simple transposition code that
one finds in spy stories in children's comics. Some of the sentences
were 'incidentally' of a personal nature, such as 'My family do not
respect my judgement', while others were neutral ('The campus
grew quite drab in winter'). Some of the neutral sentences were
preceded by the experimenter saying mildly, 'Can't you do it a little
bit faster?' The 'personal' sentences tended to be encoded more
slowly, and with more mistakes, but the worst condition of all was
when a neutral sentence was accompanied by the time pressure.
Even in such a straightforward task, where the degree of creativity
required is minimal, the exhortation to 'hurry up' is entirely counterproductive.[13]

The deleterious effect of time pressure on the quality of thinking
is also shown in a study by Kruglansky and Freund. Students were

given some personal data about a hypothetical applicant for a managerial job and asked to predict his likely success in the position. Half the students were given positive information followed by negative, and the other group were given the same information in the reverse order. Those students who received the positive information first gave significantly higher predictions of success than the others. And this tendency was exaggerated when the students were asked to make their judgements against the clock. What seems to happen is that we build up an intuitive picture of the situation as we go along, and it takes work to 'dismantle' this picture and start again. So if later information seems to be at odds with the picture so far, we may unconsciously decide to reinterpret the dissonant information, rather than radically reorganise the picture. And the more we feel under pressure, the less likely we are to make the investment of 'starting from scratch'. This tenacity is a considerable pitfall for intuition, for when we are making real-life decisions it often happens that information is not available to us all at once, but arrives piece by piece. If we 'make up our minds' quickly and intuitively, it means that later pieces of information may be ignored or downgraded if they do not happen to confirm the judgement that has already been made.[14]

Even stress that is not particularly related to the problem-solving task itself increases this rigidity. Hospital patients who are awaiting an operation give more stereotyped responses than control subjects to the Rorschach ink-blot test, and are much less fluent and creative in thinking up ways to complete similes such as 'as angry (or "interesting" or "painful") as —'. They also, incidentally, become physically more clumsy and more forgetful.[15]

One of the people who has worked most intensively on increasing the quality of intuition in practical, real-life settings is George Prince, the founder, with William Gordon, of the well-known 'Synectics' programme for enhancing creativity. Prince started out with the assumption that people needed to be trained in the art of generating more and better ideas. 'I was convinced that people tended to come to us weakly creative and leave strongly creative.' But slowly he became convinced that this was not the nub of the problem. He realised that *speculation*, the process of expressing and exploring tentative ideas in public, made people, especially in the work setting, intensely vulnerable, and that, all too frequently, in a variety of subtle (or not so subtle) ways, people came to experience their workplace meetings as unsafe.

People's willingness to engage in delicate explorations on the edge of their thinking could be easily suppressed by an atmosphere of even minimal competition and judgement. 'Seemingly acceptable actions such as close questioning of the offerer of an idea, good-natured kidding about someone's idea, or ignoring the idea – any action that results in the offerer of the idea feeling defensive – tend to reduce not only his speculation but that of others in the group.' Prince's depressing conclusion is that adults in the workplace are much more susceptible to 'hurt feelings' than we commonly admit, and that equally prevalent is the largely unacknowledged tendency for workers at all levels, and in all vocations, to see themselves as engaged in a competitive struggle to preserve and enhance a rather fragile sense of self-esteem. He concludes: 'The victim of the win–lose or competitive posture is always speculation, and therefore idea production and problem solving. When one speculates he becomes vulnerable. It is too easy to make him look like a loser.'[16]

Just as mothers-to-be may become rather particular about the conditions in which their gestation, and eventually the birth, takes place – traditionally demanding special foods or comforts that may seem to others somewhat eccentric – so too do creators, according to their own testimony, sometimes develop personal rituals and requirements that establish the conditions which are felt to be safe and conducive to intuition. Pearl S. Buck could not work without a vase of fresh flowers on her desk, and a view of the New England countryside, while Jean-Paul Sartre hated the country, and needed to look out on to the bricks and chimneys of a Parisian street. Kipling claimed to be unable to write anything worthwhile with a lead pencil. The poet Schiller liked to fill his writing desk with rotting apples, claiming that the aroma stimulated his creativity. Walter de la Mare, Sigmund Freud and Stephen Spender, along with many others, had to chain-smoke while writing. Though collective 'brainstorming' is valuable for throwing up novel ideas, the conditions for deeper insight and intuition seem most often to be solitary and free from outside pressure of any kind. Carlyle tried to build a soundproof room. Emerson would leave home and family for periods and live in a hotel room. I can do three months' good work in a fortnight in my New Zealand beach-house.

It is not only a hostile external environment that can reduce creativity. If our own belief systems are threatened, by, for example, some unanticipated implications that seem to be emerging from a seemingly innocuous line of thought, we may suppress our own intuition and speculation. What started out as an intriguing puzzle may, as we dig deeper into it, turn out to have unwanted repercussions for the way we think about and organise our lives. The more fundamental a belief is to our view of ourselves, or to a position on which we have staked our reputation, the harder it is going to be to re-examine. Sometimes this mental inertia is entirely reasonable: wholesale reorganisation of the mental household is not to be undertaken lightly. If someone suggests that we rearrange the furniture, so to speak, we might be willing to try it; but if they propose that the house would look better if we were to move the foundations a few metres to the right, they are likely to meet rather stronger resistance. Just so with fundamental changes to the structure of our knowledge.

Efraim Fischbein from Tel Aviv University comments on the inertia of science in this regard, but the same principle applies to the informal, everyday mind just as well.

> A scientist who has formulated a certain hypothesis did not formulate it by chance; it optimally suits his general philosophy in the given domain, his usual way of interpretation, his knowledge and his research methodology. He is certainly very anxious to preserve his initial interpretation not only for his own prestige – which is certainly an important factor – but chiefly because it is the hypothesis which is best integrated in the structure of his reasoning. He will be unwilling to give up this first hypothesis because by renouncing it he has to re-evaluate a whole system of conceptions.[17]

Hence what has come to be referred to as 'Planck's dictum', after the German physicist Max Planck: major advances in science occur not because the proponents of the established view are forced by the weight of evidence to change their minds, but because they retire and eventually die.

It is not only the whole class of things that we refer to as 'threats' which militates against the relaxed and hospitable mood that encourages creativity: anything that simply makes you try too hard has the same effect. Wanting an answer too much can interfere with the process of gestation. In one study, Carl Viesti asked his subjects to try to detect which of three complicated patterns was the odd one out, and looked at the extent to which their performance improved over a series of such tests. Although they were given plenty of time to examine each set, those subjects who were offered significant rewards for correct detection performed worse, and learnt less, than those who were given only a token payment. Viesti concludes that 'regardless of their size, monetary utilities [sic] do not appreciably increase performance on insight learning tasks, rather, their presence may interfere with such performance'.[18]

Interestingly, the same counterproductive effect of incentives has also been observed in the animal world. Rats and monkeys who have to learn a skill in order to get food discover less about their environment in general if they are ravenous than if they are only mildly hungry. The more pressing is the requirement to reach the goal or solve the problem, the less do animals or human beings attend to the overall patterns in their world, and the more they try to pick out just those few pointers that will get the job done. This is adaptive up to a point; but if the world then changes, so that there are new contingencies to be discovered, such an attitude is exposed as blinkered and narrow.[19] Incentives may increase routine productivity, it seems; but they do not create conditions conducive to top-quality insights and solutions. Too many carrots, as well as too much stick, are inimical to creative intuition.

The next quality which encourages creative intuition we might call 'feeling it kick'. As the seed of an idea grows, it is as if the host gradually becomes aware of the autonomous movements of new creative life inside her. And how sensitive she is to these small signals, and how she responds to them, has a significant influence on the creative process. How mental gestation turns out depends particularly on the ability to turn on to the borderlands between consciousness and the unconscious a kind of awareness that is welcoming without being predatory, and perceptive without being blinding. Crucially, skilled intuiters seem to be able to watch the emergence of their creations without chivvying them, neatening them up or trying to turn them too quickly into words.

In the 1960s poet Ted Hughes gave a series of talks for young people on the radio about writing. In one of these, he described very beautifully this quality of gentle attentiveness to one's own mind.

At school . . . I became very interested in those thoughts of mine that I could never catch. Sometimes they were hardly what you could call a thought – they were a dim sort of a feeling about something . . . [and] for the most part they were useless to me because I could never get hold of them. Most people have the same trouble. What thoughts they have are fleeting thoughts – just a flash of it, then gone – or, though they know they know something, or have ideas about something, they just cannot dig those ideas up when they are wanted.

Their minds, in fact, seem out of their reach . . . The thinking
process by which we break into that inner life . . . is the kind
of thinking we have to learn, and if we do not somehow learn
it, our minds lie in us like fish in the pond of a man who cannot
fish . . . Perhaps I ought not to call it thinking at all. I am
talking about whatever kind of trick or skill it is that enables
us to catch these elusive or shadowy thoughts, and collect them
together, and hold them still so we can get a really good look
at them.[20]

Hughes goes on to say that he is not very good at this kind of mental
fishing, but that what skill he does have he acquired not at school,
but through . . . fishing: literally coarse fishing, with a rod and a
float. When you are spending hours gazing at the red or yellow dot
in the water in front of you, all the normal little nagging impulses
that are competing for your attention gradually dissolve away, and
you are left with the whole field of your awareness resting lightly
but very attentively on the float, and on the invisible, autonomous
world of water things suspended below it, and moving – perhaps –
towards the surface, and towards your lure. Your imagination and
your perception are both working on and in the water world. Thus
fishing is an exercise which cultivates the kind of relaxed-yet-
attentive, perceptive-yet-imaginative mode of mind that fosters
intuition; and at the same time it offers a metaphor for the way in
which such a mental attitude mediates between consciousness and
the undermind.

This way of gathering and inspecting the fruits of intuition without bruising them, or avidly turning them into jams and pies, is, as
Hughes says, something which people are differentially good at, or
familiar with; and it is also an art which can be cultivated not just
through literal fishing but through any form of contemplation that
invites you to observe without interfering with the crepuscular world
that lies between consciousness and the undermind; between light
and dark; between waking and sleep. In the gloaming of the mind,
if one is quiet and watchful, one can observe the precursors of
conscious intelligence at play, and in so doing may be lucky enough
to catch the gleam of an original or useful thought. As Emerson
said in his essay on 'Self-reliance', talking of creativity: 'A man
should learn to detect and watch that gleam of light which flashes
across his mind from within . . . In every word of genius we recognise
our own rejected thoughts; they come back to us with a certain
alienated majesty.'[21] Several studies show that there are large differ-

ences between how well people are able to access these states of reverie, and that these correlate with how creative they are judged to be. People who have vivid imaginations, for example, being able to lose themselves at will in fantasy, or to recall childhood memories in great sensory detail, also score highly on standard tests of creativity.[22]

In similar vein, analytical psychologist James Hillman deplores the post-Freudian party game of 'interpreting' one's dreams. The dream, says Hillman, often has an integrity, an aura of both meaning and mystery, that is simply lost if one tries to dismember it into the familiar categories of thought. It is in the very nature of dreams to hint and allude. 'An image always seems more profound, more powerful and more beautiful than the comprehension of it.' To ask of a dream 'What does it mean?' is as misguided as to ask the same question of a painting or a poem – or of a sunset, come to that. 'To give a dream the meanings of the rational mind is . . . a kind of dredging up and hauling all the material from one side of the bridge to the other. It is an attitude of *wanting* from the unconscious, using it to gain information, power, energy, exploiting it for the sake of the ego: make it mine, make it mine.'[23] The proper attitude towards a dream, according to analytical psychology, is to 'befriend' it: 'to participate in it, to enter into its imagery and mood, to . . . play with, live with, carry and become familiar with – as one would do with a friend.' So 'the first thing in this non-interpretative approach to the dream is that we give time and patience to it, jumping to no conclusions, fixing it in no solutions . . . This kind of exploration meets the dream on its own imaginative ground and gives it a chance to reveal itself further.'

In some of the everyday problems we face, the 'goal' is clearly established in advance, and the value of the 'solution' has to be measured against predetermined criteria. If your car dies on the motorway, you want the emergency service person to get it going and fix the fault; you do not want them to start reupholstering the seats. But if a company's sales figures are declining, there is a whole range of possible 'goals' that one might pursue: advertising, customer service, market research, downsizing, product development, reorganising the structure . . . To have decided prematurely which aspect of the enterprise needs fixing may be to have missed a creative opportunity. The good intuiter is sometimes capable of delaying her decision about where she is going, even after she has set out. One of the areas where the value of this reluctance to specify the goal has been demonstrated most clearly is painting. Many artists have described the thrill of embarking on a canvas without knowing what will emerge. D. H. Lawrence, an enthusiastic amateur painter, described this vertiginous feeling.

> It is to me the most exciting moment – when you have a blank canvas and a big brush full of wet colour, and you plunge. It is just like diving into a pond – then you start frantically to swim. So far as I am concerned, it is like swimming in a baffling current and being rather frightened and very thrilled, gasping and striking out for all you're worth. The knowing eye watches sharp as a needle; but the picture comes clean out of instinct, intuition and sheer physical action. Once the instinct and intuition gets into the brush-tip, the picture happens, if it is to be a picture at all.[24]

A study of art students at the School of the Art Institute in Chicago by Getzels and Csikszentmihalyi looked in detail at the different working methods of the students as they tackled this task, and investigated whether there were any aspects of their modus operandi that correlated with the quality of the finished picture – as judged by the art tutors and practising artists. The students did indeed work in very different ways. From the large selection of objects available, from which to compose their still life, some students selected and handled as few as two, while others played with many more before settling on their selection. And some of them did 'play': they did not just pick the objects up; they stroked them, threw them in the air, smelled them, bit into them, moved their parts, held them up to the light and so on. The students also varied in the actual objects they selected. Some chose from the 'pool' those that were conventional, even clichéd, still-life subjects – a leather-bound book, a bunch of grapes. Others went for objects that were more surprising, or less hackneyed. Most interesting, though, were the differences in working practice once the students had started on their pictures. Some continued to change the composition of the objects, or even the objects themselves, for quite a long time, so that the finished structure of the picture did not emerge until rather late in the creative process. Others, once they had made their composition, stuck to it religiously, and their pictures took on recognisable form rather early.

The findings of the study were clear. The pictures produced by the students who had considered more objects, and more unusual ones, who played with them more, and who delayed foreclosing on the final form of the picture for as long as possible, changing their minds as they went along, were judged of greater originality and 'aesthetic value' than the others. What is more, when the students were followed up seven years later, of those who were still practising artists, the most successful were those who had adopted the more playful and patient modus operandi. These were clearly people who had learnt how to stay open to the promptings of their intuition, and who were comfortable setting out on a journey of discovery without the reassurance of knowing in advance where they were going. They are in good company. Picasso said of his own painting: 'The picture is not thought out and determined beforehand, rather while it is being made it follows the mobility of thought.'[25]

There is a whole variety of ways in which people differ with respect to intuition – and therefore an equal variety of ways in which we can set about trying to improve the hospitality of the conditions, both inner and outer, within which intuition can blossom. Being a 'mother of invention' is an art that we can learn. We can learn to acknowledge, and to take more seriously, the small seeds of poignancy and puzzlement that occur to us, and the gleams of thought that flash across the periphery of the mind's eye. We can discover the contexts and moods in which we are most creative and receptive, and make sure that we make time for these in our lives. We can guard against becoming too invested in a problem, and trying too hard. We can practise the art of not neatening problems up too quickly, and of not making up our minds too soon about what would count as a 'solution'. And we can cultivate patience. As the Tao Te Ching asks:

Who can wait quietly while the mud settles?
Who can remain still until the moment of action?
Observers of the Tao do not seek fulfilment.
Not seeking fulfilment they are not swayed by the desire for
    change.
Empty yourself of everything.
Let the mind rest at peace.
The ten thousand things rise and fall
While the Self watches their return.
They grow and flourish and then return to the source.
Returning to the source is stillness, which is the way of nature.

## CHAPTER 6

# Thinking Too Much? Reason and Intuition as Antagonists and Allies

> The men of experiment are like the ant; they only collect and use. The reasoners resemble spiders, who make cob-webs out of their own substance. But the bee takes a middle course; it gathers its material from the flowers of the garden and of the field, but transforms and digests it by a power of its own.
>
> *Francis Bacon*

Is it possible to think too much? Though people who cannot get to sleep for churning over their problems might answer in the affirmative, conventional wisdom sometimes seems to suggest not. In the classroom, the consulting room or the boardroom, we may operate as if the more analytical we were, the better. Or even if we do not, we may assume that we ought to; that a detailed listing and weighing up of considerations, for example, represents some kind of ideal cognitive strategy to which our actual behaviour approximates. As Benjamin Franklin wrote to the British scientist Joseph Priestley:

My way [of tackling a difficult problem] is to divide half a sheet of paper by a line into two columns, writing over the one Pro, and over the other Con. Then during three or four days' consideration, I put down under the different headings short hints of the different motives, that at different times occur to me, for or against each measure . . . I find at length where the balance lies; and if, after a day or two of further consideration, nothing new that is of importance occurs on either side, I come to a determination accordingly . . . When each [reason] is thus considered, separately and comparatively, and the whole lies before me, I think I can judge better, and am less likely to make a rash step.[1]

The rationale for such a modus operandi is, presumably, that by making our thoughts and motives explicit and orderly, we can evaluate and integrate them better, and thus make better decisions. Or, as an influential textbook on decision-making puts it: 'The spirit of decision analysis is divide and conquer: decompose a complex problem into simpler problems, get your thinking straight in these simpler problems, paste these analyses together with a logical glue, and come out with a program for action for the complex problem.'[2] Be as explicit, as articulate and as systematic as you can be, and you will be thinking in the way that generates the best decisions and solutions. Given the evidence that we have looked at so far, however, we might have cause to question this ubiquitous, common-sensical assumption. D-mode and the slower ways of knowing work together, but they can get out of balance, and lose coordination.

Jonathan Schooler from the Learning Research and Development Center at the University of Pittsburgh has conducted a number of studies over the last few years that demonstrate graphically how thinking can get in the way of a whole variety of mental functions including everyday memory and decision-making, as well as intuition and insight. These studies go to the heart of the relationship between d-mode and intuition. One concerns how people choose between several possibilities, as when we are deciding which of a range of different foods we prefer. Schooler gave his subjects five different brands of strawberry jam to taste, and asked them to rate them and say which they liked best. The jams had recently been the subject of a consumer report, and those used had been ranked 1st, 11th, 24th, 32nd and 44th by the experts. Some of the subjects were told that they would be asked to explain the reasons for their choices, and to think carefully about their reactions and preferences. The results showed that those subjects who were left to their own devices ranked the jams in a way that corresponded closely to the judgement of the experts, while those who had been instructed to analyse their reactions disagreed with the experts.

Obviously this need not be a problem: maybe thinking about your choices makes you more independent. You decide what is right for you, rather than following the herd. If thinking carefully means that your decisions are based more closely on your true values and preferences, you would expect the thoughtful subjects to be *more* satisfied with their choices, and for this satisfaction to be more long-lasting. Unfortunately the reverse is the case. In a parallel study, subjects were shown five art posters and invited to choose one to take home. Those who had deliberated most carefully turned out to be significantly *less* satisfied with their choice, a few weeks later, than those who had chosen 'intuitively'. The deliberating jam tasters were more 'individual' than their intuitive colleagues, but they were worse, not better, at making choices that reflected what they *really* liked.

Neither of these choices, you could say, has a terribly important impact on people's lives. But deciding which courses to take in college certainly does. Schooler investigated university students as they selected their second-year psychology courses. They were given full information about all the options, including comments and ratings from those students who had taken each course the previous year, and asked to say which courses they thought they would take. Again, some students were asked to reflect in detail on the information provided, and the criteria they were using to make their choices. As in the other studies, those who thought most carefully were less likely to opt for the courses recommended by their peers, and more likely to change their minds subsequently. Later, when it came to signing up, the choices of the 'deliberating' students tended to revert to those that conformed more with received opinion, and correspond with the choices of those who had chosen intuitively.

On the basis of these studies, the researchers argue that there are a number of potentially negative effects of encouraging people to be more reflective and explicit about their decisions. In choosing a picture, or a jam, or a course, there are many interwoven considerations to be taken into account, *not all of which are (equally) verbalisable*. When the decision is made in an intuitive way, these considerations are treated in a more integrated fashion, and those that are hard to articulate are given due weight – which actually may be considerable. However, when people are forced (or encouraged) to be analytical, the problem is deconstructed into those considerations that are more amenable to being put into words. Thus the way the predicament is represented to consciousness may be, to a greater or lesser extent, a distortion of the way it is represented tacitly, and decisions based on this skewed impression are therefore less satisfactory.

In particular, d-mode may exclude or downgrade those non-verbal considerations that are primarily *sensory* or *affective*. Analytic thinking therefore tends to overestimate *cognitive* factors, which may be more easily expressed, resulting in decisions which seem 'sensible', but which fail to take into account non-cognitive factors. Additionally, the more carefully one analyses the different alternatives, the more one finds that there are good and bad aspects to each, and the greater the consequent tendency for judgements to become more moderate, more similar and therefore less decisive. Hence the tendency to come to decisions that differ from those which the acknowledged experts would have advised, and to feel obscurely dissatisfied with the choice one has made. Anyone who has ever agonised over a choice while shopping, and then regretted the decision immediately they have got the item home, will be familiar with this phenomenon. It is the dislocation between conscious and unconscious decision-making that people are referring to when they say that they should have listened to their 'heart', or their 'gut feeling', or their intuition.

As usual the issue is not black and white. We might now suspect that where a problem can be adequately represented verbally, and where the solution lies at the end of a logical chain of reasoning, a predominantly d-mode approach will be effective and efficient. While where the problem is more complex, contains aspects that are hard to articulate, or demands an insightful leap, d-mode will be less successful than a more receptive, patient approach. In another series of experiments, Schooler and his colleagues explored these two types of problems, looking particularly to see where active thinking helped and where it hindered.

'Insight problems' are those where people are in possession of all the information and ability necessary to solve them, but where there is a tendency to feel blocked or 'stumped', before suddenly having a kind of 'Aha!' experience in which the solution becomes immediately or rapidly obvious. In such problems, the difficulty is often caused by the tendency to make some unconscious assumptions that get in the way, or to fail to retrieve knowledge that would actually be helpful. We have met examples of such problems before – the 'mutilated chessboard', for example. Two different puzzles of this kind are shown in Figure 6.

The figure on the left represents a triangle made up of coins. The problem is to make the triangle point downwards by moving only three of the coins. The figure on the right represents a pen containing nine pigs. The problem is to build two more square enclosures that would put each pig in a pen by itself.

Contrast these with two so-called 'analytical' problems. In the first, imagine that there are three playing cards face down on the

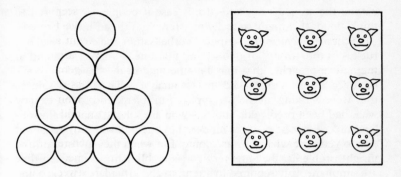

*Figure 6. Insight problems. (a) Move three coins to invert the triangle.*
*(b) Draw two squares to give each pig its own enclosure. (The answers are in*
*note 6.3 on page 236.)*

table in front of you. You are given the following pieces of information:

To the left of a queen there is a jack.
To the left of a spade there is a diamond.
To the right of a heart there is a king.
To the right of a king there is a spade.

Your job is to say what the three cards are.

In the second problem, the police are convinced that one of Alan, Bob, Chris and Dave has committed a crime. Each of the suspects in turn has made a statement, but only one of the four is true. Alan said 'I didn't do it'. Bob said 'Alan is lying'. Chris said 'Bob is lying'. Dave said 'Bob did it'. Who is telling the truth; and who did the crime?[3]

In the two analytical problems, no additional knowledge has to be supplied by the problem-solver, and it is unlikely that any assumptions will be made unwittingly that would make the problems harder to solve than they already are. All that is required is a meticulous fitting together of the pieces of information – a non-trivial but in principle straightforward task – and the answer will emerge. We might imagine that, if people were asked to think out loud while they were attempting an analytical problem, their words might track their thoughts quite easily and accurately, and would be positively related to the actual solution.

But with the insight problems, we might argue that a different kind of 'thinking' is required, one which is more of the intuitive, behind-the-scenes kind; and in this case if people are required to think aloud, this might actually interfere with the intuitive process. As Schooler says in his paper: 'Verbalisation may cause such a ruckus in the "front" of one's mind that one is unable to attend to the new approaches that may be emerging in the "back" of one's mind.' Schooler's study did in fact analyse what subjects said as they were working on the two types of problem, and found exactly what had been predicted. Subjects who are solving analytical problems are neither helped nor hindered by the demand to think about and to verbalise what they are doing. But when they are attempting insight problems, they are very considerably hampered when they are simultaneously required to attend to and articulate what is going on in their minds.

In one variation of the experiment, subjects were informed before they started that they would be working on two kinds of problems, and that one kind, the 'insight' problems, would typically lead them into an approach that did not work. They were given an example of an insight problem, and told that if they got stuck, it would probably help to try to find a different approach or a new perspective. As before, some of the subjects were told to think aloud, and others were not. There were two interesting results. The first was that this heavy hint was of no benefit at all in solving the insight problems, and did nothing to offset the decrease in performance produced by thinking aloud. It looks strongly as if the way of knowing that leads to success in the insight problems is not only outside of conscious awareness, but outside of conscious control as well. If the processes of intuition are beyond voluntary control, then there is no way in which the subjects can make use of the 'helpful hints' which they have been given.

The researchers also found that the information about insight problems did have a marked effect on the solution of the analytical problems. When subjects were thinking aloud, the hint severely *damaged* their ability to solve the logical puzzles. Just as the attempt to solve insight puzzles through exclusive reliance on d-mode is misguided, so performance in analytical problems can be impeded if you sow doubt in people's minds about their straightforwardness. The suspicion that something might be trickier than it actually is causes the confident use of d-mode to falter, as people endeavour to seek out – intuitively – complexities that do not exist. This result reinforces the point that the selection of the right cognitive mode is a matter of appropriateness, and not of the absolute superiority of one way of knowing over another.

Listening to the tapes of people's verbalisations, it became clear to Schooler and his colleagues that the *contents* of the problem-solvers' thoughts, as they tackled the two types of problem, were different. People wrestling with the analytical problems talked fluently, and most of their comments referred to the problem itself. However, when they were attempting the insight problems, subjects paused more frequently, and the pauses were longer: there were many more occasions on which there was, seemingly, nothing going on in the problem-solvers' minds. And when people doing the insight problems did verbalise, they were four times more likely to make the kinds of comments that referred not to the logic of the problem but to their own mental state. They would say things like 'There is nothing that's going through my mind that's really in any kind of ... that's in a verbal fashion'; or 'I know I'm supposed to keep talking but I don't know what I am thinking'. And this experience of 'nothing going on' was actually correlated with success on the insight problems. Those subjects who paused more solved more problems. Keeping up a running mental commentary really does interfere with the slower, less conscious processes going on at the back of the mind, and causes a drop in intelligence and creativity. We must presume that people for whom such chatter is habitual are thereby hampered when it comes to dealing with problems of greater subtlety or indeterminacy.

Jonathan Schooler's general point is of enormous significance. Some of what we know is readily rendered into words and propositions; and some of it is not. Some of our mental operations are available to consciousness; and some of them are not. When we think, consciously and articulately, therefore, we are not capturing accurately all that is going on in the mind. Rather we are selecting only that part of what we know which is capable of being verbalised; only those aspects of our cognition to which conscious awareness has access. We think what is *thinkable*; not what is 'true'. And the disposition to treat all problems as if they were d-mode problems thus skews our thoughts and our mental operations towards those that can be made explicit.

Other areas of our psychological life show similar effects. Take memory. There are many experiences that defy articulation, and our memories of them must therefore rely on non-verbal records. Our ability to recognise a huge collection of human faces, for example, with a remarkable degree of accuracy and effortlessness, attests to the power (and, I would say, 'intelligence') of these unspoken processes. What we can say about a face or an expression is a small fraction of what we can know. Thus it should come as no surprise to discover that the effort to *describe* a face so narrows our attention, and biases it towards the little that can be said, that memory is reduced quite severely. In another of his studies, Schooler gave subjects photographs of unfamiliar faces to study, and asked them to attempt to describe some of the faces but not others. These pictures were then mixed up with some new ones of rather similar-looking people, and the subjects were asked to pick out the ones they had seen before. The faces that had been described were recognised about half as well as those that had not, and this impairment was unrelated to how detailed or accurate a particular description had actually been. The same result is obtained if the stimuli are simple patches of colour.

The problem with description is twofold. First, the effort to describe the face forces one to break it up into its articulable features, and focuses attention on what can be said at the expense of what is genuinely (but non-verbally) distinctive. And secondly, at the time of recognition one may be trying to retrieve the 'written records' from memory, and match these to the pictures that one is being shown, rather than relying on the non-verbal, sensory records that have been registered. If this retrieval effect is a significant part of the problem, performance might be improved by preventing people from using the verbal 'code' while they are doing the recognition test. This could be achieved by forcing people to make their memory judgement very fast, perhaps. If you deprive them of the time it takes to *think*, they might have to fall back on the visual information which has been overshadowed by the verbal description, and thus overcome the interference. This is exactly what Schooler showed. When people had to make their recognition decisions quickly, the deleterious effect of verbalisation was removed.

In this case a 'snap judgement' is more reliable than a considered one. 'Decide first, and ask questions afterwards' may be the right strategy when dealing with non-verbal information. One can escape from the negative effects of d-mode by responding *faster* than thought, as well as more slowly.[4] The conventional wisdom that says we should always benefit from thinking and reflecting more is again seen to be in need of revision. There are interesting practical implications from this study for the handling of eye-witness testimony, identity parades, and so on. Asking witnesses to 'think carefully', and to describe what they have seen, may well interfere with their subsequent ability to recognise a face from a photograph or a line-up.

Schooler's studies have extended the range of everyday mental tasks with which articulation has been shown to interfere. In Chapter 3 we saw that learning to manage complex and unfamiliar situations, and to perform under pressure, can both be undermined by too strong a commitment to intellectual comprehension and control. Now we know that the same can be true when we are making choices and decisions, solving problems that involve insight, and even when we are simply recognising faces or other visual stimuli. However, again we should beware of falling into the trap of exaggerating the downside of d-mode. There is no value in demonising the intellect. There are many situations in life where an explicit grasp is useful or necessary. When we have to communicate our ideas to other people, in order to get practical tasks accomplished, we must obviously articulate as clearly as we can.

But we do not need d-mode just for communication. There are times when we need its analytical powers to test and refine ideas that have been thrown up by the undermind. The study of creativity in many different areas shows that what is required for optimal cognition is a fluid balance between modes of mind that are effortful, purposeful, detailed and explicit on the one hand, and those that are playful, patient and implicit on the other. We need to be able both to *generate* ideas, and also to *evaluate* them. Intuition is the primary mode of generation. D-mode is the primary mode of evaluation. Henri Poincaré summed it up when he said: 'It is by logic we prove; it is by intuition we discover.' For the scientist, intuition and contemplation may provide the vital creative insight that is both preceded and followed by the more disciplined procedures of d-mode. The chemist Kekulé, having first seen the cyclical form of the benzene carbon ring in a drowsy fantasy, concluded the report of his breakthrough to the Royal Society with the words 'Gentlemen, let us learn to dream. But before we publish our dreams, let us put them to the test of waking reason.' And Poincaré, having vaunted the need for patience, said: 'There is another remark to be made about the conditions of this unconscious work: it is possible, and of a certainty it is only fruitful, if it is on the one hand preceded and on the other followed by a period of conscious work.'[5] If we can think too much, it is also possible to think too little.

The classic formulation of scientific creativity, developed from Poincaré's observations by Graham Wallas in his 1926 book *The Art of Thought*, sees it as emerging from the interplay between four different mental modes or phases: preparation, incubation, illumination and verification. In the preparation phase, one gathers information, carries out experiments, and seeks, as hard as one can, for a satisfactory explanation – which obstinately refuses to come. D-mode is employed to the limit, and finally admits defeat. Then, as we saw in Chapter 5, the problem is put to one side to rest and incubate. If all goes well, at some unpredictable moment a new idea – novel, unexpected, but somehow full of promise – surfaces. And then, after this revelation or 'illumination', comes the return of d-mode, to apply its tests and checks, probing to see if the promise is fulfilled, and seeking ways to turn the illumination into a form which can be communicated, and which can compel the assent of others.

But it is not just scientists who value d-mode. Artists and poets too, though they are wary of its ability to strangle the creative impulse in its cradle, know that it is also a tool of which they have need. They are clear, as A. E. Housman said, that 'the intellect is not the fount of poetry, it may actually hinder its production, and it cannot even be trusted to recognise poetry when it is produced'. Yet many creative artists also speak of the value of a more deliberate, controlled, conscious mode of mind in sorting through the products of intuition, and shaping them into a finished product. Mozart distinguished between the conditions of creativity, and those of selection, when he said: 'When I am, as it were, completely myself, entirely alone, and of good cheer . . . it is on such occasions that my ideas flow best and most abundantly. Whence and how they come, I know not: nor can I force them. Those ideas that please me, I retain in memory . . .'[6] Not all the ideas that 'flow freely' are retained for future use: only those 'that please me'. John Dryden talks of intuition, or 'Fancy', 'moving the Sleeping Images of things towards the Light, there to be distinguish'd, and then either chosen or rejected by the Judgment'.[7] Even the Romantic poets such as William Wordsworth are equally clear that 'Poems to which any value can be attached were never produced . . . but by a man who, being possessed of more than usual organic sensibility, had also thought long and deeply. For our continued influxes of feeling are modified and directed by our thoughts.'[8]

Sculptor Henry Moore expressed in rather more detail the Janus-faced quality of the discriminating intellect.

It is a mistake for a sculptor or a painter to speak or write very often about his job. It releases tension needed for his work. By trying to express his aims with rounded-off logical exactness, he can easily become a theorist whose actual work is only a caged-in exposition of concepts evolved in terms of logic and words. But though the nonlogical, instinctive, subconscious part of the mind must play its part in his work, he also has a conscious mind which is not inactive. The artist works with a concentration of his whole personality, and the conscious part of it resolves conflicts, organises memories, and prevents him from trying to walk in two directions at the same time.[9]

It seems as if full-blown creativity works in a way that is not unlike biological evolution. As long ago as 1946, R. W. Gerard suggested that imagination and intuition are to ideas what mutation is to animals: they create a diversity of new forms, many of which are less viable, less well suited to the demands of the environment, than those that existed already, but some of which, perhaps only a few, contain features and properties that are adaptive as well as novel. The undermind accounts for the 'arrival' of ideas, both fit and unfit. Reason and logic then act like the environment, putting each of these candidates to the test, and ensuring that it is only the fittest that survive.[10] (More recently neuroscientist Gerald Edelman has proposed, with his idea of 'neural Darwinism', that the development of different pathways in the brain itself is determined by a similar process. Those connections that 'work' to the animal's advantage are strengthened; those that do not fade away.)[11] The analogy is limited, however, by the fact that the undermind, unlike the process of genetic mutation, generates not just random variations of what exists already, but complex, well-worked-out candidates; not just guesses but *good* guesses, *educated* guesses. The undermind is intelligent in a way that mutation, as far as we know, is not.

Several artists talk of the need for conscious thought to come to the rescue of the creative process when intuition 'stalls', as it not infrequently does. If one is lucky, like Coleridge with 'Kubla Khan', the undermind does the whole thing for you. The creative product is 'channelled', and the only role left for the conscious mind is that of scribe. However, it is not always thus. Intuition will not be managed, and sometimes it seems to down tools before the job is completed. As Amy Lowell says: 'The subconscious is . . . a most temperamental ally. Often he will strike work at some critical point and not another word is to be got out of him. Here is where the conscious training of the poet comes in, for he must fill in what the subconscious has left . . . This is the reason that a poet must be both born and made. He must be born with a subconscious factory always working for him, or he can never be a poet at all, and he must have knowledge and talent enough to "putty" up his holes.'[12]

Housman, he who is most aware of the destructive power of the critical intellect, nevertheless had to draw on it in order to get a poem finished.

> Having drunk a pint of beer at luncheon ... I would go out for a walk of two or three hours. As I went along, thinking of nothing in particular, only looking at things around me and following the progress of the seasons, there would flow into my mind, with sudden and unaccountable emotion, sometimes a line or two of verse, sometimes a whole stanza at once, accompanied, not preceded, by a vague notion of the poem which they were destined to form part of ... When I got home I wrote them down, leaving gaps, and hoping that further inspiration might be forthcoming another day. Sometimes it was, if I took my walks in a receptive and expectant frame of mind; but sometimes the poem had to be taken in hand and completed by the brain, which was apt to be a matter of trouble and anxiety, involving trial and disappointment, and sometimes ending in failure.
>
> I happen to remember distinctly the genesis of the piece which stands last in my first volume. Two of the stanzas, I do not say which, came into my head, just as they are printed, while I was crossing the corner of Hampstead Heath between Spaniard's Inn and the footpath to Temple Fortune. A third stanza came with a little coaxing after tea. One more was needed, but it did not come: I had to turn to and compose it myself, and that was a laborious business. I wrote it thirteen times, and it was more than a twelvemonth before I got it right.[13]

The creative mind possesses a dynamic, integrated balance between deliberation and contemplation. It is able to swing flexibly between its focused, analytical, articulated mode of conscious thought, and its diffused, synthetic, shadowy mode of intuition. But the mind may lose its poise and get stuck in one mode or the other. And if its balance has been disturbed in this way, it takes time and effort to free it up again.

This process of rediscovering the complementarity of the mind's different modes has been graphically charted in a study of 'women's ways of knowing' by Mary Field Belenky and others.[14] They made a detailed study of the experiences of women of a wide variety of ages and backgrounds who were studying within the formal education system, and identified five stages through which these women seemed to pass on their journey towards expanded sophistication and confidence as 'knowers'. In the early stages of this development, they claim, many women, particularly those who have previously had little successful experience of formal education, start out feeling very powerless and inept in the face of the rational, articulate way of knowing. They feel as if they have no 'voice' of their own, and are in awe of others (principally but not exclusively men) whose d-mode voices sound loud, self-confident and authoritative.

But at some point, they may realise that they *do* 'know', and that there is validity to their experiences, feelings – and intuitions. In this stage of what the authors call 'subjective knowing', their respondents feel the first stirrings of their own 'epistemological authority'; though this is associated not with their ability to be rational and explicit, but with the emergence of a new respect for 'the inner voice'. 'Truth' is discovered not through argument and articulation, but through the promptings of gut feelings. It is as if there is 'some oracle within that stands opposed to the voices and the dictums of the outside world'.

> I just know. I try not to think about stuff because usually the decision is already made up inside you and then when the time comes, if you trust yourself, you just know the answer.

> There's a part of me I didn't even know I had until recently – instinct, intuition, whatever. It helps me and protects me. It's perceptive and astute. I just listen to the inside of me and I know what to do . . . I can only know with my gut. My gut is my best friend – the only thing in the world that won't let me down or lie to me or back away from me.

This discovery is experienced as vital and welcome; but it is accompanied, for some women, by an over-reaction in which thought-out knowledge is disdained as 'remote' and 'academic', while the inner voice is accepted as inevitably right and trustworthy simply by virtue of its 'innerness'. If it 'feels right', it is impossible for it to be wrong; even for it to be questioned can be taken as a mark of disrespect or felt as a violation. The sense of self as a valid knower is so precious, and yet so tenuous, that its source has to be defended against all conceivable threats, real or imagined. Instead of there being an absolute authority which is external, now this absolute is shifted inside. The feeling that there *is* an omniscient source of certainty remains; it is just relocated. In this move, the domains of logic, articulation and science may be completely rejected. The authors comment that: 'It was as if, by turning inward for answers, they had to deny strategies for knowing that they perceived as belonging to the masculine world.'

> It is not that these women have become familiar with logic and theory as tools for knowing and have chosen to reject them; they have only a vague and untested prejudice against a mode of thought that they sense is unfeminine and inhuman and may be detrimental to their capacity for feeling. This anti-rationalist attitude is primarily a characteristic of women during the period of subjectivism in which they value intuition as a safer and more fruitful approach to truth.

For some women this attitude may become as arrogant and offensive as that which they are at pains to denounce.

> A few of the women . . . were stubbornly committed to their view of things and unwilling to expose themselves to alternative conceptions. Although they might have described themselves as generous and caring, they could be, in fact, impatient and dismissive of other people's interpretations. They easily resorted to expletives when faced with others' viewpoints – 'That's bullshit!' . . . These were women at their most belliger-ent . . . adept at turning the tables on authorities by bludgeon-ing them with wordy, offensive arguments. In the classroom, as in life, they warded off others' words and influence via ploys to isolate, shout down, denigrate and undo the other.

When women look back on this stage later, from the more balanced, integrated perspective that Belenky refers to as 'procedural knowing',

> now they argue that intuitions may deceive; that gut reactions can be irresponsible and no one's gut feeling is infallible; that some truths are truer than others; that they can know things that they have never seen or touched; that truth is shared; and that expertise can be respected . . . They have learned that truth is not immediately accessible, that you cannot 'just know'. Things are not always what they seem to be. Truth lies hidden beneath the surface, and you must ferret it out. Knowing requires careful observation and analysis. You must 'really look' and 'listen hard'.

They have realised that their love affair with the inner intuitive voice – and particularly with the brittle certainty of its 'snap judgements'

– was a vital stage on the road to establishing their confidence in their own minds, and in developing their portfolios of ways of knowing; but that it was also tainted by the fear of uncertainty, and driven by more than a little wishful thinking. The 'inner voice' can easily be interpreted as telling you that things are the way you want them to be. As Minna, emerging slowly from a bad marriage and studying to become an occupational therapist, reflected: 'I was confused about everything. I was unrealistic about things. I was more in a fantasy world. You have to see things for what they are, not for what you want to see them. I don't want to live in a dream world [any more].'

At this later stage, knowing is characterised more by a respect for plurality and relativity, complexity and patience. The women in this study seem now to be discovering a more contemplative, less impulsive, form of intuition. The forced choice between feeling-laden subjectivity and remote objectivity begins to collapse, and knowing emerges from interaction and respect. It is no coincidence that, for these women, an interest in poetry often resurfaced at this stage. One of the women, a college senior, spoke scornfully of critics who use their 'so-called interpretations' as 'an excuse to get their own ideas off the ground'. She felt that to understand a text, you had to 'treat it as you would a friend'; accept it as 'real' and as 'independent of your existence', rather than 'using it for your own convenience and reinforcement'. She had become capable, in the words of Simone Weil, of a way of knowing that 'is first of all attentive. The soul empties itself of all its own contents in order to receive into itself the being that it is looking at, just as he is, in all his truth.'[15]

## CHAPTER 7

# Perception without Consciousness

> At every moment there is in us an infinity of percep-
> tions, unaccompanied by awareness or reflection; that
> is, of alterations in the soul itself, of which we are
> unaware because the impressions are either too
> minute or too numerous.
>
> *Leibniz*

We are more in touch with, and more influenced by, the world around us than we know. In the 1960s, there was a belief that cinema audiences were being unconsciously manipulated into buying soft drinks that they didn't really want by messages flashed on to the screen too briefly to be detected consciously. Though it turns out that subliminal advertising is much less effective in persuading us to act against our best interests than we might fear, subliminal influences are indeed ubiquitous. They do not just occur when we are watching a screen or listening to a tape; they are present the whole time, and we could not manage without them. The undermind stays in continuous communication with the outside world without many of these conversations appearing in consciousness. Not only do we fail to *comprehend* what is going on in our own minds; we may not even *see* what is happening either.

Subliminal perception is hard to write about, because the relevant terms in the language are so muddled. It is symptomatic of our cultural neglect of the undermind that we have no word for 'being influenced by something of which we are unconscious'. Although it may conflict with the way some people use the word, I am going to use 'awareness' to refer to the general phenomenon of 'picking up signals' from the environment (or from the body), regardless of whether they get represented in consciousness. And I shall reserve 'consciousness', and '*conscious* awareness', for what appears before the mind's eye. Thus, in my usage, there is nothing paradoxical about using the expression 'unconscious awareness' to refer to the state of being affected or influenced by some stimulation that is not itself present in consciousness.[1]

In 1989, Thane Pittman and Robert Bornstein of Gettysburg College in Pennsylvania conducted an experiment to investigate how people decide between job applicants. Students were given the job specification – for a research assistant in the psychology department – and asked to review the applications of two young male candidates (call them Tom and Dick), and to make a recommendation as to which should be appointed. In fact, the applications differed in only one salient respect: Tom had good computing skills, but was poor at writing, while Dick was the opposite. Each application was accompanied by a photograph of a (different) young man. Before they took part in this study, the subjects had been asked to help out with another short experiment on visual perception, in the course of which they were exposed to five four-millisecond presentations of either Tom or Dick's face, accompanied by the word GOOD. An exposure of four milliseconds, under the lighting conditions used, is too short for any conscious impression to be gained. All the subjects saw was a tiny flicker of light, without any 'content'.

The students were twice as likely to choose, as the best person for the job, the candidate whose face had been subliminally presented. If it was Tom's face that had been projected, two-thirds of the students chose Tom over Dick. If it was Dick that had been flashed up, two-thirds opted for Dick. When they were asked to justify their decision, the subjects who had preferred Tom said it was because computing skills were more important for a research assistant than the ability to write well. The subjects who had chosen Dick said that anyone could learn the requisite computing skills, but to be able to communicate fluently was of vital importance.[2]

Pittman and Bornstein's subjects are unconsciously aware of the face that was flashed too briefly for conscious recognition, and of the association between that face and the positive evaluation conferred by the accompanying word GOOD. We know that they are unconsciously aware by virtue of the fact that their actual behaviour cannot be accounted for in any other way. Of course, it is true by definition that we cannot 'see' the unconscious itself directly. We can only infer its presence from its influence on things that we can observe, in just the same way that a 'black hole' cannot be photographed, yet its existence, and its properties, are implied by the fact that light behaves oddly in its vicinity. Likewise, to see how subliminal perception works, we have to look at the way people's actual behaviour, or their conscious thoughts, feelings and perceptions, are 'bent' by forces and processes that are themselves invisible.

Though there are many controlled studies of unconscious perception, we do not really need science to convince us of its ubiquity. We are constantly reacting to things that do not enter consciousness (though it takes an effort of will to notice the existence of things that one has *not* noticed). As you read these words your body is conforming to the chair or whatever you are sitting in or lying on, and you are adjusting your posture every so often in response to sensations of which you are usually unconscious. Your hands are responding to the size and stiffness of the book as you hold it and turn the pages. You may even, while reading, have the experience of realising that a clock is in the middle of striking the hour, and of being able to count up the number of unheard chimes that preceded your moment of 'awakening'.

The classic example of unconscious perception is driving a car and suddenly 'coming to'; you realise that you have driven for the last twenty minutes, absorbed in a conversation or a train of thought, without apparently – consciously – noticing anything at all of the road, the traffic or the operation of the controls. Consciousness has been absorbed in one world, while the unconscious 'automatic pilot' has been in quite another, coping very nicely on its own with roundabouts, traffic lights and pedestrian crossings. And the same ability to pursue flexible, intelligent routines, while being 'elsewhere', is manifest just as much in walking along a crowded street, doing the washing up, playing the piano, taking a shower or getting the children's tea. The ability to do things mindlessly, even cognitively quite demanding things like chatting to a friend or giving a lecture, is notorious. (You may recall the old story of the vicar who dreamt that he was delivering a sermon, and woke up to find that he was.)

The automatic pilot is sensitive to what is happening, and how things are going, just as the computer that controls the real automatic pilot in a plane is sensitive. On automatic pilot we do not just respond like stupid robots; we respond appropriately, like intelligent ones. Or we do most of the time. Sometimes we are caught out, and then we have behaved 'absent-mindedly'. You find that you are halfway to work before you remember that you were supposed to be going to the doctor's. Or, in William James' famous example, you go upstairs to change for the party, and suddenly realise that you are in your pyjamas and cleaning your teeth. Especially when 'we', that is, our conscious minds, are preoccupied, we may find ourselves pouring the hot water into the sugar jar or lighting the fire with today's newspaper.

But when consciousness is not so totally obsessed, merely entertaining itself with a fantasy or a rehearsal, then we do find that the unusual breaks through, it 'grabs our attention' and we wake up. A ball bounces out from between two parked cars just ahead, and suddenly we slow down and are on the alert for the child who may be about to dash out after it. Consciousness is re-engaged with perception and action; the conversation stops in mid-sentence. But here again there is evidence for unconscious perception, for how did we know to lock consciousness on to just *this* small detail, out of the stream of impressions that is constantly flooding in? How is it, when we impulsively turn our head, or stop and listen for a moment, that we frequently find there *is* something there to be attended to? The faint, unrecognised night noises of a friend's country cottage keep jerking me back from the edge of sleep; while the taxis that used to rattle past my London flat throughout the night left me unmoved.

The only possible explanation for these phenomena is that the undermind is keeping a continual check on what is happening below the horizon of conscious awareness, detecting what might be important or dangerous, and deciding when to butt in to consciousness with a 'news flash'. Of course it does this fallibly. Sometimes it interrupts me with false alarms – the noise which wakes me is just the beams creaking, not burglars or a fire – and sometimes it fails to alert me to what matters. But its existence is all I am trying to highlight at the moment, not its omniscience. The movements of consciousness, and the pictures it presents, reflect, like the news flash on the television, the judgements of editors, and the alertness of reporters whose existence we must infer, but whose faces we may never see.

We are constantly reacting to things not solely in terms of what they are, but in terms of what we expect them to be. We prepare ourselves, physically and mentally, for what is going to happen next on the basis of cues that frequently do not themselves enter consciousness. While this process continues in a routine and successful fashion, it usually remains unnoticed. But it reveals itself through its errors. The first few times you use a moving escalator, getting on and off feels slightly peculiar. But when you are more experienced, your body has learnt to make a subtle set of adjustments to help you keep your balance, which are automatically triggered by the surrounding visual cues as you step on to the stairs.

Now, as you approach, as travellers on the London Underground frequently do, an escalator that is stationary, the same cues trigger the same pattern, and you initiate, at just the right moment, a delicate compensation for an acceleration (or deceleration) that, disconcertingly, does not occur.[3]

An even more compelling demonstration of the same effect is provided by the room that you may be invited to enter if you visit the Psychology Department at the University of Edinburgh. The walls and ceiling of the room are actually an upside-down box that is suspended from cables just above the 'real' floor. The gap between walls and floor is too small to be noticed. You enter the room, the door closes behind you – and someone outside pushes the 'room' so that it moves relative to the floor, and to you. This produces a visual perception that normally only occurs when you yourself are walking or swaying, so, acting on this unconscious interpretation, you 'correct' your inferred movement by leaning in the opposite direction – and fall over.

Unconscious preparation may be perceptual as well as physical. Give someone a large and a small tin that weigh the same, and they will tell you that the small one is the heavier – because it is heavier than expected, given its size. Someone brings you what you unreflectingly imagine to be a cup of tea – and the first sip tastes strange, before you realise that it is actually a cup of coffee. Once your expectations are recalibrated, then its taste, the very same taste, becomes reclassified as familiar and satisfactory.

We do not see or taste or feel what is 'out there'; conscious perception is a useful fiction that misrepresents 'reality' in our own interests. As you read, your eyes are flicking along the lines of print in a succession of jumps and fixations – 'saccades' – yet what you see, consciously, is a whole, stable page of print. If you were to hold the book twice as far away from your eyes as it is now, it would hardly look any smaller, even though the image on your retina is only half the size. And though part of the page is (probably) falling on the 'blind spot' of the retina, you are not aware of a corresponding hole in the world you see. The undermind routinely makes all kinds of adjustments to the data it receives before it hands them on to consciousness – because it is usually advantageous to do so.

These everyday examples of unconscious perception often concern aspects of the world which are perfectly visible, audible and so on, but which, though they are registered by the undermind, do not make it into consciousness. They affect us, but pass unnoticed. Experimental studies such as Pittman and Bornstein's, however, have tended to use stimuli that are themselves very faint or fleeting, on the borders of what is perceptible. Such situations demonstrate very clearly the nature of unconscious perception. And they have fascinated psychologists since the very beginnings of the scientific approach to the study of the mind.

In a classic study in 1898, for example, B. Sidis showed people cards on which were printed a single number or letter – but the cards were placed far enough away that his subjects were quite unable to read what was on them. Sidis reports that 'the subjects often complained that they could not see anything at all; that even the black, blurred, dim spot often disappeared from their field of view'. However, when he asked them to name the characters on the cards, they were correct much more often than they would have been by pure guessing, even though that is exactly what they felt they were doing. Sidis concluded from his experiments that there is 'within us a secondary subwaking self that perceives things which the primary waking self is unable to get at'.[4]

Even earlier, in 1884, philosopher C. S. Pierce carried out a series of tests with his graduate student Joseph Jastrow at Johns Hopkins University in America, in which they judged over and over which of two nearly identical weights was in fact the heavier. Again, despite the fact that their subjective confidence was effectively zero, they were able to do much better than chance would dictate. Over the thousands of trials on which they made complete guesses, indicating 'the absence of any preference for one answer over its opposite, so that it seemed nonsensical to answer at all', they were in fact correct between 60 and 70 per cent of the time. What was particularly interesting about their study was their recognition that these results were not just of curiosity value; they are of real significance for the way people operate in the world, for instance how we relate to each other. They wrote that:

> The general fact has highly important practical bearings, since it gives new reason for believing that we gather what is passing in one another's minds in large measure from sensations so faint that we are not fairly aware of having them, and can give no account of how we reach our conclusions about such matters. The insight of females as well as certain 'telepathic' phenomena may be explained in this way. Such faint sensations ought to be fully studied by the psychologist *and assiduously cultivated by everyman*.[5] (Emphasis added)

Not only do they see the relevance of these findings for everyday life; they also suggest that people may increase their sensitivity to such faint sensations. Just as intuition can be educated and sharpened (as I argued in Chapter 5), so can one's ability to make use of the mass of weak impressions that underlie – and are usually neglected by – our normal ways of seeing and knowing. It is as if the mind has two thresholds, one below which it registers nothing at all, and a second above which something becomes conscious. In between the two lies the demi-monde of the undermind in which impressions are active but unconscious. And Pierce and Jastrow's suggestion is tantamount to saying that the distance apart of these two thresholds can vary, so that it is possible to increase one's *conscious* sensitivity to what had previously been going on at an unconscious level.

An early demonstration of the power of unconscious perception to influence what does appear in consciousness was given by Otto Poetzl, a Viennese neuropsychologist working with casualties of the First World War. He tested a number of soldiers who had suffered gunshot wounds to the part of the brain which processes visual stimuli, the occipital lobe at the back of each cerebral hemisphere, and discovered something rather odd. They were effectively blind in the centre of their vision, yet if they kept their eyes fixed on a picture (which they couldn't 'see'), ideas and images would begin to come into their conscious minds which were clearly related to the 'invisible' picture. Associations to the picture would start to surface in consciousness not as features of a coherent visual perception, but as more free-floating and mysterious fragments. Poetzl wondered whether he could produce the same kind of effect in people with normal vision, and devised the following test. First a detailed picture was flashed very briefly, for just one hundredth of a second, in front of the volunteers. They were asked to draw whatever they could of what they had seen – which was usually nothing or very little. Then they were told (absurd though it may sound) to go away and to have a dream that night, and to come back the next day, relate their dreams, and draw any elements of their dreams that they could. When he analysed the records of the dreams, he found that they contained many fragments and associations of the original 'invisible' picture.[6]

Though these early studies were not as tightly designed as one might wish, the essential results have been replicated recently under more stringent conditions. Cambridge psychologist Mark Price, for example, has shown that people are able to 'guess' the category to which a word belongs, even though the word itself was exposed too quickly to be detected. If subjects are flashed the word 'carrot' subliminally, they may not be able to say what the word was, but they still have a better than chance likelihood of guessing that it was a vegetable. In one of his experiments, when the subject happened to be his brother, Price inadvertently replicated the Poetzl effect. At one point he flashed the word 'camel' which his brother failed to detect. However, in the middle of the following presentation, the subject suddenly started chuckling to himself. When Mark asked him what was up, he replied that he was laughing because an absurd fantasy about camels had suddenly popped into his head 'from nowhere'.

Similar effects are obtained when stimuli are hard to detect, not because they are faint or fleeting, but because they occur in peripheral rather than central vision. John Bradshaw at Monash University in Australia has shown that we can unconsciously read words that are on the edge of the visual field, and that such subliminal perception will influence how we interpret what we are consciously attending to. He flashed people an ambiguous word like 'bank' in the middle of the screen, and simultaneously, at the edges of vision, he flashed another word, such as 'money', which suggested one of the two different meanings of the central word. Though they had no conscious awareness of the peripheral words, his subjects nevertheless tended to interpret the central word as meaning 'financial institution' rather than 'edge of a river'.[7] Information which is strong enough to exceed the threshold of awareness, but not to become conscious itself, is nevertheless able to influence what *does* appear in consciousness. As we saw with intuition, consciousness seems to demand evidence that is more definitive than does the undermind.

Another implication of the Pittman and Bornstein study discussed earlier is that, because subjects do not know *how* or even *that* they have been influenced subliminally, they are unconscious of the true source of their decision, and they are therefore unable to use consciousness to guard against the influence itself. They are susceptible to the subliminal message precisely because they do not know it is there. People may have many tendencies that they try to control or mitigate, but unless these are picked up by the 'radar' of consciousness, those controls may not be able to operate. Just as we saw with intuition, the undermind may work with a richer database than consciousness, but the tapestry it weaves may contain threads and assumptions that are false or out of date. This opens up the intriguing possibility that, by bypassing the checks and inhibitions of consciousness, a subliminal stimulus may actually have a greater effect on behaviour than one perceived clearly.

A study by C. J. Patton gives a graphic demonstration of this effect. She chose as her experimental subjects two groups of female college students, one comprising women with 'normal' attitudes to eating, and the other women who had a history of eating disorders. Clinical data suggested that many women in the latter group would have developed a very ambivalent attitude to food, on the one hand craving it as a source of comfort, and on the other disliking themselves for being so dependent, or for becoming overweight (at least in their own eyes) as a result, and struggling to control the desire to eat, especially when in company. The question was: how would these women behave if they became anxious, but were unaware of either the fact or the source of their anxiety?

Like the Pittman and Bornstein study, Patton's experiment involved two parts. In the first, all the women took what was described as a 'visual discrimination test', in which they were to try to identify sentences that were flashed on a screen. There were two key sentences: one, which has previously been shown to be capable of inducing some anxiety in many subjects, was 'Mummy is leaving me', and the other, a neutral control, was 'Mummy is loaning it'. The sentences were flashed either subliminally, for four milliseconds, or consciously, for 200 milliseconds. Half of each of the two groups was flashed one of these messages, and the other half the other. Following this, the subjects went on to take part in the second experiment, in which they were to do a taste test on three different types of biscuit. After explaining this latter task, the experimenter left each subject alone with three full bowls of biscuits to complete the test. After the supposed taste test was completed, the experimenter checked to see how many biscuits each subject had eaten.

The results showed that the neutral sentence had no effect on biscuit consumption for any of the subjects, regardless of whether it was consciously seen or not. However, after having been shown 'Mummy is leaving me', the women with the eating disorders consumed twice as many biscuits as the other women – an average of twenty – *but only when the sentence had been projected subliminally.* For the 'normal' subjects at both exposure durations, and for the eating-disordered subjects when the sentence was clearly visible, the number of biscuits eaten was the same as in the neutral, control condition. When perception is conscious, it becomes possible to override and control our impulses. When we are unconscious of having been affected, we are less able to be vigilant.[8]

A similar effect can be produced if our attention is distracted from the crucial information. J. M. Darley and P. H. Gross asked people to estimate the intellectual ability of some hypothetical children, given various kinds of information. When the only background they were given was the parents' jobs and incomes – their 'socio-economic status' – this information did not influence their judgements. However, another group of subjects were additionally shown a videotape of the children apparently performing somewhat ambiguously on an intelligence test; that is, doing well on some of the test items, and poorly on others. Under these circumstances, when people had been told that one of the children they had watched came from a poor family, they would rate that child's intellectual ability lower than that of an equivalent child whom they had been led to believe came from a better-off family. Apparently, when people are aware that they *might* be biased by such information, they can take steps to compensate. But the effort of focusing on interpreting the child's concrete performance on the test seemed to mask the subjects' awareness of the need to be vigilant about their own stereotypes, and thus their assumptions were able to sidle into their judgements unnoticed.[9]

An experiment by Larry Jacoby at McMaster University in Canada emphasises both the depth of the unconscious interpretations we make, and the power that conscious awareness has to reorganise these interpretations. His study shows that the unconscious can even mislead us about what *kind* of experience we are having: whether it is a new perception or a memory, for example. These basic categories of experience are not 'given'; they, too, are judgements or attributions about which it is possible to be mistaken. Jacoby has shown that these judgements tend to be heavily influenced simply by how easily or fluently something is processed by the undermind. We know from many previous studies that having recognised something once makes it easier to recognise again; there is a residual effect of the first recognition that facilitates the second. And it may be this relative ease of processing that underlies the judgement that something *has* occurred in the recent past. The decision to treat an experience as a memory, rather than as a fresh perception, is, at least in part, an inference based on the fact that we were able to identify and categorise it faster than might have been expected. If this is so, we might be able to trick the undermind into calling something a memory by making it easier to process. (There is clearly the basis of a psychological explanation for *déjà vu* experiences here.)

Jacoby's experiment managed to create exactly this confusion.[10] He and his colleagues showed their volunteers a list of words, and after a short delay, showed them another longer list, one word at a time, that contained the words from the first list mixed up with some new words. Subjects were asked to say, of each word on the second list, whether it had been on the first list or not: in other words, to classify their experience of each word on the second list as a 'memory' or just a 'perception'. The experimenters' cunning manipulation was to make some of the test words slightly easier to read than others by varying the clarity of the print. They found that new words in clearer type were likely to be falsely 'recognised' as having been on the original list. Unconsciously people thought that the relative ease of processing the clear words was due to having seen them before, so they judged them to be memories.

Recent laboratory studies of 'false memory syndrome' have shown that judging an experience to be a 'memory' (and therefore 'real'), rather than a fantasy, is also influenced by the nature of the experience. It may be that the more vividly you can get someone to fantasise, the more likely they are subsequently to misrepresent this experience as a true memory.[11] Certainly, it is not uncommon to experience some confusion, especially just after waking, as to whether one is recalling a lifelike dream from the night before, or a memory of an event several days ago.

There is an additional feature of Jacoby's study that is very significant. The illusion of memory is removed if people are told (or if they notice for themselves) that the visual clarity of the second batch of words is being manipulated. If they are conscious that ease of processing is being influenced by another factor, they can take this into account in their judgements. They can keep the two variables separate. But if people don't know what is going on, then the undermind bundles the two different *sources* of ease – memory and clarity – together, and comes up with some wrong answers. As with the previous examples, when we are consciously aware of an influence, we may be able to guard against it or compensate for it. We can see that what we are doing *is* making an inference or an assumption. When we are not aware of the same influence, when it comes to us subliminally and is, so to speak, already dissolved in our perception by the time it has arrived in consciousness, then we implicitly *trust* it.

In everyday life, this phenomenon of self-monitoring has a large impact on how we form, and deal with, our stereotypes. For

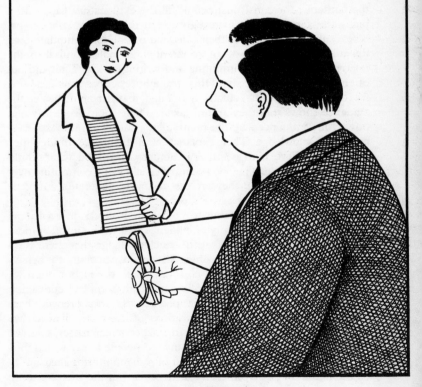

*Figure 7. What does the doctor reply?*

example, look at Figure 7 and ask yourself what the doctor might reply.

Many people think that the doctor might say something like 'You look quite slim to me', while thinking to himself 'I wonder if we have a case of anorexia nervosa here?' In fact what the doctor says is: 'Don't worry; a lot of men tend to put on weight around your age.' Even those who pride themselves on their sensitivity to gender issues may unconsciously be trapped by the picture into assuming that the man is the doctor. When the stereotype is unconsciously stirred into the perception, then we may start to try to make sense out of a seemingly puzzling situation – not realising that it is the assumption that is problematic, not the reality. But of course once we are *aware* of the assumption, the whole picture – literally – changes.[12]

There is a third feature of unconscious perception which the original experiment by Pittman and Bornstein reveals, and that is the tendency for consciousness to 'fill the explanatory gap' with a plausible story, and not to recognise that this is what it has done. Their subjects 'explained' their selection of one job candidate over the other on the basis of an apparently rational appraisal of the relative importance of computing and writing ability. They did not offer this as a *conjecture* about their thought processes, but as a bona fide account of what actually happened in their minds. Yet the evidence shows that they are mistaken. Their choice is clearly and powerfully influenced by the manipulation of which they have no conscious knowledge. Their explanation of the choice, though they genuinely believe it to be true, is actually based on what would have been plausible. They do not intrinsically value computing over writing (or vice versa); they are just trying to generate a rationale for how their preference came about.

This tendency to confabulate is not an isolated or occasional phenomenon. There is plenty of evidence that we do it much more than we think. Occasionally we may acknowledge that there is an element of surmise in our reasoning, as when we account for behaving unreasonably by saying 'I *must have been* tired'; but often we buy our own reasoning uncritically, and with complete conviction (and confidently assert that 'I snapped *because I was* tired'). There are now many experimental illustrations of the ways in which we misconstrue our own motivation. In one, a 'street trader' laid out several different pairs of tights and invited people to say which they preferred. Whichever order they were arranged in, it turned out to be the tights that were on the right-hand end of the row that were chosen most frequently. Clearly there is a statistical bias towards choosing the item in that position, regardless (within limits, obviously) of what it is. Yet when asked why they had chosen the tights they did, no one said 'because they were at the right-hand end'.[13]

In another series of studies on the so-called 'bystander' effect, people were observed in a real-life situation, waiting for a train for example, and a stooge on the platform would suddenly fall to the ground and start groaning. The question is: who goes to the person's aid? What is observed, across a variety of different conditions, is that the more people there are around, the less likely any one person is to offer help. But if you ask those who stood by *why* they didn't get involved, they will tell you all sorts of stories which make no mention of the number of other people. And if you suggest to them that the number of bystanders might have influenced them, they will dismiss the possibility out of hand. In all these cases, people do not know they are confabulating – and would be most indignant if you were to suggest they were. Their conscious interpretation *is* their reality.[14]

The more we acknowledge the existence of the undermind, and its incredible ability to register events and make connections, the less we may need to turn to magical explanations for mental phenomena that at first sight appear strange or supernatural. Take what is sometimes referred to as the 'sixth sense', the rather mysterious ability which is sometimes invoked to account for the experience of somehow 'knowing' that you are being looked at, or that there is someone else in a room that you had supposed to be empty. But is it a sixth sense that is at work, or merely a collection of unconscious impressions derived from the other five? Perhaps this form of intuition could be explained on the basis of a collection of minimal unconscious impressions derived from the five ordinary senses, each of which is too weak to impinge on consciousness itself, but which nevertheless add up to an inexplicable 'feeling'? There do not seem to be any empirical studies of this, but the possibility is effectively described in *Tender is the Night* by Scott Fitzgerald, who was himself fascinated by the activities of the cognitive unconscious.

In an inhabited room there are refracting objects only half noticed: varnished wood, more or less polished brass, silver and ivory, and beyond these a thousand conveyors of light and shadow so mild that one scarcely thinks of them as that, the tops of picture-frames, the edges of pencils or ash-trays, of crystal or china ornaments; the totality of this refraction appealing to equally subtle reflexes of the vision as well as to those associational fragments in the subconscious that we seem to hang on to, as the glass-fitter keeps the irregularly shaped pieces that may do, some time. This fact might account for what Rosemary afterwards mystically described as realizing there was some one in the room, before she could determine it.[15]

We might even venture that a heightened subliminal sensitivity to other people, or even to the contents of one's own mind, might account for some 'telepathic' phenomena. As Pierce and Jastrow suggested, 'we gather what is passing in one another's minds in large measure from sensations so faint that we are not aware of having them . . . certain telepathic phenomena may be explained in this way'. It is possible that their thinking on the matter may even have been influenced directly by a study, also reported in the 1880s by the French physician, philosopher and psychologist Theodore Flournoy, of the renowned medium Catherine Muller, or 'Helen Smith' as she was pseudonymously known. In her meetings she would fall into a trance and undergo personality changes in which she would re-enact scenes from her previous lives. She became, in turn, a fifteenth-century Indian princess, Marie Antoinette, and a visitor from the planet Mars, in which latter incarnation she was able to talk in 'Martian', and discuss the planet's landscape, vegetation and people in extraordinary detail. All her 'characters' were most convincing, and their 'messages' for her clients often highly pertinent and perceptive. Flournoy gained her trust and searched open-mindedly for a 'natural explanation'; one which would credit her with being neither a genuine space and time traveller, nor a fraud.

By meticulously researching her early life, Flournoy was able to show – just as Lowe had done for Coleridge's 'The Ancient Mariner' – that much of her material had come from books she had read as a child and had, consciously, completely forgotten. He described her behaviour under trance as 'romances of the subliminal imagination', and each character represented a reversion of her personality to a different phase of her childhood. His analysis of 'Martian', he argued, showed that it was based on the syntax of French, although the linguist Victor Henry, who also studied her, contended that much of the Martian vocabulary was derived from Hungarian – the mother tongue of Helen Smith's father.[16] In this, as in other cases, Flournoy found no evidence to convince him that her perceptions of other people could not similarly be accounted for on the basis of some combination of buried knowledge and an acute subliminal sensitivity to non-verbal, so-called 'paralinguistic' cues.

No such investigation, however scrupulous, demonstrates that genuine reversion to past lives, or contact with the spirit realm, does *not* occur. And any particular case is always arguable. But such careful and even-handed investigations do at least require us to respect the powers of the undermind, and they might, regrettably to some people, advise caution in interpreting such exotic phenomena as out-of-the-body and near-death experiences, clairvoyance and divination, and so on. There are those who see such experiences as incontrovertible evidence of supernatural powers or influences, and use them as 'proof' that there is more to life than is dreamt of in our current psychology. It is often claimed that they could not have come from 'within'; that there must be real spirits out there who are talking to us, or real powers of telepathy which defy the known laws of physics and physiology. Maybe there are. But the case for the very existence of some of these strange experiences, let alone any particular explanation, is not yet established. And at least in some cases the magical conclusion may be premature – because the role of the *unconscious* has not been fully appreciated. The implicit identification of mind with consciousness is manifested time and again in esoteric circles in the assumption that, if we cannot find an explanation for some phenomenon in conscious terms, then 'the mind' cannot have been responsible at all.

# CHAPTER 8

# Self-Consciousness

Ah, what a dusty answer gets the soul
When hot for certainties in this our life

*George Meredith*

During the course of the second year of his therapy, a middle-aged client, a man of considerable intellect and accomplishment, was talking about the negative patterns in his life. The therapist, Joseph Masling, observed: 'You seem to think you have no right to be happy.' The man immediately began to fidget almost uncontrollably, before eventually subsiding into stillness. After a long silence, he said, 'What did you say?' Another of Masling's clients, a young woman on the verge of successfully completing her graduate training programme, behaved in almost exactly the same way when Masling commented, 'Have you noticed how much easier it is for you to tell me about your failures than your successes?' She too, after a lot of squirming, had to ask him to repeat what he had said.[1]

The undermind is a layer of activity within the human psyche that is richer and more subtle than consciousness. It can register and respond to events which, for one reason or another, do not become conscious. We have at our disposal a shimmering database full of pre-conceptual information, much of which is turned down by consciousness as being too contentious or unreliable. Conscious awareness decides what it will accept as valid – and thereby misses dissonant patterns and subtler nuances. While in d-mode, consciousness tends to present to us a world that is somewhat cautious and conventional. Sometimes this is appropriate, but if we get stuck there and lose the key to the twilight world that subserves it, we mothball valuable ways of knowing which can find sense and weave meaning out of a collection of the faintest threads and scraps.

I have suggested that one way of expressing this disparity between conscious and unconscious is in terms of two thresholds, a lower one, above which the undermind becomes active, and a higher one, above which information enters consciousness. The closer together these two points are, the more 'in touch' with the unconscious we are, and the more complete is our conscious awareness of what is happening across all the mental realms. The further apart they are, the more our conscious perception is impoverished. This quantitative notion of thresholds is rather crude, but it enables us to formulate an important question: what is it that determines how near or how far apart the two thresholds are? More generally, is the relationship between conscious and unconscious forms of awareness a dynamic one, subject to change, and if so, what are the forces that control it? In cases such as Masling's, it seems clear that some information received unconsciously can cause a considerable amount of non-verbal discomfort, and that, as a result, it is gated out of consciousness. The therapist's remark occasions a very rapid raising of the conscious threshold. Perhaps it is specifically things that are threatening that cause the conscious threshold to shoot up.

Corroboration for this supposition comes from a phenomenon called 'perceptual defence', which has been known to experimental psychologists since the 1940s. In the classic version of these studies, a subject is repeatedly flashed a word very briefly, and the exposure duration is gradually increased until she is able to identify the word correctly. Some of the words used are neutral, while others are vulgar or disturbing in some way. The charged words do not become consciously visible to the subjects until they are exposed for a considerably longer duration than the neutral words. If recognition and consciousness are the same thing, this result is simply incomprehensible. How could one selectively raise the perceptual threshold for things that have not yet been recognised? Unconscious perception provides the only explanation: the taboo word *is* recognised unconsciously, and the upper, conscious threshold is immediately raised in order to try to protect consciousness from the threat or emotional discomfort that the word has generated.[2] Jerome Bruner, one of the instigators of research on unconscious perception back in the late 1940s, used to use the analogy of the 'Judas eye', the peephole used by the doorkeeper at a 'speakeasy' to distinguish between bona fide members, for whom the door opens, and undesirables, such as the police, who are shut out. Without the Judas eye, one could only tell friend from foe by opening the door – and then it was too late.

Conversely, we can demonstrate that access to information in the undermind that is dubious, not because it is directly threatening, but because it is faint or ephemeral, can be increased by making subjects feel more relaxed and 'safe'. One way to do this is to ask subjects to express this weak information without feeling that they are under pressure, or being judged in any way. Normally when people are asked to recall something previously shown, they feel that *they* are being tested.[3] Psychologists' experiments are designed to be hard enough for people to make some errors: if everybody got everything right, the data would not differentiate between different conditions, and it is these differences that tell us interesting things about how the mind is working. And nobody likes making mistakes. It may be that the normal type of memory test, for example, underestimates how much people really do know, because the feeling of being on trial makes them adopt a cautious attitude.

Experiments by Kunst-Wilson and Zajonc and others have managed to demonstrate this effect. First they show people a sequence of complicated nonsense hieroglyphics. When these are subsequently mixed up with some new shapes of the same kind, subjects are rather poor at picking out the ones they saw before. However, if, instead of asking people in the second part of the experiment to *recognise* the 'old' squiggles, they are simply asked to point out the ones they *prefer*, they tend quite reliably to choose the ones they saw before. When self-esteem is at stake, delicate unconscious forms of information and intelligence seem to be disabled or dismissed, and the way we act becomes clumsy and coarse. When we are less 'on our best behaviour', the glimmerings of knowledge from the undermind are more available to guide perception and action. Sometimes reception is good, and we are able to pick up and use the undermind's faintest broadcasts. At other times, when we are stressed, only the strongest stations get through.[4]

The same kind of release from pressure can be achieved by presenting the 'test' as if it were a guessing game, rather than a measure of achievement. When we treat something as a 'pure guess', we do not feel responsible for it in the same way. We are freed to utter things that come to us 'out of the blue', because there is no apparent standard of correctness or success against which they, or we, will be judged. One method that has been used to investigate this idea was first devised to assess the memories of people with severe retrograde amnesia. Such people seem to have completely lost their ability to remember what has happened more than a few seconds ago. If you meet one of these people, then go out of the room and come back five minutes later, they will greet you as if you were a stranger. Give them a short list of words to study, take the list away, and after a short delay ask them to recall the words, and they will look at you blankly and say 'What words?' Yet there has for many years been a suspicion that these patients *do* have some memory; it is just that they are unable to access it *deliberately*.

The nineteenth-century French physician Claparede, for example, concealed a pin between his fingers when he was introduced to one of these amnesic patients, giving him a prick as they shook hands. On leaving the room and reappearing a few minutes later, Claparede was, as expected, treated by the patient as if they had never met before – yet the patient was curiously reluctant to shake his hand. When queried about this antisocial behaviour, he rather vaguely explained that 'you never know with doctors; sometimes they play tricks on you'.[5] It is no coincidence that it was a *painful* stimulus that was unconsciously registered, for the undermind is particularly concerned with things that are of significance for our survival and wellbeing.

The suspicion that amnesiacs have more memory than it appears has recently been confirmed in the following way. Subjects are given some time to study a list of words which they are asked to remember, and then the list is removed. A little later, instead of asking the subjects to recognise or recall the words in the conventional manner, they are shown the first two or three letters of a word and asked to respond with the first word beginning with those letters that comes into their heads. As far as the patients are concerned, this a completely new exercise. But the prompts they are given have been selected so that they can be completed with one of the words on the original list – and this to-be-remembered word has been chosen to be less common in the language than some alternative words that could also be used to complete the frame. So if one of the words on the original list was CLEAT, the patients are asked to think of a word that completes the frame CLE– –. Without any memory of the list, people would respond to the cue with a more common word like CLEAN or CLEAR, but in fact the amnesic patients tend to produce the rarer word which they recently saw, but cannot 'remember'.

The words must have been recorded, but the memories only reveal themselves in the way they influence the (apparently) 'free' association. It begins to look as if the 'amnesia', in these cases at least, is more to do with an inaccessibility of memories to consciousness than with an inability to register what has happened. This same effect has now been reproduced in people with undamaged memories, using subliminal perception techniques. Subjects are presented with a number of words on a screen, one by one, too briefly for conscious perception. If they are subsequently asked to recall the words, they will, like the amnesiacs, say 'What words?' Yet, if they are tested with the same 'free association' game, the words they did not even 'see' are found to have made a significant difference to how their minds are spontaneously working. They behave exactly like the memory-impaired patients.[6]

The same kind of distrust of faint information by consciousness has recently been demonstrated by Cambridge psychologist Tony Marcel in a neat study that focused on perception rather than memory.[7] His (unimpaired) subjects were presented with very weak flashes of light, so weak that it was hard to tell whether there was anything there or not, and asked to indicate each time they thought they saw a flash. But Marcel asked them to indicate in any of three different ways: by blinking their eyes; by pressing a button; or by saying 'Yes [I see the light]'. He discovered that these three different methods of answering the same question were not equivalent but gave quite varied results. When people were blinking, they 'saw' many more of the weak flashes than when they were replying verbally, with the button-pressing somewhere in between. When the subjects were asked to respond to the light by *both* blinking and reporting, there were many occasions on which the eyes said Yes while the voice said No. (By measuring the time intervals between stimulus and response, Marcel was able to exclude the possibility that these results could have occurred simply because of reflex blinking.)

Marcel points out how these results challenge our commonsense view of the mind. Our normal assumption is that we have a unitary consciousness which registers 'what's there'. If something is 'there', you can sense it and report it, and the *way* you report it ought not to make any difference to whether you 'see' it or not. Responding is, we assume, 'downstream' of perception, and what comes later in the processing chain should not affect what comes earlier. Under ordinary conditions, this assumption seems to work. But with stimuli that are ambiguous as to whether they are 'there' or not, that model of ourselves begins to break down. The method of report now appears to have a retrospective effect on what we see. The medium that is most closely tied to normal consciousness – verbalising – turns out to be the one that 'sees' least well, while the one that is most automatic, most engrained, most *un*conscious, turns out to be the most sensitive of all. There is corroboration for the idea that *the more the self is involved, the more cautious consciousness has to be*, for fear of 'getting it wrong'. Clearly a verbal report, 'Yes, there was a flash', or 'No, there wasn't', feels like more of a personal commitment than the mere blink of an eyelid – an act that one does not usually think of as requiring close personal supervision or involving much ego investment.

There was another aspect to the experiment which directly replicated the beneficial effect of 'guessing', and its superiority over 'trying'. In each of Tony Marcel's studies, regardless of which modes of response were being used, subjects' ability to detect the weak light was less than perfect. However, when Marcel asked people not to try to report the presence of the flash accurately, but simply to guess, their performance magically shot up to almost 100 per cent! To 'try' is to have some kind of investment in the outcome. You *care*, you *bother* – and therefore you cannot help but be 'bothered' if your effort proves unsuccessful. With a pure guess, on the other hand, you feel as if you are plucking an answer out of thin air. And, as we have seen before, when the pressure is reduced, you are able to allow your choice to be guided by unconscious promptings that *are* adequate for the task, despite consciousness's lack of faith.

Twenty-five years ago, when I was beginning my graduate work in psychology at the University of Oxford, one of my fellow students was a lanky, bearded Australian called Geoff Cumming. Geoff was investigating the effects of 'backward' and 'lateral masking' on perception, using much the same kind of procedure as Tony Marcel. A faint image of a letter was projected on to a screen for a brief period, and then followed, after a short, variable delay, by another image, such as a checkerboard pattern. Geoff was looking at the effect of various characteristics of the second stimulus on people's ability to detect the first event: under some conditions the latter would wipe out conscious awareness of the former. During the course of his experiments, he noticed a rather curious phenomenon. When his subjects were responding at their own speed, they would, under particular combinations of conditions, regularly fail to detect the target letter. But when they were urged to respond as quickly as they could, under the same conditions, they would often make a fast response that correctly detected the target – but would then, a moment later, verbally apologise for having made a mistake! It would seem that they could break through their self-consciousness by responding very fast, and thus were enabled to make use, unconsciously, of the faint information from the first stimulus. But because this information had not been strong enough to make it through into consciousness, they retrospectively concluded that their response had been in error – and incorrectly corrected themselves.[8]

Some of the deleterious effects of self-consciousness on performance have been vividly demonstrated in certain types of brain disorder. Patients with neurological damage may show a dramatically increased disparity between their ability to function and what they are consciously aware of. Tony Marcel has recently reported the clinical case of a woman suffering from hemiplegia with anosognosia – she has lost the use of one side of her body as the result of a stroke, but seems curiously unaware of her deficit.[9] If she is asked to describe herself she will not mention the paralysis. If she is asked to rate her ability to do something that requires the use of two hands – catch a large beach-ball, for example – she will give herself eight or nine out of ten. When asked about herself directly, her consciousness of her condition is remarkably low. However, if questions are put to her in such a way that her self-image is not so directly involved, she then gives quite different answers. If, instead of asking 'How well can *you* do these things?', the questioner says 'If I were like you, how well would I be able to catch the ball?', she will give a rating of only one or two out of ten.

When the form of the question allows her to distance herself from her condition, to 'disown' it, she is able indirectly to acknowledge it. And this is not just a matter of 'not wanting to say', or feeling consciously embarrassed. The evidence is that her reluctance to admit her condition operates 'upstream' of consciousness, in the underground departments of the mind where the decisions about what to allow into consciousness are being made. Interestingly, one can create this 'disinhibition' of consciousness not only by allowing her to project her disability on to someone else, but by inviting her to talk in a regressive, childlike mode. If you hunker down by the side of her chair and whisper rather conspiratorially, 'Tell me, is the left side of your body ever *naughty*?', she will join in the game and whisper back, 'Yes, terribly.' We could speculate that there is a childish sub-personality that needs to exercise less tight control over what can be given access to consciousness – after all, children are used to not being in control of things. Much of the world for them is refractory, and 'naughty' is the childhood word *par excellence* for things that misbehave or will not do what they are told.

We know that, *in extremis*, people are capable of extraordinary feats of tactical unconsciousness. People suffering from 'hysterical blindness', for example, have cut off their consciousness of vision as a result of witnessing something horrific – to protect themselves from further trauma.[10] It is possible to raise the conscious threshold on one sensory modality, apparently, in a way that is both non-selective (unlike the 'perceptual defence' case) and extreme. Yet such people, though they have no visual experience, manage to navigate their way around obstacles uncannily well. Analogously, functionally deaf people show no response to sounds – they fail to show the normal startle reaction to a loud, sudden noise, for example – yet may respond 'No' when quietly asked if they can hear anything. Just as the patient with no memory nevertheless knows, at some level, not to shake hands with the doctor who has hurt them, so it is possible for people with no vision to be able to find their way about, and for people with no hearing to respond to questions.

We are all familiar with more mundane versions of the same phenomena, such as 'unconscious driving', which I used as an example in the last chapter. Though the idea of 'unconscious seeing' may strike us at first as weird or paradoxical, it does so only because it conflicts so strongly with our implicit beliefs about how the mind works, and this dissonance simply makes us unaware of how much of the time we respond appropriately in the absence of conscious awareness. If we were 'seeing unconsciously' in the absence of any conscious experience at all – if we were able magically to find our way around a strange house in pitch darkness, as it were – we would indeed be stunned. But our reliance on the undermind is conveniently masked by the fact that the 'hole' in consciousness is almost always plugged by some other content that is drawing our attention. It is only because our conscious mind is occupied with something else that we usually fail to observe the extent to which the visual information on which we have been relying has bypassed conscious awareness.

We are familiar too with the effect of 'self-consciousness' on behaviour as well as perception. Think of someone who is being interviewed for a job they desperately want, or a child who has been specifically enjoined to carry a full cup of tea very carefully. In such situations there is a sense of vulnerability, of a precarious balancing act, the successful execution of which depends on a degree of skill or control that we do not confidently possess. Thus there is anxiety and apprehension. And this leads to a coarsening of motor control, making us clumsy, in addition to the constriction of attention. Under pressure, we seize up, or 'go blank'. The interviewee fails to understand a perfectly straightforward question. The child concentrates so hard on not spilling the tea that her coordination goes, and she becomes graceless and gauche. It was the very day, in 1984, when my long-standing (and long-suffering) partner had finally finished our relationship that I dived – without any conscious suicidal intention – into the shallow end of a swimming pool and split my head open on the bottom. Losing keys, breaking plates and denting the car are similar symptoms of stress. It is when consciousness is most fiercely preoccupied, usually with a difficult and emotionally charged predicament, that it disregards information (like a large sign saying DEPTH ONE METRE) most egregiously, and siphons off from the unconscious the resources that it needs to function well.

One way in which the relationship between consciousness and the undermind can be radically transformed is through hypnosis (a phenomenon the existence of which, unlike the paranormal, is now established beyond empirical doubt). The active ingredients in hypnosis are relaxation and trust: allowing yourself to be hypnotised involves giving up your normal sense of controlling your own actions, of planning and striving, and putting yourself in someone else's hands. And in this state, for those who can attain it, the relationship between conscious and unconscious becomes unusually labile and permeable. The hypnotist is able to speak directly to the undermind, and is able to adjust, sometimes to an extraordinary extent, which aspects of the unconscious gain access to consciousness, and which do not. Under so-called 'hypnotic age regression', for example, you may gain conscious access to long-forgotten childhood memories. Or you may have compelling hallucinations that you take to be 'reality'. While on the other hand you may be rendered functionally deaf or blind, either to particular kinds of events, or across the board. In one case the conscious threshold is lowered so that normally inaccessible memories surface into conscious awareness; in the other the threshold is raised so high that even mundane experience is blocked out.

Though consciousness may be drastically altered or reduced, we can show that the undermind continues to function. The fact that you no longer have any conscious experience of hearing, or of pain, for example, does not mean that you are *really* not registering the sensations. The threshold between conscious and unconscious has simply been raised to the point where consciousness just isn't getting the usual reports of what is going on in the interior. Take the control of pain. Hypnotic analgesia is a reliably documented and effective method of pain control.[11] Studies show it compares favourably with drugs such as aspirin, diazepam (Valium) and morphine in reducing the experience of pain. Hypnotic suggestion alone produces clinically significant pain relief in as much as 50 per cent of a sample of the general population, even when the people in the sample have not been preselected for their hypnotic susceptibility.

Yet, despite the dramatic alteration of the conscious experience, some responses to the painful stimuli remain. The registration of pain can be demonstrated through physiological measures, for example. One widely used indicator of general arousal is the 'galvanic skin response', GSR, a measure of the resistance of the skin to the passage of electrical current. People who, as a result of hypnotic analgesia, show no visible reaction to a painful electric shock nevertheless show GSR reactions that are typical of the more normal response.[12] Also, it turns out that you can talk to a 'hidden part' of the person who can tell you about the pain, even though its conscious intensity is reduced or non-existent. If hypnotised subjects are asked to sit with their left hands in a bucket of iced water – which is normally quite painful – they will appear relaxed and will say that they genuinely feel little or no discomfort. However, if they are asked to respond with the other hand to a written list of questions about their general physical state, as it were inadvertently, they will report, in their answers, the pain that they do not 'feel'.

One of the clearest demonstrations of this so-called 'hidden observer' effect was recorded by Ernest Hilgard, a long-time hypnosis researcher, in a student practical class. One of the students, a suitable subject, was rendered functionally deaf: he denied hearing anything and failed to flinch at loud noises. While he was in this state, Hilgard whispered softly in his ear:

> As you know, there are parts of our nervous system that carry on activities that occur out of awareness, like circulating the blood ... There may be intellectual processes also of which we are unaware, such as those that find expression in dreams. Although you are hypnotically deaf, perhaps there is some part of you that is hearing my voice and processing the information. If there is, I should like the index finger of your right hand to rise as a signal that this is the case.

The finger rose, and the hypnotised student spontaneously commented that he felt his index finger rise, *but had no idea why it had done so*. Hilgard then released the student from the hypnotic deafness, and asked him what he thought had happened. 'I remember', said the volunteer, 'your telling me that I would be deaf at the count of three, and would have my hearing restored when you placed your hand on my shoulder. Then everything was quiet for a while. It was a little boring just sitting here, so I busied myself with a statistical problem I was working on. I was still doing that when suddenly I felt my finger lift.'

Our security is threatened by information that is painful, or predictive of pain, but it is discomfited by more than that. We possess a whole variety of beliefs, many of which are themselves unconscious or unarticulated, which specify, more or less rigidly, and in more or less detail, our character and our psychology. They define what kind of person we are, our personality or 'self image', and even how our minds are supposed to work. Could it be that our consciousness, what we are able to feel and know about ourselves, is regulated by these beliefs, as well as by the need to protect self-esteem? Is there any evidence that the threshold and the nature of conscious perception – the information in the undermind which is consciously available to us – is influenced by such assumptions?

An intriguing study by Ellen Langer at Harvard suggests that even such a basic psychological attribute as our visual acuity is determined by who we happen to believe ourselves to be. Her subjects were invited to 'become' air force pilots for an afternoon. They were dressed in the appropriate uniform and given the chance to 'pilot' a jet airplane on a flight simulator. The context was made as 'real' as possible, and the subjects were asked to try to *become* a pilot, not merely to act the part. At the beginning of the study, before the simulation had been introduced and explained, each subject was given a short physical examination, which included a routine eye test. During the flight simulation, while they were being pilots, they were asked to read the markings on the wings of another plane that could be seen out of the cockpit window. These markings were actually letters from an equivalent eye chart to the one used in the physical. It was found that the vision of nearly half of the 'pilots' had improved significantly. Other groups of subjects, who were equally aroused and motivated, but who were not immersed in the role, showed no such improvement. By changing the sense of self, more precise sensory information can become available to consciousness.[13]

The self also contains core assumptions about psychological aspects of our make-up that are generic and cultural as well as those that identify us as individuals. Some of these core beliefs concern consciousness itself: when conscious awareness occurs, what it is for, how well it can be trusted and so on. And one of these tacit assumptions is that 'What we see – consciously – is what is "there", and it is *all* that is there'. If we subscribe to this model of perception, we will unquestioningly assume that 'When I have no conscious visual experience, I cannot have registered anything about what is happening out there in the visual world'. So if by some tragic accident I were to be deprived of conscious visual *experience*, but not of unconscious visual information, it is possible that I would be prevented *by this belief* from reacting appropriately to visual *events*. Someone who is involuntarily deprived of conscious sight may thus, as a result of their fundamental belief that consciousness and perception are the same thing, handicap themselves further by cutting themselves off from their residual unconscious visual capacity.

It has recently been suggested that this is exactly what may be happening in cases of so-called 'blindsight'. These cases have now become quite celebrated – they have taken over from 'split brain' patients as the most fashionable and intriguing neurological curios. The blindsight condition results from damage to the visual area of the cerebral cortex which leaves the patient with a blind 'hole' in some part of their visual field. Despite the lack of conscious visual awareness, it has been convincingly demonstrated that these patients *can* react appropriately to stimuli that impinge within the hole – but only if they are allowed to feel that they are playing a rather bizarre kind of 'guessing game', and are not actually being asked to 'see' anything. In the original studies by Lawrence Weiskrantz at Oxford, patients were asked to indicate when they saw one of a constellation of small lights that were flashed at different locations in the visual field. Those lights that fell within the blind area were, as you would expect, not reported. However, when asked to do something absurd – to take part in the nonsensical game of pointing at the hypothetical location of the (to the patient) non-existent light – they were able to do so with remarkable accuracy and consistency. Exactly *what* these patients can respond to is still under investigation. They can certainly point accurately at flashes of light, distinguish between simple shapes such as circles and crosses, and it has been claimed that two such patients have been observed to adjust the movement of their hand appropriately as they reach out for different objects which they cannot 'see'.[14]

Though we can demonstrate that blindsight patients do possess this residual visual capacity, they seem (like Tony Marcel's experimental subjects) unable to use it to respond verbally to the flashes of light, and they do not seem to make spontaneous use of this information to further their own everyday purposes. In the 1993 CIBA symposium on 'Experimental and Theoretical Approaches to Consciousness', psychologist Nicholas Humphrey, commenting on one of the presentations, had some perceptive comments to make on exactly this issue.

> John Kihlstrom's interesting remarks about the self and its relation to unconscious processes . . . are quite well supported by some of the data from patients with blindsight . . . I worked many years ago with monkeys who had had the striate cortex [the primary visual part of the brain] removed: they retained extraordinarily sophisticated visual capacity, much better than anything which has yet been discovered in human beings with lesions of the striate cortex. One way of thinking about this is that the monkey has an advantage in that it doesn't have a particularly highly developed concept of self. Hence the monkey's non-sensory [i.e. unconscious] visual percepts are nothing like so surprising to the monkey as to the human. For a human to have a percept which isn't *his own* percept (related to himself) is very odd indeed. So human patients retreat into saying: 'I don't know what's going on' and denying their ability to see at all. For the monkey, I suspect [unconscious] perceptual information doesn't create the same sort of existential paradox, therefore the monkey is much more ready to use it . . . Interestingly, for one particular monkey I worked with for a long time, there *were* conditions under which she became unable to see again – if she was frightened or she was in pain. It was as though *anything which drew attention to her self undermined her ability to use unconscious percepts.*[15] (Emphasis added)

The more self-conscious we are – the more fragile our identity – the more we shut down the undermind. As people feel increasingly vulnerable, so their access to, and reliance on, information that is faint or fleeting declines. They become not just physically but also mentally clumsy, losing access to the subtler ways of knowing. Conversely, the less self-conscious we are, the more 'at home in our skins and our minds', the more it seems we are able to open ourselves to the undermind, and to the mental modes through which it speaks. Self-consciousness is a graded phenomenon; there are milder, more chronic and more widespread degrees of self-consciousness, in which the kinds of deleterious effects we have been discussing still occur, albeit in less intense forms. Extrapolating from the experimental studies, we might hazard the suggestion that many people (at least in busy, d-mode cultures), much of the time, are in a state of low-grade, somewhat pressurised self-consciousness, and that in this state, consciousness is edited and manipulated so that its contents are as congenial and unthreatening to the operative model of self and mind as possible. We attenuate our contact with the iridescent world of the undermind, and may thus deprive ourselves of valuable data. Though we are in fact sensitive to the shimmering reality that underlies consciousness, we act as if we were not – because we do not 'believe' in it, do not trust it, or do not like what it has to say.

Blindsight research suggests that one area in which we might expect to see the effects of 'self' on consciousness is where people are acting deliberately or intentionally, as opposed to accidentally, playfully or impulsively. The sense of self is associated most crucially, after all, with plans and actions which are designed to serve our own conscious purposes. Intentions are the conscious expressions of our valued goals – of our selves, in other words. One of the features of the blindsight syndrome seems to be the decoupling of patients' unconscious seeing from their own intentions. We can show that they have residual sight, and we can do so best under precisely those conditions when they are *not* acting on the basis of any internally generated intention: when they are not trying to achieve anything, or prove themselves in any way.[16]

This inhibiting effect of intention certainly has its parallels in everyday life. The phenomena of 'not being able to see for looking', or of 'trying too hard', are commonplace. Perhaps the presence of a strong intention locks consciousness too firmly into a predetermined framework of plans and expectations, so that other information, which could potentially be useful or even necessary, is relegated to unconscious processes of perception, where it is, in these cases, ignored. *Intention* drives conscious *attention*, to the detriment, sometimes, of intelligence. In d-mode, we are not just 'looking', we are looking *for*; and what we are looking for has to be, to an extent, pre-specified. Attention is focused and channelled by the unconscious decisions we have made about what may be 'relevant' to the solution of the problem, or the achieving of the intention. And these presumptions may be accurate, or they may not.

Sigmund Freud made exactly this point in his 'Recommendations to physicians practising psychoanalysis', published in 1912. The technique of psychoanalysis, he said,

> consists simply in not directing one's notice to anything in particular, and in maintaining the same 'evenly-suspended attention' . . . in the face of all that one hears. In this way . . . we avoid a danger. For as soon as anyone deliberately

concentrates his attention to a certain degree, he begins to select from the material before him; one point will be fixed in his mind with particular clearness and some other will be correspondingly disregarded, and in making this selection he will be following his expectations or inclinations. This however is precisely what must not be done.[17]

This line of thought suggests that threat or desire may make consciousness narrower as well as coarser, and may explain the experiments described in Chapter 5 which showed that the creativity of intuition and problem-solving is reduced by a feeling of threat or pressure. One of the major reasons why too much effort, too purposive an attitude, or a general increase in stress or anxiety is counterproductive is because it creates 'tunnel vision'. We might imagine that, at any given moment, people are shining a 'beam' of attention outward on to their environment, through the five senses, and inward, too, on to their own physiological, emotional and cognitive state. At the extremes, this beam may be tightly focused, like a spotlight, or wide and broad, like a floodlight, or the dim glow of a candle.

In principle, both forms of attention, concentrated and diffused (and all the degrees of focus in between), are useful. In a pitch-black cave, a hurricane lamp shedding a broad, dim light, which enables you to see the overall size and shape of your surroundings, is what you need first. If all you have is a torch with a fine beam, you will not be able to get your bearings so well. But once you have orientated yourself, it is useful to be able to home in on details, and now the spotlight comes into its own. The diffuse illumination gives you a holistic impression; the focused beam enables you to dissect and analyse. Both are needed, and in an optimal mental state one can flow between the two extremes, adopting a degree of concentration that is appropriate for that particular moment. This balance between perception and attention has a parallel in the equilibrium of deliberation and intuition required for optimising our creative thought processes.

It appears that being stressed, threatened or over-eager tends to narrow the beam of attention too much, whether it be to one's own internal database or to the outside world. People who are chronically anxious have been shown to have more tightly focused attention than people who are more relaxed. Several investigators have reported impaired night vision, for example, in people who are stressed or anxious.[18] When people are required simultaneously to carry out a focal task, such as trying to track a randomly moving point in the central part of a screen with a cursor, and a task that requires the use of peripheral vision, like detecting small flashes of light at the edges of the screen, then increasing the size of the reward for successful performance leads to a tighter concentration on the tracking task, and a serious fall-off in performance on the peripheral task. And if subjects had *not* been forewarned about the peripheral lights, 34 per cent of subjects who were working for large rewards failed to notice them at all, whereas only 8 per cent of those receiving small incentives failed to notice them.[19] The same kind of tunnel vision is produced if the general level of stress is increased by making people work in hot or noisy conditions.[20]

People working under pressure, whether environmental or psychological, tend to select out and focus on those aspects of the situation as a whole which they judge to be the crucial ones. And this judgement must to a certain extent, as Freud realised, be a prejudgement. You make an intuitive decision about what is going to be worth paying attention to. If this 'attentional gamble' is correct, people may learn the task, or figure out a solution, quicker, but at the expense of a broader overview. They see in terms of what they expect to see, and if this self-imposed blinkering reflects an adequate conceptualisation of the problem, time may be saved. But if it is not, or if (as in the Luchins' jars experiments) the situation changes but because of the tight focus the change is not noticed, then a commitment to the spotlight processing strategy is going to let them down. As Jerome Bruner says, reflecting on the adverse effect of motivation: 'Increase in incentive leads to a higher degree of selective attention for those parts of a complex task that subjects interpret as more important, with a concomitant tendency to pay less attention to other features of the situation.'[21] Broad, diffuse attention is precisely what is needed in non-routine, ill-defined or impoverished situations, where data is patchy, conventional solutions don't work, and incidental details may make all the difference. And that is why too much effort inhibits creativity.

An experiment by Jerome Singer demonstrates how increasing the desire for a solution can lead to the coarsening of perception. He asked subjects to estimate the size of a square placed at some distance away from them down a long corridor, by selecting one from an array of different-sized squares arranged on a stand to one side of them. Although this looks simple enough, it is in fact quite a difficult task, precisely because there is so little information with which to judge the size of the distant square. All kinds of subtle cues, such as shadow, brightness and the visual texture of the square, might be helpful. So though the square occupies a very precise point in the visual field, subjects will benefit from having a wide beam of attention in terms of the range and kinds of cues to which they are attentive. It is the sort of task, in other words, which might prove sensitive to the effects of pressure. When subjects were required not just to make their judgement but to imagine that they had a bet riding on its accuracy, their performance deteriorated – even with an imaginary as opposed to a real stake. In another version of the study, subjects were asked to spend fifteen minutes, prior to the size test, working on an insoluble problem, with the experimenter feigning surprise and disappointment at their poor performance. This was sufficient to induce a mood of anxiety and frustration which, in its turn, coarsened the perception of the cues in the distance test, and caused a deterioration of performance.

# CHAPTER 9

# The Brains behind the Operation

The Brain – is wider than the Sky –
For – put them side by side –
The one the other will contain
With ease – and You – beside –

*Emily Dickinson*[1]

We now know a considerable amount about what the intelligent unconscious can do, and the conditions under which it works best; but it remains to discover what it is – how it is physically embodied – and exactly how it works: how it makes available the slower ways of knowing and the powers of subliminal perception. We know very clearly that there is a 'brains' behind the operation, but who, or what, is it?

The brain – the three pounds of soft wrinkled tissue that occupies the skull – is the focus of intense research activity at the moment. The 1990s were designated the 'Decade of the Brain' by the US Congress. At the 1996 British Association for the Advancement of Science 'Festival of Science', the annual public showcase for the work of scientists in Britain and elsewhere, the two-day symposium on 'Brains, Minds and Consciousness' had to be moved from its original venue to the largest lecture theatre on the University of Birmingham campus in order to accommodate the audience. Scarcely a month goes by without the appearance of another book by one of the leading figures in the thriving new discipline of 'cognitive neuroscience'. In trying to understand the physical substrate of the slower ways of knowing, the brain is clearly the most fruitful place to start. If we first explore what the brain does, or what it can plausibly be supposed to do, we shall then be able to see what, if anything, is 'left over', in need of explanation by some other means.

The brain is one of three main systems that coordinate the workings of a whole animal. Together with the hormonal system and the immune system, the central nervous system, of which the brain is the headquarters, ensures that all the different limbs, senses and organs of the body act in concert.[2] The brain integrates information from the eyes, ears, nose, tongue and skin with data about the state of the inner, physiological world, and, by referring this information to the stored records of past experience, is able to construct actions that respond as effectively as possible to the current situation. The brain assigns significance, determines priorities and settles competing claims on resources, for the common good. It ties together *needs*, as signalled from the interior, *opportunities* (and threats) in the environment, as flagged by the five senses, and *capabilities*, as represented by the programmes that control movement and response. And it is able to do this, in the case of human beings, with such consummate elegance and success because it remembers and learns from what has happened before.

The brain is composed of two types of cells – glial cells and neurons – both in profusion. The glia seem to be mainly responsible for housekeeping: they mop up unwanted chemical waste, and make sure that the brain as a whole stays in optimal condition. But it is the neurons, approximately one hundred billion of them, that give the brain its immense processing power. Each neuron is like a minute tree with roots, branches – the dendrites – and a trunk called the axon. The neurons in the brain vary considerably in their actual size and shape. Some are straggly and leggy, with long axons running for some millimetres through the brain; others are short and bushy, with dense dendritic branches, but perhaps measuring only a few thousandths of a millimetre from end to end. However, all the neurons have the same function: they carry small bursts of electric current from one end to the other.

The neurons are packed together into a dense jungle, and where their roots and branches touch (at junctions called synapses) they are able to stimulate one another. Electrical activity in one – what we might call the 'upstream' neuron – influences the likelihood that its 'downstream' neighbour will become electrically active in its turn. Normally each downstream neuron needs to collect the stimulation from a number of its upstream neighbours, until it has built up enough excitation of its own to exceed some 'firing threshold'. When it fires, a train of impulses is initiated along its own axon and into its dendrites, where it can contribute to the excitation of other cells with which it is in contact. No one input will fire a downstream cell on its own, but each contributes to this general pool of activation, making the cell more or less disposed to fire in response to other inputs.

The elaboration of the story of how the neural electrical impulses originate, are conveyed along the axon, and serve to activate other neurons is one of most notable successes of twentieth-century science, and it has been told in detail on many occasions. In brief, each neuron is covered with a semi-permeable membrane that is able to retain some kinds of chemicals within the body of the cell and keep others out. Many of these chemical particles carry a small electrical charge, either positive or negative, and the membrane, in its normal state, is selectively permeable to these 'ions', in such a way that it is able to maintain an electrical gradient, a potential difference, between the inside of the cell and the fluid which surrounds it. However, under the influence of other chemicals – neurotransmitters – which may be released into the ambient fluid, the character of the membrane changes so that charged ions are allowed to flow across it, and it is this flow which initiates the chain of events that may result in a burst of electricity, an action potential, travelling from one end of the neuron to the other.

Action potentials occur spontaneously at more or less regular intervals: nerve cells are never completely quiet. Even when we are asleep their activity continues. But the pattern and frequency of firing can be dramatically altered by events at the synaptic junctions with other cells. A wave of electricity arriving at a synapse from an upstream cell causes neurotransmitters to be released into the gap between it and the downstream cell. These molecules float across the gap and attach themselves to receptors on the membrane on the downstream side, causing it, in its turn, to allow charged ions to flood into the cell and set off another action potential. The stimulation that one neuron gives to another can be inhibitory, making the next-door cell less likely to fire, as well as excitatory.

Each cell may receive stimulation from up to 20,000 different sources, so the neuronal jungle as a whole is incredibly tightly and intricately interconnected. It is estimated that there are of the order of one million billion possible connections in the outer mantle, the cortex, of the brain alone. If we could lay the brain out neatly, we would see that on one 'side' there are all the incoming 'calls' from the inner and outer senses; on the other there are the outputs and commands to all the muscles and glands of the body; and in the middle is this vast tangle of living wires, immersed in a complex, continually changing bath of chemicals, integrating and channelling

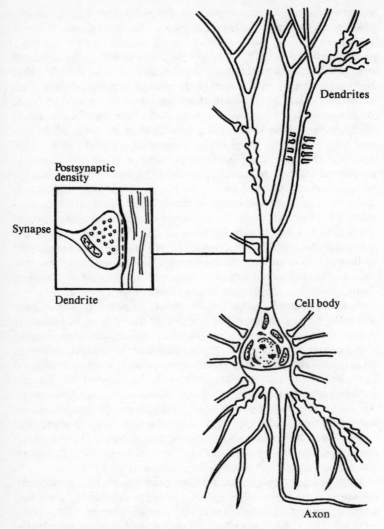

*Figure 8. A stylised neuron*
*(reproduced with permission from* Scientific American)

messages from one side of the network to the other, with many, many loops and diversions on the way.

The electrical communication between neurons can be changed permanently as a result of experience. Cells that were originally strangers, and relatively 'deaf' to each other, can develop close associations, as a result of which one only has to whisper to attract the other's attention where before it had to shout. One way in which experience affects the long-term flow of neural communication is through physical dendritic growth. It has been shown that animals who live in environments full of rich stimulation develop much bushier neurons than do those whose worlds are dull and monotonous. The total number of synapses can increase. But synapses are also capable of becoming easier for an impulse to cross – and for this we need a different process: 'long-term potentiation' (LTP).

When neurotransmitters are released into the synaptic channel between the upstream (currently active) and downstream (currently inactive) neurons, some of the pores in the downstream membrane open up easily to let the charged ions across. However, others, the so-called NMDA (N-methyl-d-aspartate) receptor sites, start out by being more tightly constricted, and will only open up if they are subjected to stimulation that is strong and long-lasting. But once they *have* been opened, they take less persuading on subsequent occasions. For some time after they have been subjected to strong stimulation, the NMDA pores will respond to a much weaker signal. This is one of the fundamental mechanisms that allows the brain to learn.[3] As one of the pioneers of brain research, Donald Hebb, wrote in his seminal book *The Organization of Behavior* in 1949, 'When an axon of cell A is near enough to excite a cell B and repeatedly and persistently takes part in firing it, some growth process or metabolic change takes place in one or both cells such that A's efficiency, as one of the cells firing B, is increased.'[4] Or, more informally, 'cells that "glow" together, grow together'.

An important characteristic of LTP is its *specificity*. Though one neuron receives inputs from a myriad of other cells, if, through LTP, it becomes more responsive to one particular upstream neighbour, it does not become indiscriminately more intimate with the others. Thus there are mechanisms in the brain which enable specific paths of facilitation to be developed between groups of neurons. When a human baby is born, there are certain genetically determined frameworks that serve to impose a general structure on the brain, but much remains unfixed. The brain is like a vast lecture theatre of students on their first day at university: full of potential friendship, but as yet strangers to each other. After a few weeks, however, each student has begun to belong to a number of different evolving circles of acquaintance: study groups, sports teams, neighbourhoods and clubs. Just so, every neuron in the brain comes to belong to a variety of developing clusters, each of which is bound together in such a way that stimulation of one member, or a small group of members, is likely to lead to the 'recruitment' of the others. And the major reason why they tend to become associated is simply that they are active at the same time. Communication becomes more selective, and information begins to flow, throughout the neural community as a whole, along more stable channels.

In order to begin to make links between our understanding of mind and brain, we need to think in terms of the behaviour of these clusters or assemblies of neurons, not of the individual nerve cells themselves. Though we need the molecular level to help us understand how assemblies of cells are formed, and how they behave, there are properties of large assemblies of neurons that are not reducible to their biochemistry – just as the way a hockey team functions cannot be understood from even the most detailed observation of a single player. Nor could the behaviour of an individual student in the context of her hockey practice be predicted or explained on the basis of her performance in the chemistry laboratory, or the college bar. Though there is some direct evidence about how assemblies of neurons behave, it is much more difficult to gather, and still relatively sparse compared with what we know about individual cells. We have to reach out beyond established fact into the realm of plausible hypothesis; to use what we know as a basis on which to build more holistic images of what the brain is like and how it works.

Suppose you are shown dozens of photographs of the same person, Jane, in different moods, wearing different clothes, in different company, engaged in different activities. Some of Jane's features will stay the same across all the photos: the colour of her eyes, the shape of her nose, and more hard-to-pin-down constellations of the way her features are put together. You may not be able to say what it is about the pattern of Jane's face, but after a while you would recognise her anywhere. The neural clusters that correspond to these 'core features', those that recur every time you see Jane, will always be co-active, and it is they that will therefore bond most tightly together. They become the nub of your concept of Jane. Others of her features, such as her ready smile or her penchant for floppy hats, while not critical, become closely associated with her representation, so that, in the absence of any definite information to the contrary, you may automatically fill in these default characteristics when you think of Jane.[5]

More loosely associated still, there are a whole variety of features and associations that have been connected with Jane, but which are less characteristic or diagnostic of her: the time she was photographed with chocolate ice cream all round her mouth; the scarlet suit she wore to Tim and Felicity's wedding. These memories, we might say, form a neural penumbra that is activated a little when we are reminded of Jane, but not – unless the context demands it – strongly enough for them to fire in their own right. Thus the overall neural representation of Jane is not clear-cut. It is composed of an ill-defined, fuzzy collection of features and associations, some of which are bound very tightly together at the functional centre of her neural assembly, while others are more loosely affiliated, and may on any particular occasion form part of the activated neural image of Jane, or may not. Some of these features, whether core, default or incidental, may be distinguishable, or nameable, in their own right – 'nose', 'smile' – while others may comprise patterns that are not so easy to dissect out and articulate.

Depicting the concept of 'Jane' as a vast collection of more or less tightly interwoven neurons does not commit us to supposing that there is a single 'Jane' neuron anywhere in the brain to which all the others lead, or even that the 'extended family' of neurons must all be found in the same location within the brain. There is plenty of evidence that such families of neurons can be – indeed usually are – widely distributed across the brain as a whole. Even if we focus only on the world of sight (and concepts are generally multi-sensory) we find that different aspects of vision – colour, motion, size, spatial location – are processed in quite widely separated areas. Current estimates suggest that there are at least thirty to forty discrete areas of visual processing in the brain, and that these different systems of neurons are themselves interconnected in highly intricate ways. Add in the other senses, as well as memory, planning and emotion, and there will be traces of 'Jane' in every corner of the brain. Just as, in the modern world, geographical proximity is only loosely indicative of the strength of people's relationships, so the intimacy between neurons is reflected in their functional, and not necessarily their physical, closeness.

Since before birth, experience has been constantly binding the brain's neurons together into functional groupings which act so as to attract and 'capture' the flow of neural activation. And these centres of activity in their turn become strung together to form pathways along which neural activation will preferentially travel, so that the brain as a whole develops a kind of functional topography. To explore the consequences of this idea, we might imagine that a concept like 'Jane' or 'cat' or 'student' forms an activation 'depression' towards which neural activity in the vicinity will tend to be drawn, as water finds its way into a hollow. Experience wears away bowls and troughs in the brain which come to form 'paths of least resistance', into and along which neural activity will tend to flow.

At the bottom of a hollow are the attributes that are most characteristic of its concept: those by which, whether we know it or not, the concept is recognised. On the sides of the valley lie the default properties, and higher up are those associations that are optional or incidental. Experience erodes and moulds the mass of neurons into a three-dimensional 'brainscape' where the 'vertical' dimension indicates the degree of functional interconnectedness, the mutual sensitivity and responsivity, of the neurons in that conceptual locality. The deeper the dip, the more tightly bound together the neurons; the more 'deeply engrained', we might say, that concept, that way of segmenting reality, is. But hollows vary in their steepness of incline as well as their depth and their size. Valleys that have steep sides are those where the concept being mapped is well defined: it has relatively few associations that are not criterial. Gentle slopes indicate a wider range of looser connections and connotations.

It is technically impossible to examine in a living brain how such large-scale, distributed collections of neurons come to be associated in the course of everyday kinds of learning. However, it is possible to write computer programs that simulate the properties of neurons, and explore the learning that relatively small numbers of such artificial neurons can accomplish. It turns out that these so-called 'neural networks' are remarkably intelligent. They can, for instance, mimic very closely the kinds of learning that we discussed in Chapters 2 and 3, where complex sensory patterns are picked up and transformed into expertise, without any conscious comprehension or explanation of what has been learnt.

Take as an example the problem of using echo-sounding equipment such as Asdic and sonar to detect mines at sea. The need to discriminate between underwater rocks and sunken mines is obviously a pressing and practical one, both during naval warfare and in the clean-up operations that follow it. Yet it is not an easy problem to solve, for a number of reasons. Echoes from the two types of object can be indistinguishable to the casual ear. And the variations *within* each class are massive – both rocks and mines come in a wide variety of sizes, shapes, materials and orientations – much greater, apparently, than the differences *between* the two. If there are any consistent distinctions, they are almost certainly not to be made in terms of single features, such as the strength of a signal at a specific frequency, but will involve a variety of patterns and combinations of such features.

Suppose we were to analyse any particular echo into thirteen frequency bands, and to measure the amplitude of the signal in each of these bands. Call the bands A, B and so on up to M; and say, for the sake of argument, that the signal strength in each band could range from 0 to 10. It is unlikely that any one of these bands on its own would provide us with a decisive fingerprint. The solution to the problem of discrimination is not as simple as saying that all rocks score more than 7 in band H and all mines score less than 7. It is not even as simple as saying 'If the echo came from a rock, then its strength in band C will be between two and three times its strength in band J'. The only kind of pattern that might conceivably distinguish rocks from mines will be something like: 'An echo probably comes from a mine if *either* the total value on bands A, D and L exceeds the product of the value on bands E and F by more than a factor of six, at the same time as H minus K is less than half of J divided by B; *or* the total of G, H, K and L is more than 3.5 times the total of A, B and C divided by the difference between I and M.' To make the discrimination successfully, if it can be done at all, will require the detection of patterns of this degree of complexity: ones that are hard to describe, let alone to discover.

In fact, human operators can become quite accurate at making these judgements, though, like the subjects in the 'learning by osmosis' experiments, they cannot articulate what it is they know. (You may remember the finely tuned 'intuition' of the sonar operator in the film *The Hunt for Red October*.) Nevertheless, human beings are less than perfect at it, and mistakes are potentially costly. To learn to discriminate between rocks and mines poses an interesting real-life challenge for a simulated brain.

A neural network comprising only twenty-two different 'neurons' has learnt to perform this discrimination surprisingly well. The neurons are arranged together in three 'layers' (as shown in Figure 9). The first layer of thirteen 'sensory' neurons corresponds to the thirteen frequency bands into which the sound spectrum of the echo is divided. They are tuned to detect the signal strength

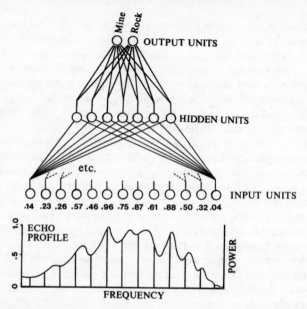

*Figure 9. A simple neural net for distinguishing rocks from mines*

within their particular band, and to emit a signal, like a real neuron's burst of action potentials, that is proportional to this strength. All of these sensory neurons send their signals to each of the seven neurons that are arranged in the next layer. And each of these seven starts out by sending a copy of its output to each of two units in the final layer, the output of one of which corresponds to the decision 'It is a rock' and the output of the other signals that 'It is a mine'. This simplified brain is not able to grow any more connections, but it is able to adjust the selective sensitivity of every neuron to each of the inputs that it receives, in exactly the way that real nerve cells do.

The 'job' of the network is gradually to adjust these sensitivities, in the light of experience, so that the flow of activity through the connections reliably activates the 'rock' neuron whenever it is given a rock echo, and the 'mine' neuron when it is given a mine. Neither the programmer, nor certainly the computer, knows at the outset what the requisite sensitivities are, nor even whether a set of sensitivities that will solve the problem actually exists. The best the programmer can do is to get a large and varied set of genuine sonar echoes which she *knows* arose from either a rock or a mine, and to feed these, one by one, into the network, telling it, after it has generated a decision, whether it was correct or not. In this 'training phase' the computer is given some relatively simple 'learning rules' which tell it how in general to adjust the sensitivities of the neurons as a function of its success or failure. For example, the network may be programmed to adjust all the sensitivities after each trial on the basis of their history of being associated with a correct response. Those units that have a better 'track record' are adjusted very little; those that have a poorer track record are adjusted by a larger amount. Finally, after the 'brain' has been given a large number of such feedback sessions, where it is 'told' whether it was right or wrong, it can be tested with a new set of echoes it has never met before to see what judgement it now makes.

In this example the network behaves exactly like the human subjects in the 'learning by osmosis' experiments. Simple neural networks turn out to be excellent models for this kind of learning. The network starts out 'guessing' and making many mistakes; but gradually its performance improves until finally it is capable of distinguishing quite accurately between rock and mine echoes which it has never heard before. These simulations show convincingly that brains can do what people do; that is, detect intricate, unverbalised patterns that are embedded within a wide range of seemingly diverse experiences, and use these to guide skilful action. Neither the real-life human being nor the artificial brain 'knows' what it is doing, nor on what basis it is doing it. Their 'knowledge' – successful, sophisticated knowledge – is contained in small adjustments in the way the neurons of the brain respond to each other; adjustments which simply direct the flow of activation along different channels, and combine it in different ways. All the brain needs is a diet of training experiences, some feedback, and clear, unpremeditated, unpressurised *attention* to what is happening; its intrinsic operating characteristics will do the rest.

It is worth noting that, in the rocks and mines simulation, the artificial brain came to make the discrimination with a degree of accuracy that even surpassed that of experienced human sonar operators on a long tour of submarine mine-sweeping duty. The neural network, despite its simplicity, outclasses a human expert – not because the computer is 'cleverer', but just because we have not been equipped by evolution with ears sensitive enough to divide sonar echoes into so many frequency bands. We might confidently suppose that if the same kind of problem were to use, instead of a range of metallic 'pings', human babies' cries denoting either 'hunger' or 'wind', mothers would outperform the computer comfortably. Conversely we might expect to find that a dolphin could be trained to beat both the computer and the human operator on the rocks and mines problem.

The imperfect performance of human beings reminds us that there are, of course, limits to the complexity that the unencumbered brain can handle. The world must contain many subtle contingencies that even the fine tuning of the human brain cannot pick up – especially those that have not in the past been directly relevant to survival, or which embody new technological, pharmaceutical or sociological patterns which the biological receiving apparatus was not designed to detect. And also there are many situations which we might *like* to master where there simply is no useful information, no pattern, to be picked up. But what is clear is that *the fundamental design specification of the unconscious neural biocomputer enables it to find, record and use information that is of a degree of subtlety greater than we can talk or think about.* If we let our view of the mind as predominantly conscious and deliberate blind us to the value, or even the existence, of unconscious ways of knowing, we are the poorer, the stupider, for it.

The brain works by routing activity from neural cluster to neural cluster according to the pattern of channels and sensitivities that exist at any moment, and that is all it does. Just as a pebble thrown into a pond starts an outward movement of concentric ripples, so activity in one area of the brain forms what Oxford neuroscientist Susan Greenfield calls an 'epicentre' from which activity spreads out, interacting with other flows of activation, and triggering new epicentres, as it goes. One can literally watch it happening. Studies by Frostig, Grinvald and their colleagues in Israel have used special dyes that can be introduced into cortical neurons, and which fluoresce when the cell becomes electrically active. If a spot of light is flashed into an animal's eye, a neuronal cluster can be seen to form instantaneously and may double its size in a matter of 10 milliseconds. After 300 milliseconds there may be a very large group of active cells distributed over a wide area.

The distributed nature of the neural clusters has been demonstrated by Wolf Singer in Germany. Singer has found that neurons that are widely separated across the visual cortex can nevertheless synchronise their firing patterns in response to a stimulus. Thus, as I have already suggested, the flow of activity is not literally from place to place, but between distributed patterns that continuously segue into one another. The brainscape is, as I have argued, delineated functionally, not physically. If we were able to track the brain's activity, and simplify it, we should see something that looked not like a brightly lit train travelling at night, but more like a luminous kaleidoscope being continually shaken. But to show these iridescent patterns shimmering across the brain is beyond our technical capacity, not least because they move so fast. Ad Aersten and George Gerstein have shown that neuronal groups are highly dynamic, forming and reforming within periods as short as a few dozen milliseconds. And what is more, the same neuron may take part in different patterns from moment to moment. Despite the huge technological problems, there is already some direct evidence for the existence and the properties of these neural patterns.[6]

As well as long-term, 'structural' changes in the brain, there is a variety of shorter-term influences on its responsiveness as well. The topographical 'erosion' of the brainscape is heavily modulated by much more transient influences. Brain responses are affected by the state of need, for example. If an animal is hungry, thirsty, sexually aroused or under threat, groups of neurons tend to adopt the same pattern of firing – to work in synchrony, in other words – more than when the animal is relaxed and sated. Heightened arousal seems to encourage groups of neurons to bind more tightly together into functional teams, and this, Susan Greenfield argues, has a number of interesting consequences, in addition to making each such group more excitable.

Given that neurons are linked together by both excitatory and inhibitory connections, increased arousal can have a mixed effect, causing some neighbours of an active cell to fire more readily while effectively suppressing others. In particular, there tends to be what is called *reciprocal inhibition* between a group of neurons that is currently active, and others that lie outside the group, and the extent of this inhibition makes for a more or less competitive relationship between different centres of activity. When one cluster is creating a strong inhibitory surround, it will tend to suppress other potential epicentres, and at the same time it also tends to sharpen the borders of its own pattern. Instead of priming a broad range of its associates, to varying extents, inhibition makes for a clearer cut-off, and the neural repercussions of any centre of activation thus become more limited. In the presence of a drug – bicuculline – that is known to block mutual inhibition between neurons, a pool of activity can be enlarged as much as tenfold. Thus when arousal is lower, several different centres may be activated simultaneously, because the competition is less fierce; and at the same time patterns of activation that started from different centres can flow into each other like watercolours on wet paper. And from these effects, Greenfield argues, may arise a third consequence of arousal: precisely because of the greater competition, any current 'winner' is more unstable, more likely to be toppled by the next emerging epicentre at any moment, and thus the train of thought may gather greater speed.

We know some of the chemical mechanisms that underlie this 'neuromodulation' effect. The brain stem, the oldest part of the brain, forms a bulge at the top of the spinal cord, and from it bundles of neurons project up into the midbrain, and thence into the cortex. It is these neurons that underlie the role of need, mood and arousal in varying the way cortical neurons behave. They can release chemicals called amines into the cellular milieu which make synapses transiently more or less sensitive. These amines include serotonin, acetylcholine, dopamine, norepinephrine and histamine. Acetylcholine, for example, inhibits one of the normal 'braking' mechanisms that causes a neuron to turn itself off after it has been active for a while. In general, neurons and neural clusters can be primed or sensitised by the influence of amines, so that they are more on a 'hair trigger'.

The dynamics of the brain can therefore vary in a number of different ways. The *direction* of activity flow is influenced both by the sensitivity of the long-term connections between cells, but also by the extent to which different areas are primed. A weak pathway may be temporarily boosted to the point where it is preferred over one that is normally stronger – and thus the 'points' can be switched so that the train of activation is diverted on to a less familiar branch-line. The *breadth* or degree of focus of activation of a concept may be reduced or expanded, so that a familiar conceptual hollow can be made to function as if it were either more or less clearly delineated – as more stereotyped, or more flexible, than its underlying set of structural interconnections would suggest. In one mood, a pattern may have boundaries that are clear and sharp; in another, its influence may spread more widely and taper off more gradually. The number and *variety* of different epicentres that can be simultaneously active also vary. In a state of high arousal, a single chain of associations that is more conscious and more conventional will tend to be followed. In a state of relaxation, activity may ripple out simultaneously from a range of different centres, combining in less predictable ways. And finally the *rate* of flow can vary. In a state of low arousal, a weak pool of activity may remain in one area of the network for some time before it moves on or is superseded.

Under greater arousal – when threatened or highly motivated – activity may flow more rapidly from concept to concept, idea to idea.

## CHAPTER 10

# The Point of Consciousness

In the beginning, man was not yet aware of anything
but transitory sensations, presumably not even of himself. His unconscious brain-mind did all the work.
Everything man did was without understanding.

*Lancelot Law Whyte*

Interesting intuitions occur as a result of thinking that is low-focus,
capable of making associations between ideas that may be structur-
ally remote from each other in the brainscape. Creativity develops
out of a chance observation or a seed of an idea that is given time
to germinate. The ability of the brain to allow activation to spread
slowly outwards from one centre of activity, meeting and mingling
with others, at intensities that may produce only a dim, diffuse
quality of consciousness, seems to be exactly what is required.

There is direct evidence that creativity is associated with a state
of low-focus neural activity. Colin Martindale at the University of
Maine has monitored cortical arousal with an encephalograph, or
EEG, in which electrodes attached to the scalp register the overall
level and type of activity in the brain. When people are more
aroused, these 'brainwaves' are of a higher frequency, and are more
random, more 'desynchronised'. When they are relaxed (but still
awake), their brainwaves are slower and more synchronised: the
so-called 'alpha' and 'theta' waves. Martindale recorded the EEGs of
people taking either an intelligence test – one that required analytical
thought – or a creativity test – one which asked people to discover
a remote associate that linked apparently disparate items, or to generate a wide range of unusual responses to a question such as 'What
could you use an old newspaper for?' Using a standardised questionnaire, the subjects had first been divided into those who were generally creative and those who were not. Cortical arousal was seen
to increase equally for both groups when they were taking the
intelligence test, relative to a relaxed baseline. When subjects

were working on the creativity test, the EEG of the uncreative subjects was the same as for the intelligence test; but the arousal level of the creative people was *lower* even than their baseline control readings.

In a follow-up study, Martindale divided the creative task into two phases: one in which people were required to think up a fantasy story, and a second in which they wrote it down. He argued that the first stage, which he called the phase of 'inspiration', would rely on creative intuition, while the second, the 'elaboration' phase, would involve a more conscious, focused attempt to work out the implications of the storyline and arrange them into some coherent sequence. As predicted, those subjects who were judged to be less creative showed the same high level of arousal in both phases, while the creative subjects showed low arousal during inspiration, and high during the elaboration. In Chapter 6 I argued that the productive use of intuition required a variable focus of attention; the ability to move between the concentrated, articulated processes of d-mode and a broader, dimmer, less controlled form of awareness. Martindale's results show that this fluidity is mirrored in the physiological functioning of the brain. Creative people are those who are able to relax and 'let the brain take the strain'.[1]

The classic description of creativity divides it into four phases: preparation, incubation, illumination and verification. (Martindale's 'inspiration' corresponds to incubation and illumination, while his 'elaboration' corresponds to what in a more scientific context would be called verification.) During the preparation phase, information is gathered and analysed through focused attention or d-mode, in which the brain acts as if the neuronal clusters were relatively sharply demarcated, and trains of associations unfold in a relatively conscious, relatively conventional manner. If the problem in hand is routine this mode will suffice to generate a solution. However, if the problem is more unusual, this way of knowing will result in a series of blind alleys. Activity rushing through tightly delineated channels will not be able to spread out broadly or slowly enough to make simultaneously active the remote associations on which the creative solution rests.

But if someone is able to move into the incubation phase, the sharp inhibitory surrounds which d-mode employs to keep activity focused and corralled, and which tend to turn gentle valleys into functional canyons, are relaxed, and the wider distribution of activity across the brainscape allows a greater number of different foci to become active at the same time. The pattern of activation in the low-focus brain resembles more the one produced by a handful of gravel flung scattershot across the surface of a still lake, than the linear sequence of epicentres created by the 'skimming' of a smooth flat stone. Now if the residual activation from the earlier, preparatory stage remains – if the problem has been put to the back of the mind without being forgotten entirely – the neural clusters that correspond to the problem specification will still be primed. (Uncreative people, as well as having lost the knack of entering the low-focus state, may also be unable to retain this background level of priming: they do not know how to put something 'on the back burner' without it falling down behind the cooker.)

As creative people go about their business, the normal exigencies and incidents of daily life will keep activating thousands of concepts and clusters throughout the brain. If one of these should inadvertently facilitate a link between previously unconnected, but primed, parts of the network, there may be just enough added activation to make an image or a metaphor exceed its threshold and shoot into consciousness – producing an 'insight', an illumination. Finally, during the phase of elaboration or verification, the focus of activation may narrow again in order to explore in more detail the implications that have been opened up.

If creativity is associated with forms of brain activity that are dim and diffuse, and if this is because, in such a state, a greater number of different foci can be simultaneously active, then we might expect that forms of unconscious awareness – subliminal perception – would show the same increase in diversity of associations. Specifically, if an idea is activated unconsciously, its associative ripples may extend out more widely than if its activation is concentrated to a degree that produces clear consciousness. There is some evidence that this is the case. Recall the study by Bowers and his colleagues (in Chapter 6) in which subjects tried to find a single word that was a remote associate of each of fifteen words presented in a cumulative list. Spence and Holland have used the same type of materials to examine the effect of unconscious perception. Their subjects were given a list of twenty words to memorise, of which ten were remotely associated with a single word such as, in the Bowers example, 'fruit'. The other ten words were of the same general type and familiarity, but were not linked together in this minimal way.

Prior to learning the list, some of the subjects were exposed to the word 'fruit' presented subliminally; some saw 'fruit' presented consciously; and some were shown only a blank screen. The results showed that the subjects who perceived 'fruit' unconsciously recalled more of the associated words on the list than did either of the other two groups. Spence and Holland interpret this result to mean that having an object clearly in conscious awareness reduces the range of associates that are active in memory to its 'immediate family'; while a stimulus that does not quite reach the focus or intensity required for consciousness subliminally primes a wider circle of associations. Focal consciousness, we might surmise, is associated with the concentration of a limited pool of 'activity' within a smaller area of the memory network.

I showed in Chapter 4 that incubation also supports better thinking through allowing time for false starts and erroneous conceptualisations to fade away, and to be superseded by a different approach. We can now see how the brain makes this process of 'reappraisal' possible. Imagine that activity in the neural network is flowing along a channel and comes to a point of choice – a T-junction. Which way is it to go? Under normal circumstances, we can assume (for purposes of illustration) that all the activity has to follow the most well-established route. If one arm of the T is worn deeper, and/or is more highly primed, than the other, then that will be the path that is preferred. We can represent the relative facilitation of each pathway, as in Figure 10, by the thickness of its line. At each junction, the activity has to 'choose' the thicker line.[2]

Now suppose the starting point for thinking about a particular problem, given the way it has been initially construed, is at point 'A'; and where you need to get to, the 'solution', is at point '!'. If you follow the thickness of the lines, you will see that the nature of this bit of the network is such that you can never get from A to !. You just keep going round in a circle. However, if for some reason you were to stop thinking about the problem in terms which require you to start at A, and were instead, by accident, to access the same bit of circuitry via another point, B, you would 'magically' find that now you *are* able to get from B to ! very easily. 'BING!', as Konrad Lorenz would say. You have an 'insight'. When you drop out of d-mode, and just let the mind drift around all kinds of 'irrelevant' or even 'silly' associations, you may well, by chance, find yourself thinking not about A but about B – and the recalcitrant solution suddenly becomes obvious. Thus it is, for example, that the most effective antidote to the 'tip-of-the-tongue' state is to stop *trying* to recall the word that stubbornly refuses to come to mind, and to allow yourself to drift off, or to do something else. And then, at some unpredictable moment, the word comes to you, out of the

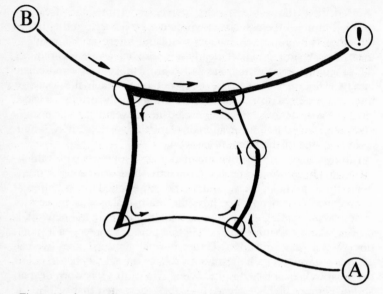

*Figure 10. A map of neural pathways, showing how a change of entry-point can make an insoluble problem soluble. Activation has to follow the thicker line at each junction.*

blue. The head-on assault contains the seeds of its own destruction, while the sideways approach, in which you allow the undermind to sneak up on the item you want, wins the day. That is why brainstorming and daydreaming are – as creative people have always known – effective ways of knowing: they capitalise on the brain's biochemistry.

This neural model also makes it clear why creativity favours not just a relaxed mind, but also one that is well- but not over-informed. In those parts of the neural network that represent the most familiar or routine areas of life, the continual repetition of patterns in experience may have carved out mental canyons and ravines so steep-sided that even when excitation is generally increased and inhibition relaxed, the course of activation flow will still be set. We cannot but construe the world in terms of concepts that are so engrained. However, where the brainscape is contoured enough to formulate an interesting problem, but not so deeply etched that a single approach is inescapable, then moving to the broad focus mode may well reveal novel associations.

It is widely assumed that the total amount of the brain that can be active at any one time varies only within a circumscribed range: there can be only so much activation to 'go round'. At higher levels of arousal the pool may be increased somewhat, but it is clear that if activation were allowed to proliferate unchecked, cognition would lose any sense of direction or definition at all. We might experience an entertaining psychedelic firework display of ever-expanding associations and allusions, but we would rapidly become swamped by them, unable to discriminate the useful and pertinent from the random and trivial. (This is exactly what happens in certain kinds of brain disorder. A famous case was described by the celebrated Russian neurologist A. R. Luria in his book *The Mind of a Mnemonist*.)[3] As patterns of activity move through the brain, there must be inhibitory mechanisms that 'turn off the lights behind them'.[4]

The assumption of limited resource helps to explain why thinking in words can impede non-verbal, more intuitive or imaginative kinds of cognition, and how it is possible to become clever at the expense of being wise. Some of the conceptual hollows in the brainscape are labelled; they have been given names, like 'Jane' or 'breakfast' or 'cat'. Names naturally pick out and focus attention on those features and patterns of the concept that are most familiar and essential: they tend to be associated with the nub of features at the bottom of the hollow, rather than with those that are on the slopes. Slightly fancifully, we might extend the landscape metaphor by planting a tall flagpole at the centre of such articulated concepts, at the top of which flutters a flag bearing the concept's name. This image will serve provided we remember that the 'flag' represents another set of neural patterns, to which the concept is linked, corresponding to the way the word sounds, looks, is spoken and written.

As a child learns language, the flags proliferate, and themselves become connected together into strings of linguistic bunting that begin to create a 'wordscape' that overlays the experientially based brainscape. Words can be combined to 'name' concepts that have no underlying reality, no direct conceptual referent, in that person's experience. Such verbal concepts are heavily influenced by the categories of a particular culture, and conveyed, moulded, through both formal and informal tuition. Different languages carve up the world of experience in different ways. The Inuit famously have dozens of words for 'snow'. English has no concept that even remotely resembles the Japanese *bushido*, the warrior code that combines fighting skill, considerable cruelty and aesthetic and emotional

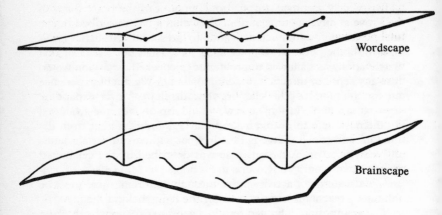

*Figure 11. Brainscape and wordscape. Some concepts have no labels; and some labels are not directly underpinned by concepts.*

sensitivity. The topography of each 'plane', the brainscape and the wordscape, and the relationship between them, represents an evolving compromise between the erosion of the brainscape by first-hand experience, and the dictates of a language about what segments and groupings are to be named.[5]

This model generates a brain-based account, for example, of the fact (referred to in Chapter 6) that describing faces can make them harder to recognise. Instead of simply focusing on the face as a unique whole, and allowing a rich pattern of connections between neuron clusters to become associated together in the brainscape, energy has to be put into making the face conform to general physiognomic concepts that have verbal flags. One is forced to construct a representation that is based on stereotypes ('bulbous nose', 'bushy eyebrows'); is motivated by what can be said rather than what is there; and is an accumulation of fragments rather than a holistic impression.

When a neural cluster, a concept, is activated diffusely, the intensity of activity at its focus may not be sufficient to trigger the verbal label that is attached to it. We have seen that it is possible to be aware of something, both consciously and unconsciously, without being able to retrieve its name. But when an adequate pool of excitation has concentrated at the epicentre of the concept, that activation may flow into, and activate, the representation of the name, which may, in turn, set off a chain of verbal descriptions and associations. And when verbal propositions become activated, the total reservoir of activation is depleted. The fact that the total pool of activation is limited means that activation of a portion of the wordscape must be at the expense of other possible activations and movements within the non-verbal parts of the brainscape. So the more activation that has been 'syphoned up' into the wordscape to subserve an attempt to construct conscious theories and explanations, the less is left to activate other areas of the brainscape. Activation that is focused and verbal tends to support concepts and ways of thinking that are more highly abstract, lacking broader detail and resonance.

In particular, widespread, low-intensity activation incorporates into the representation of a situation more of its personal resonances and connotations. The more the ripples of association spread out, the richer the pattern of meanings that is activated. Objects of perception or thought are imbued with greater significance, because their representation is shot through with a person's felt concerns: their hopes, fears, plans and interests. Situations therefore 'make more sense'; we know more clearly where we stand when the sensory impressions they generate are grounded and glued together by feeling. Because attention is broadly and dimly distributed, these elements of feeling may not themselves emerge into the bright light of consciousness, but nevertheless their activation ensures that perception is suffused with significance.

Felt meaning is embodied. When we sense significance deeply, it affects us physically. We do not simply understand: we are 'touched' or 'moved'. Much of the warp of significance that is woven into the fabric of sensation consists of such bodily connotations. In general, the inner as well as the outer senses contribute to the overall flow and pattern of the brain's activity, and when activity is diffused, the state of the viscera and musculature, any bodily felt *emotions, needs or threats*, are incorporated within the representation as a whole.[6]

Conversely, as the focus of activation tightens, the image of the world that is created becomes more abstract, more intellectualised, and less rich in meaning and feeling. David Gelernter, who has reviewed much of the evidence for the link between focus and emotion, concludes:

As we inch upwards, gradually raising or tightening our focus . . . thinking starts unmistakeably to grow numb. We are less and less able to *feel* our recollections; we merely witness them

... Thought loses its vividness ... In the end we are left to
the cold comfort of logic alone to peg together a powerful and
penetrating – [but] numb and pale – thought-stream and drive
it forward.

The trains of thought that may be stimulated in the wordscape are
also likely to be more rigid, more stereotypical and more defined
by the conventions of the linguistic culture at large, than the patterns
of the brainscape. Thus it may well be harder – as many creative
people have argued – to be original in propositions than in intuitions,
or to unearth and question cultural assumptions that are embodied
in the very way the wordscape is constructed.

One crucial aspect of the functioning of the brain remains to be
explored: how and where and why does it create consciousness?
Even to pose the question in this way makes one important assumption: that consciousness *is* a product of the brain, rather than, for
example, a universal property of all matter, or a signal from elsewhere that is picked up by the brain in the same way that a broadcasting channel is detected and transformed by a television receiver.
Both of these views have lengthy histories in philosophy and religion,
but I shall join the neuroscientific consensus in seeing consciousness
as a correlate of certain kinds of activity that occur only in nervous
systems of a particular kind and degree of complexity. The prima
facie evidence for this starting point is, after all, overwhelming. We
know through direct experience that members (or, if we are being
absolutely precise, at least one member) of one complex species –
*homo sapiens* – possess consciousness, whereas the idea that amoebas, daffodils or pebbles are conscious is at best a conjecture or a
projection. And we know that it is damage to the central nervous
system – rather than to the liver or the lungs, for example – that
results in alterations to or even loss of consciousness.

But if consciousness is a property of brains, we can be sure that
it is not a property of individual neurons. We see and think in terms
of concepts and images that are associated with large groups of
interwoven neurons, not single cells. And we may also discard the
idea that consciousness can be localised within some particular area
or structure of the brain. Despite many attempts to find such a
specific anatomical substrate for consciousness, none has been discovered. Nor should we really expect it to, given that we know that
a single neural cluster is itself distributed widely throughout the
brain. Descartes thought that the pineal gland, situated right at the

centre of the brain, was the 'seat of the soul' and, as Daniel Dennett describes it, the door into the mental cinema where consciousness is projected on to a screen. But we now know without doubt that there is no localised 'headquarters' in the brain to which all inputs are referred, and from which all orders are issued.[7]

We have to think of consciousness, therefore, as associated with *states* of the nervous system rather than places. So our question has to be: what are the conditions which are necessary and/or sufficient for the brain to make its activities conscious? The short answer is that nobody knows for sure. Finding an answer to this question is the 'holy grail' of neuro- and cognitive science at present – as it has been, recurrently, for philosophers and theologians over the centuries. But there are some clues. First, consciousness is associated with *intensity*. The strength or the concentration of activation in a cluster of neurons seems to play a role. Signals that are loud or bright or shocking grab our conscious attention, and stimuli which are being processed unconsciously can be 'boosted' into consciousness if their magnitudes are increased.

For example, even blindsight patients can become directly conscious of activity in the blind field if a stimulus suddenly gets brighter or starts to move faster. In another neurological disorder called prosopagnosia, people are specifically unable to recognise faces that they do in fact know well. Show them Princess Diana and they will not know that they know the face, nor be able to put a name to it. But if these patients have previously been shown a picture of Prince Charles – which they have not been able to identify either – the chances of their consciously recognising Diana are increased. Even though Charles did not himself exceed the threshold of consciousness, his picture has been unconsciously recognised, and this is sufficient to send some activation to prime the underlying Diana network. This, added to the activation from Diana's picture, may be strong enough for the image to reach consciousness.[8]

But intensity alone cannot account for consciousness. Even very strong stimuli can, with time, be ignored. And there is evidence that some neurons, paradoxically, are *more* strongly stimulated by subliminal stimuli than by conscious ones. There are cells in the visual area of the brain, for example, that give a more vigorous response to a light that is shone in the eye of a fully anaesthetised animal than one that is wide awake.[9]

A more crucial condition for consciousness seems to be the *persistence* of neural activity. Benjamin Libet at the University of California at San Francisco has found, by directly stimulating the part of the brain that is responsible for the sense of touch, that even quite strong stimuli must reverberate in the brain for a minimum period of about half a second before they become conscious, whereas reactions to unconscious processing can occur much faster, and with much briefer stimuli.[10] If there is such a minimum time condition for consciousness, then it is possible that people who are asked to move from an unconscious to a conscious mode of responding might show not a smooth increase in their reaction time, but a jump as they move from one mode to the other. Sure enough, the time it takes for people to push a button, as quickly as they can, in response to the onset of a light, is about 200 milliseconds. But if they are asked to slow down their responding by the tiniest possible amount, there is a quantum jump of about half a second, giving total response times of nearly three-quarters of a second. It is as if there is no halfway house: either you are responding instinctively, or you have to wait for consciousness to develop (like a photograph) and then respond.[11]

There are a number of different reasons why the neural effect of a stimulus might last long enough for consciousness to develop. One, obviously, is the duration of the stimulus event itself. But strong stimuli may become conscious, even if they are objectively too short, simply because intensity causes neural activity to reverberate for a longer time – just as the sound of a well-struck gong hangs in the air. (In this way we can subsume the 'intensity' condition under the 'persistence' condition.) And such reverberation may come about not just as a result of the strength of a stimulus. Semir Zeki at University College London has proposed, in the case of vision, that two different areas of the visual cortex must be able to 'sustain a dialogue' if consciousness is to occur. Francis Crick and Christof Koch at the Salk Institute in California have made a somewhat similar suggestion: that what is required to subserve consciousness is a reverberatory loop linking the thalamus in the midbrain to the neocortex.

A third possibility, and one which ties in with some of the experimental evidence we looked at in Chapter 8, is that it is the involvement of the 'self' that extends neural activity to the point of consciousness. We have just seen that asking people to respond self-consciously, rather than instinctively, seems to bump them into a qualitatively different, slower mode of processing. And we have also seen the reverse effect: if people are asked to respond very fast to weak signals – to jump from a careful reliance on the cautious criteria of consciousness to a fast, unchecked way of responding –

they give correct answers which consciousness, when it catches up, countermands. From this point of view consciousness may be *sui generis* self-consciousness.[12] We become conscious of stimuli (except when they are in themselves strong or persistent) because they are being referred to a special part of the neural network which corresponds to our self-image, to see if they fit comfortably with the sense of who we are, and with the ongoing life story in which we see ourselves as taking part. This checking process takes time, and thus in itself tends to fulfil the conditions required for neural activity to generate consciousness (though other censoring processes may quickly come into play – as we saw with the phenomenon of 'perceptual defence' – if it turns out that the information is judged, on this analysis, to be threatening or uncongenial).[13]

If a stimulus is either too short or too weak, or does not fit with my model of who I am and what is going on, it can still play its part in determining how the brain is reacting to events, but it will do so covertly rather than explicitly. As John Kihlstrom, one of the leading researchers on the cognitive unconscious, sums it up:

> When a link is made between the mental representation of self and the mental representation of some object or event, then the percept, memory or thought enters into consciousness; when this link fails to be made, it does not. Nevertheless, unconscious percepts and memories, images, feelings and the like can still influence ongoing experience, thought and action . . .[14]

This association between the self and consciousness, and the idea that consciousness demands a time-consuming resonance or reverberation between different circuits of the brain, raises the intriguing possibility that areas of the brain might be getting on with their business at an unconscious level, without bothering to wait for consciousness to develop. A pool of neural activation may split into two, one part resonating with the self, and thus subserving the emergence of conscious awareness, while the other carries on with further processing such as planning a response. This makes sense, particularly if time is short. It may pay you to continue with the preparations for building an extension on to your house at the same time as you are waiting for formal 'planning permission' – on the assumption that the permission will be forthcoming in the end. If permission is in the event refused, you can abort the plan before you have started the actual process of construction. Provided you have not physically 'jumped the gun', nothing, except the planning time, is lost. If the brain were capable of operating in this dual-track manner, we would have to rethink the function of consciousness. Far from being the instigator of action, the *source* of orders and decisions, consciousness could, at least under some conditions, simply be receiving notification of what was in fact being decided elsewhere.

Another study by Ben Libet has demonstrated that this bifurcation of the mind does occur.[15] He asked people to hold out a hand and, whenever they felt like it, to flex one of their fingers. While they were doing this simple task, he recorded three points in time. Firstly, by virtue of electrodes attached to the person's head, he was able to pinpoint the moment at which the precursors of the action were discernible in the brain's patterns of electrical activity (the EEG). Secondly, Libet asked people to indicate, by registering the position of a spot on a rotating clock in front of them, the moment at which they were first aware of the intention to make the movement. And finally, by recording activity in the muscles of the finger, he was able to note when the physical movement itself began. He found that the voluntary action began to develop in the brain about 350 milliseconds – a third of a second – *before* the appearance of the conscious intention, which occurred, in turn, some 200 milliseconds before the start of the action itself. These results indicate clearly that it is the unconscious brain which decides what to do, and when; and that what we experience as an intention is merely a *post hoc* confirmation of what has already been set in motion. Consciousness receives a kind of corollary 'despatch note', and then presents this as if it were the original order.

'Will', or even 'free will', on this evidence, seems to belong to the brain, rather than to consciousness. But this does not mean that consciousness is left without any function at all. If conscious awareness is associated with the process of checking a situation for concealed threats to self, it may also be instrumental in inhibiting actions and experiences that are adjudged to be risky, rather than in routine instigation and construction. The detection of some irregularity or threat, real or imagined, may result in a veto that can block the execution of the evolving plan before the 'point of no return'. As psychologist Richard Gregory has speculated, it may be that we – that is, consciousness – do not have 'free will', but we do have 'free won't'.

This brings us to the question: what is consciousness for? We are most conscious of those things which might threaten us (except when the feeling of threat is so great, so threatening in itself, that the very experience is inhibited, as in hysterical blindness, traumatic amnesia or psychopathological forms of repression). Where the brain's initial unconscious diagnosis declares the situation to be safe and familiar, there is no need to dwell on it – no need to refer it to the self for further tests. The flow of activation moves on too quickly for the persistence condition for consciousness to be met. But where there is some initial doubt, then the flow of activation is arrested and the predicament is allowed to resonate with the priorities and checks of the self, so that further data may be collected, or a wider pool of associations activated. If this results in the 'all clear', action can proceed unhindered. If a threat is uncovered, then censorship and self-control can save the day. Consciousness is for self-protection.

From this perspective, conscious awareness emerges as fundamentally quizzical and questing. We become conscious of that which is being actively probed for its significance. As many cognitive scientists have recently argued, focused, conscious awareness, the spotlight of the mind, is primarily associated with states of disruption or emergency, and with the activities of investigation, detection and resolution that follow.[16] Some segment of the world poses a puzzle, and high levels of activation concentrate in the corresponding areas of the neural network, so that the nature of the predicament – 'What *is* it, out there (or in here)?' – can be discerned more fully, and an appropriate response discovered. In an emergency, or a state of utter absorption, the proportion of the brain's (limited) resources that are appropriated by this enquiry may be so great that other competing activities are temporarily shut down – we freeze, we stop breathing. Consciousness, it has been suggested, originally made its evolutionary appearance in the context of this focused, arrested response to threat or uncertainty.

This view of the mind is very different from the commonsense one, which sees consciousness as the executive boardroom of the mind, and as the theatre in which 'reality' is displayed. Specifically, the evidence seems to suggest two rather radical conclusions. First, common sense tells us that consciousness is what we can trust: that the world is as it appears to be. But brain research indicates that consciousness manifests not what is certain, but what is in question. Focused consciousness is associated with those aspects of the mind's activity *which are currently being treated as problematic*. Whatever occupies the centre of conscious attention is there precisely because its meaning, its significance, its interpretation, is in doubt. Through dwelling on something we may be led to a richer, 'truer' understanding of it, but that is the result of being conscious, not the prerequisite.

The second conclusion is that consciousness *per se* does not actu- ally *do* anything. Consciousness accompanies, and is therefore symptomatic of, a particular mode of operation of the brain-mind *as a whole*: one in which ongoing action is arrested, careful attention to the probable source of the interruption is being paid, all the sub-systems are listening carefully for new information, priorities are being revised, and new plans being laid. These are the circumstances under which the brain-mind 'generates' conscious awareness. The intense pooling and resonating of activation creates the conditions in which consciousness appears. Consciousness accompanies a very particular and very useful mode of mind, though it does not itself possess any executive responsibility. Although I earlier suggested that we, qua consciousness, may possess 'free won't', even this formulation credits consciousness with power that it does not intrin- sically possess; for the process of vetoing action or editing experience is itself carried out by the brain. Even the 'self' turns out to be just one sub-system among many within the overall neural economy of the brain – the system that defines what is to count as 'threat' or 'desire', and which examines experience through these filters.

The idea that the brain, unsupervised by the conscious intellect, does smart things on its own, and that consciousness *per se* does not carry out any cognitive function, can be discomfiting, because it seems to leave 'us' at times with nothing to do. Yet this disconcerting feeling of redundancy may be the price we have to pay if we wish greater access to the slow ways of knowing. Certainly there are many neuroscientists who firmly believe that the physical brain is *all* we need to account for human intelligence. One of the clearest spokespeople for this point of view is the self-styled 'neuro- philosopher' Patricia Churchland, who has written:

The cardinal principle for the [neuroscientist] is that . . . there is no little person in the brain who 'sees' an inner television screen, 'hears' an inner voice . . . weighs reasons, decides actions and so forth. There are just neurons and their connec- tions. When a person sees, it is because neurons, individually blind and individually stupid neurons, are collectively orches- trated in the appropriate manner . . . In a relaxed mood we still understand perceiving, thinking, control and so forth, on the model of a self – a clever self – that does the perceiving, thinking and controlling. It takes effort to remember that the cleverness of the brain is explained not by the cleverness of a *self* but by the functioning of the neuronal machine that is the brain . . . In one's own case, of course, it seems quite shocking that one's cleverness should be the outcome of well-orchestrated stupidity.[17]

Perhaps any discomfort that these ideas cause is merely a product of having identified ourselves too closely with the habits and values of d-mode, and that all that is required to dispel the unease is an expansion of this shrunken definition of intelligence to reincorporate the brain.

# CHAPTER 11

# Paying Attention

One day a man of the people said to Zen master Ikkyu:
'Master, will you please write for me some maxims of
the highest wisdom?' Ikkyu immediately took his brush
and wrote the word 'Attention'. 'Is that all?' asked the
man. 'Will you not add something more?' Ikkyu then
wrote twice running 'Attention. Attention'. 'Well,'
remarked the man rather irritably, 'I really don't see
much depth or subtlety in what you have just written.'
Then Ikkyu wrote the same word three times running:
'Attention. Attention. Attention'. Half-angered, the
man demanded: 'What does that word "Attention"
mean anyway?' And Ikkyu answered gently: 'Attention
means attention.'

*Philip Kapleau Roshi*

In d-mode, perception is diagnostic. Its role is to sample the
information that is arriving through the senses until it can recognise,
categorise and label what is 'out there' – a 'traffic jam', a 'politician'
– or 'in here' – 'sadness', a 'headache'. Once perception has come
up with its diagnoses, its job is done, and interest shifts downstream
to what can be inferred, and done, about the situation thus
described. If the snap diagnosis is accurate and adequate, thought
builds on a firm foundation. But there is always a risk that such a
skimpy approach to perception may neglect information that does
not, on first sight, *seem* to be significant, but which, had attention
been less precipitate, might have revealed its relevance and its worth.
D-mode determines the way in which attention is to be deployed,
and it is not always the best way. If we get stuck in d-mode's
particular way of attending, we may prematurely and unwittingly
discard just what we need. Sometimes a slower, more meticulous
approach to perception can lead to a richer mental image of what
is happening, and hence to a better way of knowing. If we are to

know as well as we can, we sometimes need to switch from the high-speed scanning of d-mode into a contemplative perceptual stance in which the world is allowed to speak more fully for itself. This chapter explores four different ways of paying attention, or 'slow seeing': detection, focusing on inner states, poetic sensibility, and mindfulness.

The habit of attending closely and patiently to the evidence, even – sometimes especially – to tiny, insignificant-looking shreds of evidence, is characteristic of skilled practitioners of a variety of arts, crafts and professions, prototypically the hunter. From a bent twig, a feather or a piece of dried excrement the expert hunter can recreate an animal, its age and state of health; and he does so in an apparently leisurely fashion in which these scraps of information are allowed to resonate, largely unconsciously, with his mental stock of lore and experience. You can't rush a tracker. Each detail, slowly attended to, is allowed to form a nucleus, an epicentre in the brain, around which associations and connotations gradually accrete and meld, if they will, into a rich, coherent picture of the animal and its passage. As Carlo Ginzburg, author of a fascinating essay on 'Clues', has surmised, the hunter squatting on the ground, studying the spoor of his quarry, may be engaged in the oldest act in the intellectual history of the human race.[1] Many other feats of vernacular connoisseurship – telling an ailing horse by the condition of its hocks, an impending storm by a change in the wind, a run of salmon by a scarcely perceptible ripple on the river, a hostile intent by a subtle narrowing of the eyes – are of the same kind. Each is an act of high intelligence, bringing to bear on the present a complex body of past knowledge, and accomplished by the eye, with little if any assistance from deliberate thought.

In this process of attentive resonance, knowledge does not become the object of explicit thought; rather it implicitly dissolves itself in a gathering sense of the situation as a whole. There is an apocryphal story of a venerable factory boiler that broke down one day, and of the old man who was called to fix it. He wandered around among its convoluted pipework, humming quietly to himself and occasionally putting his ear to a valve or a joint, and then pulled a hammer out of his toolbag and banged hard on one small obscure corner. The boiler heaved a deep sigh and rumbled into life again. The old man sent in a bill for £300, which the manager thought excessive, so he sent it back with a request that it be itemised. When it came back, the old man had written:

for tapping with hammer: 50p
for knowing where to tap: £299.50p.

Similarly, the painter J. M. Whistler, at the trial of John Ruskin, was asked by the judge how he dared ask £350 for a 'Nocturne' that had taken him only a few hours to paint. Whistler replied that the fee was not for the painting, but for 'the knowledge of a lifetime'.

In the late nineteenth century, three new professions came into being that explicitly relied on the ability to read clues: the authentication of artworks, police detection, and psychoanalysis. In the mid-1870s, Giovanni Morelli developed a method for discriminating original paintings from copies and fakes, based not on overall composition or draughtsmanship but on the execution of such tiny details as earlobes and fingernails. He argued that it was precisely in these unimportant details, when both the 'master' and the copyist were 'off guard', that differences of technique would manifest themselves most clearly. As with a casual signature, rather than a self-conscious script, it was in these inadvertent trivia that personality would reveal itself – but only to the eye which understood this to be the case. Like the hunter, one had to be alert to the presence of meaning in the scraps and marginalia.

Morelli directly influenced the development of the 'science' of detection, which was to be dramatised by the emerging writers of detective fiction, such as Gaboriau in France in the late 1870s, and, a little later, most famously, by Sir Arthur Conan Doyle in his Sherlock Holmes stories. Gaboriau, in one of his 'Monsieur Lecoq' adventures, contrasts the novel approach of the detective Lecoq with the 'antiquated practice' of the old policeman Gevrol, 'who stops at appearances, and therefore does not succeed in seeing anything'.[2] While in the Sherlock Holmes story called 'The Cardboard Box', which begins with the mysterious arrival, at the home of 'an innocent maiden lady', of a box containing two severed ears, Holmes literally 'morellises'. Dr Watson reports: 'Holmes paused, and I was surprised, on glancing round, to see that he was staring with singular intentness at the lady's profile.' And Holmes later explains:

You are aware, Watson, that there is no part of the body which varies so much as the human ear . . . I had, therefore, examined the ears in the box with the eyes of an expert, and had carefully noted their anatomical peculiarities. Imagine my surprise then, when, on looking at Miss Cushing, I perceived that her ear corresponded exactly with the female ear which I had just inspected . . . I saw at once the enormous importance of the observation. It was evident that the victim was a blood relation, and probably a very close one.[3]

Sigmund Freud too was influenced, in his developing formulation of the psychoanalytic method, by Morelli, and quite possibly by Conan Doyle as well. Freud is recorded as speaking of his fascination with the Sherlock Holmes stories to one of his patients (the so-called 'wolf-man'). Certainly he had become intrigued by the techniques of Morelli at least ten years before he began to develop his ideas about psychoanalysis in print. In a retrospective essay, 'The Moses of Michelangelo', published in 1914, Freud writes of this influence thus:

> Long before I had any opportunity of hearing about psycho-analysis, I learnt that a Russian art-connoisseur, Ivan Lermolieff [a pseudonym of Morelli's], [was] showing how to distinguish copies from originals . . . by insisting that attention should be diverted from the general impression and main features of a picture, and he laid stress on the significance of minor details . . . which every artist executes in his own characteristic way . . . It seems to me that his method of inquiry is closely related to the technique of psycho-analysis. It, too, is accustomed to *divine secret and concealed things from unconsidered or unnoticed details, from the rubbish heap, as it were, of our observations.*[4] (Emphasis added)

It is interesting to observe, in this context, the changing approach to medical diagnosis over the course of the last two hundred years. The process of detection and identification of disease these days is often devoid of this leisurely resonance of attentive observation with the working knowledge of a lifetime's experience. The modern general practitioner makes a succession of snap decisions as to either the nature of the disorder with which she is confronted, or what further objective, 'scientific' tests to order. She is now so rushed, and so enchanted (as we all are) by technology, and technological ways of thinking, that she generally prefers to trust a read-out from a machine over a considered clinical judgement. An instrument gives us 'real knowledge' about the patient, whereas the poor doctor on her own can offer nothing more substantial than an 'opinion'. Reliance on informed intuition seems increasingly 'subjective', risky and old-fashioned. As medical historian Stanley Reiser says:

> Without realising what has happened, the physician in the last two centuries has gradually relinquished his unsatisfactory

attachment to subjective evidence ... only to substitute a devotion to technological evidence ... He has thus exchanged one partial view of disease for another. As the physician makes greater use of the technology of diagnosis, he perceives his patient more and more indirectly through a screen of machines and specialists; he also relinquishes control over more and more of the diagnostic process. These circumstances tend to estrange him from his patient *and from his own judgement*.[5] (Emphasis added)

Yet throughout the history of medicine, the doctor has functioned more like the tracker or the detective than a technician. And even today there are striking examples of this attentive, resonant intuition at work. There is the much-retailed account of the day the Dalai Lama's personal physician, Yeshi Dhonden, visited Yale Medical School, for example. He gave a demonstration to the assembled group of sceptical Western doctors of traditional Tibetan medical diagnosis by examining a woman patient with an undisclosed illness. On approaching the woman's bed, Yeshi Dhonden asked her no questions, but simply gazed at her for a minute or so before taking her hand and feeling for her pulse. Richard Selzer was one of the physicians present:

> In a moment he has found the spot, and *for the next half-hour* he remains thus, suspended above the patient like some exotic golden bird with folded wings, holding the pulse of the woman beneath his fingers, cradling her hand in his. All the power of the man seems to have been drawn down into this one purpose. It is palpation of the pulse raised to the state of ritual ... his fingertips receiving the voice of her sick body through the rhythm and throb she offers at her wrist. All at once I am envious – not of him, not of Yeshi Dhonden for his gift of beauty and holiness, but of her. I want to be held like that, touched so, *received*. And I know that I, who have palpated a hundred thousand pulses, have not felt a single one.

Finally Yeshi Dhonden laid the woman's hand down. He turned to a bowl containing a sample of her urine, stirred it vigorously, and inhaled the odour deeply three times. His examination was over. He had still not uttered a single word. His diagnosis, whatever it was, would be based solely on his protracted attention to the appearance, the feel and the smell of the woman's sick body. Back in the conference room Yeshi Dhonden, through his young interpreter, delivered his verdict in curiously poetic terms. 'Between the chambers of the heart, long, long before she was born, a wind had come and blown open a deep gate that must never be opened. Through it charge the full waters of her river, as the mountain stream cascades in the springtime, battering, knocking loose the land, and flooding her breath.' Finally the woman's consultant disclosed his diagnosis: 'congenital heart disease: interventricular septal defect with resultant heart failure'. Unless he was very lucky, or had been secretly primed, we may conclude, with the originally sceptical Selzer, that Yeshi Dhonden was 'listening to the sounds of the body to which the rest of us are deaf'. Having stilled his mind through the practice of meditation, he looks, listens, feels and smells *without thinking*, without trying to make any sense, allowing all his sensory impressions to seep at their own speed into the furthest corners of his vast, largely inarticulate storehouse of knowledge, and to deliver back to him, in consciousness, images and figures that make sense of the whole.[6]

This kind of detection comes into its own under certain conditions. It needs a problem that can be clearly formulated – how long since the horses passed?; who planted the bomb?; what is causing the fever? – but to which the answer is not obvious. It requires 'clues': pieces of information whose significance, or even presence, is not immediately apparent. It works with a mind that has a rich database of potentially relevant information, much of which is tacit or experiential rather than articulated. And this kind of detection requires a particular mental mode in which details can be dwelt upon, at first without knowing what their meaning may be, so that slow ripples of activation in the brain may uncover any significant connections there may be. Without this patient rumination, the clue, the problem and the database will not come into the fruitful conjunction that reveals the ways in which they are related.

The successful detective trains her awareness on the outside world, in order to find meaning in the minutiae of experience. The second fruitful way of paying attention is similar, except awareness is now directed inward, towards the subtle activities and promptings of one's own body. The ability to 'listen to the body' is very useful in gaining insight into a whole variety of personal puzzles and predicaments. This ability has been dubbed *focusing* by the American psychotherapist Eugene Gendlin. Back in the 1960s, Gendlin and colleagues at the University of Chicago were involved in large-scale research project designed to discover why it was that some people undergoing psychotherapy made good progress while others did not, no matter who the therapist was or what she did. After analysing thousands of hours of tape-recorded sessions, Gendlin uncovered the magic ingredient, which could be picked up even in the first one or two sessions, and which would predict whether the client would make progress or not. It was not anything to do with the school or the technique of the therapist, nor, apparently, with the content of what was talked about. It was the clients' spontaneous tendency to relate to their experience in a certain way. If they did, they would make progress; if they did not, they wouldn't.[7]

The successful clients were those who spontaneously tended to stop talking from time to time; to cease deliberately thinking, analysing, explaining and theorising, and to sit silently while, it seemed, they paid attention to an internal process that could not yet be clearly articulated. They were listening to something inside themselves that they did not yet have words for. They acted as if they were waiting for something rather nebulous to take form, and groping for exactly the right way of expressing it. Often this period of silent receptivity would last for around thirty seconds; sometimes much longer. And when they did speak, struggling to give voice to what it was they had dimly sensed, they spoke as though their dawning understanding was new, fresh and tentative – quite different from the tired old recitation of grievance or guilt which frequently preceded it.

Gendlin called this hazy shadow which they were attending to, and allowing slowly to come to fruition, a *felt sense*, and it was quite different both from a string of thoughts and from the experience of a particular emotion or feeling. It seemed to be the inner ground out of which thoughts, images and feelings would emerge if they were given time and unpremeditated attention. It appeared that many people lacked the ability, and perhaps the patience, to allow things to unfold in this way. Instead they would, in their haste for an answer, pre-empt this process of evolution, creating a depiction of the problem which told them nothing new, and which gave no sense of progress or relief.

Gendlin discovered that the felt sense will form not in the head, but in the centre of the body, somewhere between the throat and the stomach. The awareness is *physical* and when it has been allowed to form, has been heard, and accurately captured in a phrase or an image, there is a corresponding physical sense of release and relaxation. It is as if some inarticulate part of the person, almost like a distressed child, feels understood, and has responded with a sigh of relief: 'Yes. That's exactly how it is. You understand. Thank you.' When this 'felt shift' happens, then the previous feeling of blockage eases, and by going back again patiently to the felt sense, people find that it is ready to tell them something further; to unfold a little more.

In focusing one takes an issue to consider, asks oneself 'What is this whole thing about?', and then *shuts up*. Over the course of half a minute or so, by holding awareness in the body, a physical sense of 'the whole thing' begins to form in a way that, at first, is unsegmented, and therefore inarticulable. The normal d-mode-dominated tendency to leap to conclusions, to construct a clear and plausible narrative as quickly as possible, is reversed. Answers from d-mode, which tend to come quickly and with a veneer of 'this-is-obviously-the-way-it-is' certainty, are ignored.[8] You know you are doing 'focusing' right, according to Gendlin, when you are not sure if you are doing it right – because you cannot yet *say* what is there. 'The body is wiser than all our concepts', he says, 'for it totals them all and much more. It totals all the circumstances we sense. We get this totalling if we let a felt sense form in inward space.[9]

Because this 'way of knowing' had not previously been identified as one of the main active ingredients in successful therapy, many therapists were unaware of the need to cultivate the client's ability in this regard. Yet, Gendlin discovered, once it was recognised it could be 'taught' quite directly. Anybody, with practice, could learn how to do it, and could benefit from it, not just in dealing with the kinds of problem that took people into therapy in the first place, but in a whole variety of situations in everyday life. To begin with, focusing feels strange, because it really is a different way of knowing from the one with which people are most familiar. As with a medical student learning to read X-rays, it takes time to 'see' what is there, and to stabilise these unfamiliar, shadowy objects of attention. But the tentative, exploratory 'feel' of focusing soon becomes unmistakeable. In one session in which I took part, the focuser said: 'I feel kind of scared, but I don't know what of. Inside it's like an animal that's totally alert, ears pricked . . . It's like something's coming, and some part of me has picked it up and is getting ready for it, but "I" don't know what it is yet.' It is this sense of the imminence of meaning not yet revealed that characterises focusing. The fruit of the felt sense is often an image or an evocative phrase, rather than a fully fledged story – such as the image quoted above of a startled animal, sensing danger, or the unknown, but not yet able to identify it. The first form that the emerging meaning takes is often poetic or symbolic, rather than literal and transparent.

Focusing is not, of course, a new discovery (though turning it into a technology certainly fits with the Promethean spirit of the age). It is very akin, for example, to the Japanese concept of *kufū*, which D. T. Suzuki in *Zen and Japanese Culture* describes as:

> not just thinking with the head, but the state when the whole body is involved in and applied to the solving of a problem . . . It is the intellect that raises a question, but it is not the intellect that answers it . . . The Japanese often talk about 'asking the abdomen', or 'thinking with the abdomen', or 'seeing or hearing with the abdomen'. The abdomen, which includes the whole system of viscera, symbolises the totality of one's personality . . . Psychologically speaking, [*kufū*] is to bring out what is stored in the unconscious, and let it work itself out quite independently of any kind of interfering consciousness . . . One may say, this is literally groping in the dark, there is nothing definite indicated, we are entirely lost in the maze.[10]

It may also have been Gendlin's 'felt sense' which was referred to as *thymos* by the classical Greeks. Located in the *phrenes*, again the central part of the body – lungs, diaphragm, abdomen – *thymos* is that part of a person which 'advises him on his course of action, it puts words into his mouth . . . He can converse with it, or with his "heart" or his "belly", almost as man to man . . . For Homeric man the *thymos* tends not to be felt as part of the self: it commonly appears as an independent inner voice.'[11] It appears that, in other cultures and other times, 'thinking with the abdomen' was a routine and familiar way of knowing. It is only in our contemporary European d-mode culture, dominated by the idea that thinking is the quick, conscious, controlled, cerebral manipulation of information, that the ability to think with the body has to be isolated, repackaged and taught as a novel kind of skill.

With focusing one has, as with detection, a predetermined agenda – a problem to solve or clarify – and the process of dwelling on the details is therefore circumscribed and channelled by a purpose. There is openness and patience, but there is also a background monitoring of progress and relevance. However, the third way of paying attention I want to consider, *poetic sensibility*, has the ability to reset or create our agenda; to uncover issues and reveal concerns, perhaps in unexpected quarters, or surprising ways. By allowing ourselves to become absorbed in some present experience without any sense of seeking or grasping at all, we can be reminded of aspects of life that may have been eclipsed by more urgent business, and of ways of knowing and seeing that are, perhaps, more intimate and less egocentric. As we gaze out to sea or up at a cloudless sky, listen to the sound of goat-bells across a valley or to a Beethoven quartet, we may sense something that lies beyond the preoccupations of daily life. We feel perhaps a kind of obscure wistfulness, a bitter-sweet nostalgia for some more natural, more simple facet of our own nature that has been neglected.[12]

Returning home from a day in the country, people commonly feel calmer, more whole, more balanced. We may not have understood anything, not arrived at any insights or answers, yet we may feel somehow transformed, as if something healing or important has been intimated, but not revealed. In some moods it is possible to gain glimpses of what seems to be knowledge or truth of a sort – of a rather deep sort, perhaps – which is *not* an answer to a consciously held question; and which cannot be articulated clearly, literally, without losing precisely that quality which seems to make it most valuable. There is a kind of knowing which is *essentially* indirect, sideways, allusive and symbolic; which hints and evokes, touches and moves, in ways that resist explication. And it is accessed not through earnest manipulation of abstraction, but through leisurely contemplation of the particular.

When we lose ourselves in the present, we do just that: lose our selves. As the linguist and philosopher Ernst Cassirer put it, the mind 'comes to rest in the immediate experience; the sensible present is so great that everything else dwindles before it. For a person whose apprehension is under the spell of this attitude, the immediate context commands his interest so completely that nothing else can exist beside and apart from it. The ego is spending all its energy in this single object, lives in it, loses itself in it.'[13] One slips away from self-concern and preoccupation into the sheer presence of the thing, the scene, the sound itself, until, as Keats said:

> Thou, silent form, dost tease us out of thought
> As doth eternity . . .

The ego, or the 'self', is essentially a network of preoccupations: a set of priorities that must be attended to in the interests of our survival, our wellbeing, or even our comfort. When the ego is in control of the mind, we act, perceive and think as if a wide variety of things – reputation, status, style, knowledgeability – mattered vitally, and as if their antitheses – unpopularity, ignorance and so on – constituted dire threats. When we are lost in the present, these conditioned longings fall away, and anxious striving may be replaced by a refreshing sense of peaceful belonging. Unskewed by hope or fear, perception is free simply to register what is there. As Hermann Hesse wrote in his essay 'Concerning the soul' in 1917: 'The eye of desire dirties and distorts. Only when we desire nothing, only when our gaze becomes pure contemplation, does the soul of things (which is beauty) open itself to us.'

By its very nature, this more dispassionate, yet more intimate, way of knowing cannot be brought about by an effort of will. It arises, if it does at all, spontaneously. The experience is like that of seeing the three-dimensional form in a 'Magic Eye' image. If you look intently at such an image with the normal high-focus gaze, scanning it for its 'meaning', all you will see, for as long as you look, is a flat field of squiggly shapes. You see plenty of detail, but it does not cohere. However, if you give up 'trying to see what's there', relax your eyes so that they gaze softly *through* the image, and stay for a while in this state of patient incomprehension, then the details begin to dissolve and melt into one another, and a new kind of seeing spontaneously emerges, one which reveals the 'hidden depths' in the picture. There is no doubt when this revelation has occurred: it has a visceral impact which cannot be forced or feigned – just as the 'getting' of a joke is a spontaneous, bodily occurrence that cannot be engineered. Someone who 'thinks' they see the image, like someone who 'understands' a joke, simply has not got it.

Though poetic sensibility cannot be commanded, it can, as with the three-dimensional visual image, be encouraged. One can make oneself prone to it by cultivating the ability to wait – to remain attentive in the face of incomprehension – which Keats famously referred to as 'negative capability': 'when a man is capable of being in uncertainties, mysteries, doubts, without any irritable reaching after fact and reason.' To wait in this way requires a kind of inner security; the confidence that one may lose clarity and control without losing one's self. Keats's description of negative capability came in a letter to his brothers, following an evening spent in discussion with his friend Charles Dilke – a man who, as Keats put it, could not 'feel he had a personal identity unless he had made up his mind about everything'; and who would 'never come at a truth so long as he lives; because he is always trying at it'.[14]

The domination of culture and education by d-mode seems to have created a whole society of Charles Dilkes: to have estranged people from a way of knowing that is, perhaps, part of their cognitive and aesthetic birthright. It certainly appears as if children may have more ready access to poetic sensibility than adults. Young children have been found to be very 'poetic' in their way of knowing in at least one respect: they are much better than older children and adults at producing and using metaphors. Psychologists Howard Gardner and Ellen Winner have found that three- and four-year-old children produce many more appropriate metaphors for a situation than do seven- and eleven-year-olds, and all children are much more fluent users and creators of spontaneous metaphor than college students.[15] And Wordsworth, in his 'Ode to Immortality', famously bemoans the loss of his childhood ways of knowing.

> There was a time when meadow, grove and stream,
> The earth, and every common sight,
> To me did seem
> Apparelled in celestial light,
> The glory and the freshness of a dream.
> It is not now as it hath been of yore; –
> Turn whereso'er I may,
> By night or day,
> The things which I have seen I now can see no more.

It may well have been the child's ability to be lost in the present that prompted the following exchange:

> 'Come along!' the nurse said to Félicité de la Mennais, eight years old, 'you have looked long enough at those waves and everyone is going away'. The answer: 'ils regardent ce que je regarde, mais ils ne voient pas ce que je vois', was no brag, but merely a plea to stay on.[16]

Though it is often lost by the time one reaches adulthood, the knack of absorption can be recaptured. One can cultivate the requisite attitude of receptivity, of allowing oneself to become quietly immersed in things – and then to wait and see. As Jacques Maritain, author of the monumental *Creative Intuition in Art and Poetry*, has said of 'poetic intuition':

> It cannot be improved in itself; it demands only to be listened to. But the poet can make himself better prepared for or available to it by removing obstacles and noise. He can guard and protect it, and thus foster the spontaneous progress of its strength and purity in him. He can educate himself to it by never betraying it.[17]

Many writers and artists have commented on the quality of knowing that emerges from patient absorption. Kafka, in his 'Reflections', says: 'You do not need to leave your room. Remain sitting at your table and listen. Do not even listen, simply wait. Do not even wait, be quite still and solitary. The world will freely offer itself to you to be unmasked, it has no choice, it will roll in ecstasy at your feet.'[18] T. S. Eliot in 'East Coker' enjoins us to 'be still, and wait without hope/for hope would be hope for the wrong thing'.[19] Martin Heidegger's *Discourse on Thinking* puts it very clearly.

> Normally when we wait we wait *for* something which interests us, or can provide us with what we want. When we wait in this human way, waiting involves our desires, goals and needs. But waiting need not be so definitely coloured by our nature. There is a sense in which we can wait without knowing for what we wait. We may wait, in this sense, without waiting for anything; for anything, that is, which could be grasped and expressed in subjective human terms. In this sense we simply wait, and waiting may come to have a reference beyond [ourselves].[20]

Rainer Maria Rilke, in his *Letters to a Young Poet*, has this advice for his self-appointed poetic apprentice:

> If you hold to Nature, to the simplicity that is in her, to the small detail that scarcely one man sees, which can so unexpectedly grow into something great and boundless; if you have this love for insignificant things and seek, simply as one who serves, to win the confidence of what seems to be poor: then everything will become easier for you, more coherent and somehow more conciliatory, not perhaps in the understanding, which lags wondering behind, but in your innermost consciousness, wakefulness and knowing.[21]

Poetic sensibility is available to everyone. It is not the special preserve of Poets with a capital P: people who deliberately create those forms of words called 'poems'. To be a Poet it is necessary to see 'poetically': necessary, but not sufficient. In addition, the Poet must be able to use language in such a way that the reader of the poem is invited not just into the Poet's world, but into the same mental mode, the same slow, poetic way of knowing, that gave rise to the poem in the first place. When we look at things in their own right, without referring them immediately to our own self-interest – which is what the poet invites us to do – then we are in a mode of sensing, knowing and learning that can reveal to us aspects of the world that lie outside the perimeter of our intentions and desires. In fact it can give us self-knowledge by situating our concerns within a wider context that is normally obscured. By allowing the poem to suck us in, we are drawn into a mode of perception that is situated upstream of our usual habits of conceptualization and self-reference. Simultaneously we know the world, and we know ourselves, differently. A poem, viewed thus, is a device for inducing a specific kind of sensibility in the reader. In Paul Valéry's terms, a poem is 'a kind of machine for producing the poetic state of mind by means of words'.

The Poet achieves her effect by doing two things at once. She paints a picture that invites our interest, our engagement, and our identification. And she does this with language that hampers our habitual ways of construing. We cannot see through our own system of categories and concerns without grossly violating the poet's words and thus we hang motionless for a moment in the presence of something made strange and new. George Whalley, writing about the 'teaching' of poetry in school, emphasises the vital necessity of 'experiencing' the poem, by which he means 'paying attention to it as though it were not primarily a mental abstraction, but as though it were designed to be grasped directly by the senses, inviting us to function in the perceptual mode'.[22] If we present a poem, especially to young minds, as something to be 'interpreted' and 'explained', as a kind of mental problem to be solved, like an extended crossword puzzle, we have missed the point. Reading poetry is an exercise in 'holding cognitive activity in the perceptual mode'. One must not *search* for meaning, but marinate oneself in the poem, so to speak, and let meaning come. If one does not treat the poem respectfully, as if it had a life and an integrity of its own, one ends up constructing a surrogate poem as a plausible substitute for the real one: one that disconcerts you less, and merely gives you back your own familiar code of conduct and comprehension.

A poem that is grasped intellectually generates a certain cerebral satisfaction. But a poem with which one is really engaged creates a bodily frisson of undisclosed import; a visceral and aesthetic response, and not just a mental one. Just as with the process of focusing, the body feels something that the mind may not understand. A. E. Housman illustrates the physicality of poetry, as we might expect, with power and humour:

Poetry indeed seems to me more physical than intellectual. A year or two ago I received from America, in common with others, a request that I would define poetry. I replied that I could no more define poetry than a terrier could define a rat, but that I thought we both recognise the object by the symptoms which it provokes in us ... Experience has taught me, when I am shaving of a morning, to keep watch over my thoughts, because, if a line of poetry strays into my memory, my skin bristles so that the razor ceases to act. This particular symptom is accompanied by a shiver down the spine. There is another which consists in a constriction of the throat, and a precipitation of water to the eyes. And there is a third which I can only describe by borrowing a phrase from one of Keat's last letters, where he says, speaking of Fanny Brawne, 'everything that reminds me of her goes through me like a spear.'[23]

Benedetto Croce, writing in the early years of this century, attempts in his *Aesthetic* to make the response in terms of beauty the linchpin of his approach to intuition.[24] For Croce, beauty is not a property of objects or of nature, but of an intuitive response. For the viewer of a painting or a sculpture or a dance, as much as for the reader of a poem, the aesthetic response is the felt manifestation of a certain way of seeing, or knowing, which that object has succeeded in inducing. That which is seen just as it is, fully attended to, not subsumed by categories or reduced to labels, is beautiful. One must learn to recognise, tolerate, enjoy and eventually value this intrinsic ambiguity and impenetrability; in Louis MacNeice's phrase, 'the drunkenness of things being various'.[25] Poetic sensibility and intuition are richer, fuller and subtler than everyday language. There are forms of knowledge that defy articulation. Impressions speak and resonate as vibrant wholes, undismembered. In this way of knowing, beauty, truth and ineffability come together. Argentinean writer Jorge Luis Borges, for example, adumbrating some of the natural 'attractors' of the poetic mode of mind, suggests that:

> Music, states of happiness, mythology, faces belaboured by time, certain twilights and certain places try to tell us something, or have said something we should have missed, or are about to say something: *this imminence of a revelation which does not occur is, perhaps, the aesthetic phenomenon*.[26] (Emphasis added)

Exquisite though the poetic way of knowing may be, we should not be seduced into desiring it as a permanent replacement for mundane reason. It remains one mental mode among many, and to be trapped in the poetic mode would be as disastrous as to be trapped in d-mode. Neurologist Oliver Sacks, in *The Man Who Mistook his Wife for a Hat*, recounts the moving story of one of his patients who was in just this position.[27] Rebecca, at nineteen, was unable to find her way around the block, could not confidently use a key to open a door and sometimes put her clothes on back to front. She had difficulty understanding straightforward sentences and instructions and could not perform the simplest calculations. Yet she loved stories and especially poetry, and seemed to have little difficulty following the metaphors and symbols of even quite complex poems. 'The language of feeling, of the concrete, of image and symbol formed a world she loved and, to a remarkable extent, could enter.' She performed appallingly on standard neurological tests, which are, as Sacks perceptively notes, specifically designed to deconstruct the whole person into a stack of 'abilities'. And just because of this, the tests gave no inkling of 'her ability to perceive the real world – the world of nature, and perhaps of the imagination – as a coherent, intelligible, poetic whole'. In the domain of conscious, deliberate intelligence she was severely handicapped. In the pre-conceptual, unreflective world, she was healthy, happy and competent.

At first, Sacks suggested she should attend classes to try to improve some of her basic 'skills', but they were of no use as they inevitably fragmented her. As Rebecca herself said: 'They do nothing for me. They do nothing to bring me together . . . I'm like a sort of living carpet. I need a pattern, a design, like you have on that carpet. I come apart, I unravel, unless there's a design.' And indeed, when she was spontaneously absorbed in an activity that engaged all of her, she was a different person. She was moved from the 'remedial classes' to a theatre workshop – which she loved and where she blossomed. She became *composed*, complete, and played her roles with poise, sensitivity and style. Sacks concludes his account: 'Now, if one sees Rebecca on stage, for theatre soon became her life, one would never even guess that she was mentally defective.' Lost in the particular, Rebecca became fluent and complete. In the world of the abstract, she was shattered and lost.

Skimping on perception runs two risks. Not only may one overlook aspects of the inner and outer world that are informative or even inspiring; one may inadvertently stir into perception as it develops assumptions and beliefs that are not justified or required. What is finally served up to consciousness may be simultaneously impoverished and elaborated, even adulterated. The mind in a hurry tends to see what it expects to see, or wants to see, or what it usually sees. One of the problems with the name 'Guy' is that I am forever reacting to calls that were not meant for me. People shouting 'Hi!' or 'Bye!' in crowded streets are likely to find me looking at them expectantly – before I detect and correct my perceptual mistake. Leaping to conclusions in this way is a gamble. By setting the threshold for the recognition of my name on a hair trigger, I make sure I react quickly – but I also make a lot of 'false positive' errors. By assuming that what usually happens is what did happen, I save processing time, but at the cost of misdiagnosing the situation when it is unusual. The fourth manner of paying attention which I want to describe in this chapter is a way of seeing through one's own perceptual assumptions. It is called mindfulness.

The extent to which the world-as-perceived is a mirror of our preconceptions and our preoccupations (and therefore the extent to which our subsequent thoughts, feelings and reactions are assimilated by these assumptions) is easy to underestimate. It takes an effort to see what is happening, because our beliefs are dissolved in the very organs which we use to sense. Take, as a trivial example, saliva. Be aware, for a moment, of the saliva in your mouth. Collect a little and roll it around. Feel how it lubricates your tongue as it slides over your teeth. Now get a clean glass, spit some of this saliva into it – and drink it. Notice how your perception of, and attitude towards, the same substance has miraculously changed. What was 'clean' and 'natural' has, through its brief excursion beyond the body, turned into something 'dirty' and 'distasteful'. The spit has not changed; only the interpretation.

One of the major contributions of experimental psychologists this century has been to keep providing us with new and telling demonstrations of what they call the 'theory-ladenness' of perception (just as it has been the function of the poets throughout history to keep showing us that the world is more 'various', more open to reinterpretation, and more inscrutable than we normally suppose). Much of the work on visual illusions shows this clearly. In the Kanizsa figures below, for example, we see – literally see – shapes that are not 'really there', because it seems plausible to the mind to suppose that they are.[28] We are used to seeing as 'whole', objects parts of which are occluded by other objects in front of them. And this expectation can drive us, if it 'makes sense' to do so, to hallucinate an intervening shape, even creating visible edges for it, adding impressions of depth and contrasts in brightness, to make the interpretation more convincing. Such tinkering with reality goes on

*Figure 12. Illusory shapes and contours, after Kanizsa (1979)*

all the time, and at levels of mind that are way below conscious intention or control.

A less stylised example is provided by the concept of 'old age'. Being 'old' is not just a biological phenomenon; how one goes about 'being old' depends on one's (largely unconscious) *image* of what it is like, what it means, to be old, and this in turn reflects a whole raft of both cultural assumptions and individual experiences. Ellen Langer and colleagues at Harvard University have examined the effect on elderly people of their own vicarious experiences, as children, of ways of being old. They reasoned that children may unconsciously pick up images of old age from their own grandparents – which they might then recapitulate as they themselves get older. Specifically, they surmised that the younger their grandparents were when children first got to know them, the more 'youthful' would be the image of old age that the children would unconsciously absorb, and the more positively they would therefore approach their own ageing.

In order to test this idea, they interviewed elderly residents of nursing homes in the vicinity of Boston to find out if they had lived with a grandparent as they were growing up and, if so, how old they were when the grandparent first moved in. When they were independently evaluated by nurses who knew nothing about the research, it was found that those elderly people who had lived with a grandparent when they themselves were toddlers were rated as more alert, more active and more independent than those whose first experience of living with a grandparent had not occurred till they were teenagers. While further research is needed to clarify the interpretation of these results, it does look as if the ways in which different people age depends quite directly on the assumptions and beliefs they have picked up in their own childhoods about what it is to be old.[29]

The unconscious assumptions that people stir into their experience are often rather hard to alter, but sometimes they can be changed just by a suggestion, especially if it comes from some kind of authority figure. The experience of pain, for instance, can be dramatically altered, in normal conscious subjects, simply by telling them to think of it differently. When a group of people who had volunteered to suffer some mild electric shocks were told to think of the shocks as 'new physiological sensations', they were less anxious, and had lower pulse rates, than those who were not so instructed.[30] In another study, hospital patients who were about to undergo major surgery were encouraged to realise how much the experience of pain depends on the way people interpret it. They were reminded, for example, that a bruise sustained during a football match, or a finger cut while preparing dinner for a large group of friends, would not hurt as much as similar injuries in less intense situations. And they were shown analogous ways of reinterpreting the experience of being in hospital so that it was less threatening. Patients who were given this training took fewer pain relievers and sedatives after their operations, and tended to be discharged sooner, than an equivalent group who were untrained.

These experiments demonstrate how other people may be able to rescue us from what Langer refers to as 'premature cognitive commitments' – help us become aware of the assumptions that we had dissolved in perception, and contemplate alternative ways of construing the situation. Helping others to change not the circumstances of their lives but their interpretations of those circumstances is a widespread therapeutic technique called 'reframing'. R. D. Laing, for example, in a classic case, helped a man who was desperate about his 'insomnia' to reconstrue his extra hours of wakefulness as a boon. 'Just think of all those people out there who are suffering from "somnia", forced to spend as much as eight or nine hours every day doing nothing,' Laing observed. When the 'problem' that we are facing is created by our own unconscious additives, no amount of good thinking or earnest effort will bring a solution. Such contortions only compound the original mistake. The only way out of the trap is to *see through* the interpretation which one had been making; to see it *as* an interpretation. Only with such self-awareness, or 'mindfulness', is it possible to be released from the pernicious belief.

Mindfulness involves observing one's own experience carefully enough to be able to spot any misconceptions that may inadvertently have crept in. There are a number of ways in which this quality of mindfulness towards the activity of our own minds can be cultivated, though all involve slowing down the onrush of mental activity, and trying to focus conscious awareness on the world of sensations, rather than jumping on the first interpretation that comes along and hurtling off in the direction of decision and action. Mindfulness can be taught directly, as a form of secular meditation, for example. Jon Kabat-Zinn, Director of the Stress Reduction Program at the University of Massachusetts Health Center, gives a clear idea of what is involved:

> The essence of the state is to 'be' fully in the present moment, without judging or evaluating it, without reflecting backwards on past memories, without looking forward to anticipate the future, as in anxious worry, and without attempting to 'problem-solve' or otherwise avoid any unpleasant aspects of the immediate situation. In this state one is highly aware and focused on the reality of the present moment 'as it is', accepting and acknowledging it in its full 'reality' without immediately engaging in discursive thought about it, without trying to work out how to change it, and without drifting off into a state of diffuse thinking focused on somewhere else or some other time ... The mindful state is associated with a lack of elaborative processing involving thoughts that are essentially *about* the currently experienced, its implications, further meanings, or the need for related action. Rather mindfulness involves direct and immediate experience of the present situation.[31]

There is now good evidence for the efficacy of such mindfulness training in helping people with all kinds of distresses and diseases. Kabat-Zinn's programme has enabled hundreds of people with painful and upsetting conditions to release the secondary fears and anxieties that such conditions invariably create. Even the painfulness of pain itself, as we have just seen, can be reduced through mindfulness.

One particularly compelling demonstration of the practical value of mindfulness comes from clinical psychologist John Teasdale in Cambridge, who has been working on ways to prevent relapse in people who suffer from chronic depression.[32] To simplify a complex story: in many types of depression people suffer some upsetting experience or feeling, but instead of just dealing with it as best they can, and moving on, a set of negative assumptions is activated which then triggers a downward spiral of pessimistic thoughts, memories, feelings and interpretations. Once this process has taken hold, people come to see the world and themselves through increasingly critical glasses, and this makes it all the more likely that they will attend to just those features of their experience that validate and exacerbate their feelings of inadequacy or hopelessness. It becomes impossible to remember, or even to notice, anything positive or encouraging. The conscious mind may become obsessed with 'personal goals that can neither be attained nor relinquished'.

Teasdale argues that the way to stop this vicious spiral from getting going is not to try to prevent experiences of disappointment or uncertainty: that is not a practical option. There will always be upsets. Rather the solution is to get people to practise new habits of thinking and attending which will stop the self-destructive patterns from gaining control of the mind. And mindfulness can do this by preventing you from leaping to conclusions, and then carrying on as if those conclusions were solid and true: first of all by keeping you closer to the 'bare facts' and enabling you to see molehills *as* molehills, rather than automatically inflating them into mountains; and secondly, as you become more attentive to the movements of the mind, you relearn your attitude towards them. The conclusions that present themselves to consciousness are not seen any more as 'valid descriptions of who I really am' – 'worthless', for example – but as 'thoughts produced by the mind'. You reinterpret them as 'mental states', or 'events in the field of awareness', not as 'reflections of reality'. So even when negative interpretations and conclusions do bubble up, mindfulness enables you to refuse the lure and question their validity. It is no longer 'me' who is forced to defend myself; it is the content of consciousness that now appears dubious. The tables are turned.

John Teasdale's conclusions from his research with depressive people may 'ring bells', perhaps, with a wider population.

Depressive relapse often seems to occur when patients fail to take appropriate remedial or coping activity at an early stage of incipient relapse, when control over depression is likely to be relatively easy to obtain. Patients may defer recognition or acknowledgement of problems to a later stage in the relapse process, where a more full-blown depressive syndrome may be much more difficult to deal with . . . Mindfulness training . . . in 'turning towards' potential difficulties, rather than 'looking away' from them, is likely to facilitate early detection of signs . . . and so to increase the chances that remedial actions will be implemented at a time when they are likely to prove most effective.

Though pharmacological approaches to depression continue to play a vital role in its amelioration, the research shows that Teasdale's and Kabat-Zinn's approach is at least as effective as administering conventional anti-depressant drugs (and there are, of course, fewer negative side effects).

Daniel Goleman, in *Emotional Intelligence*, has documented the role that mindfulness can play in preventing 'emotional hi-jacking'.[33] When couples begin to get into a marital 'fight', for example, things can easily go from bad to worse if either or both of the partners falls into a self-reinforcing pattern of negative thinking. Mindfulness increases the likelihood that such a pattern can be spotted and neutralised before it has done too much damage. Goleman gives the example of a wife who feels in the heat of the moment that 'he doesn't care about me or what I want; he's always so selfish', but who, on catching herself in the act of 'demonising' her husband, is able to remind herself that 'There are plenty of times when he has been caring – even though what he did just now was thoughtless and upsetting.' Through the moment of mindfulness she is able to neutralise the exaggerated thought that, if accepted, would have justified a negative reaction that would only have inflamed the situation further.

The value of slowing down the mind is evident in dozens of everyday situations. Take the example of a divorcing couple arguing over custody of their child. In such an emotionally intense situation, it is very easy for an impoverished perception to lead to a rigid response: one in which it looks as if only one partner can 'win', and the other therefore must 'lose'. Any more subtle analysis of the actual predicament, and especially of what it is that each party is actually trying to achieve, is sacrificed in favour of a knee-jerk adherence to a one-dimensional view. Underneath the dogmatism, however, there may be a whole host of other factors and values, consideration of which might make it possible for everybody – including the child – to win. Is each parent really seeking the full-time physical presence of the child, or is it a quality of relationship they want to preserve? Are they trying to use the custody issue as a way of punishing their partner, or of asserting a need for control that they feel they lost in the marriage? Could there not be advantages to being a part-time parent that have been overlooked? And what might be best for the child itself? Increasingly it is the role of counsellors and mediators to try to ease people out of their entrenched, antagonistic positions, and to see the situation more fully. To cultivate mindfulness is to be able to adopt that role for oneself.

The cultivation of mindfulness does not require instruction in any kind of formal meditation, though it may help. Our own culture possesses many venerable and effective activities – or inactivities – that encourage the mind to shift out of doing-and-thinking mode into a mode that is relaxed and spacious, yet alert to its own meanderings. Coarse fishing, for example, as Ted Hughes noted, is a meditation in everything but name; a perfect excuse to gaze at the float while the mind wanders free, enjoying the shimmer of the light on the water or the soft touch of the rain. It is not unknown for fishermen to experience mild resentment when their reverie is disturbed by the inconvenient attentions of fish to hook. Rhythmic activities such as knitting, weeding and swimming may all encourage mindfulness of simple body sensations, sounds or smells, drawing attention away from problem-solving and back into the perceptual world. Watching a relatively unimportant county cricket game live is good practice. On TV you can drift off too much, knowing that the action replays will show you the highlights. At the ground, if you don't maintain awareness you miss the action when it occasionally happens. Yet you cannot spend all day focused and concentrated. You have, gradually, to develop the quality of attention of a cat: relaxed and watchful at the same time. You feel the spontaneous 'pulsing' of awareness that we spoke of in Chapter 10.

If perception samples experience only in order to categorise it, and to decide whether it is potentially useful or harmful, the conscious image it creates is likely to be rather flat and dull. Having lost perceptual vividness, we seek to put ourselves in extreme situations where the outside world startles us strongly enough for perceptual intensity to return. Thus the entertainment 'industry', whether in the form of violent or pornographic films, terrifying theme-park rides, raves or cocaine, becomes geared to providing transient experiences of aliveness that our habitual mode of mind prevents us from having for, and by, ourselves. Greater mindfulness makes conscious experience of life richer and more vivid. Rediscovering the ability to dwell in perception gives it back its charm and its vitality.

## CHAPTER 12

# The Rudiments of Wisdom

The wisdom of a learned man cometh by opportunity of leisure; and he that hath little business shall become wise.

Ecclesiasticus 38:34

Summerhill, the progressive English school founded by A. S. Neill, is run by a council, the 'moot', which meets once a week. Every member of the school, from the newest five-year-old to the oldest teacher, has a single, equal vote. The moot decides everything: school rules, bedtimes, sanctions to be applied in particular cases. It is the mid-1970s. The founder and his wife sit quietly in the meeting listening to a complaint being brought by two girls against one of the boys who has, allegedly, been irritating them by, among other things, flicking them with towels. The mood of the meeting is against the defendant. Student after student denounces him. A harsh penalty looks likely. Both Neill and his wife sit with their hands raised, waiting to be invited to speak by the thirteen-year-old girl who is chairing the moot. Eventually Mrs Neill has her turn. 'Just think how dull your lives would be if you didn't have these boys to annoy you,' she says, with a twinkle. The meeting laughs. Then Neill speaks in a gruff, laconic voice, as if he is raising a point of procedure. 'I don't think the meeting has any right to interfere in a love affair,' he says. Again the meeting laughs. The boy and one of the girls grin sheepishly at each other. The meeting moves on.[1]

The exploration of 'knowing better by thinking slower' eventually brings us to a consideration of wisdom. The dictionary tells us that wisdom is 'the capacity to judge rightly in matters relating to life and conduct; sound judgement especially in practical affairs; making good use of knowledge'. But that does not get us very far. What does it mean to 'judge rightly', or to have 'sound judgement'? Who is to decide what is right or sound? What sort of knowledge does one need, and how does one learn to make good use of it? All the interesting questions are begged. Our study of the complex and sometimes troubled relationship between the hare brain and the tortoise mind can help us to get a better handle on this most elusive, but most important, of concepts.

The Neills' reactions demonstrate some of the qualities of wisdom. Above all, wisdom is practical, dealing directly with 'matters relating to life and conduct'; with 'practical affairs'. It is also creative and integrative. The Neills 'reframe' a polarised situation in a way that skilfully avoids taking sides. Where the protagonists are stuck in a world-view in which one must 'lose' if the other 'wins', the wise counsellor finds a perspective that integrates and transcends the opposing positions. Apparently stark choices are magically transformed into common purposes. A classic example of this creative reframing occurred during one of the many riots in nineteenth-century Paris, when the commander of an army detachment was ordered to clear a city square by firing at the *canaille* – the rabble. He commanded his soldiers to take up their firing positions, their rifles levelled at the crowd, and, as a ghastly silence descended, he drew his sword and shouted at the top of his lungs: 'Mesdames et messieurs, I have orders to fire at the *canaille*. But as all I can see from here are a great number of honest, respectable citizens peacefully going about their lawful business, may I request that they clear the square quietly so that I can safely pick out and shoot the wretched *canaille*.' The square was emptied in a few minutes, with no loss either of life or face.[2]

Wisdom often involves seeing through the apparent issue to the real issue that underlies it. Where the Summerhill students saw only conflict, Neill saw a much more complex dynamic that included affection and playfulness in addition to the superficial disgruntlement; while his wife gently hinted at a longer-term perspective within which enacting and resolving such minor conflicts constituted an important and proper part of the 'curriculum' of growing up. The girls are reacting from a level of irritation which is genuine but also incomplete, less than the whole story, and Mrs Neill gently reminds them of a larger set of values which they share, but have temporarily forgotten. There is a part of the girls that would indeed be disappointed if the boys were to leave them in peace. On a grander scale, Nelson Mandela, in his famous inaugural speech as president of South Africa in 1994, sought to reframe the fears, and the aspirations, of his countrymen and women. He attempted to peel away one layer of understanding to reveal another.

Our deepest fear is not that we are inadequate. Our deepest
fear is that we are powerful beyond measure. It is our light,
not our darkness, that most frightens us. We ask ourselves,
who am I to be brilliant, gorgeous, talented and fabulous?
Actually, who are you not to be? . . . There's nothing enlightened about shrinking so that other people won't feel insecure
around you . . . As we let our own light shine, we unconsciously
give other people permission to do the same. As we are liber-
ated from our own fear, our presence automatically liberates
others.

We might say that wise people are able to act and judge 'rightly'
because they see through the complicated intermediate layers of
value in which people sometimes become enmeshed to the simple
truths and concerns that animate almost everyone: to feel safe; to
express oneself without fear; to understand one's place and purpose
in the world; to act with integrity; to belong somewhere; to love
and be loved. As the French psychologist Gisela Labouvie-Vief has
concluded from her study of wisdom, 'What makes the artist, the
poet or the scientist wise is not expert technical knowledge in their
respective domains but rather knowledge of issues that are part of
the human condition. Wisdom consists, so to say, in one's ability
to see through and beyond individual uniqueness and specialisation
into those structures that relate us to our common humanity.'[3]

Wise judgements take into account not just ethical depth but the
social and historical repercussions that may ensue. An expedient
solution may follow from a partial analysis of a problem that represents only one point of view, or excludes a long-term perspective.
Those attempting to conduct ethical business, for example, try to
make decisions that benefit a constituency of 'stakeholders' that
includes employees and their families, customers and local residents,
as well as the shareholders, and which respect the interests and
rights of future generations. Wisdom works with 'the big picture',
one that accurately incorporates the moral, practical and interpersonal detail, however inconvenient, and tries to find a solution that
fits and respects this complexity as well as possible. Wisdom does
not search the rule-book for templates and generalities that the
situation can be forced to fit. It tends to go back to the moral and
human basics and custom-build a response that reconciles as many
of the constraints and desiderata as possible.

Wisdom is uncompromising about fundamental values, but flexible and creative about the means whereby they are to be preserved

or pursued – sometimes surprisingly or even shockingly so. A Zen master astounded his monks by burning statues of the Buddha to keep warm. Jesus cut through a convoluted moral predicament by telling his confused and angry followers to obey the law, but keep their spirits free, by 'rendering unto Caesar the things that are Caesar's, and unto God the things that are God's'. A wise action may seem to disregard convention, or even rationality. In desperate situations, where all other avenues are blocked, it may be wise to do something apparently absurd.

When in 1334 the Duchess of Tyrol encircled the castle of Hochosterwitz she knew that the fortress, built on a steep rock rising high above the valley floor, was impossible to attack directly, and would yield only to a long siege. And so it proved. Eventually both the defenders and the Duchess's troops were on the verge of giving up. The defenders were down to their last ox and their last two bags of barley. The attacking soldiers were becoming bored and unruly, and there was pressing military business elsewhere. The commander of the castle, at this point, seemed to lose his sanity. He ordered the ox to be slaughtered, its carcass to be stuffed with the barley, and the body thrown over the ramparts, whence it rolled down the cliff and came to rest in front of the enemy camp. Upon receiving this disdainful message, and assuming that anyone who could afford such an extravagant gesture must be well provisioned and in good heart, the discouraged Duchess gave up the siege and moved on.

If a predicament can be solved by d-mode, it does not need wisdom. Wisdom has been defined as 'good judgement in hard cases'.[4] Hard cases are complex and ambiguous; situations in which conventional or egocentric thinking only results in heightened polarisation, antagonism and impasse. In hard cases personal values may conflict: to choose the course of honesty is to risk the sacrifice of popularity; to choose adventure is to jeopardise security. As with the 'bystander studies' which I mentioned earlier, publicly to go to someone's aid may risk your being late for your appointment, getting your clothes dirty, or looking a fool when the situation turns out to be a student prank. Hard cases are those where important decisions have to be made on the basis of insufficient data; where what is relevant and what is irrelevant are not clearly demarcated; where meanings and interpretation of actions and motives are unclear and conjectural; where small details may contain vital clues; where the costs and benefits, the long-term consequences, may be difficult to discern; where many variables interact in intricate ways.

The conditions in which wisdom is needed, in other words, are precisely those in which the slow ways of knowing come into their own. To be wise is to possess a broad and well-developed repertoire of ways of knowing, and to be able to deploy them appropriately. To be able to think clearly and logically is a constituent of wisdom, but it is not enough on its own; many unwise decisions have been made by clever people. One needs to be able to soak up experience of complex domains – such as human relationships – through one's pores, and to extract the subtle, contingent patterns that are latent within it. And to do that, one needs to be able to attend to a whole range of situations patiently and without comprehension; to resist the temptation to foreclose on what that experience may have to teach. (The poet and critic Matthew Arnold, during his time as an inspector of schools, used to tell of a colleague who boasted of thirteen years' experience – whereas, Arnold would comment, it was perfectly clear to anyone who knew the man that he had had nothing of the sort. He had had one year's experience thirteen times.) And one must be able to take one's time: to mull over a problem and to dwell on details and possibilities. In short, to be wise one needs the tortoise as well as – perhaps even more than – the hare.

Allowing oneself time to be wise is vital in the context of caring professions such as counselling and psychotherapy. Robin Skynner, co-founder of both the Institute of Family Therapy and the Institute of Group Analysis, and author, with comedian John Cleese, of the books *Families and How to Survive Them* and *Life and How to Survive It*, has talked of his perennial confusion on working with a new group or a new family.[5] Even with more than forty years' experience, it regularly happens, he says, that a few minutes into the first consultation he feels lost. Suddenly his accumulated knowledge and skill appear to desert him. It seems as if he has no precedents on which to draw. He may wonder what he is doing there, or may even feel fraudulent. Nothing wise occurs to him to say or do. Yet, Skynner says, one of the major benefits of his vast experience is the courage not to flee from this barren state. It remains, after all this time, uncomfortable, yet he now recognises it to be an essential 'winter' phase, in which nothing seems to be growing, which precedes the arrival of spring. After half an hour or so, some tentative inklings and intuitions begin to form, and gradually a new sense of being able to work with this unprecedented situation emerges. Skynner's knowledge of interpersonal dynamics does not manifest itself as fast and certain prescriptions: far from it. It appears through the courage to wait, and to notice and trust the fragile shoots of understanding that eventually start to appear.

The way of knowing that generates wisdom is a curious one, for it seems to transcend conventional dualities. It is at once subjective and objective; both involved, caring and affectionate, and yet dispassionate and unclouded by personal sentiment or judgement. There is what I have called 'poetic sensibility', in which the object of attention is known intimately, even 'lovingly', but without projection: no hopes or fears intrude to obscure the clarity of perception. If the 'object' is another person, someone in distress, for example, the wise counsellor is touched by their predicament, and yet untouched by it. She feels the situation as a human, and not just as a technical, one; but her empathy does not dissolve into mere sympathy or, worse, collusion. She is able to see – with her mindfulness – beliefs and opinions, both her own and others', *as* interpretations, and not – as they may appear to the sufferer – as the transparent truth.

Psychotherapist Carl Rogers described empathy as:

> entering the private perceptual world of the other and becoming thoroughly at home in it . . . It means temporarily living in his/her life, moving about in it delicately without making judgments . . . as you look with fresh and unfrightened eyes at elements of which the individual is fearful . . . To be with another in this way means that *for the time being* you lay aside the views and values you hold for yourself in order to enter another's world without prejudice.[6]

To perform this delicate balancing act, the wise person needs to be mindful not only of the other's world, but of his own as well. As in focusing, he needs to be able to 'tune in' to his own inner state to ensure that no judgements or projections are slipping unnoticed into his own interpretation of the situation. Only if his perception is clean and full will his judgement be subtle, fair and trustworthy. That is why the so-called 'counter-transference' in psychotherapy – projections of affection or even sexual attraction, for example, on to the client by the therapist – is such an important issue (and why doctors, in the United Kingdom at least, are not permitted to treat members of their own families). But this dispassionate yet kindly vantage point is not easily achieved. As the Danish philosopher Kierkegaard said: 'The majority of men are subjective towards themselves and objective towards all others – terribly objective sometimes

– but the real task is in fact to be objective towards one's self and subjective towards all others.'[7]

There is little empirical research on wisdom itself; but there is some information about what people consider wisdom to be. Yale psychologist Robert Sternberg summarised the general view of the wise individual thus:

> [She or he] listens to others, knows how to weigh advice, and can deal with a variety of different kinds of people. In seeking as much information as possible for decision-making, the wise individual reads between the lines . . . The wise individual is especially able to make clear . . . and fair judgements, and in doing so, takes a long-term as well as a short-term view of the consequences of the judgement made . . . [She or he] is not afraid to change his or her mind as experience dictates, and the solutions that are offered to complex problems tend to be the right ones.[8]

The ability which mindfulness brings, and which wisdom seems to presuppose – to see one's own knowledge, as well as that of others, as a personal and social construction, capable of being interrogated, reframed or reconstrued – is not easily developed, nor does it come without cost. It requires a considerable sense of personal security to give up the belief in certain knowledge. It is not just the admission that one's knowledge is always incomplete – that there is always more that one could consider – which is required, but the recognition that knowledge itself is essentially unsure, equivocal, open to question and reinterpretation. As Harvard educationalist Robert Kegan has recently reminded us in his book *In over our Heads: the Mental Demands of Modern Life*, this perspective is only gained at the cost of 'a human wrenching of the self from its cultural surround'.[9] Adult educators, for example, who are demanding this reflective, critical ability from their students, are asking

> them to change the whole way they understand themselves, their world, and the relation between the two. They are asking many of them to put at risk the loyalties and devotions that have made up the very foundation of their lives. We acquire 'personal authority', after all, only by relativizing – that is, only by fundamentally altering – our relationship to public authority. This is a long, often painful voyage, and one that, for much of the time, may feel more like mutiny than a merely exhilarating expedition to discover new lands.

To be wise requires the development of a mode of mind which can accept the relative nature of knowledge without tipping into rampant subjectivity or solipsism. One must be able to live with Voltaire's dictum: 'doubt is an uncomfortable condition, but certainty is a ridiculous one'. Yet this doubt must leave freedom to act – sometimes quickly and decisively. The wise person walks a narrow line between the twin perils of rigid dogmatism and paralysing indecision. As psychologist John Meacham has put it: 'one abandons both the hope for absolute truth and the prospect that nothing can be known; in wisdom, one is able to act with knowledge while simultaneously doubting.'[10] Meacham makes his point with an example in which this awareness of fallibility is conspicuously absent. In the film *The Graduate*, the young man Ben (Dustin Hoffman) is taken aside at his graduation party by his father's friend Mr Maguire. 'Come with me for a minute,' says Maguire, 'I want to talk to you. I just want to say one word to you. Just one word.' 'Yes, sir,' Ben replies. 'Are you listening?' insists Maguire. 'Yes, sir, I am,' says Ben. 'Plastics,' says Maguire. There is a long pause while they look at each other. Finally Ben asks, 'Exactly how do you mean that, sir?' The humour of this laconic scene turns precisely on Maguire's unwise conviction that his knowledge, so generously offered, is absolutely incorrigible. Meacham points out that an epistemological milieu, whether it be school, university or workplace, that requires one to appear certain militates against the development of wisdom. On the contrary, 'An intellectual climate hostile to ambiguity and contradiction is one that encourages easy solutions such as stereotyping and intolerance.'

Wisdom arises from a friendly and intimate relationship with the undermind. One must be willing, like Winnie the Pooh, to 'allow things to come to you', rather than, like Rabbit, 'always going and fetching them'. D-mode clings to lines of thought that are clear, controlled, conventional and secure: precisely those to which 'hard cases', by their very nature, will not succumb. Wisdom comes to those who are willing to expand their sense of themselves beyond the sphere of conscious control to include another centre of cognition to which consciousness has no access, and over which there seems to be little intentional jurisdiction. As Emerson puts it: 'A man finds out that there is somewhat in him that knows more than he does. Then he comes presently to the curious question, Who's who? Which of these two is really me? The one that knows more or the one that knows less; the little fellow or the big fellow?'

Those sages and seers who represent the most clear-cut embodiments of wisdom tend to give two different answers to Emerson's question. Those that belong to the theistic religious traditions are inclined to retain their personal identification with the 'little fellow', and to assume that the 'big fellow' is some external source of authority who, through its grace and mercy, has chosen to speak 'through' them. In the Judeo-Christian tradition, for instance, the broadcasting authority is referred to as God. Yet even within these religions there have been dissenting voices which have insisted on seeing the source of wisdom as immanent. The 'big fellow' is still called God, or the Godhead, but is now construed as an inscrutable force or process that is located within. In the so-called 'apophatic' tradition within Christianity, for example, there are many mystics and sages who have expressed their rediscovered intimacy with the undermind in these terms. The founder of the apophatic tradition, the sixth-century Dionysius the Areopagite, described the mystic as one who 'remains entirely in the impalpable and the invisible; having renounced all knowledge [he] is united to the Unknowable – to God – in a better way, and *knowing nothing, knows with a knowledge surpassing the intellect*'.[11] To Dionysius, 'The most godly knowledge of God is that which is known by unknowing.'

Eckhart von Hochheim, 'Meister Eckhart' to history and to his followers, is acknowledged to be perhaps the greatest of the Christian mystics, though in his time his works were condemned by a papal commission as heretical and dangerous. He died only just in time to avoid being burnt at the stake, it appears. For Eckhart, 'A really perfect person will be so *dead to self*, so lost in God, so given over to the will of God, that his whole happiness consists in being *unconscious of self and its concerns*, and being conscious instead of God.' The goal of spiritual practice is to find the inner place 'where never was seen difference, neither Father, Son nor Holy Ghost, where there is no one at home, yet where the spark of the soul is more at peace than within itself'.

Johannes Tauler, a Dominican monk and disciple of Meister Eckhart who lived and taught in the German Rhineland in the mid-fourteenth century, was one of the first of the apophatics to adopt an explicitly psychological interpretation of his religious experience. His practical instruction on the contemplative life, delivered mostly in his sermons at the convents and monasteries which he visited, offers very clear advice to the nuns and monks on how to pursue their devotions inwardly. For him it was self-evident that human beings yearned for inwardness; for 'personal renewal through a submersion in the divine ground from which all creatures have arisen'.[12] And this divine ground was none other than the Kingdom of God.

> This Kingdom is seated properly in the innermost recesses of the spirit. When the powers of the senses and the powers of the reason are gathered up into the very centre of the man's being – *the unseen depths of his spirit, wherein lies the image of God* – and thus he flings himself into the divine abyss . . . where [everything] is so *still, full of mystery and empty*. There is nothing there but the pure Godhead. Nothing alien, no creature, no image; no form ever penetrated there.

Tauler's method for approaching this Kingdom relied on the deliberate, methodical cultivation of passivity, a way of turning oneself over to forces and impulses that did not originate from the sphere of the conscious self, which he, following Eckhart, and anticipating Heidegger, referred to as *Gelassenheit*, 'letting be'. Of the pragmatic benefits of this attitude Tauler was in no doubt: 'In this way nature and reason become purified, the head strengthened, and the individual more peaceful, more kind and more restful.'

Tauler's psychological interpretation of the Church's symbolism must also have seemed, to his more conventional contemporaries, to border on the blasphemous. In his view of the Trinity, for instance, no longer is God the Father a transcendent figure. He becomes the Godhead, the innermost source, the unconscious mystery. The Son represents the perpetual birth of 'something' out of this divine 'nothing', the amazing coming into being of conscious experience and physical acts, continually gushing forth from the impenetrable fountainhead. And the Spirit is the transformation of being, the wisdom, the 'peace of God which passeth all understanding', which is available to all those who are willing, as Eckhart put it, to 'naught themselves', to put their faith in the inner God who 'moves in a mysterious way'.

There is a surprising coming together of the direct insight of these sages (and one could quote from dozens more in similar vein) with the new science of the intelligent – even wise – unconscious. Perhaps it would encourage people to see that their own capacity for wisdom is amenable to cultivation if this confluence of understanding were to be more widely known. As Lancelot Law Whyte, one of the first to trace the history of the unconscious back into pre-Freudian times, concludes:

Today faith, if it bears any relation to the natural world, implies
faith in the unconscious. If there is a God, he must speak
there; if there is a healing power, it must operate there . . . The
conscious mind will enjoy no peace until it can rejoice in a
fuller understanding of its own unconscious sources.[13]

Of all the major world religions, though, it is Buddhism that most
clearly and consistently identifies the source of wisdom with the
undermind. Indeed, Buddhism goes so far as to say that wisdom
resides in the recognition that *all* the activities and contents of consciousness are merely manifestations of unconscious processes.
Even our most rational and well-considered trains of thought are
not created by the conscious 'I', but are merely displayed in consciousness, like images and text on a computer screen. The screen
has no intelligence of its own; it merely portrays the results of a
certain kind of activity within the unobservable world of the micro-
chips. The Buddhist project, we might say, is to bring about a
shift in our identity's 'centre of gravity' from consciousness to the
mysterious undermind. Through the meticulous form of attention
that is cultivated by mindful meditation, we become more fully
aware of the passing details of experience, and the wayward, ephemeral and suspect nature of the conscious mind begins to become
more evident. One has to be able to adopt a more sceptical attitude
towards consciousness itself, reconceptualising it as a drama, a passing show, rather than as a reliable star to steer by. In the words of
a contemporary Tibetan *dzogchen* master:

> Whatever momentarily arises in the body-mind . . .
> has little reality.
> Why identify and become attached to it,
> passing judgement on it, and ourselves?
> Far better to simply
> let the entire game happen on its own,
> springing up and falling back like waves.[14]

The relocation of the centre of both identity and intelligence to the
undermind is expressed, within Buddhism, most clearly in the Zen
tradition. The contemporary Japanese teacher Shunryu Suzuki
Roshi, who founded and for many years directed the Zen Centre
in San Francisco, used to say: 'In Japan we have the phrase *shoshin*,
which means "beginner's mind". The goal of practice is always to
keep our beginner's mind . . . If your mind is empty, it is always
ready for anything; it is open to everything. In the beginner's mind

there are many possibilities; in the expert's mind there are few.'[15]
Another contemporary Zen teacher, the Korean Seung Sahn Sunim,
who has also made his home in the United States, instructs his
students thus:

> I ask you: What are you? You don't know; there is only 'I
> don't know'. Always keep this don't-know mind. When this
> don't-know mind becomes clear, then you will understand. So
> if you keep don't-know mind when you are driving, this is
> driving Zen. If you keep it when you are talking, this is talking
> Zen. If you keep it when you are watching television, this is
> television Zen. You must keep don't-know mind always and
> everywhere. This is the true practice of Zen.[16]

But, as in Christianity, the discovery of the value of the hidden
layers of the mind is not a modern achievement. As long ago as the
seventh century, Chinese teacher Hui-Neng, the Sixth Patriarch of
Zen, presumed author of the influential *Platform Sutra*, was encouraging his followers to take note of the activities of the undermind.[17]
A bluff, down-to-earth fellow, by all accounts, he devoted much
of his energy to trying to combat the prevalent idea that spiritual
realisation involved cutting off your thoughts, and that if you medi-
tated earnestly and frequently enough, 'enlightenment' would be
your inevitable reward. Spiritual practice, for Hui-Neng, was not
about calming or emptying the mind; it was about noticing, in
any and every moment, in whatever you were up to, the dynamic
relationship between conscious and unconscious.

> O friends, if there are among you some who are still in the
> stage of learners, let them turn their illumination upon the
> source of consciousness whenever thoughts are awakened in
> their minds . . . The [conscious] mind has nothing to do with
> thinking, because *its fundamental source is empty*. . . It is called
> 'ultimate enlightenment' when one has awakened to the source
> of the mind.

Buddhist scholar D. T. Suzuki explains very clearly what this awareness of the interface between conscious and unconscious means, in
terms of the key Buddhist concept of *prajna*, usually translated as
'wisdom'.

> Prajna points in two directions, to the Unconscious, and to a
> world of consciousness which is now unfolded . . . When we
> are so deeply involved in the outgoing direction of conscious-
> ness and discrimination as to forget the other direction of

Prajna, [wisdom] is hidden, and the pure undefiled surface of
the Unconscious is now dimmed . . . Ordinarily the apperceiv-
ing mind is occupied too much with the outgoing attention,
and forgets that at its back there is the unfathomable abyss of
the Unconscious. When its attention is directed outwardly, it
clings to the idea of an ego-substance. It is when it turns its
attention within that it realises the Unconscious.

The undermind for Hui-Neng, like the Godhead for Tauler, is the
'nothing' that constantly brings forth the 'somethings' of the mind.
The miracle, as D. T. Suzuki puts it, is that 'It is in the nature of
Suchness [the Unconscious] to become conscious of itself . . . In
the self-nature of Suchness there arises consciousness . . . Psychologically we can call [Suchness] the Unconscious, in the sense that
all our conscious thoughts and feelings grow out of it.' And: 'To
see self-nature means to wake up in the Unconscious.'

When Ma-tsu and other Zen leaders declare that 'this mind is
the Buddha himself', it does not mean that there is a kind of
soul lying hidden in the depths of consciousness; but that a
*state of consciousness . . . which accompanies every conscious and
unconscious act of mind is what constitutes Buddhahood.*

## CHAPTER 13

# The Undermind Society: Putting the Tortoise to Work

The great spectre that recurrently haunted many of the most sensitive men of the last two hundred years is that there may eventually come a time when all the richness and amplitude of Creation will simply pass through the eyes of a man into his head and there be turned by the brain into some sort of formula or equation.

Nathan Scott

In the ballroom of the Washington Hilton, 1,500 top educators from around the world sit rapt while Doug Ross describes the future. Ross, then Assistant Secretary of Labor for Employment and Training in the Clinton administration, is delivering the closing address to the 1994 international conference 'A Global Conversation about Learning'. He is talking about the practical steps that the multi-billion-dollar agency he oversees is taking to bring about the 'learning society'. Plans already exist, he tells the audience, to create a system of tax incentives to encourage people to become lifelong learners: to invest in developing their own 'cognitive capital' throughout their working lives. Taking out a loan for learning should be as easy, and as attractive financially, as taking out a mortgage. New 'school to work' programmes are being designed to break down the traditional divide between academic and vocational study, and to bring sophisticated theory to life in the workplace.

Ross makes no bones about the fact that the new 'poor', the new marginals in society, will be those who cannot or will not learn. Learning and earning are already inextricably entwined, and becoming more so by the day. The work of the future will be overwhelmingly *mind* work. Manual and blue-collar workers, Ross claims, will constitute just 10 per cent of the workforce by early in the next century. Economic and vocational forces are placing unprecedented pressure on individuals throughout society continually to learn new knowledge and skills – and to develop their *ability* to learn: their confidence and their resourcefulness as learners, and their skill at managing their own learning lives.

The concern with learning is not just American: it is worldwide. In Britain a massive 'Campaign for Learning', supported by the government and big business, and coordinated by the Royal Society of Arts, was launched in 1996 to encourage people to, in the words of its chairman Sir Christopher Ball, 'care about their learning in the same way that we are all gradually learning to care about the environment or about our own personal health'. The aim of the campaign is 'to help create a learning society in the UK in which every individual participates in learning, both formal and informal, throughout their lives. This means boosting people's desire to learn, highlighting existing ways of learning and proposing new ones'. The campaign arose as a response to a large-scale survey which revealed that, while more than 80 per cent of the population believed that learning was important for them, less than a third had plans to do anything about it. For only £2.50 you can receive from campaign headquarters a pack of material that will enable you to create your own Personal Learning Action Plan, and many other initiatives are in the pipeline.

The pressure on individuals to become learners is not just a consequence of changes in the job market, however. If employers and governments can no longer offer everyone 'jobs for life', many other cultural sources of stability and authority have also been weakened and undermined over the course of the twentieth century. It is already something of a cliché to talk of 'the collapse of certainty', whether it be in terms of the disappearance of traditional communities, the rise of geographical mobility, the explosion in information and communication technologies, the interpenetration of different ethnic cultures, the appearance of a bewildering variety of new spiritual movements and leaders to challenge the authority of the orthodox religions, or the freedom to develop personal preferences and lifestyles that may bear no relationship to those into which one was born. For many people in Western society it is now not only possible to choose, to a considerable extent, how, where and with whom they are going to live, and who they are going to be: it is incumbent upon them to do so. Whether the freedom to invent oneself is experienced as welcome or unwelcome, the onus is on individuals to learn and grow, in all kinds of ways, as never before.

In the midst of these uncertainties and opportunities, it is, therefore, of paramount importance that people possess not just the confidence but the know-how to be able to learn *well*. Governments can create incentives and campaigns can exhort, but if people feel unsafe or unsupported, or are unskilled in the craft of learning itself, they may shy away from the learning opportunities that they encounter – even when such learning would clearly help them to pursue their own valued goals and interests. Learning, whether it involves mastering a new technology or recovering from a divorce, is a risky business, and a lack of either the tools or the self-assurance to pursue it results in stagnation.

In this context, it is all the more significant that cognitive science is currently drawing our attention to the curious fact that we have forgotten how our minds work. As we have seen, the modern mind has a distorted image of itself that leads it to neglect some of its own most valuable learning capacities. We now know that the brain is built to linger as well as to rush, and that slow knowing sometimes leads to better answers. We know that knowledge makes itself known through sensations, images, feelings and inklings, as well as through clear, conscious thoughts. Experiments tell us that just interacting with complex situations without trying to figure them out can deliver a quality of understanding that defies reason and articulation. Other studies have shown that confusion may be a vital precursor to the discovery of a good idea. To be able to meet the uncertain challenges of the contemporary world, we need to heed the message of this research, and to expand our repertoire of ways of learning and knowing to reclaim the full gamut of cognitive possibilities.

This will not be easy, for the grip of d-mode on late twentieth-century culture is strong, and reflects a trend in European psychology that has its origins right back in classical Greece. For Homer, the seat of human identity and intelligence was emotional rather than rational, and opaque rather than transparent. The undermind, *psyche*, was a living reality, experienced in the body, and interpreted either, as Julian Jaynes has suggested, as 'the voice of the gods', or as a vital ingredient of the human personality.[1] But by the time of Plato the centre of gravity of a person's being had drifted upwards into the head, and begun to become associated with reason and control. The undermind still existed as an emotional or intuitive force, but it had come to be seen as secondary and subversive, something wayward, primitive and unreliable. It was in reason that people were most truly, most nobly, themselves.[2]

This ambivalent relationship with the undermind was to continue for more than a millennium. Conscious reason might be the apotheosis of the self, but the acknowledgement of its more mysterious, sometimes more inspired but less controllable shadow remained. Plotinus, the third-century neo-Platonist, commented that 'feelings can be present without awareness of them'; and that 'the absence of a conscious perception is no proof of the absence of mental activity'. A century later, St Augustine famously wrote: 'I cannot totally grasp all that I am. The mind is not large enough to contain itself: but where can that part of it be which it does not contain?' Aquinas, in the thirteenth century, noted that 'there are processes in the soul of which we are not immediately aware'.[3]

Shakespeare clearly recognised the whole variety of unconscious influences on conscious life. He touches on the inability to see the source of one's own experience, or to comprehend its true meaning, in many of the plays, most famously, perhaps, in Antonio's lament at the beginning of *The Merchant of Venice*:

> In sooth, I know not why I am so sad:
> It wearies me; you say it wearies you;
> But how I caught it, found it, or came by it,
> What stuff 'tis made of, whereof it is born,
> I am to learn;
> And such a want-wit sadness makes of me,
> That I have much ado to know myself.

In *The Comedy of Errors* he notes the power of subliminal influences on perception, when he speaks of:

> . . . jugglers that deceive the eye,
> Dark-working sorcerers that change the mind.

While in *A Midsummer Night's Dream* he sketches a theory of creativity that anticipates by 300 years the insights that we encountered in Chapter 4:

> . . . as imagination bodies forth
> The form of things unknown, the poet's pen
> Turns them to shapes, and gives to airy nothing
> A local habitation and a name.

Prior to the sixteenth century, people's sense of their own minds was both deeper and broader than it was later to become: deeper, in embracing with equanimity the existence of internal forces that were beyond their ken; and broader, in accepting, for the most part rather uncritically, external sources of knowledge and authority.

'Mind' was not such an individual possession; it was embedded in and distributed across society. Over the course of the next two hundred years, however, both of these facets of the mind were to change significantly. First, it became normal, rather than abnormal, for individuals to 'make up their own minds' about things. Even by the year 1600, according to Lancelot Law Whyte, 'the person thinking for himself ceased to be a social freak inhibited by his difference from others, and began to claim the opportunity to realise himself and to guide the community'. And in the seventeenth century 'we can recognise the germ of a new experience and a new way of living which in our own time has become a social commonplace: the existentialist complaint that there is no tradition which makes life bearable ... From then onward every sensitive and vital young person had to make his own choice.'[4] By the eighteenth century, the inclination and the ability to think for oneself was becoming firmly accepted as the goal of development, and the essential characteristic of maturity.

This increasing tendency to look for authority not outwards, to the received traditions and mythologies of Church or state, but inwards, to one's own mental workings, was accompanied by the shrinkage of 'mind' to admit only conscious reason, and to deny the legitimacy or even the very existence, of aspects of the mind that were not open to introspection. One came to conclusions, one *knew*, by working things out for oneself, and the way one did that was by conscious, deliberate thinking. That was the only kind of 'thinking', the only cognitive activity, there was. This outrageous, but as it turned out compelling, claim is usually attributed to René Descartes who, as Jacques Maritain puts it, 'with his clear ideas divorced intelligence from mystery'; though in truth Descartes' contribution was to encapsulate and give powerful expression to an intellectual movement that already possessed considerable momentum. Descartes wrote to his friend Mersennes: 'nothing can be in me, that is, in my mind, of which I am not conscious; I have proved it in the *Meditations*', and his insistence that the intelligent unconscious simply did not exist carried such conviction that it seeped into the culture at large, and became 'common sense'. All intelligence is conscious, and conscious intelligence – reason – is who, essentially, I am.

By 1690, John Locke was simply stating the obvious when he said that 'It is altogether as intelligible to say that a body is extended without parts, as that anything thinks without being conscious of it.' And in his *Essay Concerning Human Understanding* he neatly encapsulates what was fast becoming psychological orthodoxy:

> [A person] is a thinking, intelligent being, that has reason and
> reflection, and can consider itself as itself, the same thinking
> thing, in different times and places, which it does only by that
> consciousness which is inseparable from thinking, and, as it
> seems to me, essential to it . . . Consciousness always accom-
> panies thinking, and it is that which makes every one to be
> what he calls self.

The assumption that conscious reason was the core of human identity, and the highest achievement of mental evolution, fed the growth
of empirical science and the plethora of technological miracles to
which it gave rise. From technology, it was a short step to the
effective cultural takeover which we see today, and which Neil Postman, as we have seen, refers to as 'technopoly': a world in which
every adverse or inconvenient circumstance is construed as a technological problem to be fixed through purposeful, rational analysis
and invention.[5] In such a world, cognition becomes synonymous
with this kind of busy, intentional mental activity – d-mode – and the
very idea of 'letting things come', of waiting, becomes paradoxical or
ridiculous. Thinking *is* the conscious manipulation of information
and ideas preferably (now) with the aid of spreadsheets and
graphics; and if solutions do not come, that simply means you are
not thinking hard or clearly enough, or you need better data. Just
as the invention of the printing press created 'prose', but by the
same token made poetry appear exotic and élitist, and the balladeers
redundant, so the hegemony of modern information technology has
vastly increased the speed and complexity of data-handling, while
making rumination and contemplation look hopelessly inefficient
and old-fashioned.

Tools are not ideologically or psychologically neutral. Their very
existence channels the development of intelligence (as well as a host
of other facets of culture, such as vocational prestige), opening
up and encouraging certain cognitive avenues, and simultaneously
closing down and devaluing others. We are fashioned by our tools,
and none more so than the computer. For 'the computer redefines
people as "information processors" and nature itself as information
to be processed. The fundamental metaphorical message of the
computer, in short, is that we are machines – thinking machines,
to be sure, but machines nonetheless.'[6] Computers epitomise the
definition of 'intelligence' as fast, explicit and clear-cut, based on
objective data and under tight control.

The computer makes possible the fulfilment of Descartes' dream of the mathematization of the world. Computers make it easy to convert facts into statistics and problems into equations. And whereas this can be useful . . . it is diversionary and dangerous when applied indiscriminately to human affairs. So is the computer's emphasis on speed and especially its capacity to generate and store unprecedented quantities of information. In specialised contexts, the value of calculation, speed and voluminous information may go uncontested. But the 'message' of computer technology is comprehensive and domineering. The computer argues, to put it baldly, that the most serious problems confronting us at both personal and public levels require technical solutions through fast access to information otherwise unavailable.[7]

This implication is both false and pernicious. Our most serious problems are not technical, nor do they arise from inadequate information. If a war breaks out, or a family falls apart, it is not (usually) because of inadequate information. Yet to speak up for the virtues of mental sloth, for thinking that is elliptical and allusive, where even the goals may not be clearly known, is to be an information technology heretic. As they say in the trade, 'it does not compute'. Computers know nothing of the value of confusion, or the virtues of torpor. Their quality is assessed in terms of the size of their memory and the speed of their processor.

It is not the computer that is at fault so much as the computational frame of mind. Martin Heidegger, in his famous speech commemorating the birth of the German composer Conradin Kreutzer in 1955, drew a distinction between 'calculative' and 'meditative' thinking, and expounded powerfully the risks of leaning exclusively on the former.[8]

Man today will say – and quite rightly – that there were at no time such far-reaching plans, so many inquiries in so many areas, [so much] research carried on as passionately as today. Of course. And this display of ingenuity and deliberation has its own great usefulness. Such thought remains indispensable. But – it also remains true that it is thinking of a special kind. Its peculiarity consists in the fact that whenever we plan, research and organize, we always . . . take [conditions] into account with the calculated intention of their serving specific purposes . . . Such thinking remains calculation even if it neither works with numbers nor uses an adding machine or computer.

Man finds himself in a perilous position ... A far greater danger threatens [than the outbreak of a third world war]: the approaching tide of technological revolution in the atomic age could so captivate, bewitch, dazzle and beguile man that *calculative thinking may someday come to be accepted and practised as the only way of thinking*. What great danger then might move upon us? Then there might go hand in hand with the greatest ingenuity in calculative planning and inventing, indifference toward 'meditative' thinking, total thoughtlessness. And then? Then man would have denied and thrown away his own special nature – that he is a meditative being. Therefore the issue is the saving of man's essential nature. Therefore the issue is keeping meditative thinking alive. (Emphasis added)

The mind itself comes to be evaluated in calculative terms. Despite the tentative introduction of project and course work, students continue to be largely assessed, at school, college and university, on their ability to manipulate data under pressure of time. The so-called GMAT, the Graduate Management Admission Test, which is almost universally used in the States, and increasingly elsewhere, to select graduates for admission to management and business schools, consists of nine sections, seven of which contain a sequence of multiple-choice questions that is too long to be completed in the carefully stipulated time. The questions are designed to test 'basic mathematical skills and understanding of elementary concepts, the ability to reason quantitatively, solve quantitative problems, and interpret graphic data; the ability to understand and evaluate what is read; and the ability to think critically and communicate complex ideas through writing'.[9]

There is no reason to believe that these skills are anything other than useful or even vital for the manager of the future; but the implicit assumption that they cover *all* the important abilities seems, in the light of the research which I have been describing, staggeringly myopic. On the contrary, the ability to be innovative, or to detect the meaning in a snippet of information (the beginning of a consumer trend, for example) – abilities which companies frequently claim are in desperately short supply – requires expertise in slow and hazy, rather than fast and clear, ways of knowing. The GMAT seems designed to discard right from the start people with contemplative or aesthetic dispositions. Those with the potential and the inclination to become virtuoso ruminators need not apply. In general, all such tests of 'general ability', or 'intelligence' in the narrow sense, favour those who are able to think (a) fast, (b) under pressure, (c) on their own, about (d) abstract, impersonal problems, which are (e) clearly defined, have (f) a single 'right' answer, and have (g) been formulated by unknown other people.[10] Sometimes this ability is just what a practical situation demands. Many predicaments need something quite different. It is not surprising that tests of IQ correlate so poorly with measures of real-life, on-the-job performance. (As we saw in Chapter 2 the reasoning of professional horse-racing handicappers is based on a highly sophisticated (but largely unarticulated) mental model that includes up to seven independent variables – yet their individual levels of performance are completely unrelated to their IQs.[11])

In the business world particularly, the idea that the quality of thought depends on the amount and up-to-dateness of such information has completely taken over. It is all but impossible to resist the prevailing idea that all thinking worth its salt involves the deliberate manipulation of data – which boils down to 'facts and figures'. Everything is purposeful, explicit, calculated and well informed. Tom Peters, one of the most revered of management gurus, quotes, with apparent approval, the following.

> *Gentry* magazine, June–July 1994. A multipage advertisement for Silicon Valley realtor Alain Pinel includes, yes, each agent's Internet address. For example, you can reach Mary S. Gullixson at mgullixs@apr.com. One friend, who works for the firm, tells me she arrives at work each morning to about 100 e-mail messages![12]

> When Harriet Donnelly is on a business trip, she religiously carts along her NCR Safari 3170 notebook computer, her SkyWord alphanumeric pager and her AT&T cellular phone. She told *Fortune* magazine . . . that after a day of meetings: 'The first thing I do when I get back to my hotel is . . . return any messages I can, using voice mail. Next I plug my computer into the telephone and download [my] e-mail . . . I also get messages on my pager.'[13]

> One morning I stood on a wobbling dock [in Bangkok] waiting for a sooty old commuter 'express' boat to arrive. Next to me was a crisply starched Thai businessman . . . As he skilfully shifted his weight to keep his balance, he placed a string of

calls on his portable cellular phone, talking excitedly. Wherever I went, I saw that kind of hustle, that kind of entrepreneurship, that kind of fervour . . . Frankly, it felt as if America had opted out of any effort to compete.[14]

If even Mary Gullixson and Harriet Donnelly – whose habits have rapidly become the business norm – aren't 'competing', it is clearly because American business is not quite 'up with the play' in terms of speed and information. The possibility that work depends on ideas, that ideas differ in their *quality*, as well as their up-to-dateness, and that quality takes time to mature, seems to be almost universally dismissed.

There are grumblings in the business world about the perils of rampant d-mode, though no one yet seems to see how deeply the problem runs. Roy Rowan, in his book *The Intuitive Manager*, talks about a business type which he calls 'the articulate incompetent': full of good ideas, immaculately presented, which lack substance and don't work. The verbal fluency of the articulate incompetents makes them persuasive; they are clever enough to be able to make an impressive-sounding case for whatever they have come up with. They tend to build an imposing superstructure of justification on a minimal foundation of observation. (And they may be so committed to being 'right' that they refuse to give their position up, even when it becomes apparent that there are considerations which they have not taken into account.) Rowan quotes interviews with some CEOs who tend to blame the phenomenon of articulate incompetence on schools. The inadequacy of the GMAT, and the 'education' that builds upon it, are noted by Robert Bernstein, chairman of giant publishers Random House, for example. Bernstein says: 'That's what frightens me about business schools. They train their students to sound wonderful. But it's necessary to find out if there's judgement behind their language.'[15]

Alternatively, there is a negative type, the 'articulate sceptic', whose cleverness manifests itself as a reflex need to show how bright he is by criticising whatever anyone else has proposed. As Edward de Bono has pointed out: 'The critical use of intelligence is always more immediately satisfying than the constructive use. To prove someone else wrong gives you instant achievement and superiority. To agree makes you seem superfluous and a sycophant. To put forward an idea puts you at the mercy of those on whom you depend for evaluation of the idea.'[16] It may be safer, for the essentially evaluative d-moder, to be seen to be reactive rather than proactive

– to respond to a presented problem, rather than to take a fresh look at a situation and reconceptualise what the problems are. Being generative, which is creative and intuitive, is bound to be riskier than being evaluative.

The need for creative responsiveness to changing conditions is now widely recognised in the pressurised cabins of business board-rooms. Grand strategic plans are fine in a stable world but, as Rowan says, talking of the senior manager: 'the farther into an unpredictable future his decisions reach . . . the more he must rely on intuition.'[17] As we have seen, it is not that in uncertain conditions we have to 'make do' with intuition, as if we were clutching at straws. It is that well-developed, tentatively used intuition is actually the best tool for the job; while the apparent solidarity of a rational, strategic plan offers nothing more than a comforting illusion. Henry Mintzberg, professor of management at McGill University in Canada, in his classic *The Rise and Fall of Strategic Planning*, demonstrates once and for all the insufficiency of d-mode as a way of knowing for the business world.[18] 'A good deal of corporate planning . . . is like a ritual rain-dance. It has no effect on the weather that follows, but those who engage in it think it does . . . Moreover, much of the advice related to corporate planning is directed at improving the dancing, not the weather.'[19]

Not only does inflexible attachment to a plan (which it has taken a lot of time, effort and money to create) make a company unrespon-sive; such plans, Mintzberg shows, tend to be based on only those considerations that can be clearly articulated and – preferably – quantified: 'hard data'. They therefore fail to take into account precisely that 'marginal' information – impressions, details, hunches, 'telling incidents' and so on – which provide the vital 'straws in the wind' on which prescient decisions can be based – and on which intuition thrives. Because consciousness demands information that is tidy and unequivocal, it can never be as *richly* informed as intuition. If you wait until a market trend is clear, you will have lost the edge. In business, as elsewhere, when decisions depend on the use of faint clues in intricate situations, the tortoise outstrips the hare.

But the calibre of intuition varies enormously. If intuition is merely a panicky, impulsive reaction to the failures of d-mode, it will be unreliable. In a world where TQM stands, more often than not, for 'Terrible Quality of Mind', intuition needs cultivating and nurturing. To be positive, we need to give some attention to specify-ing those conditions which facilitate the production of top-quality intuitions. The first requirement is a climate in which the value of intuitions, and the nature of the mental modes that produce them, are clearly understood by all. The second is leadership which models and acknowledges the value of intuition: managers who 'walk their talk' as far as slow knowing is concerned, encouraging the contribution of ideas that are judged on their merits, and not on how slickly or persuasively they are initially put across. Business leaders need to be open to 'the germ of an idea' that may seem unconventional at first sight, or which may be expressed in terms of analogies or images.

De Bono, in his many books on 'lateral thinking', has provided a wealth of telling illustrations of ideas that look 'silly' or 'childish' to start with, but which turn out on closer inspection to contain the seeds of highly creative and appropriate solutions.[20] A child drew a wheelbarrow that had the wheel at the back, near the handles, and the legs at the front. This 'silly mistake', which could easily have been 'corrected' by a conventional teacher, in fact produces a barrow that is ideal for manoeuvring round tight corners (on the same principle as a dumper truck that has the steering wheels at the back).

An apocryphal firm that sold mail-order glassware was suffering an unacceptably high level of breakages in transit, and they could not find a way to construct or label their parcels which made any difference. However much padding they put in, the glass was still getting broken. Then someone suggested that they simply stick the address label on the glass, and send it completely unprotected. Though this was rejected as being too risky, the idea behind it – that if the postal handlers could clearly *see* that the goods were fragile, rather than being told so by stickers which they tended to ignore, they might naturally take more care – was a good one, and it led, via a second creative leap, to the design of packaging that had a glassy finish, which had the desired effect.

A creative workplace needs to encourage people to engage with their work mindfully and to think about what they are doing. The development of such a working environment is stimulated by giving individuals, and especially teams, genuine responsibility for planning and carrying out meaningful pieces of work, and for deciding how their goals are to be best achieved. A recent official report on *Fostering Innovation* by the British Psychological Society, having surveyed all the relevant research, concludes that: 'Individuals are more likely to innovate where they have sufficient autonomy and control over their work to be able to try out new and improved ways of doing things' and where 'team members participate in the setting of objectives'.[21] The more people feel that they have some stake in their work, the more likely they are to be interested in spontaneously looking for improvements, and in keeping their thoughts, impressions and ideas simmering quietly on the back burners of their minds, both when they are 'on the job' and off it.

We saw in Chapter 5 that some daily routines and physical surroundings are more conducive to the germination of ideas than others. Giving workers some control over the work environment may be helpful – though individuals vary considerably in what works best for them, and many conducive conditions do not fit easily within the structure of conventional employment. I work best by the sea. Some people sink into the right frame of mind with the help of certain pieces of music. Often people are at their most receptive when they are able to spend a great deal of time in silence. Some people jog; some swim; some meditate. Descartes, it is said, did much of his best thinking lying in bed till late in the day. Encouraging people to note what works best for them is a practical first step. In a corporate setting it is not out of the question that – with a modicum of ingenuity – some of these conditions could be created, at least for some of the time.

People are obviously more inclined to be innovative and intuitive when they feel safe being so. *Fostering Innovation* suggests:

> One of the major threats to innovation is a sense of job insecurity and lack of safety at work . . . Where individuals are threatened they are likely to react defensively and unimaginatively . . . They will tend to stick to tried and tested routines rather than attempt new ways of dealing with their environments . . . People [are] generally more likely to take risks and try out new ways of doing things in circumstances where they feel relatively safe from threat as a consequence. It is the thrust of this document that the revolution in management practices in the 80s has now to be paralleled by a new revolution in the 90s and into the next century, which emphasises . . . psychological safety at work and practical support for the development and implementation of new and improved ways of doing things.

It is not just that people are bolder about trying things out when they feel relaxed and secure; threat creates a mindset of anxiety and entrenchment in which awareness is constricted and focused on the avoidance of the threat, rather than the spacious, open attitude that the slow ways of knowing require to work. People need to feel that they can say 'This may sound silly, but . . .'; or 'Can I just think aloud for a moment . . .' Where half-baked ideas are immediately torn to shreds, people rapidly learn to wait until d-mode has delivered a position that is polished and watertight (but quite possibly over-cautious and already out of date) – or not to contribute at all.

Finally there is time itself. The slow ways of knowing will not deliver their delicate produce when the mind is in a hurry. In a state of continual urgency and harassment, the brain-mind's activity is condemned to follow its familiar channels. Only when it is meandering can it spread and puddle, gently finding out such uncharted fissures and runnels as may exist. Yet thinking slowly, paradoxically, does not have to take a long time. It is a knack that can be acquired and practised. The mind needs to be *given* time; but its ingenuity also depends on the cultivation of a disposition to *take* one's time, as much as there is. One can learn to access and use these other ways of knowing more fluently. One might even suggest that managers – and their workforces – might try meditation; though, as a preliminary, they would need to understand what that means, and how it helps.

However, those who try to manage nations and corporations – ministers and executives of all persuasions – may be panicked by the escalating complexity of the situations they are attempting to control into assuming that time is the one thing they have not got. Their fallacy is to suppose that the faster things are changing, the faster and more earnestly one has to think. Under this kind of pressure, d-mode may be driven to adopt one shallow nostrum, one fashionable idea after another, each turning out to have promised more than it was capable of delivering. Businesses are re-engineered, hierarchies are flattened, organisations try to turn themselves into *learning* organisations, companies become 'virtual'. Meetings proliferate; the working day expands; time gets shorter. So much time is spent processing information, solving problems and meeting deadlines that there is none left in which to think. Even 'intuitive thinking' itself can easily become yet another fad that fails – because the underlying mindset hasn't changed.

Although it is important to think about how to encourage the *appearance* of slow knowing, it is even more important to think about the conditions that equip people with the longer-term *dispositions*, the *personal qualities* and the *capabilities* to make full use of their varied ways of knowing, regardless of the messages of the particular setting they happen to find themselves in. Even when the culture is implicitly directing them to use d-mode, people need to know how to make use of slow knowing, and when it is appropriate. This, at the current point in history, must surely be the true function of education.

In any school or college, there is not one curriculum but two. The first we might call the *content curriculum*: it is the body of knowledge and know-how that people are there to learn – sums, French, philosophy, dentistry, whatever. Both students and teachers are, all being well, clear about what the subject is and how progress is to be gauged. If this were the only curriculum, teachers would be free to use whatever means they could to make learning easier, quicker, more pleasant and more successful. But it isn't. Underneath every concern with content lies another curriculum, less visible but just as vital – the *learning curriculum* – which is teaching students about learning itself: what it is; how to do it; what counts as effective or appropriate ways to learn; what they, the students, are like as learners; what they are good at and what they are not. And if US Labor Secretary Doug Ross and 'Campaign for Learning' chairman Sir Christopher Ball are right, and the future of both social and personal wellbeing depends on people's confidence and competence *as learners*, then this second curriculum simply cannot be ignored. The learning society requires, above all, an educational system which equips all young people – not just the academically inclined – to deal well with uncertainty.

People come to any learning experience with a set of learning-related abilities and attitudes which will determine – in conjunction with the learning task and the learning culture which surrounds it – how learning goes. These learning skills and dispositions are themselves changed by experience: people's capabilities as learners, their fortitude in the face of difficulty, their implicit understanding of what learning entails, and their images of themselves as learners will all have been altered somewhat (even if only confirmed or consolidated) by any learning event. It is with the cumulative nature of such changes that the learning curriculum is concerned. It asks: how can this succession of learning challenges and encounters be designed and presented so that, over the long term, people's learning power changes in a positive direction?

The learning curriculum does not compete, or alternate, with the content curriculum: it follows it like a shadow. To be concerned with the education of young people as learners, generically, does not mean that we give up teaching them specific subjects. Like all human qualities, those of the 'good learner' develop in the course of engaging in appropriate activities. There has to be some content, something to learn *about*. Questions about the content curriculum remain to be answered. But the criteria which are used to determine the selection of subject matter, the methods of teaching and learning that are encouraged, and the focus of assessment: these are altered by the acknowledgement of the second curriculum. Whatever the topic, part of educators' attention remains on the mastery of the specific skills and materials; but another part now has to rest on the long-term residue of learning dispositions and capabilities which is accruing from those encounters.

We cannot simply assume that what is good in terms of one curriculum is necessarily good in terms of the other. If we wanted a swimmer to improve on her personal best time, we could threaten her with dire consequences if she did not. We could even, if all else failed, tow her rapidly up and down the pool. Neither of these ploys, however, would do much for her long-term performance. The fear of failure would probably make her tense and hostile. The towing, if we persisted with it, would succeed only in weakening her muscles. There is direct evidence that the two curricula do sometimes shear apart in just this way.

Carol Dweck of the University of Illinois has explored in a detailed series of experiments the quality of 'good learning' that we might call *resilience*: the ability to tolerate the frustrations and difficulties that inevitably occur in the course of learning, without getting upset and withdrawing prematurely. Dweck found that people's resilience varies enormously, right across the whole educational age spectrum from pre-school to graduate-level study. Some people, when they hit difficulty, would quickly start to become distressed, and instead of persisting with the task would have to find ways of shoring up their self-esteem. For them, the experience of finding learning hard was aversive. Her results showed that, of all the students she tested, it was the 'bright girls', those who were doing relatively well on the content curriculum, who were most likely to show this lack of resilience: to be 'failing', in other words, on the learning curriculum. Dweck speculated that the bright girls are the group with the most fragile grasp on learning precisely because they are the students with the least experience of sticking with difficulty, and (sometimes) succeeding. Being 'bright', they find learning relatively easy, so they encounter difficulty more rarely than a less 'able' child. And being girls, they are likely, when they *do* get stuck – especially when they are young – to be comforted or offered an alternative activity by a teacher, whereas a boy is more likely, so the research shows, to be encouraged to 'stick with it'. Thus the bright girls are the students who have had the fewest opportunities to build up their 'learning muscles', and their stamina consequently remains weak.[22]

Any learners who lack resilience will be fine while learning is going smoothly, but will be prone to fall apart when it gets rocky, and this vulnerability leads them, Dweck discovered, to make conservative learning choices. Those lacking resilience will choose tasks that they know how to handle, and become anxious when a teacher changes the rules, or offers a different, less familiar kind of learning experience. They will, as a result, tend to develop only a narrow range of learning skills, those that offer the highest probability of success. If doing well in a school is defined in terms of the ability to articulate and explain, then only d-mode will be exercised and developed. The risk is that without resilience, some young people, the 'successes', will overdevelop one side of their learning potential and neglect the other; while a different set of young people, the relative 'failures', may hardly develop as learners at all, or even go backwards. At worst, there is an unenviable choice between stunting and lopsidedness.

The learning curriculum, therefore, must first and foremost be committed to the strengthening of resilience, and this requires conveying to young people an accurate view of the many faces of learning, of the mind, and of themselves. Learning sometimes involves confusion, for example, so it makes no sense to create a content curriculum which systematically deprives young people of the opportunity of getting used to being confused and of learning how to deal with it productively; nor to create, or allow to be created, a learning culture which implicitly construes confusion as aversive or presumes that learning can and/or should happen without confusion; nor one in which learners come to feel that their sense of self-worth is contingent on clarity, and subverted by ignorance. If children learn to feel threatened by ignorance, their resilience will be weakened; and likewise if they learn to feel threatened by failure or frustration.

Carol Dweck has shown that resilience is also undermined by a false view of 'intelligence'. She distinguished between two general views of intelligence or 'ability'. In one view – the more accurate one which I am using here – ability is seen as a kind of expandable toolkit of ways of learning and knowing. As one learns, so one can also be learning how to learn; becoming a better learner. The other, which unfortunately tends to permeate educational discourse, sees 'ability' as an innately determined endowment of general-purpose brainpower which places a ceiling on what you can expect, or be expected, to achieve. To say of a child 'Sally is tall, has brown eyes, and is very bright' is implicitly to be subscribing – and encouraging Sally to subscribe – to the latter view. In such a view, no amount of effort on Sally's part will change her height, her eye colour or her 'ability'.

Dweck's dramatic discovery is that a lack of learning resilience is frequently underpinned by this latter, deterministic view of the mind. A child who agrees with the idea that 'it is possible to get smarter' is likely to be persistent and adventurous in her learning. A child who disagrees, who thinks that ability is fixed, is more likely to get upset when she fails. Why? Because the fluid 'theory' of intelligence encourages a child to stretch herself: by doing so she might become cleverer. That is a possibility for her. For her less fortunate classmate, it is not. She feels herself to be possessed of a certain immutable amount of cleverness, so that failure or confusion, for her, can only be construed as evidence that her ability is inadequate. And the more discomfited she is by this, the more impelled she is to withdraw, hide, defend or attack. The practical lesson of this research for the learning curriculum is that teachers themselves must understand the fluid and variegated nature of learning ability, and use language that conveys this view to children.

Once this view of the mind as expandable is established in principle, the next demand of the learning curriculum is that it should offer the opportunity to practise the whole gamut of ways of knowing and learning. On the foundation of resilience can be built greater *resourcefulness*. D-mode must be developed and refined, but so must the powers of intuition and imagination, of careful, non-verbal observation, of listening to the body, of detecting (without harvesting them too quickly) small seeds of insight, of basking in the mythic world of dream and reverie, of being moved without knowing why. If they were more aware of both the possibility and the value of doing so, teachers would be able to find a host of opportunities to vary more widely the learning modes which their lessons encouraged or demanded, and this is especially true of teachers in secondary, further and higher education. Many of the ways of knowing which we have been exploring are familiar to the primary school child and her teacher. But conventional content curricula tend to make the mistake of seeing them as 'childish', to be supplanted, as quickly as possible, by more explicit, more articulate forms of cognition. This attitude is profoundly misguided. The slower ways of knowing do not need to be replaced. They need to be cultivated and nurtured, right on into adulthood; and they need to be supplemented, not overshadowed, by the more formal ways of knowing that start later.

Intuition, for example, can readily be honed by including it explicitly within the learning context. Recent research on the learning of science shows that children develop a much richer understanding of how to *do* science, a much firmer, more flexible grasp of scientific thinking, if they are encouraged to bring their intuitions about how the world works into the laboratory with them: to share them, explore them and test them out. As we saw in Chapter 4, intuition is a vital way of knowing in scientific research. By working with their intuitions rather than ignoring them, children are learning not just science as a body of knowledge, but to think like scientists.[23]

Cultivating a relaxed attitude of mind, in which one can 'let things come', is also something that education could address. Some young people have picked up the knack for themselves; others may need a little coaching. Archbishop William Temple was clearly one of the former.

> When I was a boy at school I used to be set the task of composing poetry in Latin, which was, as you know, rather difficult. However, I was working by candle-light, and whenever I got 'stuck' and couldn't find the right phrase, I would pull off a stick of wax from the side of the candle and push it back, gently, into the flame. And then the phrase would simply come to me.[24]

Likewise there is good evidence for the value of imagination as a learning tool throughout the lifespan. Whether it be in learning physical skills such as sports, in preparing oneself for difficult encounters, or in sorting out one's own values and beliefs, active imagination and visualisation can often prove much more effective than rational self-talk.[25] Imagination and fantasy are areas in which young children are naturally expert, and in which their learning power can easily be refined and developed as they grow up. Conversely, if their imaginative birthright is allowed to atrophy through progressive neglect, learning power will be narrowed and reduced.

What about mindfulness? If we want young people to become successful on the content curriculum, we can afford to teach them as if knowledge were certain. In this kind of teaching, it may well be more efficient to adopt the 'textbook' approach: to act as if knowledge, and the appropriate methods for dealing with it, were (for the most part) agreed and secure. 'Why risk unsettling children with fancy talk about the "social construction of knowledge"?' is a perfectly fair question. But on the learning curriculum students must be helped to develop a greater sense of ownership of knowledge and the knowledge-making process, and this means presenting knowledge as more equivocal; a human product that is always open to question and revision. If we want to start children out on a journey that heads, however remote the destination, in the direction of wisdom, then we may need to run the risk of creating some epistemological insecurity. In fact it turns out that the risk is not so great.

Ellen Langer at Harvard has conducted a series of studies with both high school and college students in which different groups were presented with the same information in different ways. For example, in one study undergraduates were given a paper to read which described a theory about the evolution of urban neighbourhoods. For one group of students the paper was written as if the theory were the simple truth. For another group, it was presented *as* a theory, using phrases such as 'You could look at the data this way', or 'It may be that . . .' When tested on their ability to use the knowledge they had gained, Langer found that, though retention was the same for the two groups, the 'could be' group was much better at using the information in flexible and creative ways. She concluded that the fear of making children insecure about knowledge was groundless, provided teachers present its provisional status as an intrinsic feature of knowledge, and not as personal indecisiveness. 'Children taught conditionally [in this way] are *more* secure, because they are better prepared for negative or unexpected outcomes.'[26]

How does the distinction between the content and the learning curricula bear on the old debate between traditional 'chalk-and-talk' teaching and 'discovery learning'? From the point of view of the content curriculum, it can seem terribly inefficient to allow children to flounder around 'reinventing the wheel', when there are so many different 'wheels' that have to be learnt. If the important thing is the wheels, this objection is entirely valid. But within the learning curriculum, what matters most is not the wheel but the inventing – and the strengthening of the powers of invention which occurs through being allowed and encouraged to invent. Time spent discovering things for yourself, even though someone could have simply told you the answer or given you the information, may be time well spent if the outcome is greater confidence and competence as an explorer. Discovery learning both draws upon and develops the power of 'learning by osmosis', and like intuition and imagination, this ability to extract patterns from experience, without necessarily being able to say what they are, continues to be of inestimable use throughout life. Both the acquisition of knowledge and skill *and* the development of learning power are important. Learners, whether children or adults, can flounder unproductively if they are given neither the tools nor the knowledge that they may need to get started on a piece of learning. The real enemy of the learning curriculum is dogmatism, whichever side it takes.

To be resourceful, in these terms, is to have at your disposal the full range of learning resources – different ways of knowing – and to have developed a good intuitive sense of the kinds of problem which each is good for, and the kinds of knowledge that each delivers. The resourceful learner is able to attend to puzzling situations with precision and concentration, and also with relaxed diffusion. She is able to 'let things speak', to see what is actually there, and not, as Hesse put it, to observe everything in 'a cloudy mirror of your own desire'. She is able to make good use of clues and hints. She is able to analyse and scrutinise, but also to daydream and ruminate. She is able to ask questions and collaborate, but is also able to keep silent and contemplate. She is able to be both literal and metaphorical, articulate and visionary, scientific and poetic: to know as Madame Curie, and to know as Emily Dickinson. To be a resourceful learner is to have had the opportunity to play, explore and experiment with each of these ways of knowing and learning, so that their power, their precision and their pertinence have all been uncovered.

The resourceful learner has also to develop the ability to be a good 'manager' of her own learning projects: to be able to judge when an approach to a problem appears not to be working, when to persist, when to change tack, and when to give up. Good learning requires the ability to be *reflective*; to take a strategic, as well as a tactical, perspective on one's learning and knowing; to be aware of 'how things are going', and of what alternative approaches there might be. Thus the learning curriculum demands that learners take on, at an appropriate rate, some genuine responsibility for deciding what, when and how they will learn, and for evaluating their own efforts. Knowing *how* to know develops through the discovery of the strengths, weaknesses and limitations of different learning styles and strategies, as they are applied across a range of real-life learning settings.

One cannot, as I have said, treat 'learning' as a new 'subject' to be added to the content curriculum. The evidence of 'study skills' programmes, for example, shows that learning strategies cannot successfully be taught directly, and that any benefits that do accrue tend not to transfer from explicit lessons into spontaneous use.[27] Learning power grows through experience; it cannot be reduced to formulae and transmitted into someone's head by instruction. Thus where the content curriculum might demand tight scheduling and supervision, the learning curriculum suggests that students be given some time, freedom and encouragement to explore. On the content curriculum, it is important that learners are told how well they are doing by being measured against 'objective' criteria: such feedback informs them of their progress, and may 'motivate' them if it doesn't demoralise them. On the learning curriculum, however, it is vital that learners are given some responsibility, encouragement and assistance to reflect upon the value of their own efforts, because only by doing so will they develop an intuitive 'nose' for quality; the ability to tell for themselves, in terms of their own (largely tacit, and certainly not quantifiable) values, what is 'good work'.

On the content curriculum, it may be seen as damaging if students are set problems that are too easy – they will be bored – or too hard – failure will dent their self-esteem. On the learning curriculum, there is less need to protect students from difficulty, or from 'biting off more than they can chew', for learning power is strengthened and broadened by the attempt to chew, and much of value may be learnt by pondering Eliot's 'Ash Wednesday' when you are ten (if you want to), just as it may from going fishing with an elder sister, even though you are too small to lift the rod, or from 'helping' your mother with the crossword, even though you solve no clues. If you are always fed a diet of problems that have been neatened up and graded, you are deprived of the opportunity to develop those slow, intuitive ways of knowing that are designed precisely to work best in situations that are untidy, foggy, ill conceived.

If the learning society is to evolve, practical changes to workplace ethos and educational methods, of the kind I have been sketching, need to be encouraged. But at a deeper level we are being asked nothing less than to conceive of the human mind in a new way. Descartes' legacy to the twentieth century is an image of the mind as 'the theatre of consciousness', a brightly illuminated stage on which the action of mental life takes place; or perhaps as a well-lit office in which sits an intelligent manager, coolly weighing evidence, making decisions, solving problems and issuing orders. In this executive den, human intelligence, consciousness and identity come together: they are, in effect, one and the same thing. 'I' am the manager. 'I' work in the light. I have access to all the files that comprise my 'intelligence'. What I cannot see, or see into, either does not exist, or it is mere 'matter', the dumb substance of the body that can do nothing of any interest on its own. It may manage certain menial operations like digestion, respiration and circulation without supervision; but to do anything clever it has to wait for instructions from head office.

This image continues to animate and channel our sense of our own psychology, of our potentialities and resources, and it is wrong in every regard. The naïve mind–body dualism on which it rests is philosophically bankrupt and scientifically discredited. *Unconscious* intelligence is a proven fact. The need to wait for inspiration rather than to manufacture it – to envisage the conscious self as the recipient of gifts from a workplace to which consciousness has no access – is likewise undeniable. We need now a new conception of the unconscious – one which gives it back its intelligence, and which reinstalls it within the sense of self – if we are to regain the ways of knowing with which it is associated. Highlighting the ways of knowing that are associated with consciousness, control and articulation enabled the extraordinary explosion of scientific thinking and technological achievement of the last two centuries; but the cost was a disabling of other faculties of mind that we cannot afford to be without. As Lancelot Law Whyte puts it:

> The European and Western ideal of the self-aware individual confronting destiny with his own indomitable will and skeptical reason as the only factors on which he can rely is perhaps the noblest aim which has yet been accepted by any community . . . But it has become evident that this ideal was a moral mistake and an intellectual error, for it has exaggerated the ethical, philosophical and scientific importance of the awareness of the individual. And one of the main factors exposing this inadequate ideal is the [re-]discovery of the unconscious mind. That is why *the idea of the unconscious is the supreme revolutionary conception of the modern age.*[28] (Emphasis added)

This conception of the unconscious – which I have been calling the undermind – is very different from the notions of the unconscious that twentieth-century European culture generally admits – such as the Freudian subconscious, the sump of the mind into which sink experiences, impulses and ideas too awful or dangerous to allow into consciousness. This representation of the unconscious is pathological and repressed. It accepts the basic Cartesian premise that consciousness is intelligent and controlled, and therefore the corollary that the unconscious must be other than, and opposed to, consciousness: emotional, irrational, wild and alien. The unconscious cannot be 'I'; it has to be 'it' – '*das Es*', as Freud originally called it, before it was gratuitously mystified by its translation into English as 'the Id'.

Clinical practice and the development of psychotherapy in the nineteenth and twentieth centuries have shown that we do indeed require this sense of the unconscious to explain aspects of human experience and behaviour – but the Cartesian image is left basically unchallenged if we make the mistake of assuming that this dark, subversive corner of the mind is the *only* part that lies outside conscious awareness. Even if we add on, as Arthur Koestler said of Jung, a kind of exotic 'mystical halo' to this fundamentally pathological picture, the core alliance of consciousness, intelligence and identity survives. The images of the unconscious that have resurfaced and survived in contemporary culture are merely elaborations of, or footnotes to, an image that continues to control the way we think about ourselves.

Yet throughout the last 350 years there *has* been a succession of other voices, demanding that the undermind be returned to its central place in our view of the mind. Less than twenty years after the publication of Descartes' *Meditation* we have Blaise Pascal reminding us that 'the heart has its reasons of which reason itself knows nothing'. Before the end of the seventeenth century Cambridge philosopher and scientist Ralph Cudworth was writing: 'It is certain that our human souls themselves are not always conscious of whatever they have in them; for even the sleeping geometrician hath, at that time, all his geometrical theorems some way in him. [And] we have all experience of our doing many . . . actions nonattendingly.' And Sir William Hamilton, one of the first philosophers writing in English to be influenced by the rise of the German Romantic movement, was lecturing on the proposition that 'The sphere of our conscious modifications is only a small circle in the centre of a far wider sphere of action and passion, of which we are conscious only through its effects.'

In 1870 the French historian and critic H. A. Taine wrote an essay, '*Sur l'Intelligence*', in which he deliberately elaborated the image of the 'Cartesian theatre' to include its unconscious background.

One can therefore compare the mind of a man to a theatre of indefinite depth whose apron is very narrow but whose stage becomes larger away from the apron. On this lighted apron there is room for one actor only. He enters, gestures for a moment, and leaves; another arrives, then another, and so on . . . Among the scenery, and on the far-off backstage there are multitudes of obscure forms whom a summons can bring on to the stage or even before the lights of the apron, and unknown evolutions take place incessantly among this crowd of actors of every kind to furnish the stars who pass before our eyes one by one.

Here, finally, we may be able to turn the Cartesian image of the mind against itself, for the bare image of a 'thought in a spotlight' hardly does justice to our understanding of 'theatre'. It leaves out at least two things without which theatre simply isn't theatre: the wings, and the nature of drama itself. The action on the stage only makes sense in terms of entrances and exits. Actors are not borne on stage; they *arrive*, and then, after a while, they *leave*. And we know they arrive from somewhere, and go to somewhere. If we do not know – even when we are not consciously thinking about it – that there *is* a 'behind the scenes' of dressing-rooms, technicians, props and paraphernalia, and a hidden world of rehearsal and discussion in which interpretations and performances are much more fluid and tentative than those that finally appear, in costume, in front of the footlights, then we don't understand what 'theatre' is. The visible performance presupposes an enormous amount of invisible apparatus and activity.

Likewise we are liable to become very confused if we don't understand the distinction between the actor and the role; between drama and 'real life'. What we watch in the theatre is a simulation, a fiction, that is designed to resemble 'real life', but also to *dramatise* it; to distort, highlight, doctor and if necessary mislead, in order to make a point or create an effect. If you forget that what is happening on stage is not 'real', then you will find yourself cowering in your front-row seat when the villain pulls a gun, or scrambling up on to the stage in order to save the heroine from a fate worse than death. A good 'play' may engage your attention and your sympathies, you may 'lose yourself' in its world for a while, but if ultimately you cannot tell the difference between play and reality, you will be in trouble. To assume that consciousness is showing and telling us the complete and literal truth is to make precisely that mistake. Thus once we begin to analyse the metaphor of the theatre, it starts immediately to unravel and subvert itself. We find that we cannot do without the wings of the undermind; and we cannot take what is going on in consciousness at face value.

An image such as the expanded theatre can help to convey a feeling for the new relationship between conscious and unconscious that we are seeking, but, in a d-mode culture, such images do not carry much weight. The voices of philosophy, poetry and imagery are relatively weak in a world that largely assumes that only science and reason speak with true authority. Thus, paradoxically, it is only science itself that can bring credible tidings of unscientific ways of knowing. One must speak to d-mode in its own language if it is to entertain the idea that it may itself be limited. The empirical research on the slower ways of knowing, and on the cognitive capacity of the undermind, can contribute significantly to the creation of the much-needed shift in our understanding of the mind. As this research gathers further momentum, it will, it must be hoped, seep into the culture at large, and encourage educators, executives and politicians to use mental tools more suited to the intricate jobs that confront them. The hare brain has had a good run for its money. Now it is time to give the tortoise mind its due.

# *Notes*

## Chapter 1

1. See Fensham, P. J. and Marton, F., 'What has happened to intuition in science education?', *Research in Science Education*, Vol. 22 (1992), pp. 114–22. This paragraph comprises a collection of assertions which I shall unpack and justify as the argument unfolds.

2. Norberg-Hodge, Helena, *Ancient Futures: Learning from Ladakh* (Shaftesbury: Element, 1991).

3. Postman, Neil, *Technopoly* (New York: Knopf, 1992).

4. The demonstration that there is a source of 'unconscious intelligence' in the mind is crucial to the argument of the book, and so that we can talk about the processes and properties of this source, it needs a name – or names. Sometimes I shall just call it 'the unconscious', and distinguish it from the Freudian repository of repressed memories by referring to the latter as the 'subconscious'. Where the contrast needs making more strongly, I shall use the expressions 'the intelligent unconscious' or 'the cognitive unconscious'. For variety, and where I feel there may be a danger that using the term 'the unconscious' may import unhelpful connotations, I shall also use my own coinage, 'the undermind'. When we go on to discuss how unconscious intelligence is exemplified in and generated by real flesh-and-blood human beings, I shall refer to the source as 'the unconscious biocomputer', or sometimes just as the 'brain-mind'. My intention is that, by using a range of different terms, I shall be able to build up a composite picture of 'the intelligent unconscious' that does justice to its many faces and functions.

5. There is presently an explosion of interest, both popular and scholarly, in the subject of consciousness. Julian Jaynes' *The Origin of Consciousness in the Breakdown of the Bicameral Mind*, Daniel Dennett's *Consciousness Explained*, Roger Penrose's *Shadows of the Mind* and Robert Ornstein's *The Psychology of Consciousness* are just four of the dozens of books that have appeared in the last few years on the nature, the evolution and the function of the conscious mind. My book is partly a reflection of this wave of enthusiasm, and partly a reaction against it. I certainly have things to say about what consciousness is and what it is for, but I also argue that we cannot

understand the nature of the conscious mind without having a better image of the dark, inaccessible layers – the minds behind the mind – that underlie it, and from which it springs. Consciousness can only be understood in relation to the unconscious. If we persist in trying to make sense of consciousness in and on its own terms, we shall continue to see those modes of mind that are most associated with consciousness as pre-eminent; and continue to ignore or undervalue those that are less conscious, less deliberate, or which require a different image of mind if they are to become visible, and to make sense. Fascinating though it is, much of this wave of research and speculation on consciousness must be seen as symptomatic of our cultural obsession with the conscious intellect, and not a corrective to it. None of these books has anything to say about the practical effect on our psychology of this reconceptualisation of the relationship between conscious and unconscious.

6. It should be clear that what I am talking about here is not a thinly disguised version of the distinction between the 'right' and 'left' brain, which was popular a few years ago as a way of thinking about the brain's lost capacities. Though it persists in the literature of 'pop psychology', this distinction has run out of steam. First, in all but a few unfortunate people, the brain works as a whole. Functionally it does not have two separate halves. To ask people to switch from a 'left-brain' to a 'right-brain' way of thinking would be like insisting that people switched from driving an 'engine car' to a 'steering-wheel car'.

Secondly, people's aspirations for the 'right brain' ran wildly out of control: far ahead of what the scientific research could possibly justify. Certainly there is a greater linguistic ability, for most right-handed people, in the left cerebral hemisphere of the brain than in the right; but the research shows that there is language on the left side, just as there are 'holistic' properties on the right, as well. Michael Gazzaniga, one of the scientists, along with Nobel laureate Roger Sperry, whose work on so-called 'split-brain' patients was the inspiration of the 'left brain, right brain' distinction, wrote despairingly as long ago as 1985: 'How did some laboratory findings of limited generality get so outrageously misinterpreted? . . . The image of one part of the brain doing one thing and the other part something entirely different was there [in the popular literature], and [the fact] that it was a confused concept seemed to make no difference.' (Gazzaniga, M., *The Social Brain: Discovering the Networks of the Mind*, New York: Basic Books, 1985).

When I talk here about 'tortoise mind', the 'undermind' or the 'intelligent unconscious' I am not talking about a new *bit* of the brain. I am talking about a set of different modes of mind that, above all, require a less busy, less purposeful, less problem-orientated mental ambience.

## Chapter 2

1. Gardner, Howard, 'The theory of multiple intelligences'. Presentation to the Annual Conference of the British Psychological Society Division of Educational and Child Psychologists, York, (January 1996).

2. Goleman, Daniel, *Emotional Intelligence* (New York: Bantam, 1995).

3. Rozin, Paul, 'The evolution of intelligence and access to the cognitive unconscious', in Sprague, J. M., and Epstein, A. N. (eds), *Progress in Psychobiology and Physiological Psychology*, Vol. 6 (New York: Academic Press, 1976). Rozin was one of the first to use the term 'cognitive unconscious', and I have borrowed other arguments and examples in this section from his seminal paper.

4. See Wooldridge, D., *The Machinery of the Brain* (New York: McGraw Hill, 1963).

5. Studies by Aronson (1951), reported by Rozin, op cit, p. 252.

6. See Smith, Ronald, Sarason, Irwin and Sarason, Barbara, *Psychology: the Frontiers of Behavior*, 2nd edition (San Francisco: Harper and Row, 1982), p. 273.

7. The research that confirms these differences has been recently summarised by Reber, Arthur, *Implicit Learning and Tacit Knowledge: an Essay on the Cognitive Unconscious* (Oxford: OUP, 1993).

8. Some of the studies that lead to this conclusion will be discussed in Chapter 8.

9. Carraher, T. N., Carraher, D. and Schliemann, A. D., 'Mathematics in the street and in schools', *British Journal of Developmental Psychology*, Vol. 3 (1985), pp. 21–9. Ceci, S. J. and Liker, J., 'A day at the races: a study of IQ, expertise and cognitive complexity', *Journal of Experimental Psychology: General*, Vol. 115 (1986), pp. 255–66.

10. Berry, Dianne C. and Broadbent, Donald E., 'On the relationship between task performance and associated verbalizable knowledge', *Quarterly Journal of Experimental Psychology*, Vol. 36A (1984), pp. 209–31. See also the recent overview of these results in Berry, Dianne and Dienes, Zoltan, *Implicit Learning* (London: Lawrence Erlbaum Associates, 1992).

11. Watzlawick, Paul, Weakland, John and Fisch, Richard, *Change: Principles of Problem Formation and Problem Resolution* (New York: Norton, 1974).

12. A great body of this work is summarised in Lewicki P, Hill, T. and Czyzewska, M. 'Nonconscious acquisition of information', *American Psychologist*, Vol. 47 (1992), pp. 796–801.

13. Reber, op cit.

### Chapter 3

1. Developmental psychologist Annette Karmiloff-Smith makes the same observation in her book *Beyond Modularity: a Developmental Perspective on Cognitive Science* (Cambridge, MA: MIT, 1992).

2. These studies are comprehensively reviewed in Berry and Dienes, op cit.

3. In fact there is research to show that our faith in articulation, as a measure of competence, is misguided in the 'real world'. Medical students' performance in written examinations, for example, is quite unrelated to their clinical skill and judgement. Yet our implicit (in the sense, this time, of 'unquestioned') faith in good old exams is such that we continue to put generations of students through them. See Skernberg, R. T., and Wagner, R. K. (eds), *Mind in Context* (Cambridge: CUP, 1994).

4. See Wason, Peter, and Johnson-Laird, Philip, *The Psychology of Reasoning: Structure and Content* (London: Batsford, 1972).

5. Coulson, Mark, 'The cognitive function of confusion', paper presented to the British Psychological Conference, London (December 1995).

6. Lewicki *et al*, op cit.

7. Master, R. S. W., 'Knowledge, knerves and know-how: the role of explicit versus implicit knowledge in the breakdown of a complex skill under pressure', *British Journal of Psychology*, Vol. 83 (1992), pp. 343–58.

8. The polar planimeter, as a metaphor for know-how, is described by Runeson, Sverker, 'On the possibility of "smart" perceptual mechanisms', *Scandinavian Journal of Psychology*, Vol. 18 (1977), pp. 172–9.

9. Bourdieu, Pierre, *In Other Words: Essays towards a Reflexive Sociology* (Stanford, CA: Stanford University Press, 1990).

10. Huxley, Aldous, *Island* (London: Chatto, 1962).

11. Korzybski was the founder of the intellectual movement known as 'general semantics', influential in the 1940s and 1950s, that explored the relationship between language and human experience: see, for instance, his book *Science and Sanity* (New York: W.W. Norton, 1949).

### Chapter 4

1. From Spencer, Herbert, *An Autobiography*, quoted in Ghiselin, Brewster (ed), *The Creative Process* (Berkeley, CA: University of California Press, 1952).

2. This sense of 'intuition' is associated with a long tradition of religious

and philosophical writers, which includes pre-eminently Spinoza, and later Henri Bergson, who see intuition as being the royal road to some higher or 'spiritual' truth. In Spinoza's use of the term, intuition refers to a kind of profound, unmediated understanding of the 'nature of things' which arises through a deep contemplative communion with people and objects. This 'intuition', according to Spinoza, is inevitably accurate; it carries with it an unquestionable certainty and authority, and only makes its appearance after the purposeful questings of reason have exhausted themselves.

3. All these first three problems can be solved by filling the largest jar, and from it filling the middle-sized jar once and the smallest jar twice. You are then left with the desired volume in the big one. For an account of these studies, see Rokeach, Milton, 'The effect of perception time upon the rigidity and concreteness of thinking', *Journal of Experimental Psychology*, Vol. 40 (1950), pp. 206–16.

4. All you have to remember from your school maths is that the circumference of a circle is 6.28 times as big as the radius (i.e. twice the product of the radius and the constant 'pi', which is 3.14).

Suppose the radius of the smoothed-out Earth is R.

Then the original length of string is 6.28 times R.

If the size of the gap we are interested in is called 'r', then the new total radius is R+r, and the new total length of string is therefore 6.28 times (R+r).

But this is the original length, 6.28R, plus 2m, or 200cm. So

$$6.28(R+r) = 6.28R + 200.$$

Take away the = 6.28R from both sides of the equation, and divide both sides by 6.28. This leaves

$$r = 200/6.28, \text{ or about 32cm.}$$

5. This tension between reason and intuition has been known since antiquity. It is recorded, for example, that the Athenian general Nicias, at the siege of Syracuse, decided to follow an intuitive interpretation of a lunar eclipse and, against his 'better' – i.e. rational – judgement, to postpone a tactical retreat, his faith in intuition leading to a decisive defeat.

6. See, for example, McCloskey, M., 'Intuitive physics', *Scientific American*, Vol. 248 (1983), pp. 114–22.

7. Ceci, S. J. and Bronfenbrenner, U., 'Don't forget to take the cupcakes out of the oven: strategic time-monitoring, prospective memory and context', *Child Development*, Vol. 56 (1985), pp. 175–90.

8. This example actually relies on 'learning by osmosis' rather than the kind of intuition that is the main focus of this chapter. But it makes graphically a point which applies to both sorts of slow knowing.

9. Kahneman, Daniel and Tversky, Amos, 'Intuitive prediction: biases and

corrective procedures', in Kahneman, D., Slovic, P. and Tversky, A. (eds), *Judgement under Uncertainty: Heuristics and Biases* (Cambridge: CUP, 1982).

10. These quotations come from a survey carried out in 1992, in which eighty-three Nobel laureates in science – physics, chemistry and medicine – answered the question: 'Do you believe in scientific intuition?' The vast majority were in no doubt that a vital stage on the road to their discoveries was listening to hunches about their results, or promptings about the direction to take, which were quite incapable of rational defence or explanation. See Fensham, Peter, and Marton, Ference, 'What has happened to intuition in science education?', *Research in Science Education*, Vol. 22 (1992), pp. 114–22.

11. Spencer Brown, George, *Laws of Form* (London: Allen and Unwin, 1969).

12. Noddings, Nel and Shore, Paul, *Awakening the Inner Eye: Intuition in Education* (New York: Teachers' College Press, 1984).

13. Lowe, John Livingston, *The Road to Xanadu* (Boston: Houghton Mifflin, 1927).

14. Quoted by Gerard in Ghiselin, op cit.

15. Coleridge, Samuel Taylor, 'Prefatory note to Kubla Kahn', quoted in Ghiselin, op cit.

16. Quotation from Woodworth, R. S. and Schlosberg, H., *Experimental Psychology* (1954), quoted in Smith, S. M. and Blankenship, S. E., 'Incubation and the persistence of fixation in problem-solving', *American Journal of Psychology*, Vol. 104 (1991), pp. 61–87. See also Smith, S. M., Brown, J. M. and Balfour, S. P., 'TOTimals: a controlled experimental method for studying tip-of-tongue states', *Bulletin of the Psychonomic Society*, Vol. 29 (1991), pp. 445–7; and Smith, S. M., 'Fixation, incubation and insight in memory and creative thinking', in Smith, S. M., Ward, T. B. and Finke, R. A. (eds), *The Creative Cognition Approach* (Cambridge, MA: Bradford/ MIT Press, 1995).

17. Yaniv, I. and Meyer, D. E., 'Activation and metacognition of inaccessible stored information: potential bases for incubation effects in problem-solving', *Journal of Experimental Psychology: Learning, Memory and Cognition*, Vol. 13 (1987), pp. 187–205.

18. These studies are reported in Bowers, K. S., Regehr, G., Balthazard, C. and Parker, K., 'Intuition in the context of discovery', *Cognitive Psychology*, Vol. 22 (1990), pp. 72–110; and Bowers, K. S., Farvolden P. and Mermigis, L., 'Intuitive antecedents of insight', in Smith, S. M. *et al* (eds), *The Creative Cognition Approach*, op cit. The solutions to the picture puzzles are: 1A shows a camera; 2A shows a camel.

19. The words in 1A are all associates of 'candle'; the words in 2B are all associates of 'carpet'; the words in 3A are all associates of 'pipe'.

20. The common associate to all fifteen words is 'fruit'. If you feel inclined to argue with some of these associations, bear in mind that they were all statistically defined as low-frequency associative responses to the cue word from a large-scale survey of American students.

## Chapter 5

1. Skinner, B. F., 'On "Having" a Poem', reprinted in *Cumulative Record* (New York: Appleton-Century-Croft, 1972).

2. Quoted in Ghiselin, op cit.

3. These last two examples were used to illustrate a television programme, broadcast on 1 September 1996 on Channel 4, called 'Break the Science Barrier with Richard Dawkins'.

4. James, Henry, 'Preface to *The Spoils of Poynton*', in Ghiselin, op cit.

5. Canfield, Dorothy, 'How Flint and Fire Started and Grew', in Ghiselin, op cit.

6. Simonton, D. K., *Genius, Creativity and Leadership: Historiometric Inquiries* (Cambridge, MA: Harvard University Press, 1984).

7. The example is discussed by Schooler, Jonathan and Melcher, Joseph, 'The ineffability of insight', in Smith, Stephen *et al* (eds), *The Creative Cognition Approach* op cit. A general survey of the evidence of an 'inverted U'-shaped relationship between knowledgeability and creativity is provided by Simonton, op cit.

8. Westcott's 'successful intuitives' are similar to the character type identified by C. G. Jung as 'introverted intuitive'. These people are not only socially somewhat introverted, as Westcott's intuitives were; they are also those who, according to Jung, have the closest relationship with their own unconscious. See, for example, Jung's classic *Psychological Types*, translated by H. G. Baynes (London: Routledge, 1926). For Jung, intuition is one of four basic mental functions, the other three being 'thinking', 'feeling' and 'sensing'. It is a rather nebulous faculty which detects possibilities and implications in a holistic fashion, at the expense of details. Jung's view is that every person has one of these four modes which is more developed than, and used in preference to, the others. One has a disposition to be a thinking or an intuitive 'type', for example. In addition to these four functions, Jung proposed that we also differ in the extent to which our basic orientation is towards the external or the internal world – whether one is an 'extravert' or an 'introvert'.

Jung, as is well known, also distinguishes between two layers of the unconscious mind, the personal and the collective. In the personal unconscious lie those memories and perceptions which are too weak ever to make it over the borderline into consciousness, or which have been repressed. The collective unconscious contains the *archetypes*, forms of universal human-species knowledge derived from one's whole ancestral lineage, and reborn in each individual's brain structure. The collective unconscious makes itself known through our innate understanding of ubiquitous human situations and relationships, and through universal systems of symbols. 'Introverted intuitives', in Jung's scheme, do, in a sense, have access to a 'higher' knowledge than the other types by virtue of their more intimate relationship with the symbolic life and fundamental knowings of the collective unconscious.

Jung's typology (and the variety of personality tests, such as the Myers-Briggs Type Inventory, based upon it) now looks rather coarse in the light of our better, more empirically based, understanding of the way in which the unconscious itself generates intuitions. (Jung held to a view of intuition as a way of seeing *into* the unconscious, rather than as a product *of* it.) And there is a second important way in which Jung's pioneering work has been superseded. He, rather fatalistically, tended to talk of the four basic personality types as if they were constitutional, and therefore largely unalterable. We now know that the intuitive way of knowing is educable, capable of being enhanced and sharpened. It is for these reasons that Jung's ideas receive less attention in this book than some readers might have expected.

9.  Westcott, Malcolm, *Toward a Contemporary Psychology of Intuition* (New York: Holt, Rinehart & Winston, 1968).

10.  Schon, Donald, *The Reflective Practitioner: How Professionals Think in Action* (New York: Basic Books, 1983).

11.  Rokeach, op cit.

12.  Cowen, Emory L., 'The influence of varying degrees of psychological stress on problem-solving rigidity', *Journal of Abnormal and Social Psychology*, Vol. 47 (1952), pp. 512–19.

13.  Combs, Arthur and Taylor, Charles, 'The effect of perception of mild degrees of threat on performance', *Journal of Abnormal and Social Psychology*, Vol. 47 (1952), pp. 420–4.

14.  Kruglansky, A. W. and Freund, T., 'The freezing and unfreezing of lay inferences: effects on impressional primacy, ethnic stereotyping and numerical anchoring', *Journal of Experimental Social Psychology*, Vol. 19 (1983), pp. 448–68.

15.  Wright, Morgan, 'A study of anxiety in a general hospital setting', *Canadian Journal of Psychology*, Vol. 8 (1954), pp. 195–203.

16.  Prince, George, 'Creativity, self and power', in Taylor, I. A. and Getzels, J. W. (eds), *Perspectives in Creativity* (Chicago: Aldine, 1975).

17. Fischbein, Efraim, *Intuition in Science and Mathematics* (Dordrecht, Holland: Kluwer, 1987), p. 198.

18. Viesti, Carl, 'Effect of monetary rewards on an insight learning task', *Psychonomic Science*, Vol. 23 (1971), pp. 181–3.

19. Bruner, Jerome, Matter, Jean and Papanek, Miriam, 'Breadth of learning as a function of drive level and mechanisation', *Psychological Review*, Vol. 62 (1955), pp. 1–10.

20. Hughes, Ted, *Poetry in the Making* (London: Faber, 1967).

21. Emerson, R. W., 'Self-reliance', in *The Collected Works of Ralph Waldo Emerson*, Vol. II (Cambridge, MA: Belknap Press, 1979).

22. Lynn, Steven and Rhue, Judith, 'The fantasy-prone person: hypnosis, imagination and creativity', *Journal of Personality and Social Psychology*, Vol. 51 (1986), pp. 404–8; and Bastick, Tony, *Intuition: How We Think and Act* (Chichester: Wiley, 1982).

23. Hillman, James, *Insearch: Psychology and Religion* (Dallas, TX: Spring, 1967); and *Archetypal Psychology: A Brief Account* (Dallas: Spring, 1983).

24. Lawrence, D. H., 'Making pictures', in Ghiselin, op cit.

25. Quoted by Zervos, C., 'Conversation with Picasso', in Ghiselin, op cit.

## Chapter 6

1. From Goodman, N. G. (ed), *A Benjamin Franklin Reader* (New York: Crowell, 1945). Quoted in Wilson, Timothy and Schooler, Jonathan, 'Thinking too much: introspection can reduce the quality of preferences and decisions', *Journal of Personality and Social Psychology*, Vol. 60 (1991), pp. 181–92. The experiments discussed in the first part of this chapter come from several papers by Schooler and his associates, including this one, and: Schooler, Jonathan and Engstler-Schooler, Tonya, 'Verbal overshadowing of visual memories: some things are better left unsaid', *Cognitive Psychology*, Vol. 22 (1990), pp. 36–71; Schooler, Jonathan, Ohlsson, Stellan and Brooks, Kevin, 'Thought beyond words: when language overshadows insight', *Journal of Experimental Psychology: General*, Vol. 122 (1993), pp. 166–83; and Schooler, Jonathan and Melcher, Joseph, 'The ineffability of insight', in Smith *et al* (eds), *The Creative Cognition Approach*, op cit.

2. Raiffa, H., *Decision Analysis* (Reading, MA: Addison Wesley, 1968). Quoted in Wilson and Schooler, op cit.

3. The solutions to the two 'insight' problems are:

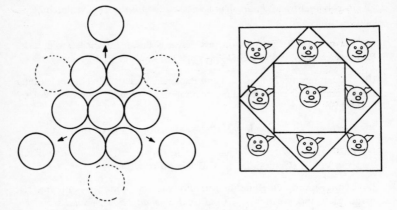

*Figure 13. Solutions to coin and pig-pen problems, page 89.*

The solutions to the two 'analytical' problems are:
a) The three playing cards, left to right, are:
   Jack of Hearts; King of Diamonds; Queen of Spades
b) Bob is telling the truth. Alan committed the crime.

4. Things, as always, are more complicated than we would like them to be. Shifting from d-mode to responding 'impulsively' also has its risks, especially in situations that are emotionally charged. Daniel Goleman in *Emotional Intelligence* (op cit) has pointed out that a habit of impulsiveness leaves us open to being 'emotionally hi-jacked by reflexes that may have served us well in the jungle, but which can now be extremely dangerous and counter-productive.'

5. Henri Poincaré, quoted in Ghiselin, op cit.

6. Mozart, 'A letter', quoted in Ghiselin, op cit.

7. Dryden, John, 'Dedication of "The Rival-Ladies",' quoted in Ghiselin, op cit.

8. Wordsworth, William, 'Preface to Second Edition of Lyrical Ballads', quoted in Ghiselin, op cit.

9. Moore, Henry, 'Notes on sculpture', quoted in Ghiselin, op cit.

10. Gerard, R. W., 'The biological basis of the imagination', *Scientific Monthly* (June 1946).

11. Edelman, Gerald, *Neural Darwinism: The Theory of Neuronal Group Selection* (New York: Basic Books, 1992). I shall postpone any further discussion of brain mechanisms till Chapter 9.

12. Lowell, Amy, quoted in Ghiselin, op cit.

13. Housman, A. E., 'The Name and Nature of Poetry', quoted in Ghiselin, op cit.

14. Belenky, Mary Field, Clinchy, Blythe McVicker, Goldberger, Nancy Rule and Tarule, Jill Mattuck, *Women's Ways of Knowing: The Development of Self, Voice, and Mind* (New York: Basic Books, 1986).

15. Weil, Simone, *Gravity and Grace* (London: Routledge, 1972). Quoted by Belenky *et al*, op cit, p. 99.

## Chapter 7

1. The existence of unconscious perception is almost universally accepted now by cognitive scientists. There are one or two die-hards, such as Douglas Holender, who wrote a long review article in 1986 attempting to find methodological flaws in every study that claimed to show unconscious perception. Norman Dixon, the doyen of the field of subliminal processing, concluded his response to Holender's paper by saying: 'the most interesting phenomenon to which Holender's paper draws attention is the extraordinary antipathy some people still have toward the idea that we might be influenced by things of which we are unaware. Would it be putting it too strongly to say it reminds one of the skepticism of "flat earth theorists" when confronted with the alarming theory that the world is round?' See Holender, D., 'Semantic activation without conscious identification in dichotic listening, parafoveal vision and visual masking: a survey and appraisal', *Behavioral and Brain Sciences*, Vol. 9 (1986), pp. 1–23; and Dixon, N.F., 'On private events and brain events', *Behavioral and Brain Sciences*, Vol. 9 (1986), pp. 29–30.

2. This study is described by Pittman, Thane, 'Perception without awareness in the stream of behavior: processes that produce and limit nonconscious biasing effects', in Bornstein, R. F. and Pittman, T. S. (eds), *Perception without Awareness: Cognitive, Clinical and Social Perspectives* (New York: Guildford Press, 1992).

3. This phenomenon is discussed in a paper by Simpson, Brian, 'The escalator effect', *The Psychologist*, Vol. 5 (1992), pp. 462–3.

4. Sidis, B., *The Psychology of Suggestion* (New York: Appleton, 1898), quoted by Merikle, P. M. and Reingold, E. M., 'Measuring unconscious perceptual processes', in Bornstein and Pittman, op cit.

5. Pierce, C. S. and Jastrow, J., 'On small differences in sensation', *Memoirs of the National Academy of Science*, Vol. 3 (1884), pp. 75–83. Quoted by Kihlstrom, J. F., Barnhardt, T. M. and Tataryn, D. J., 'Implicit perception', in Bornstein and Pittman, op cit.

6. Both the Poetzl studies and the more recent follow-ups are described

in Ionescu, M. D. and Erdelyi, M. H., 'The direct recovery of subliminal stimuli', in Bornstein and Pittman, op cit.

7. Bradshaw, John, 'Peripherally presented and unreported words may bias the perceived meaning of a centrally fixated homograph', *Journal of Experimental Psychology*, Vol. 103 (1974), pp. 1200–2.

8. Patton, C. J., 'Fear of abandonment and binge eating: a subliminal psychodynamic activation investigation', cited in Masling, Joseph, 'What does it all mean?', in Bornstein and Pittman, op cit. The fact that it may be hard consciously to accept that the subliminal perception of 'Mummy is leaving me' can have such a dramatic effect on behaviour is itself further evidence of Patton's effect.

9. Darley, J. M. and Gross, P. H., 'A hypothesis-confirming bias in labeling effects', *Journal of Personality and Social Psychology*, Vol. 44 (1983), pp. 20–33.

10. Whittlesea, B. W., Jacoby, L. L., Girard, K. A., 'Illusions of immediate memory: evidence of an attributional basis for feelings of familiarity and perceptual quality', *Journal of Memory and Language*, Vol. 29 (1990), pp. 716–32.

11. Schacter, Daniel (ed.), *Memory Distortions: How Minds, Brains and Societies Reconstruct the Past* (Cambridge, MA: Harvard University Press, 1995).

12. I shall say more about the way in which we can protect ourselves from our own unconscious assumptions in the discussion of 'mindfulness' in Chapter 11.

13. See Nisbett, R. and Wilson, T., 'Telling more than we know: verbal reports on mental processes', *Psychological Review*, Vol. 84 (1977), pp. 231–59.

14. Latane, B. and Darley, J. M., *The Unresponsive Bystander: Why Doesn't He Help?* (New York: Appleton-Century-Crofts, 1970).

15. Fitzgerald, F. Scott, *Tender is the Night* (New York: Scribner, 1934).

16. The case of Flournoy and Helen Smith is discussed by Ellenberger, Henri, *The Discovery of the Unconscious* (New York: Basic Books, 1970), p. 316.

## Chapter 8

1. Masling, Joseph M., 'What does it all mean?', in Bornstein and Pittman, op cit.

2. Bruner, Jerome and Postman, Leo, 'Emotional selectivity in perception and reaction', *Journal of Personality*, Vol. 16 (1947), pp. 69–77.

3. It is surprising how frequently well-educated adults in our society fear that any kind of psychological trick or test is liable to expose something unwelcome about their mental powers. Television quiz shows, for example, both reflect and promote the ridiculous assumption that rapid retrieval of trivia is a valid index of 'intelligence' – though schools, with somewhat more pretension, may fall into the same trap.

4. A similar interpretation of the Zajonc studies has been offered by Reber, op cit.

5. Quoted by Reber, op cit, p. 18.

6. These studies of 'implicit memory', as it is called, are comprehensively reviewed by Schacter, Daniel, 'Implicit memory: history and current status', *Journal of Experimental Psychology: Learning, Memory and Cognition*, Vol. 13 (1987), pp. 501–18.

7. Marcel, Tony, 'Slippage in the unity of consciousness', in CIBA Symposium 174, *Experimental and Theoretical Studies of Consciousness* (Chichester: Wiley, 1993).

8. Cumming, Geoff, 'Visual perception and metacontrast at rapid input rates', DPhil thesis, University of Oxford (1971).

9. Marcel, op cit.

10. The best discussions of 'functional blindness' are still those provided by P. Janet in his classic text *The Major Symptoms of Hysteria* (New York: Macmillan, 1907). A marvellous fictionalised description is provided by William Wharton in *Last Lovers* (London: Granta, 1991).

11. See Wall, Patrick, in CIBA Symposium 174, op cit.

12. Sutcliffe, J. P., '"Credulous" and "skeptical" views of hypnotic phenomena: experiments in esthesia, hallucination and delusion', *Journal of Abnormal and Social Psychology*, Vol. 62 (1961), pp. 189–200.

13. Langer, E., Dillon, M., Kurtz, R. and Katz, M., 'Believing is seeing', unpublished paper, Harvard University, referred to in Langer, Ellen, *Mindfulness: Choice and Control in Everyday Life* (London: Harvill, 1991).

14. For a review of blindsight studies, see Weiskrantz, Lawrence, *Blindsight: A Case Study and Its Implications* (Oxford: Clarendon, 1986).

15. Humphrey, Nicholas, comments in discussion, in CIBA Symposium 174, op cit, p. 161.

16. I once asked Tony Marcel whether a thirsty blindsight patient would spontaneously reach for a glass of water that was within the blind field. In practice, he pointed out, this would be an almost impossible test to carry out, as none of these patients is 'blind' in all areas of the visual field, so when their eyes are free to move about, as they normally are, any significant

objects in their world would rapidly be picked up through the areas of normal sight. But (for what it's worth) his strong intuition, having worked with such patients for some time, is that they would not.

17. Freud, Sigmund, 'Recommendations to physicians practising psychoanalysis', in Strachey, J. (ed. and trans.), *The Standard Edition of the Complete Psychological Works of Sigmund Freud*, Vol. 12 (London: Hogarth Press, 1958/1912).

18. Granger, G. W., 'Night vision and psychiatric disorders', *Journal of Mental Science*, Vol. 103 (1957), pp. 48–79.

19. Bahrick, H. P., Fitts, P. M. and Rankin, R. E., 'Effect of incentives upon reactions to peripheral stimuli', *Journal of Experimental Psychology*, Vol. 44 (1952), pp. 400–6.

20. Bursill, A. E., 'The restriction of peripheral vision during exposure to hot and humid conditions', *Quarterly Journal of Experimental Psychology*, Vol. 10 (1958), pp. 113–29.

21. Bruner, J. S., Matter, J. and Papanek, M. L., op cit.

## Chapter 9

1. Dickinson, Emily, 'The Brain', in *Complete Poems* (Boston: Little, Brown, 1960), reprinted in Mitchell, S. (ed), *The Enlightened Heart* (New York: Harper & Row, 1989).

2. We now know that the three 'systems' are so tightly integrated with each other that it is more accurate to see them as three aspects of what is in effect a single system. Indeed, if we are to appreciate the physical underpinnings of the slow ways of knowing we need to reinstall the brain in its bodily context, and we shall do so in Chapter 10. But it makes sense to start with the brain on its own.

3. So far the cells that have shown LTP, in an area of the midbrain called the hippocampus, do tend to revert to their original recalcitrant state eventually, so they themselves cannot be responsible for the memories of a lifetime. It will not be long before some similar but even more permanent mechanism is found that will weld together cells in the cortex. But for the moment this remains just beyond the leading edge of what it is technically possible to investigate.

4. Hebb, D. O., *The Organization of Behavior* (New York: McGraw Hill, 1949).

5. See for example Minsky, Marvin, *The Society of Mind* (London: Picador, 1988).

6. This evidence is reviewed in Greenfield, Susan, *Journey to the Centers of*

*the Mind* (Oxford: Freeman, 1955). My image of brain organisation, a preliminary version of which was first published in my *Cognitive Psychology: New Directions* (London: Routledge, 1980) is in many respects similar to Susan Greenfield's, a coincidence that may be not unrelated to the fact that we were graduate students together in Oxford in the early 1970s. The major differences are that my model tries to find a place for language; and our views on the role of arousal are somewhat divergent.

## Chapter 10

1. This evidence is reviewed in Martindale, Colin, 'Creativity and connectionism', in Smith, Ward and Finke, op cit.

2. This illustration is a development of one used by Edward de Bono in *The Mechanism of Mind* (Harmondsworth: Penguin, 1971).

3. Luria, A. R., *The Mind of a Mnemonist* (Harmondsworth: Penguin, 1975).

4. There is some direct evidence for this obvious assumption: see Grossberg (1980), cited in Martindale, op cit; Kahneman, D., *Attention and Effort* (Englewood Cliffs, NJ: Prentice Hall, 1973); Baddeley, A. D. and Weiskrantz, L. (eds), *Attention: Selection, Awareness and Control* (Oxford: Clarendon, 1993).

5. The influences of culture and experience cannot in practice be separated in the neat way that this picture might imply. Much if not all of a child's direct experience with the words is both mediated by 'agents' of the culture, and saturated with cultural assumptions. Parents, older siblings and teachers are continually guiding a young person's attention, implicitly instructing her as to what is worth attention, and what significance these selected experiences are to be assigned. (Children are quick to pick up value judgements – such as phobias, for example – from observing the reactions of their seniors.) And even if no agent is physically present, the child's world is full of objects and experiences that embody the values and assumptions of the culture: toys, games, artifacts and rituals of all kinds. Even the buildings which she inhabits, and the physical landscape through which she moves, are repositories of cultural meaning.

The two-plane model which I am using here is a simple version of the kind of 'hybrid' model that many neural network theorists are currently exploring. See for example Churchland, P. S. and Sejnowski, T. J., *The Computational Brain* (Cambridge, MA: Bradford/MIT Press, 1992).

6. I am drawing on some of Gelernter's ideas in the paragraphs that follow. See Gelernter, David, *The Muse in the Machine: Computers and Creative Thought* (London: Fourth Estate, 1994).

7. Dennett, Daniel, *Consciousness Explained* (London: Viking, 1992).

8. Young, A. W. and De Haan, E. H., 'Face recognition and awareness after brain injury', in Milner, A. D. and Rugg, M. D. (eds), *The Neuropsychology of Consciousness* (London: Academic Press, 1992).

9. Research quoted by Greenfield, op cit.

10. Libet, Benjamin, 'The neural time factor in conscious and unconscious events', in CIBA Symposium 174, op cit.

11. Experiment by Jensen (1979), quoted by Libet, op cit, p. 126.

12. I have elaborated this argument in my *Noises from the Darkroom: the Science and Mystery of the Mind* (London: HarperCollins, 1994).

13. For an elaboration of this argument, see my article 'Structure, strategy and self in the fabrication of conscious experience', *Journal of Consciousness Studies*, Vol. 3 (1996), pp. 98–111.

14. Kihlstrom, John, 'The psychological unconscious and the self', in CIBA Symposium 174, op cit, p. 152.

15. Libet, op cit.

16. I have argued this point of view in more detail in my *Noises from the Darkroom: The Science and Mystery of the Mind*, op cit; as also has Velmans, Max, 'Is human information processing conscious?', *Behavioral and Brain Sciences*, Vol. 14 (1992), pp. 651–726; and Mandler, George, *Mind and Emotion* (New York: Wiley, 1975). The argument is very similar to Keith Oatley's, in *Best Laid Schemes* (Cambridge: CUP, 1992).

17. Churchland, Patricia, *Neurophilosophy* (Cambridge, MA: MIT Press, 1986).

## Chapter 11

1. Ginzburg, Carlo, *Myths, Emblems, Clues* (London: Hutchinson Radius, 1990). I am grateful to Alan Bleakley for putting Ginzburg's work my way.

2. Cited in Ginzburg, op cit, p. 211.

3. Conan Doyle, Sir Arthur, 'The Cardboard Box', first published in the *Strand* magazine, Vol. 5 (1893), pp. 61–73. Quoted by Ginzburg, op cit.

4. Freud, Sigmund, 'The Moses of Michelangelo' (1914), in *Collected Papers* (New York: Hogarth Press, 1959). Quoted by Ginzburg, op cit.

5. Reiser, Stanley, *Medicine and the Reign of Technology* (Cambridge: CUP, 1978). Quoted in Postman, op cit.

6. From Seltzer, Richard, *Mortal Lessons* (New York: Simon and Schuster, 1974). Reprinted in Feldman, Christina and Kornfield, Jack (eds), *Stories of the Spirit, Stories of the Heart* (San Francisco: HarperCollins, 1991).

7. This research, as well as details of the focusing process, are described in Gendlin, Eugene, *Focusing* (New York: Bantam, 1981).

8. I can vouch both for the effectiveness of focusing, and for its subtle, slippery quality, as I have taken two training courses in it. Some people find it easier and quicker to grasp than others, and it needs coaching, feedback and modelling, as well as direct tuition, if one is to get the hang of it. Learning to 'focus' is of the same order of difficulty as any other form of delicate perceptual learning – wine-tasting, reading X-rays or animal tracks, and so on.

9. Gendlin, op cit.

10. Suzuki, D. T., *Zen and Japanese Culture* (Princeton, NJ: Princeton University Press, 1959), pp. 104–5, 109, 157.

11. Dodds, E. R., *The Greeks and the Irrational* (Berkeley, CA: University of California Press, 1951). See also Onians, R. B., *The Origins of European Thought* (Cambridge: CUP, 1951).

12. This feeling, referred to as *yugen*, is much prized by the Zen-inspired painters and poets of Japan. The poet Seami says *yugen* is 'To watch the sun sink behind a flower-clad hill, to wander on and on in a huge forest with no thought of return, to stand on the shore and gaze after a boat that goes hid by far-off islands, to ponder on the journey of wild geese seen and lost among the clouds.' To which Alan Watts, in *Nature, Man and Woman* (London: Thames and Hudson, 1958), adds: 'But there is a kind of brash mental healthiness ever ready to rush in and clean up the mystery, to find out just precisely where the wild geese have gone . . . and that sees the true face of a landscape only in the harsh light of the noonday sun. It is just this attitude which every traditional culture finds utterly insufferable in Western man, not just because it is tactless and unrefined, but because it is blind. It cannot tell the difference between the surface and the depth. It seeks the depth by cutting into the surface. But the depth is known only when it reveals itself, and ever withdraws from the probing mind.'

13. Cassirer, Ernst, *Language and Myth* (New York: Harper, 1946).

14. Quoted by Scott, Nathan, *Negative Capability: Studies in the New Literature and the Religious Situation* (New Haven, CT: Yale University Press, 1969).

15. Gardner, Howard and Winner, Ellen, 'The development of metaphoric competence: implications for humanistic disciplines', in Sacks, S. (ed.), *On Metaphor* (Chicago: University of Chicago Press, 1979).

16. Dimnet, Ernest, quoted in de la Mare, Walter, *Behold this Dreamer!* (London: Faber & Faber, 1939), p. 647.

17. Maritain, Jacques, *Creative Intuition in Art and Poetry* (London: Harvill, 1953).

18. Quoted by Scott, op cit.

19. Eliot, T. S., *Four Quartets* (London: Faber & Faber, 1959).

20. From John Anderson's introduction to Heidegger's *Discourse on Thinking* (New York: Harper and Row, 1966)

21. Rilke, Rainer Maria, *Letters to a Young Poet*, translated and introduced by R. Snell (London: Sidgwick, 1945).

22. Whalley, George, 'Teaching poetry', in Abbs, Peter (ed), *The Symbolic Order* (London: Falmer Press, 1989), p. 227.

23. Housman, A. E., quoted in Ghiselin, op cit.

24. Croce, Benedetto, *Aesthetic*, translated by Ainslie Douglas (New York: Noonday/Farrar, Straus, 1972).

25. MacNeice, Louis, 'Snow', reprinted in Allott, Kenneth (ed), *The Penguin Book of Contemporary Verse* (Harmondsworth: Penguin, 1962). Interestingly, in the context of the present discussion, Allott comments of MacNeice: 'He is too eager and impatient to accept his subject quietly and try to understand it. He grabs it, pats it into various shapes, and varnishes any cracks in the quality of his perception with his prestidigitatory skill with words and images.'

26. Borges, Jorge Luis, *Labyrinths* (Harmondsworth: Penguin, 1970).

27. Sacks, Oliver, 'Rebecca', in *The Man who Mistook his Wife for a Hat* (London: Duckworth, 1985) pp. 169–77.

28. Kanizsa, G., *Organisation of Vision: Essays in Gestalt Psychology* (New York: Praeger, 1979).

29. This and several of the other illustrations in this chapter are taken from Langer, Ellen, *Mindfulness*, op cit.

30. Holmes, D. and Houston, B. K., 'Effectiveness of situation redefinition and affective isolation in coping with stress', *Journal of Personality and Social Psychology*, Vol. 29 (1979), pp. 212–18.

31. Teasdale, John, Segal, Zindel and Williams, Mark, 'How does cognitive therapy prevent depressive relapse and who should attentional control (mindfulness) training help?', *Behavioral Research and Therapy*, Vol. 33 (1995), pp. 25–39.

32. Teasdale *et al*, op cit.

33. Goleman, Daniel, *Emotional Intelligence*, op cit.

## Chapter 12

1. This incident occurs in a film made about Summerhill in the 1970s by the Canadian Film Board.

2. This story is told in Watzlawick, Paul, Weakland, John, and Fisch, Richard, *Change: Principles of Problem Formation and Problem Resolution*, op cit.

3. Labouvie-Vief, Gisela, 'Wisdom as integrated thought: historical and developmental perspectives', in Sternberg, R. J. (ed), *Wisdom: its Nature, Origins and Development* (Cambridge: CUP, 1990).

4. Kekes, J., quoted by Kitchener, Karen and Brenner, Helene, 'Wisdom and reflective judgement', in Sternberg, op cit.

5. Robin Skynner discussed this in a seminar with Fritjof Capra at Schumacher College, Devon, in June 1992.

6. Rogers, Carl, *A Way of Being* (New York: Houghton Mifflin, 1981).

7. Kierkegaard, Soren, quoted by Pascual-Leone, Juan, 'An essay on wisdom: toward organismic processes that make it possible', in Sternberg, op cit.

8. Sternberg, Robert J., 'Implicit theories of intelligence, creativity and wisdom', *Journal of Personality and Social Psychology*, Vol. 49 (1985), pp. 607–27.

9. Kegan, Robert, *In over our Heads: the Mental Demands of Modern Life* (Cambridge, MA: Harvard University Press, 1994).

10. Meacham, John, 'The loss of wisdom', in Sternberg, *Wisdom*, op cit.

11. In this and the other quotations in this section, the emphases have been added to the originals.

12. Quotation and details of Tauler's life taken from Moss, Donald M., 'Transformation of self and world in Johannes Tauler's mysticism', in Valle, R. S. and von Eckartsberg, R. (eds), *The Metaphors of Consciousness* (New York: Plenum Press, 1981).

13. Whyte, Lancelot Law, *The Unconscious before Freud* (London: Julian Friedmann, 1978), p. 10.

14. Excerpts from *Free and Easy: A Spontaneous Vajra Song* by Lama Gendun Rinpoche.

15. Suzuki, Shunryu, *Zen Mind, Beginner's Mind* (New York: Wetherhill, 1970).

16. Sahn, Seung, *Dropping Ashes on the Buddha*, S. Mitchell (ed) (New York: Grove Press, 1976).

17. Quotations drawn from Suzuki, D. T., *The Zen Doctrine of No Mind* (London: Rider, 1969); and Yampolsky, Philip, *The Platform Sutra of the Sixth Patriarch* (New York: Columbia University Press, 1967).


**Chapter 13**

1. Jaynes, Julian, *The Origin of Consciousness in the Breakdown of the Bicameral Mind* (Boston: Houghton Mifflin, 1976).

2. Dodds, op cit.

3. These quotations are drawn from Whyte, op cit.

4. Ibid, pp. 41–2.

5. Postman, op cit.

6. Ibid. p. 111.

7. Ibid. pp. 118–19.

8. From Heidegger, Martin, *Discourse on Thinking*, op cit.

9. Description of the GMAT in the 1996–7 GMAT Bulletin, published by the Graduate Management Admission Council, Princeton, NJ.

10. See the American Psychological Association review of 'Intelligence: knowns and unknowns', chaired by Ulric Neisser, published in *American Psychologist*, Vol. 51 (1996), pp. 77–101.

11. Ceci and Liker, op cit.

12. Peters, Tom, *The Pursuit of Wow! Every Person's Guide to Topsy-Turvy Times* (New York: Vintage, 1994).

13. Peters, Tom, 'Too wired for daydreaming', *Independent on Sunday*, 13 February 1994.

14. Peters, *The Pursuit of Wow!*, op cit.

15. Rowan, Roy, *The Intuitive Manager* (Boston: Little, Brown, 1986).

16. De Bono, Edward, *De Bono's Thinking Course* (London: BBC, 1985).

17. Rowan, op cit.

18. Mintzberg, Henry, *The Rise and Fall of Strategic Planning* (New York: The Free Press, 1994).

19. Quinn, Brian, quoted by Mintzberg, op cit.

20. See de Bono, op cit.

21. West, Michael, Fletcher, Clive and Toplis, John, *Fostering Innovation: A Psychological Perspective* (Leicester: British Psychological Society, 1994).

22. For a summary of Dweck's work, see Chiu, C., Hong, Y. and Dweck, C. S., 'Toward an integrative model of personality and intelligence: a general framework and some preliminary steps', in Sternberg, R. J. and Ruzgis, P. (eds), *Personality and Intelligence* (Cambridge: CUP, 1994).

23. For further examples and discussions of science education, see Claxton, Guy, *Educating the Inquiring Mind: The Challenge for School Science* (Hemel Hempstead: Harvester/Wheatsheaf, 1991); Claxton, Guy, 'Science of the times: a 2020 vision of education', in Levinson, R. and Thomas, J. (eds), *Science Today: Problem or Crisis?* (London: Routledge, 1996); Cosgrove, Mark, 'A study of science-in-the-making as students generate an analogy for electricity', *International Journal of Science Education*, Vol. 17 (1995), pp. 295–310; and Osborne, Roger and Freyberg, Peter (eds), *Learning in Science* (Auckland and London: Heinemann, 1985).

24. Archbishop William Temple, quoted in Watts, Alan, *In my Own Way* (New York: Vintage, 1973).

25. See for example Gallwey, Timothy, *The Inner Game of Tennis* (London: Cape, 1975); Clark, Frances Vaughan, 'Exploring intuition: prospects and possibilities', *Journal of Transpersonal Psychology*, Vol. 3 (1973), pp. 156–69.

26. Langer, E., Hatem, M., Joss, J. and Howell, M., 'Conditional teaching and mindful learning: the role of uncertainty in education', *Creativity Research Journal*, Vol. 2 (1989), pp. 139–50.

27. See Nisbet, J. and Shucksmith, J., *Learning Strategies* (London: Routledge, 1986).

28. Whyte, op cit.

# Index

Numbers in italics refer to diagrams.

ability 217–18
absorption 175
accumulated clues task 65–6
acetylcholine 146
activation 148, 149–50, 152, *152*, 153, 155, 157, 159, 161, 162
adaptation 16
Aersten, Ad 145
*Aesthetic* (Croce) 178
affect 74
allusions 153
alpha waves 148
amines 146
amnesia 118–19, 161
Amsler, Jacob 39, 40
analytical problems 88–90, 91
'Ancient Mariner, The' (Coleridge) 59, 114
animals
    counterproductive effect of incentives 80
    and fast intuitions 51
    and intelligence 15–16, 44
    and mutation 95
    and neurons 137
anosognosia 122
anxiety 96, 130, 132, 213
apophatic tradition 196
Aquinas, St Thomas 204
Arnold, Matthew 192
art 4, 8
    and d-mode 8–9, 95
    modus operandi 82–4
*Art of Thought, The* (Wallas) 94
articulate incompetence 210
articulate sceptics 210–11
associations 148, 149, 151, 153, 155
attention 129–32, 143, 149, 162, 164
    detection 165–9
    focusing on inner states 165, 169–72
    mindfulness 165, 180–87, 198
    poetic sensibility 165, 172–9
Augustine, St 204
automatic pilot 102–3
awareness 100, 107, 169
axon 134, 135, 137

babies
    brain 137–8
    and intelligence 16
    and unconscious operations 20
backward masking 121
Bacon, Francis 85
Ball, Sir Christopher 202, 215
Bankei 3
beauty 174, 178
beginner's mind (*shoshin*) 198–9
behaviour
    behavioural rigidity 75–6, 77
    effect of self-consciousness on 123–4
Belenky, Mary Field 97–9
Berg, Paul 57, 70
Bernstein, Robert 210
Berry, Dianne 22, 23, 26, 30, 31, 32–3
bicuculline 145
blindness
    functional 124
    hysterical 122–3, 161
'blindsight' patients 127–9, 157
Borges, Jorge Luis 178
Bornstein, Robert 101, 104, 107, 108, 111
Bourdieu, Pierre 41
Bowers, Kenneth 63, 65, 66, 150
Bowers' degraded images 63–4, *64*
Bradshaw, John 107
brain
    breadth of activation of a concept 146
    composed of two types of cells 134
    and consciousness 156–7
    direction of activity flow 146
    dual-track operation 159–60
    and epicentres 144, 145, 146
    focus of current intense research activity 133
    and information 134
    low-focus 149–50
    rate of flow 146–7
    responses affected by the state of need 145
    routes activity from neural cluster to neural cluster 144

brain – *cont.*
  ties together needs, opportunities and
    capabilities 134
brain damage 19, 122
brain stem 146
brain-mind
  and consciousness 162
  and effective action 20
  enquiring 19
  function of 18–19
  as plastic 18, 19
  powers of observation and detection 25
  starved of perceptual data 33
  and thinking slowly 214
brainwaves 148
'Brains, Minds and Consciousness'
    symposium (University of
    Birmingham) 133
brainscape 140, 144–5, 148, 149, 152–5,
    *154*
brainstorming 13, 78, 152
Brawne, Fanny 178
British Association for the Advancement of
    Science 133
British Psychological Society 212
Broadbent, Donald 22, 23, 26, 30, 31,
    32–3
Bronfenbrenner, U. 54
'Bronze Horses, The' (Lowell) 60
Brown, Michael 57, 67
Bruner, Jerome 72, 117, 131
Buck, Pearl S. 78
Buddha 191, 200
Buddhism 198
business world
  and d-mode 210–11, 214
  and information 209–10
  and slow way of knowing 213–14
'bystander' effect 112, 191

'Campaign for Learning' 202, 215
Canfield, Dorothy 70–71
capabilities 134
'Cardboard Box, The' (Conan Doyle)
    166–8
Carlyle, Thomas 78
Cassirer, Ernst 173
Ceci, S.J. 54
central nervous system 134, 156
chaos 11
character attributes experiment (Lewicki)
    36–8
children
  custody battles 185–6
  development of ability to ruminate 44–6
  and intelligence 16–17
  and production of metaphors 174–5
  and unconscious operations 20
Chow, Yung Kang 71–2
Christianity 196–7, 199
Churchland, Patricia 162–3

CIBA 127
clairvoyance 114
Claparede 119
Cleese, John 192
'Clues' (Ginzburg) 165
coaching 33–4, 42, 219
coarse fishing 81, 186
Cocteau, Jean 58–9
cognition
  d-mode and 16, 21, 49, 206
  earnest, purposeful 14
  and language 10
  relaxed 10
cognitive neuroscience 133, 157
cognitive science 3–4, 38, 67, 157, 203
Cohen, Stanley 57, 67
Coleridge, Samuel Taylor 59–60, 95, 114
'collapse of certainty' 202
Combs, Arthur 76
*Comedy of Errors, The* (Shakespeare) 204
common sense 31, 205
complexity, language and 11–12, 25, 30
computers 206–7
Conan Doyle, Sir Arthur 166–7
'Concerning the soul' (Hesse) 174
confusion 203, 207, 217
conscious awareness 100–101, 103, 107,
    109, 116, 121, 123, 126, 151, 159–62
conscious, and unconscious 19–20, 37, 63,
    72, 116–17, 124, 199–200, 204, 226
consciousness 100–101
  and an automatic pilot 102–3
  associated with intensity 157
  in d-mode 116
  'disinhibition' of 122
  focal 151
  focused 162
  implicit identification of mind with 115
  inaccessibility of memories to 119
  and information 117, 120, 211
  and intuition 211
  lack of any executive responsibility 162
  manifests what is in question 161
  and perception 106, 126–7
  and persistence of neural activity 158
  as a property of brains 156–7
  scepticism towards 198
  and self-consciousness 128–9, 159
  for self-protection 161
  and the self's involvement 120
  tendency to confabulate 111–12, 225
  and thinking 206
  and the TOT effect 62–3
  and the unconscious 64, 66, 223–4
  and the undermind 37–8, 81, 107, 116,
    124
contemplation 4, 47, 49, 93, 96, 174, 206
content curriculum 215–22
corporate planning 211
cortex 135, 144, 146, 158
Coulson, Mark 34–5

counter-transference 193
*Creative Intuition in Art and Poetry*
  (Maritain) 175
creativity 47, 63, 81
  and age 71
  asssociated with a state of low-focus
    neural activity 148–9
  counterproductive effect of incentives
    79–80
  enhanced when people are forced to slow
    down 52, 75
  and evolution 95
  favours a relaxed, well-informed mind
    152
  four phases of 94, 149–50
  icons of 4
  and the nature of the 'incubator' 69
  scientific 94
  seeing through invisible assumptions
    71–2
  Shakespeare and 204
  and the slow mind 3
  and the 'Synectics' programme 77
  and threatened belief systems 78–9
  and verbalisation 91
  and vivid imagination 82
creativity test 148–9
Crick, Francis 158
Croce, Benedetto 178
Csikszentmihalyi 83
Cudworth, Ralph 224
culture
  d-mode and 41, 203, 226
  good at solving analytic and technological
    problems 6
  inner culture 5–6
  learning 215, 217
  neglect of the unconscious 6–7
  outer culture 6
Cumming, Geoff 121
Curie, Madame Marie 221
curiosity 19, 69
custody battles 185–6

d-mode
  abandoning 34, 35
  adopted as the 'default mode' of the
    Western mind 4
  as an evolutionary and cultural parvenu
    21
  and analytical/insight problems 90–91
  and art 8–9, 95
  and attention 164
  and the business world 210–11, 214
  and cognition 16, 21, 49, 206
  and common sense 31
  consciousness in 116
  defined 2
  and education 174, 217, 218
  escaping the negative effects of 92–3
  and Euclidean geometry 40

given exclusive credence 7
grip on late twentieth-century culture 203
and the incubation phase 149
and intuition 8, 53–5, 57, 86, 149
likes explanations and plans that are
  'reasonable' and justifiable, rather than
  intuitive 8
maintains a sense of thinking as being
  controlled and deliberate 10
and map-reading 46–7
more interested in finding answers than in
  examining the questions 7
neither likes nor values confusion 8
operates at the rates at which language
  can be received, produced and
  processed 10
operates with a sense of urgency and
  impatience 9
overshadows learning by osmosis 27
as precise 9–10
and the preparation phase 94, 149
as the primary instrument of technopoly 7
as the primary mode of evaluation 93
as purposeful and effortful rather than
  playful 9
relies on language that appears to be
  literal and explicit 10
response when disconcerted 32
as the right tool 35–6, 88, 93, 94
and the Rubik cube 29–30
and scientific creativity 94
seeks and prefers clarity 8
sees thought as the essential problem-
  solving tool 7
sees understanding as the essential basis
  for action 7
speaking to it in its own language 226
treats perception as unproblematic 7
used when the problem is easily
  conceptualised 3
values explanation over observation 7–8
and wisdom 191, 195
works with concepts and generalizations
  10
works well when tackling problems which
  can be treated as an assemblage of
  nameable parts 10–12
as the wrong tool 76, 88
Dalai Lama 4, 168
Darley, J.M. 109
Darwin, Charles 56
daydreaming 152
de Bono, Edward 210
de la Mare, Walter 78
deafness, functional 123, 124, 125
'Decade of the Brain' (1990s) 133
decision-making 86
*déjà vu* 109
deliberation 96
dendrites 134
Dennett, Daniel 157

depression 184–5
Descartes, René 7, 13, 157, 205, 207, 213, 222, 224
detection 165–9
development
  of the ability to ruminate 44–5
  'stage theory' of 21
Dhonden, Yeshi 168–9
Dickinson, Emily 133, 221
Dilke, Charles 174
Dionysius the Areopagite 196
'Discourse on Thinking' (Heidegger) 176
divination 114
doctors 167–9
Donnelly, Harriet 209, 210
dopamine 146
dreams
  dream game 34
  'interpreting' 82
  remembering 63
Dryden, John 94
Dweck, Carol 216, 217, 218

'East Coker' (Eliot) 176
Ecclesiasticus 188
echo-sounding equipment 140–44, 142
Eckhart, Meister (Eckhart von Hochheim) 3, 196, 197
Edelman, Gerald 95
education
  and d-mode 174, 217, 218
  and intuition 21, 218, 219
  privileges one form of conscious, intellectual intelligence 21, 26
  promotion of 'book learning' and formal education 41–2
  resourcefulness 218, 221
  'study skills' programmes 221–2
  see also content curriculum; learning curriculum
EEGs see encephalographs
ego 173
Einstein, Albert 4, 56, 57, 70
elaboration 149, 150
Eliot, George (Mary Ann Evans) 48, 49
Eliot, T.S. 176, 222
Emerson, Ralph 78, 81, 195, 196
Emotional Intelligence (Goleman) 185
emotions 155
empathy 193
encephalographs (EEGs) 148, 160
enlightenment 199
entertainment 'industry' 186–7
epicentres (in the brain) 144, 145, 146
errors of omission 72
Essay Concerning Human Understanding (Locke) 205–6
Euclidean geometry 39, 40–41, 43
evaluation 93
evolution 17, 19, 44, 56, 95

'Experimental and Theoretical Approaches to Consciousness' (CIBA symposium, 1993) 127
explanation
  d-mode values over observation 7–8
  dislocation between expertise and 31
  intuitive 8
  'reasonable' and justifiable 8

factory task (Berry and Broadbent) 22–3, 30, 32–3
factory task (Coulson version) 34–5
false memory syndrome 110
Families and How to Survive Them (Skynner and Cleese) 192
fantasy 82, 149, 219
  Kekulé and 57, 93
felt sense 170, 171, 172
'Festival of Science' (British Association for the Advancement of Science) 133
Fischbein, Efraim 79
Fitzgerald, Scott 113
five senses 113, 134
'Flint and Fire' (Canfield) 70–71
Flournoy, Theodore 113, 114
focusing on inner states 165, 169–72, 193
Fortune magazine 209
Fostering Innovation (British Psychological Society) 212, 213
Franklin, Benjamin 85–6
'free association' game 119, 120
free will 160–61
Freud, Sigmund 13, 78, 129–30, 131, 167, 224
Freund, T. 76
Frostig 144
functional fixedness 44

Gaboriau, Émile 166
galvanic skin response (GSR) 125
Gardner, Howard 16, 175
Gelassenheit ('letting be') 197
Gelernter, David 155–6
Gendlin, Eugene 169–71, 171, 172
general relativity theory 56
Gentry magazine 209
geometry 39, 40–41, 43, 53
Gerard, R.W. 67, 95
Gerstein, George 145
gestation
  cannot be controlled or hurried 68
  child-bearing 68–9, 78
Getzels 83
Ginzburg, Carlo 165
Giotto 40–41
glial cells 134
'Global Conversation about Learning' conference (Washington, 1994) 201
GMAT (Graduate Management Admission Test) 208–9, 210
'godhead' 3, 196, 197, 200

Goleman, Daniel 16, 185
Gordon, William 77
Graduate Management Admission Test
    (GMAT) 208–9, 210
*Graduate, The* (film) 195
Greenfield, Susan 144, 145, 146
Gregory, Richard 160–61
Grinvald 144
Gross, F.H. 109
GSR *see* galvanic skin response
'guessing' 118, 121, 127
Gullixson, Mary 209, 210

hallucinations 124, 180
Hamilton, Sir William 224
Hebb, Donald 137
Heidegger, Martin 3, 176, 197, 207–8
Henry, Victor 114
Hesse, Hermann 174, 221
'hidden observer' effect 125–6
Hilgard, Ernest 125
Hillman, James 82
histamine 146
Hoffman, Dustin 195
Holland 150–51
Homer 203
*House at Pooh Corner, The* (Milne) 48
Housman, A.E. 69–70, 94, 96, 177–8
Hughes, Ted 80–81, 186
Hui-Neng 199, 200
Humphrey, Nicholas 127–8
*Hunt for Red October, The* (film) 141
Huxley, Aldous 46
hypnosis 124
hypnotic age regression 124
hypnotic analgesia 124–5

Id, the 224
ideas
    gradual formation and development of
        58–9
    and intuition 49
    and the mental womb 69
    necessary to generate and evaluate 93
    problems of speculation in the workplace
        77–8
    thinking and 206
    and the undermind 95
illogical tasks 34, 35
illumination 94, 149
illusory shapes (Kanizsa) 180, *181*
imagery 10, 59, 226
imagination 220
    creating a diversity of new forms 95
    as a learning tool 219
    and perception 81
    subliminal 114
    vivid 82
imaginative seeds
    allowing oneself to be impregnated 69
    artists' sensitivity to poignant trifles 70

need to make contact with a 'body of
        knowledge' of the right kind 71
scientists stimulated by an unexplained
        detail or incongruity 70
sensitivity to growth of 80
*see also* ideas; intuition
*In over our Heads: the Mental Demands of
        Modern Life* (Kegan) 194
incubation period 60, 61, 62, 94, 149, 151
information
    accessing information in the undermind
        117–18
    and the brain 134
    and the business world 209–10
    and consciousness 117, 211
    discovering new patterns or meanings
        within 49
    dissonant 77
    distraction from crucial 109
    insistence on high-quality 72
    making judgements/decisions using
        inadequate 72–5
        wild guessers 73, 74
    'marginal' 211
    as not always an asset 33
    and the preparation phase 94, 149
    thinking and 206
    unconsciously drives perceptions and
        reactions 38
    *see also* knowledge
Information Super-Highway 14
information technology 206, 207
inner states, focusing on 165, 169–72, 193
innovation 212, 213
insight 49, 60, 63, 86, 93, 105
insight problems 88, *89*, 90–91, *236*
inspiration 59, 71, 96, 149, 223
intellect *see* d-mode
intelligence
    computers and 206
    and curiosity 19
    emotional 16
    helps animals to survive 16–17
    and learning 17–19
    'multiple intelligences' 16
    practical *see* know how
    resurgence of interest in the concept 16
    sensorimotor 21
    unconscious 20–21, 44, 223
    and verbalisation 91
intelligence test 148
intention 129
intolerance 195
intuition 47, 49, 220
    consciousness and 211
    creating conditions conducive to 78, 84,
        211–14
    creating a diversity of new forms 95
    d-mode and 8, 53–5, 57, 86, 149
    defined 50
    disparaged 50

intuition – *cont.*
   Earth example 52–3
   fast 51, 53, 55
   the language of 57
   and learning by osmosis 67
   mistaken 35–6, 55
   optimum conditions for 75
   and patience 49, 57–8
   as provisional 50
   refuses to be managed 95–6
   schools and 21, 218, 219
   in science 56–8, 219
   in shadowy, intricate or ill defined
      situations 56, 72, 96
   slow 55–6
   and thinking 86, 148, 214
   and the undermind 50, 53
   *see also* ideas; imaginative seeds
Intuitive Manager, The (Rowan) 210
IQ (intelligence quotient) 20, 209

Jacoby, Larry 109, 110
James, Henry 70
James, William 102
jar puzzle *see* Luchins
Jastrow, Joseph 105, 106, 113
Jaynes, Julian 203
Jesus Christ 191
'Judas eye' analogy 117
judgement 188, 190, 191, 193, 194
Jung, Carl Gustav 224
justification, d-mode's concern with 8

Kabat-Zinn, Jon 183, 185
Kafka, Franz 175–6
Kahneman, Daniel 55
Kanizsa, G. 180, *181*
Karmiloff-Smith, Annette 45
Keats, John 172, 174, 178
Kegan, Robert 194
Kekulé von Stradonitz, Friedrich 57, 93
Kierkegaard, Sören 193–4
Kihlstrom, John 128, 159
Kipling, Rudyard 78
know how
   'bundling' of 43
   and confidence in ability 30
   'formatted' differently to knowledge 39,
      41
   good for different kinds of purposes 39
   implicit 20–27
   and IQ 20
   and knowledge 19, 36, 39, 41
   and learning 203
   and measures of 'conscious intelligence'
      20
   as not articulated 43–4
   and the polar planimeter 40
   relationship with conscious
      comprehension 30
   and slow knowing 214–15

   tied to particular domains 42–3, 46
knowing 171–6, 178, 223
   and perception 164–5
   procedural 98
   slow 3, 4, 6–7, 11–14, 176, 203,
      212–15, 218–19, 222, 226
   subjective 97
knowledge
   conceptual 33–4
   explicit 30, 31
   implicit 30, 31, 36
   and know how 19, 36, 39, 41
   practical 42
   subliminal 62
   taught as if certain 219
   taught as open to question and revision
      220
   the undermind acquires 37
   working 33–4
   *see also* information
'Knowledge, knerves and know how'
   (Masters) 38
Koch, Christof 158
Koestler, Arthur 224
Korzybski, Alfred 47
Kreutzer, Conradin 207
Kruglansky, A.W. 76
'Kubla Khan' (Coleridge) 59–60, 95
*kufū* 172
Kunst-Wilson 118

'labour-saving' devices 5
Labouvie-Vief, Gisela 190
Ladakh 5
Laing, R.D. 182
Langer, Ellen 126, 181, 182, 220
Langer, Suzanne 3
language 10
   and complexity 11–12, 25, 30
   of d-mode 46
   imposes a particular timeframe on
      cognition 10
   incomprehensible sentences 11–12, 25,
      30
   learning 45
   liberation by 46
   and 'reality' 46
   and rigidity 46
   and the world of experience 153–4
   the world seen through 10–11
lateral masking 121
lateral thinking 212
Lawrence, D.H. 83
*Laws of Form* (Spencer Brown) 58
learning
   d-mode and 8, 42
   discovery 220–21
   and earning 201
   emerging in a gradual, holistic way 8
   explicit 38–9
   good 221

growth of learning power through
  experience 222
imagination as a learning tool 219
intuitive 38–9
  language 45
'learning beyond success' 45–6
resilience 216–17, 218
as a risky business 203
slow 8–9
as a survival strategy 17–18
and uncertainty 6, 17–18
worldwide concern with 202
learning by osmosis 20–27, 49
d-mode and 21, 34, 35
and discovery learning 220
and intuition 67
limitations of 42–3
nature of 26–7
need for 20
and neural networks 143
and slow intuition 56
learning curriculum 215–22
learning society 201, 202, 215, 222
Leibniz, Gottfried 3, 100
*Letters to a Young Poet* (Rilke) 176
Levi-Montalcini, Rita 56–7
Lewicki, Pawel 23, 25, 26, 36–8
Lewicki experiments 23–5, *24*, 30, 46
lexical decision task 62
Libet, Benjamin 158, 160
*Life and How to Survive It* (Skynner and
  Cleese) 192
Locke, John 205
logic 12, 93, 95
logical tasks 34, 35
long-term potentiation (LTP) 137
Lorenz, Konrad 57–8, 151
Lowe, John Livingston 59, 114
Lowell, Amy 60, 67, 96
LTP *see* long-term potentiation
Luchin, Abraham 51
Luchins, Edith 51
Luchins jar puzzle 51, 52, 71, 76, 131
Luria, A.R. 153

MacNeice, Louis 178
'Magic Eye' images 174
*Man Who Mistook his Wife for a Hat, The*
  (Sacks) 179
Mandela, Nelson 189–90
Maori *marae* 4
map-reading 46–7
Marcel, Tony 120, 121, 122, 127
Maritain, Jacques 175, 205
Martindale, Colin 148, 149
Masling, Joseph 116, 117
Masters, R.S. 38–9
Meacham, John 195
medicine 167–9
meditation 4, 186, 198, 213, 214
*Meditations* (Descartes) 205, 224

memory 9–10
amnesiacs and 119, 123
childhood 82, 124
degraded 19
and *déjà vu* 109
face recognition 92
false memory syndrome 110
and incomprehensible sentences 11
memory tests 118
retention of ideas 94
and Rubik's cube 30
thinking and 86
and the TOT state 61, 62
*Merchant of Venice, The* (Shakespeare) 204
Meredith, George 116
metaphor 10, 175, 179
Meyer, D.E. 62
midbrain 146, 158
*Midsummer Night's Dream, A* (Shakespeare)
  204
Milne, A.A. 48
mind
becomes stuck in one mode or the other
  96–7
beginner's (*shoshin*) 198–9
current image of the mind 222–3
expanded theatre image 225
historic changes in view of 204–5
implicit identification with consciousness
  115
shift in understanding of the 226
slow *see* slow mind
swinging flexibly between conscious
  thought and intuition 96, 99
as 'the theatre of consciousness' 222,
  224–5
three different processing speeds 1–2
*see also* undermind
*Mind of a Mnemonist, The* (Luria) 153
mind-body dualism 223
mindfulness 165, 180–87, 193, 194, 219
Mintzberg, Henry 211
mnemonics 13
Moore, Henry 95
Morelli, Giovanni 165, 167
'Moses of Michelangelo, The' (Freud) 167
motivation 131
Mott, Sir Neville 57
Mozart, Wolfgang Amadeus 94
Muller, Catherine (known as Helen Smith)
  113–14
mutation 95
'mutilated chessboard' 35–6, *36*, 55, 88

*Native American Medicine Cards* 1
near-death experiences 114
needs 17, 134, 145, 155
negative capability 174
Neill, A.S. 188, 189
Neill, Mrs 188, 189
neocortex 158

neural Darwinism 95
neural networks 140, 141–3, *142*, 151, 152, 161
neural pathways 151, *152*
neuro-transmitters 135, 137
'neuromodulation' effect 146
neurons *136*
  and amines 146
  in the brain 134
  change in electrical communication bewteen 137
  clusters of 138, 144, 145, 146, 149, 150, 154–5, 157
  cortical 144
  electrical impulses 134–5
  'extended family' of 139
  groups working in synchrony 145
  reciprocal inhibition 145
  response to subliminal stimuli 157–8
*New York Times* 71
Newton, Sir Isaac 58
NMDA (N-methyl-d-aspartate) receptor sites 137
Noddings, Nel 58
Norberg-Hodge, Helena 4–5
norepinephrine 146

observation 49, 210
  d-mode values explanation over 7–8
  developing powers of 218
  limits to powers of 25
  as a major learning vehicle 33
'Ode to Immortality' (Wordsworth) 175
'old age' 181–2
'On "Having" a Poem' (Skinner lecture) 68–9
opportunities 134
*Organization of Behavior, The* (Hebb) 137
out-of-body experiences 114

pain, relief of 124–5, 182, 183–4
paralinguistic cues 114
Pascal, Blaise 224
passivity 197
past lives, reversion to 113–14
patience
  cultivating 84
  focusing and 172
  and intuition 49, 57–8
Patton, C.J. 108
perception(s) 182, 193
  coarsening of 131–2
  and consciousness 106, 126–7
  d-mode treats as unproblematic 7
  degraded 19
  as diagnostic in d-mode 164
  effects of 'backward' and 'lateral masking' on 121–2
  and imagination 81
  inability to articulate 44
  information and 38

  and intuition 130
  power of subliminal forces on 204
  seeking perceptual intensity 186–7
  skimping on 179–80
  subliminal 100, 101, 107, 108, 117, 119–20, 150–51
  unconscious 100–115, 117, 150
  visual 101, 104
perceptual defence 117
Peters, Tom 209
philosophy 3, 4, 226
Piaget, Jean 21
Picasso, Pablo 84
Pierce, C.S. 105, 106, 113
pineal gland 157
Pinel, Alain 209
Pittman, Thane 101, 104, 107, 108, 111
Planck, Max 79
Planck's dictum 79
plasticity 18, 19
*Platform Sutra* 199
Plato 203
Plotinus 204
poetic sensibility 165–9, 193
poetry 4, 99, 176–9, 206
  Coleridge and composition 59–60
  d-mode and 10, 94–6, 226
  Housman on 69–70
  and intelligence 17
  Lowell on composition 67
  and slow knowing 3, 176
Poetzl, Otto 106, 107
Poincaré, Henry 60, 93–4
polar planimeter 39–40, *40*, 43
Postman, Neil 5, 206
practice, as a major learning vehicle 33
*prajna* (wisdom) 199–200
premature cognitive commitments 182
preparation
  and creativity 94, 149
  unconscious 104
Price, Mark 106–7
Priestley, Joseph 85
Prince, George 77, 78
prosopagnosia 157
*psyche* 203
psychoanalysis 129–30, 166, 167
psychology 9
psychotherapy 3, 9, 169–70, 193, 224

reactions, quick 51
reality
  conscious interpretation as 112
  hallucinations 124, 180
  language and 46
  a play and 225
reason 95, 203, 204, 205, 226
Reber, Arthur 26
rebus problems 60–61
receptivity 57, 170, 175
recognising faces 92, 93

'Recommendations to physicians practising psychoanalysis' (Freud) 129–30
'Reflections' (Kafka) 175–6
reflexes 17
reframing 182, 189, 194
Reiser, Stanley 167–8
repression 161
resilience, learning 216–17, 218
resourcefulness 218, 221
retrograde amnesia 118–19
Rilke, Rainer Maria 176
*Rise and Fall of Strategic Planning, The* (Mintzberg) 211
Rogers, Carl 193
Rokeach, Milton 52, 75–6
Rorschach ink-blot test 77
Rosli, Philop Kapleau 164
Ross, Doug 201–2, 215
Rowan, Roy 210, 211
Royal Society of Arts 202
Rubik cube 28–30, *29*
rules of thumb 37
rumination 45–6, 47, 49, 206, 209
Ruskin, John 166

saccades 104
Sacks, Oliver 179
Sahn Sunim, Seung 199
Sartre, Jean-Paul 78
Schiller, Johann von 78
scholastic fallacy 41
Schon, Donald 75
Schooler, Jonathan 86, 88, 90–93
science 226
    the creative stimulus 70
    the imaginative stimulus 70
    inertia of 79
    intuition in 56–8, 219
    and slow learning 9
*Scientific Monthly, The* 67
Scott, Nathan 201
self
    being unconscious of 196
    changing the sense of self 126
    conscious 223
self-consciousness 116–32, 158–9
self-image 126, 159
self-monitoring 110
'Self-reliance' (Emerson) 81
Selzer, Richard 168, 169
sensory deprivation 19
serotonin 146
Shakespeare, William 204
*shoshin* (beginner's mind) 198–9
Sidis, B. 105
Singer, Jerome 131
Singer, Wolf 144
sixth sense 113
Skinner, B.F. 68
Skynner, Robin 192–3
slow mind

associated with creativity or 'wisdom' 3
decline of slow thinking 4–7
defined 2
used for intricate, shadowy or ill defined situations 3
as vital 2
slow ways of knowing *see under* knowing
Smith, Helen (Catherine Muller) 113–14
Smith, Steven 60, 61–2, 63
'snap judgements' 51, 92
*Social Statics* (Spencer) 48
soul, the 99, 157, 174, 204, 224
special relativity theory 56
speculation 77–8
Spence 150–51
Spencer Brown, George 58
Spencer, Herbert 48–9, 50, 56
Spender, Stephen 71, 78
Spinoza, Benedict 3
'split brain' patients 127
*Spoils of Poynton, The* (James) 70
Stein, Gertrude 68
stereotypes 109, 110–11, *111*, 195
Sternberg, Robert 20, 194
stress 76, 130–31
'study skills' programmes 221–2
subconscious 223
subliminal advertising 100
subliminal perception *see under* perception
Summerhill school 188, 189
*Sur l'Intelligence* (Taine) 224–5
Suzuki, D.T. 172, 199–200
Suzuki Roshi, Shunryu 198–9
synapses 134, 135, 137, 146
'Synectics' programme 77

Taine, H.A. 224–5
*Tao Te Ching* 3, 84
Tauler, Johannes 196–7, 200
Taylor, Charles 76
Teasdale, John 184–5
technology 223
    and a sense of urgency 9
    and technopoly 5
technopoly 5, 7, 206
telepathy 105, 113, 114
Temple, Archbishop William 219
*Tender is the Night* (Fitzgerald) 113
thalamus 158
Theories of Everything 6
theta waves 148
thinking
    calculative 207, 208
    and consciousness 206
    drawbacks of analytical thinking 86–8
    and ideas 206
    and information 206
    and intuition 86, 148, 214
    lateral 212
    low-focus 148
    meditative 207, 208

thinking – *cont.*
  scientific 223
  slow 5, 214
  Spencer's mode of 48–9
  thinking fast 52
  thinking freshly 51
  thinking too little 94
  thinking too much 85, 94
  thinking what is thinkable 91
thought(s)
  d-mode and 7, 10
  fleeting 80
  and intuition 96
  logical 42
  and the mind's three different processing
      speeds 1–2
  pattern of thought gradually forming itself
      48–9
  perversion of 49
  sudden insight 56–7
threats 17, 134, 145, 155, 161, 213
*thymos* 172
time
  changing conception of and attitude
      towards 4–5
  and creativity 52, 75, 76–7
  'saving' 5
'tip-of-the-tongue' (TOT) state 61–2, 64,
    151
trial and error 32, 42
tunnel vision 130
Tversky, Amos 55
Tyrol, Duchess of 191

'Unborn' 3
uncertainty
  and intuition 74–5
  learning and 6, 17–18
unconscious
  accomplishment of tasks 4
  and conscious/consciousness 18–19, 37,
      63, 72, 116–17, 124, 199–200, 204,
      226
  and consciousness 64, 66, 223–4
  history of the 197
  misleading about an experience 109
  neglect of 6–7
  new conception of 223
  regarded as wild and unruly rather than
      as a valuable resource 7
  relationship with one's 3
unconscious perception 100–115, 117,
    150
unconsciousness, tactical 122
undermind
  ability to register events and make
      connections 113
  accessing information in the 117–19
  acquires knowledge 37
  ambivalent relationship with 203–4
  becomes active 116

concerned with survival and wellbeing
    119
  and consciousness 37–8, 81, 107, 116,
      124
  and *déjà vu* experiences 109
  evidence for 67
  and ideas 95
  and information 25
  and intuition 50, 53
  key to the 13–14
  makes adjustments to the data it receives
      104
  neglected by modern Western culture 7
  news flashes by 103
  and problem-solving 63
  promptings of the 72
  relocation of the centre of identity and
      intelligence to 198
  research on cognitive capacity of 226
  respecting the powers of the 114
  and self-consciousness 128
  and subliminal knowledge 62
  and wisdom 195
understanding
  'articulated' 43
  d-mode and 7
  and language complexity 11–12
University of Birmingham ('Brains, Minds
    and Consciousness' symposium) 133
University of Edinburgh Psychology
    Department 104

Valéry, Paul 28, 177
verbalisation 90, 91, 92, 120
verification 94, 149, 150
Viesti, Carl 79–80
vision
  impaired night vision 130
  peripheral 131
  visual acuity, and sense of self 126
visualisation 13, 219
Voltaire 195

Wallas, Graham 94
Wason, Peter 31–2
Watzlawick, Paul 23
Weil, Simone 99
Weiskrantz, Lawrence 127
Westcott, Malcolm 72–5
Whalley, George 177
Whistler, J.M. 166
Whitehead, A.N. 15
Whyte, Lancelot Law 148, 197–8, 205, 223
Winner, Ellen 175
wisdom
  acting with knowledge while doubting 195
  allowing oneself time 192–3
  Buddhism and 198
  concept of *prajna* 199–200
  and d-mode 191, 195
  defined 188, 191

and fundamental values 190–91
icons of 4
seeing through the apparent issue to the real issue underlying it 189, 190
and the slow mind 3
and the undermind 195
the wise individual 194
works with 'the big picture' 190
'wits' 2, 6, 21, 30, 43
Wittgenstein, Ludwig 52
*Women's Ways of Knowing* (Belenky *et al*) 97

word association 64–5
wordscape 153–6, *154*
Wordsworth, William 94–5, 175

Yale Medical School 168
Yaniv, I. 62

Zajonc 118
Zeki, Semir 158
*Zen and Japanese Culture* (Suzuki) 172
Zen Buddhism 191, 198–9

All Fourth Estate books are available from your local bookshop,
or can be ordered direct from:

Fourth Estate, Book Service By Post, PO Box 29,
Douglas, I-O-M, IM99 1BQ

*Credit cards accepted.*

Tel: 01624 836000    Fax: 01624 670923

Or visit the Fourth Estate website at: www.4thestate.co.uk

*Please state when ordering if you do **not** wish to receive further
information about Fourth Estate titles.*

# RADIO MY WAY

## FEATURING CELEBRITY PROFILES FROM JAZZ, OPERA, THE AMERICAN SONGBOOK AND MORE

*By*

## Ron Della Chiesa

*With*

## Erica Ferencik

**PEARSON**

Boston   Columbus   Indianapolis   New York   San Francisco   Upper Saddle River
Amsterdam   Cape Town   Dubai   London   Madrid   Milan   Munich   Paris   Montréal   Toronto
Delhi   Mexico City   São Paulo   Sydney   Hong Kong   Seoul   Singapore   Taipei   Tokyo

**Editorial Director:** Daryl Fox
**Editor-in-Chief, Communication:** Karon Bowers
**Editor:** Ziki Dekel
**Development Editor:** Kay Ueno
**Assistant Editor:** Stephanie Chaisson
**Associate Managing Editor:** Bayani Mendoza de Leon
**Executive Marketing Manager:** Wendy Gordon
**Marketing Manager:** Phil Olvey
**Project Coordination, Text Design, and Electronic Page Makeup:** Integra
**Manufacturing Buyer:** Mary Ann Gloriande
**Senior Cover Design Manager/Cover Designer:** Nancy Danahy
**Cover Images:** Top, © Charlies Giuliano; bottom and author photo on the flap, © Fred Collins; co-author photo on the flap © Nina Huber

Credits and acknowledgments borrowed from other sources and reproduced, with permission, in this textbook appear on page 290.

**Library of Congress Cataloging-in-Publication Data**

Della Chiesa, Ron.
  Radio my way/Ron Della Chiesa; with Erica Ferencik.
    p. cm.
  Includes bibliographical references.
  ISBN-13: 978-0-205-19078-2
  ISBN-10: 0-205-19078-2
    1. Della Chiesa, Ron.  2. Radio broadcasters—United States—Biography.
  3. Radio and music—United States.  I. Ferencik, Erica.  II. Title.
  ML429.D46A3 2011
  791.44092—dc23
  [B]                                                    2011031836

1 2 3 4 5 6 7 8 9 10—STP-Courier—15 14 13 12

ISBN-10:     0-205-19078-2
ISBN-13: 978-0-205-19078-2

*To Joyce: you're all the world to me.*

92
DELLA
CHIESA

# table of contents

*Foreword    ix*

**CHAPTER 1** Radio Kid    1

**CHAPTER 2** Finding My Voice    7

**CHAPTER 3** Steppin' Out    15

**CHAPTER 4** Nice Work If You Can Get It    31

**CHAPTER 5** A World of Joyce    45

**CHAPTER 6** I Miss You in the Afternoon    60

**CHAPTER 7** Fly Me to the Moon    71

**PROFILES    89**

**I. The World of Opera    89**
    1. Luciano Pavarotti: *King of the High C's    89*
    2. Placido Domingo: *The Incomparable    92*
    3. Carlo Bergonzi: *Viva Verdi!    95*
    4. James McCracken: *Esultate!    98*
    5. Ben Heppner: *Hip Heldentenor    101*
    6. Marcello Giordani: *La Dolce Vita    105*
    7. Norman Kelley: *"Knew Them All!"    108*
    8. Richard Cassilly and Patricia Craig: *The Opera Couple    110*
    9. Jose Van Dam: *The Singing Actor    113*
    10. Robert Merrill: *Star Spangled Baritone    115*
    11. Jerome Hines: *Devils and Demons    118*
    12. Eleanor Steber: *West Virginia Diva    121*
    13. Eileen Farrell: *The Feisty Soprano    124*
    14. Aprile Millo: *After the Golden Age    128*

## II. All That Jazz  130

1. Dizzy Gillespie: *The Diz*  130
2. Benny Carter: *Gentleman Ben*  134
3. Illinois Jacquet: *Jacquet's Got It*  138
4. Stan Getz: *Hot, Cool, and Swinging*  142
5. Buddy Rich: *Super Drummer*  146
6. Joe Venuti: *The Irascible*  150
7. Lionel Hampton: *A Lifetime of Swing*  154
8. Ruby Braff: *Boston's Own*  157
9. George Shearing: *It's Shearing You're Hearing!*  161
10. Dave McKenna: *Three Handed Swing*  165
11. Dick "Spider" Johnson: *Born to Play*  169
12. Joe Williams: *Boss of the Blues*  172

## III. The Great American Songbook  176

1. Tony Bennett: *The John Singer Sargent of Singers*  176
2. Bobby Short: *The Embodiment of Elegance*  184
3. Frank Sinatra, Jr.: *His Own Way*  189
4. John Pizzarelli: *Style's Back in Style*  195
5. Mel Torme: *"Don't Call Me the Velvet Fog!"*  199
6. Nancy Wilson: *The Thrush of Columbus*  203
7. Rosemary Clooney: *Blue Rose*  207
8. Sammy Cahn: *A Song's Best Friend*  212

## IV. Conductors and Composers  217

1. Arturo Toscanini: *The Maestro*  217
2. Gunther Schuller: *"Where the Word Ends"*  224
3. Harry Ellis Dickson: *Wild About Harry!*  229
4. David Raksin: *Grandfather of Film Music*  234
5. Andre Previn: *Musical Wunderkind*  238

## V. Comics, Actors, and Everybody Else  242

1. Jean Shepherd: *A Voice in the Night*  242
2. Mort Sahl: *If You're For It, I'm Against It*  250
3. Ernest Borgnine: *The Importance of Being Ernest*  256
4. Chuck Jones: *What's Up, Doc?*  262
5. Robert B. Parker: *Dean of American Crime Fiction*  269

## Fabulous People/Simple Food  274

by Joyce Scardina Della Chiesa

*Acknowledgments  281*
*Endnotes  285*
*Credits  290*

Let me begin with a warning: if you can plow past this foreword, you won't want to put this book down. Here he is, for once without a microphone, as breezy, informative, and witty as you've heard him all these years. If you like Ron Della Chiesa, and who doesn't, you're going to love this book.

I've known Ron Della Chiesa since I began as a production assistant at WGBH in 1985. I remember the first time I was asked to work with him. It was to direct fundraising on *MusicAmerica*; and with this assignment came a warning: he doesn't take direction, and he doesn't follow the fundraising mandates—including time limits on pitching segments. Somehow I sensed, correctly as it turned out, that there was more of a problem with the fundraising directive than with the host in question.

Having experienced successful fundraising drives at other stations, I knew that the "accepted" techniques didn't always fill the coffers. With Ron I sensed the opportunity to go for a humorous approach. In no time we had worked up a shtick where I was giving "updates" from Fundraising Central (our term), with a recording of an old-style teletype chomping in the background. Even though I was in the same room with him, we'd pretend that it was difficult for us to hear each other, or to know if I was on the air. What was the goal of the fundraising campaign? Were we close? How much closer? We were breaking each other up and having a ball. The phones would not stop ringing with callers eager to pledge large donations.

It may be difficult for some to believe that Ron's enthusiasm is genuine. But it is. Countless times he's burst out of the sound booth after a Boston Symphony broadcast pouring out his excitement, even more than what he conveyed on the air. He loves what he hears, his passion undiminished even after listening to thousands upon thousands of hours of musical performances from every genre imaginable. Most of us would be jaded over time. That is not Ron Della Chiesa.

Many books of this nature contain more than a dollop of correcting past wrongs for posterity. That isn't Ron Della Chiesa either. If you're reading this now only to see if you are the worthy victim of a score-settling, you may as well put this back down. It isn't here.

It's a good bet that if you are mentioned, you have a place of honor. And if you have a section of this book devoted to you, you have a vaunted place in

Ron's vast pantheon of musicians and personalities that have enriched his generation, and I trust, generations to come. My congratulations!

## —Brian Bell

*Brian Bell produced the Boston Symphony radio broadcasts beginning in the fall of 1991, and wrote every script Ron Della Chiesa read for those programs, as well as concert presentations by the Boston Pops, the Handel and Haydn Society, the Boston Philharmonic, and OperaBoston.*

# Radio Kid

THERE ARE TIMES WHEN I LOOK BACK AT MY LIFE TO DISCOVER where this passion for radio began, only to realize there's no mystery at all. My beginnings were storybook-perfect to grow the seeds of an obsession for broadcasting. Ever since that first day I sat in awe in front of a radio, I wondered how I could become part of the magic of the voices inside the box.

I was born on February 18, 1938, in Quincy, Massachusetts, a city known as the birthplace of two presidents (John Adams and John Quincy Adams, both of whom had lived just a mile from our home on High Street) and for its stone quarries, where my grandfather toiled many long hours. In my neighborhood, the two most important things to kids after school were sandlot baseball and getting home in time to listen to their favorite radio programs. After all, this was the

First dip in the ocean, Nantasket Beach, 1939.

Author's collection.

1

Author's collection.

Golden Age of Radio, a time when radio was the focus of entertainment right up through the advent of television in 1948.

We would race home to listen to the *Lone Ranger, Superman, Captain Midnight, Duffy's Tavern, Tom Mix and the Ralston Straight Shooters,* and *The Shadow.* These wondrous programs filled our afternoons and evenings with entertainment. Everything happened in our minds: that was the magic and the mystery of radio. Our charged imaginations kept us hooked day after day. Who did Captain Midnight look like? What about his evil enemy, Ivan Shark? And Tom Mix, the cowboy? We could send away for a picture, but we were too busy building our own movie sets in our minds, conjuring the faces of our heroes, their loves, and their arch enemies. Each program would end on a cliffhanger, leaving us breathless from one day to the next. I'd lie in bed staring in the dark worrying about Superman: Would kryptonite trip him up and keep him from springing Lois Lane from the evil clutches of those who were terrorizing the city?

Most of the shows were sponsored by drinks or cereals: Ovaltine sponsored *Captain Midnight,* Ralston and Farina backed *Tom Mix.* So many labels from Ovaltine got you the Captain Midnight secret decoder ring. But when you finally got a ring, the message was always the same: "Buy Ovaltine!"

I confess I still have my ring and an embarrassing number of premiums in an envelope somewhere. To this day I get excited when I see a lumpy envelope in the mail! Hell, it might be a decoder ring or prize. A pair of glow-in-the-dark Tom Mix spurs, anything's possible. It boiled down to scoring boxtops, labels, and premiums;

so I had my mom, a very Catholic woman, stealing Ovaltine labels in the store. But that's what it was like being an only child for me.

My parents would go to hell and back to make me happy.

■ ■ ■

My mother, Florence, and my father, Aldo, had been married a good thirteen years before I was born. As an only child, I became the focus of their life. They both grew up in Quincy and became schoolteachers. My earliest memories of my father, however, really begin at the end of World War II, in 1946. I didn't see much of him before that time because he was called up from the reserves shortly after Pearl Harbor in 1941.

My mother described the day my father caught the train to report for active duty. We went with him to South Station. This was in 1942, and I was just four years old. Crying, I begged my father not to leave and ran after him into the train. It was not an easy day. A kind-hearted couple who'd been watching approached and offered to drive us home or help us in whatever way we needed, but my mom declined the offer and just thanked them. We went home, just the two of us, where we had to build our lives together until he returned.

Those early days were a bittersweet mix of missing my dad and enjoying a quintessential New England childhood that was tinged with all that World War II brought to the populace. People tightened their belts; Rosie the Riveter rolled up her sleeves. I remember lying in bed listening to the sounds of destroyers being built in the Quincy shipyard nearby, a rhythmic pounding that went on all night long. I'd gaze out my window at a sky lit ghostly white from the factories below. We'd ration stamps, buy savings bonds, and save scrap metal. A star was displayed in the windows of the families who had a brother or a father serving overseas.

My mother, Florence Della Chiesa, 1922.

Author's collection.

We sat in the dark for air raids, waiting for the all-clear to turn our lights back on again. I remember my photograph being taken in a little army uniform, then gazing at it the next week in the newspaper where my mother had proudly sent it.

Though times were hard for my mom, she definitely knew how to take care of business. She handled all the practical matters: the finances, the shopping, everything required of a World War II housewife. I think she tried to keep things as "normal" as possible for me, but we both missed Dad.

Meanwhile, I'd listen to the radio.

■ ■ ■

Our radio was the centerpiece, the beating heart of the living room, in fact, of the whole house. At seven years old I became a regular listener of the Metropolitan Opera on Saturday afternoons. My mother would be cooking in the kitchen on snowy winter days when you couldn't get out, while thundering, passionate voices filled the rooms, warming us. As young as I was, I would have given anything to see an actual opera.

The voice of Milton Cross was magic to me. He had a way of telling the story of the opera and describing the action on stage that brought it fully alive. He would capture the colors of the costumes, the way the great gold curtain parted to magnificent scenery, the way the singers moved, and the reaction of the audience. The bravos, the pageantry, all the romantic or wrenching drama were forever etched in the theater of my mind. I thought at the time that it would be the best sort of life to be an opera singer. I eventually realized that I didn't have the voice. But from those earliest childhood days, I felt that someday I just had to be on the radio.

I was so inspired by radio programs that one day I built a fake radio station in my bedroom. I had a turntable, 78 rpm records, and a little cardboard microphone. I'd write my own scripts, create my own shows, commercials, and man, did I impress myself!

I had the radio bug, and I had it bad.

■ ■ ■

My dad came home in 1946; I was eight years old. It was then that I really began to get to know him. The truth was, my father was a dreamer. A graduate of Massachusetts College of Art in the twenties, he thought only of painting, travel, and the beautiful places in the world he wanted to visit someday. For him, it was all about aesthetics. My father really was one of these inherently gifted Renaissance men; he certainly wasn't someone out to make a lot of money or change the world. He lived in his own world of art, music, drawing, paintings, sunsets, and beauty in general. When he wasn't teaching art and mechanical drawing at Braintree High School, he'd be painting or drawing or indulging his love of art by visiting museums. He told me he once spied John Singer Sargent walking along Huntington Avenue and followed him all day, though he was so in awe of the man that he couldn't bring himself to approach him or say a word.

My father, Aldo Della Chiesa, 1920.

As young as I was, it was hard to understand the notion of my father at war and his necessary transition to civilian life. I now know that part of his recovery from the trauma of what he had seen those past four years was to immerse himself back into his world, especially that of music and painting. He also couldn't wait to be a father and husband again.

On Saturday afternoons the radio would always be tuned to the Metropolitan Opera, but now my dad added a whole new dimension to the experience. Not only could he paint, he could sing all the arias from the operas even though he couldn't read a note of music. He had a collection of 78 rpm records that he played on our wonderful cherrywood Victrola; his parents had brought it with them from Italy. To this day I listen to Caruso recordings on that very same turntable—incredible how the sound fills the room!

Dad would sing while he was shaving, sing at the breakfast table, sing as he got dressed for work. He'd put on a record and belt out arias from Verdi, from Puccini, and sing along with all the great tenor voices from Italian opera: Caruso, Gigli, and Di Stefano. In a way he was like my dear friend Tony Bennett: when he wasn't painting he would sing; when he wasn't singing he was sketching or painting.

My father began to take me into Boston on Sundays. First stop was always art supplies, his brushes and paints, followed by a museum, after which I would beg him to take me to radio stations. I was obsessed with wanting to know how a radio station looked, how it was all laid out, who did what, and how everything came together to create shows. Usually he'd give in to my pleading, and we'd go visit one of them. Many of the stations at that time were in hotels. WBZ was in the Hotel Bradford in the theater district; WNAC was in the Hotel Buckminster in Kenmore Square near Fenway Park. A radio station in a hotel was such a romantic notion; tourists would be checking in on one floor, while the floor above would be dedicated to a station. A plate glass window was the only thing separating an orchestra with a live studio audience from the broadcast booth, an announcer, and gaggles of wide-eyed tourists and visitors.

It was my father who took me to my first opera, *La Boheme*, in 1946, at the old Boston Opera House on Huntington Avenue, not far from Symphony Hall. It was a beautiful building, patterned after the great opera houses in Europe. It has since been torn down, but there remains a little street called Opera Place, its original home.

This is where it all started for me, on that hot June afternoon in Boston, where the sounds that moved me while listening to the radio and my father's records finally became wed to a live performance. We sat way up and far in the back on one of the several balconies that hugged a horseshoe-shaped stage, peering down at the performers through binoculars. My ears vibrated from the sheer power of their voices; even at that young age I was stirred by the dramatic emotion behind the words. I sat in awe as love, betrayal, suffering, and joy seemed to explode all around me.

Soon after this day we took a trip to New York City, venturing past the old Metropolitan Opera house where so many greats had sung—Lilly Pons, Beniamino Gigli, and Rosa Ponselle. From there we went backstage on 7th Avenue where all the scenery was stored. The back door to the stage was open, and I turned to my dad and asked if we could go in. He found one of the stage hands and got permission to walk out on the stage. We passed through a dark, musty hallway to the back entrance. Finally, I stood there on the well traveled blond wood—it seemed to go on endlessly until it disappeared into darkness at the back of the house. It was magic. My father grinned at me and said, "Come on, Ronnie, why don't you sing a few notes?"

My heart pounded as I gazed up at the famous golden horseshoe, but I think I sang a little bit of Pagliacci. The great hall seemed to hold my voice in its giant hands, echoing it softly back at me. Now I can say that I sang on the stage of the old Metropolitan Opera house, the same stage graced by the great Caruso.

What a nice way for a Quincy boy to begin a relationship with opera.

## CHAPTER 2

# Finding My Voice

**F**ANTASIZING ABOUT A CAREER IN BROADCASTING COULD DISTRACT me for hours, but eventually I had to get outside and play ball, literally. And I found that real life was different out on the sandlot. Don't get me wrong—I had all the baseball cards, and I loved the Red Sox. The truth was, I was a terrible baseball player. It wasn't for lack of trying, or for lack of love of everything about the game, especially the getup for the catcher, but baseball greatness was like

Can't run, can't hit, love Red Sox.

Author's collection.

7

girls: unattainable. I loved nothing better than the crack of the bat, skidding into bases, sun in my eyes and the mound under my feet, but you see, I lacked speed. I'd field a groundball, but without fail it bounced back up and hit me in the face. It drove me flipping crazy. As a kid, my arm just wasn't strong enough, and there was no one around to really show me the ropes. Remember, there was no Little League back then. I recently complained about my problem to an old friend, former Red Sox great Sam Mele. He said, "How'd you like to have a ball coming at your head at ninety-eight miles an hour? No, seriously, you need someone to coach you, end of story. Every day I was out there I had my brothers to teach me." In any case, I could count on one hand the number of times I got a piece of that baseball. And I didn't have a clue about football—except that football players really did get all the girls.

How to get the girls: now *there* was an enigma that took years to decipher...

As I was screwing up in right field or brooding in the dugout, I'd picture one of my favorite Red Sox players, Dom DiMaggio, Joe DiMaggio's younger brother. I liked him because he seemed as if he should have nothing to do with playing baseball. He was this scholarly looking guy they called "The Little Professor," not at all rough and tumble or even very big at 5'9", playing out there in center field with his glasses and glove, holding his own, and a great ball player. He had real style. He actually answered a letter I'd sent him when I was ten years old, saying how much *he* appreciated hearing from *me*.

It was a day I'll never forget.

■ ■ ■

Childhood memories become especially precious for most of us as we grow older. Mine flit through my mind, often unbidden:

The comfortable cape house my dad built for the grand sum of five thousand dollars; the steam engines that chugged by the playing fields after school; the pet parakeet that got a weird growth and died but sang his tail off for the five years we had him; the canary that didn't sing, but for some reason we named Perry Como; bowling at Kippy's Candlepins in Quincy; building a tree house with stolen lumber and installing a makeshift stove with a pipe; dropping a sheet from the branches and showing movies with a five-cent "entrance fee"; being confounded by algebra and geometry but entranced by Richard Halliburton's *Book of Marvels*; starting a "hamster club" and, unfortunately, baking a few of them in the hot summer sun; eating my mom's great desserts: apple pie, corn muffins, lemon meringue pie, fruitcake, Toll House cookies; and my mom calling me home in the evenings for dinner. Yes, they really did stuff like that back then, just as in the *Henry Aldrich* radio show. The thing was to play it at least a little bit cool and not run back home right away....

As it turned out, I can't mourn my days on the baseball field too much. I mean, I was a different sort of a kid. How many eleven-year-olds in my neighborhood had seen their first opera and daydreamed about singing arias in front

First business venture, July 1947. L–R: John Ross, David Maglio, Charles Ross, me.

of thousands? I was seriously out of the mainstream on that count. What sort of pre-teen hangs out at the Public Library Music Room listening to Puccini, Verdi, Bjoerling, and Lanza for hours on end? I used to go to the one in Quincy. They had two or three booths where I could dig in and listen to all the opera and classical recordings we didn't have in our collection at home. I'd sit there and pour over the notes on the back of the album: the history of the composer, the story of the opera, the Libretto. Often I'd sitting there in the dark, gorgeous voices soaring in my head, until someone tapped me on the shoulder and told me the library was closing and it was time to go home.

Even my friendships were based on a shared love of music. By my teens, a small circle of friends had developed and drew quite close. Bob Buccini and I were thirteen, hanging at his place when we heard Paul Antonelli belt out this mindblowing rendition of "Old Man River." Here was this kid our age with this mature, full baritone operatic voice. I was slackjawed. His dad, Guido, was a teacher and prominent violinist who played Broadway music in pit orchestras, very well known in Boston music circles. We bonded over a passion for music, his Italian-American background, and the fact that he was an only child too, also with musical ambitions. Bob, who at age twelve had already lost both parents and lived with an elderly Italian aunt, used to come over to our house all the time to hang out and listen to music and, more often than not, sit down to dinner with us. My mother and father became a sort of surrogate set of parents for him. The three of us boys not only huddled around the Victrola listening to opera, but we did the usual kid things: get in trouble and chase girls. Even today we're in touch;

though I have to say they can find me, but I can't always find them. They'll hear me on the radio or see me on TV, and they'll call or email me. So wonderful to hear from friends I've been blessed with for over sixty years.

■ ■ ■

In the 1940s and '50s it was all about discovery and "firsts." There are so many I will never forget: my first animated film, *Pinocchio*; and my first live action, *Lassie Come Home*, which I saw with my mom at Camp Lee in Virginia where my Dad was based. I bawled my eyes out watching that film, because of Lassie, of course, but also no doubt because it was such an emotional time seeing my dad between his deployments all over the world.

Movies were huge in the early forties, especially since this was before television. Not only was it a big deal when a movie came out, but each showing was a major production. After you settled in with your popcorn and soda, things would kick off with a cartoon or newsreel, then a B movie, the coming attractions, and *finally* the feature. Imagine seeing all of this for just a quarter. It was a whole afternoon or a whole morning's worth of entertainment. You could really lose yourself in there; I know I did. At that time Quincy had four movie theaters: the Adams, the Strand, the Capitol, and the Art. We loved the Adams Theater most of all because that's where the weekly serials ran, each episode leaving you with a cliffhanger: someone jumping out of a building or holding a knife to the hero's neck. You could hardly wait to return the following week for the next episode.

Watching *Fantasia* was a profound experience for me, mainly because of the unprecedented way Disney matched great animation with music. *The Three Caballeros* (1944) took things even further: it was one of the first films that mixed animation with live action. Donald Duck himself ventured down to Brazil and danced the Samba with Conchita. *Song of the South* impressed me the same way.

Disney characters inspire and delight me to this day, possibly because the very first album given to me for Christmas was the soundtrack to *Pinocchio*. A lot of great music came out of that, including "When You Wish Upon A Star." I became enamored of movie musicals: *Singing in the Rain, Bandwagon, Oklahoma!,* and *Carousel,* as well as the tough guy films. Who could forget Bogart in *Casablanca, Key Largo,* and *Treasure of Sierra Madre?* I discovered sound tracks for movies, as well as film composers such as Bernard Herrmann, who cranked up the suspense in 1960s movies like Hitchcock's *Psycho, North by Northwest,* and *Rear Window.* There was Max Steiner, who almost painted the brooding skies above Tara with his score for the 1930s film *Gone with the Wind,* and Franz Waxman's heart wrenching music for *A Place in the Sun.* Dimetri Tiomkin created the moving score for *The Robe,* and Eric Wolfgang Korngold composed the rollicking music for swashbuckler films such as *The Sea Hawk,* starring Errol Flynn. Many of these composers had fled Europe because they were Jewish, and the Nazis were coming into power in the 1930s. The youngest at that time was André Previn, who came to Hollywood when he was a teenager.

■ ■ ■

My father took me to my first concert at Boston's Symphony Hall when I was thirteen. Charles Munch was conducting, and I was blown away by the sound. Dad also introduced me to my first Boston Pops concert with the legendary Arthur Fiedler, a quintessential conductor and charismatic man who conducted the Pops for fifty years. I remember one day walking outside of Symphony Hall when Fiedler himself stepped out of his car. Two kids riding by on their bikes screeched to a stop and pointed, saying, "Look, it's Beethoven!"

Little did I dream that one day I would be up in the broadcast booth as the voice of the Boston Symphony, or have the chance to interview not only André Previn, but John Williams, who would become the most popular film composer of all time.

Another first from that era was my introduction to New York City through visits with my aunt and uncle who lived in Scotch Plains, New Jersey. My lifelong love affair with New York began with those one-hour trips from their home into the city with my dad. In the late forties the city had a whole different look: automats that dropped a wax-paper-wrapped sandwich down a chute for a quarter, women and men in sharp suits, and everyone in hats. A bus tour took us by the old Met opera house on 39th and Broadway, through the Bowery, uptown to Harlem, and midtown to the Empire State Building, the tallest building in the world at the time. A visit to the live television show *Broadway Open House,* hosted by Jerry Lester, really opened my eyes, as did the RCA–NBC tour, where I stood at the very podium where the great Toscanini conducted concerts.

It was on our way home from one of these trips that I begged my dad for a TV. There weren't too many television sets around in Quincy in 1948, when they were just starting to be commercially available, but I just had to have one. He finally took me to a trade show put on by RCA Victor in the old Mechanics Hall on Huntington Avenue in Boston, where you could see an actual television. We walked around the hall gaping at these big black and white boxes with screens only seven or ten inches wide. It was magical. The first person we saw on television was in fact Arturo Toscanini, all dressed in black, with his mane of white hair flying as he conducted the NBC Symphony Orchestra.

Two years later, after untold begging, we finally got a TV, a twelve-inch GE. The first show I saw was the *Ed Sullivan Show*, which I loved, but I really howled watching *The Show of Shows* with Sid Caesar and Imogene Coca, who brilliantly satirized Italian opera and movies by mixing Italian with all kinds of nonsensical words. *Playhouse 90* showed live plays, and it was quite a revelation to see what radio stars Jack Benny, George Burns, and Gracie Allen actually looked like.

TV got me so excited that I got together with a group of friends and built a mini-television studio in a friend's garage in Quincy. Empty cartons morphed into video cameras; soup cans became camera lenses. We made a boom micro-phone with a long stick and a can rubberbanded on. I'd sit behind a desk and announce the news of the day, introduce guests, pitch toilet paper and Supersuds. My buddy, cameraman Glen McGee, claims it all started in that old garage. We even took the show on the road—to the baseball park where we did a play-by-play in front of all sorts of confused passersby.

Girls, as I recall, were not intrigued by our efforts. This may have had some-thing to do with my entertaining the thought of becoming a priest, especially in my mid-teen years.

■ ■ ■

I was confirmed by Cardinal Cushing, an Irish, Boston-born man with a good heart. I admired him, and I became an impassioned Catholic, though I never went to Catholic school or was an altar boy. When I was much younger I even made a little shoebox with curtains with a cup inside, crushed bread for wafers. With a prayer book off to the side, I'd genuflect to my heart's content. When I look back on it, I realize I was in love with the whole ritual of it: the mass, the mystique of the altar, the tabernacle, the chalice, how the priest would raise it to the heavens, but not so much the theology. It was that primo acting role that really got me: the ritual of the bells and the incense. Certainly the chalice-raising scene in *Parsifal* hovered somewhere in my subconscious mind. Regardless, to this day it all intoxicates me each time I visit the Cathedral of the Holy Cross, a magnificent Vatican-like edifice in the South End.

The teen years—around the time hormones really hit—can be a rocky time. I for one didn't quite know what to do with them. I certainly felt different from a lot of my peers. Instead of being out in the streets stealing hubcaps like other teens, I was home listening to Enrico Caruso. Not a great way to meet girls, plus there was my physique: I was on the skinny side, no matter what I ate.

I remember preparing myself for weeks to ask a stunning Sandra Dee look-alike to dance at a high school record hop. How you can actually plan for some-thing like that I don't know, but I practiced every verbal and logistic scenario humanly possible before that night, hoping to make failure an impossibility by leaving nothing to chance. "Would you like to dance?" "How does a dance sound?" "Feel like dancing?" Or pretty much just yank her up onto the floor. This girl was *mine.* Arriving early, I scouted the room, finding the best possible places to hang out, look casual, and maintain my Tony Curtis hairstyle.

And there she was, sitting alone, which was perfect. I didn't want to have to ask her in front of her giggling, smirking friends. She looked gorgeous under the glittering lights of the high school gymnasium. But still I waited; I wanted a nice, slow ballad so I could make my intentions clear.

The moment came. The song was "The Great Pretender," by The Platters. I approached and she gave me this coy look. I thought I even detected a smile creeping up one side of her face.

"Would you like to dance?"

"No," she said, and looked away.

All the air left my body and my face froze. I walked away out of the gym into the beautiful spring night. What the hell did I do wrong? I felt like an ass. I must be wearing the wrong kind of jacket. My tie was stupid. The pricey suede shoes looked suddenly ridiculous.

It all put me into a deep funk for a good three weeks: taking apart those moments and putting them back together, hearing that "No" in my head a thousand times over, my undignified, silent exit. It didn't help that in the following week I got into a fist fight with this kid who'd been bullying me for a year, losing in a big way. To this day I'll indulge in the occasional daydream about getting even.

■ ■ ■

Between striking out with girls and baseball, I started to get obsessed with swing music and playing the trumpet. My friend Dick Hayes had a Zenith console radio with a turntable that played 33 1/3 rpms—a huge deal at the time. We'd hang at his house listening for hours to Benny Goodman, Les Brown, and Ted Heath, but mostly to the Glenn Miller orchestra. "In the Mood," "String of Pearls," "Chattanooga Choo Choo," "Moonlight Serenade"—we couldn't get enough of that big band sound. We also watched *Sun Valley Serenade* and *Orchestra Wives* more times that I care to recall. I loved the big brassy sound and became convinced I could play the trumpet.

I really gave it my best shot. How could I not try to play an instrument when I was steeped and stewed in music, both through my own hours of listening but also with my dad belting out arias over breakfast or jumping up at a party to sing.

My cousin gave me a trumpet, and I started to take lessons from an old big band trumpeter in Quincy named Joe DiBona who charged just two bucks a lesson. I started my research in a record store, also in Quincy, called Jason's. They had booths where you could listen to the records before you bought them. I wallowed in Miller, Goodman, Count Basie, and Duke Ellington, segueing into Sinatra, Mario Lanza, Peggy Lee, and Ella Fitzgerald. I bought what I could and played along with the records at home.

I was starting to develop my sound, but after many months realized I didn't quite have it. I wanted to play so badly, but I pushed to make it happen faster than I could. Impatient, I guess. So I quit, even though the high school band director asked me to join the band. I guess I didn't have the confidence. Little did I know that later in life that brief acquaintance with the trumpet would come in very handy. As far as my own future as a musician, it didn't look promising, but I knew I had to be involved with music somehow. So I went back to my cardboard broadcasting booth. That's where I really felt comfortable—writing my own commercials, delivering them, and playing DJ: "Here is a wonderful recording by Mario Lanza. He's got a big hit with 'Be My Love'..."

■ ■ ■

As much fun as I was having playing broadcaster, I had a growing awareness that I didn't know what the hell I sounded like. I needed a tape recorder, which wasn't cheap back in the early fifties. They cost around fifty dollars—a lot of money for my parents, both teachers, to spend on anything, much less something so strange as a tape recorder. Undaunted, however, I kept begging my dad for one. After about the tenth or eleventh "no," I realized I needed a new tactic.

We were driving to Wollaston Beach in Quincy on a beautiful summer day.

"Dad," I said, "I just gotta have a tape recorder."

My mother rolled her eyes and folded her arms, staring at the beach out the window.

"We've talked about this," my dad said.

"You hear me upstairs, I know you do. Talking into a cardboard mic."

"That is a problem."

"Look, I can't hear what I sound like. Really hear my voice. I don't know what I'm doing wrong, I'm just talking to myself like an idiot. Talking to the wall. I have to be able to record my own radio shows and play them back—"

"Christmas is coming," he replied, as the hot July winds blew.

"I can't wait till Christmas. This can be my Christmas present, okay? I don't care about anything else."

He pulled over to the side of the road, got out of the car, slammed the door, and started walking, steam pouring out of the top of his head. A little way off he stood with his arms folded. My mom said, "Well, now you've done it."

I got out of the car and ran over to him. "Dad—"

"Ron, I can't get you everything you ask for."

"I know."

"It's not possible now. Wait a few months."

"But Dad, you know how you feel about your painting? And your singing?"

He looked at me for the first time since this discussion began. The look in his eyes made me remember the kind of pain he was in when he first got back from the war, and I almost couldn't continue.

"That's how I feel about being in broadcasting someday. Being on the radio, having my own show."

He looked away again. "Why don't you just go back to the car."

Defeated, I did what he said, calculating all the way how many months or even years it would take me to save that much money. Nobody said a word to each other all the way home.

The next day, I'd stowed away the fake microphone and was lying on my bed, daydreaming about serenading my latest crush with my trumpet, then quickly substituting that with a fantasy tenor saxophone.

Dad appeared in the doorway. "Okay, let's go get you your tape recorder."

## CHAPTER 3

# Steppin' Out

My FUTURE IN RADIO TRULY BEGAN WITH THAT FIRST, RELATIVELY gigantic tape recorder. It was a reel-to-reel VM, or "Voice of Music." At first I was ecstatic, following my parents around and recording them, my cat, my friends, crazy stuff off the radio. That feeling soon ebbed. I started taping my own shows, one after the other, playing them back for hours, and trying to pin down what was wrong. After a while it became pretty obvious: I'd been cursed with a horrible speaking voice. I couldn't believe how high and whiny I sounded, but to add to my own personal hell, there was something else terribly, eerily wrong that I couldn't quite put my finger on.

WJDA, established in 1947, was the first radio station in Quincy. So, of course, I was obsessed with it. When I was ten years old I drew a sketch of what I thought the inside of the station looked like and sent it in. My uncle Amilio, the first Italian Mayor of Quincy, did his nephew a favor and called the manager and owner of the station, James D. Asher. He told Mr. Asher that I was interested in radio and wanted one of my taped shows to be critiqued. I was invited for a tour of the place, even did a little spot on a kid's show. Still no word from Mr. Asher about my tape, however, so I started bugging my dad to follow up with another phone call. In the end, I got my wish.

I don't think I've ever labored so hard on anything as I did on that demo tape, and I don't think I've ever sweated quite so much in anyone's office as I did in James Asher's office on that sweltering September day.

After directing me to what seemed like an enormous leather chair opposite him, he took my precious tape, put it in his recorder, and pressed "play." He put his chin in his hands and stared at the machine as we both listened. Heart pounding, I watched his face as if it were the shroud of Turin and someone had told me they'd seen it move.

He frowned and turned off the recorder. "Son," he said, "I don't think you're quite ready to work in radio yet."

"Yet." I was hanging on to the "yet" for dear life. "But, what do you—"

"You're not bad, but you have a real diction problem."

"Diction?"

"You've got a real heavy Boston accent."

"I do?"

"You don't pronounce your 'r's."

"I don't?"

He rewound the tape and played it through. Not one "r" in the entire reel. Talk about a light bulb going off. I thanked the man and couldn't get out of there fast enough to get home and listen to my tapes with fresh ears.

I listened to Milton Cross introduce the Metropolitan Opera with his perfect diction, then I played myself doing the same thing. Holy *paahk yaah caah in Haahvahhd Yahhd*—did I sound like a Bostonian!

I knew that in order to be on the radio I had to get rid of my Boston accent; I had to have a neutral accent. Beyond all that, Mr. Asher stressed before I left that day that I would need to know how to use phrases and inflection to deliver copy and sell, because in radio you're always selling. I became keenly aware that I hoped my audience would be the entire United States, not just Boston. I listened incessantly to all the announcers: Hugh James, Don Wilson (the announcer on the Jack Benny show), Don Amici and his brother Jim; Orson Welles, and Ben Grauer who used to do the NBC broadcast with Toscanini.

I sat down and made a list of all the problem words I could think of. All the poor, helpless words I was flattening and murdering with my accent: car, bar, later, never, over, dark; or as I was saying them: *cahh, bah, latah, nevah, ovah, dahk*. The list grew daily, and daily I repeated the words over and over until I had officially welcomed the letter "r" into my life and made peace with it. But changing how I spoke wasn't easy, especially when I was hanging out with friends who wouldn't have known what to do with me if I started speaking in a completely different way. They may have even started kicking my ass. Thankfully I was able to speak one way "upstairs" as well as "normally" in real life.

All in all it took me years to lose that accent. In fact, I still get called on it, most recently during a BSO broadcast. Sometimes I'll slip after a couple of beers with friends who won't hesitate to clue me in. But keep in mind, I live in Dorchester, a wonderful neighborhood, but one where r's are nonexistent. It's so easy to just relax, walk out my door, and fall back into an "r"-less life.

■ ■ ■

My first job was at the Alhambra Tea Room in Quincy Center. It was a restaurant with a soda fountain, and they made their own candy in the basement. The place was a landmark; my mother had worked her way through high school there in the twenties. When I was sixteen, she introduced me to the owner, a kindly Greek gentleman. I'll never forget her saying, "This is my son. Hire him, you'll like him."

The next day the manager, Al, taught me the finer points of making sodas, sundaes, and sandwiches. My first day on the job is still seared into my brain: a blur of ice cream and blenders, take-out orders, fries, burgers, and milkshakes—a thousand different things happening all at once. I was at the register, learning how to make change for the very first time, when my second or third customer, a woman in a scarf and sunglasses, bought coffee and skittered out of the store. Minutes later she burst back in and charged right up to the manager, saying, "That young man at the register, he cheated me! Look, I gave him twenty dollars for coffee, and I only got two back!" Well, I knew she didn't give me twenty dollars, and I got very upset, close to hysterical. My newness was so obvious—it was written all over me. Al stared at me as I insisted on my innocence, then turned to the woman. With a stone-cold look he told her he'd check the register and receipts. After a minute that felt like days, he looked up from the register and told her to get lost.

Things smoothed out after that, and I really got into the rhythm of being a soda jerk; in fact I worked there straight through high school to support my pricey LP habit. There were other benefits to being the guy behind the counter slinging hash and fizzing the drinks: I started to feel comfortable in my own skin. I began to realize that I had a personality, a sense of humor. Kids who'd knock me around as soon as look at me showed new respect or at least acknowledged me, sometimes even tried to pal up to me. Besides, the Alhambra was not just a social hangout, it was *the* place to be. On top of all that, the jukebox was jumping with the greats all day long: Nat King Cole, Ella Fitzgerald, Stan Getz, Billie Holiday, whatever you wanted to hear.

■ ■ ■

Around that time I stumbled on some muscle mags. I still wasn't happy with my build. Steve Reeves on the cover of *Strength and Health* with his 28" waist, 45" chest, chiseled features, and bulging biceps looked like a god to me. He was one great specimen of physical prowess. I asked my dad if he'd help me set up a bench in the garage with a mirror. He said sure, and that he'd tack a couple pictures of my hero on the wall to boot.

I ordered my weights from the York Barbell Company in York, Pennsylvania, and the mailman nearly blew his knees out lugging them up the steps. I flipped through the pictures of Steve, posing and puffing, then looked at my thin arms, back to the weights. I have to say I had some serious doubts at that point, but then I thought to myself, just open the box and set the thing up, for crying out loud.

I did the first curl and felt the blood rushing through every single vein and capillary in my arm. It was this whoosh of feeling alive, this reminder that I had a body. I looked through the book of exercises that came with the weights: bench, leg, and overhead presses, deltoid pulls, and bicep curls. I was out of control—it felt so good. As the days and weeks went by I'd look in the mirror and think, shazam! Something very good is happening here. I ordered more weights and set new goals, but after a while I needed more weights than it made sense to buy, so I joined Jack Donovan's Health Club.

At this point, I was looking forward to getting a little more one-on-one attention. Jack must have been in his sixties, but he looked fantastic, maybe because he never ever stopped lifting weights. A legend in the Quincy bodybuilding world, Jack was an old-time musclehead who ran the gym on his own. I'd come to lift and he'd be alone in the gym benching two-fifty. As I let myself in, he'd squeak out,

"Come on in and get set up, kid, you can do it. I'll talk you through it."

I'd say, "Jack, how much should I lift?"

"Yaaarghh...arghhh...two-twenty!" Clang bang of the barbell back in the rack.

So I'd go ahead and do my thing as he did his, gathering pointers here and there as he deadlifted, straining and squatting and groaning all around me.

As I finished up he'd call out, "Yeeeaaaaaah! Lock that door on the way out, wouldja, buddy?"

Sometimes I'd come in and he'd be locked in this little steam cabinet with only his head sticking out, a towel wrapped around his thick neck. I guess he got into the thing through a secret side entrance of some kind; in any case he was always smiling in there and I used to wonder what he was doing with his hands, although I never asked and I never got in the thing myself.

The point was, I wasn't scrawny any more, my voice was changing, and it felt like things were coming together. Girls weren't exactly all over me, but there were hopeful signs here and there.

■ ■ ■

I made my second foray into the real world of radio when I was sixteen years old at WPLM in downtown Plymouth, a station that had only been on the air for two years. Again I sat there sweating through every pore in my body as station owner Jack Campbell, a WWII vet, listened to my tape. Again the serious look and snap of the button as he turned off the player. Except this time I heard, "When would you like to start?"

My first challenge was how to get to work. I'd copped a ride with my dad to the station for the interview, but I didn't have my driver's license yet, and there was no way he could drive me there and back each day. All so strange because there I would be, fifty years later, broadcasting on WPLM! Regardless, at the time it was a much needed boost to my ego.

First a radio job, and then, out of nowhere, I fell in love for the very first time. On my way to the Alhambra I'd have to walk by Dorothy Murial's Bakery, and each time I did my heart would do a flip in my chest. This Kim Novak look-alike would be at the counter wrapping up a cake, fixing me with these doe eyes, an enigmatic little smile on her lips. Each time I went in to see her she lit right up; she made my teenage boy head spin. Gerri Robbins from Hough's Neck; I'll never forget her!

Luckily Gerri loved music, too. In fact, her dad turned me on to Sinatra; up until that time it had been strictly jazz, big band, opera, and classical. He put "Songs for Swinging Lovers" on the turntable and that was it for me. Gerri and I used to hit the Neponset Drive-in movies on the weekend, steaming up the windows so bad we had to turn the windshield wipers on. There were even a couple of times we drove away practically dragging the whole drive-in with us because we forgot to unlatch the speakers from the windows...

I may have been a teenager in love, but I didn't feel like a teenager when I took Gerri to the Newport Jazz Festival in Newport, Rhode Island, the granddaddy of jazz festivals. That's where we first saw Duke Ellington's Orchestra, Count Basie, Dinah Washington, Erroll Garner, and Billie Holiday.

Norman Granz was the gentleman who created Jazz at the Philharmonic, or JATP, as it came to be known. Granz initially organized desegregated jam sessions in Los Angeles in the early 1940s. These jam sessions evolved into the JATP concerts, which eventually went on tour. He was committed to the notion that black and white performers should be able to play together and that they deserved equal pay. From this belief grew the innovative idea of bringing together and taking on tour some of the greatest jazz musicians of the time, artists who up until then played mostly in smoke-filled clubs and dives, and often under segregated circumstances. Granz would open the evening with a jam session, after which each artist would emerge and play his favorite tune.

When the JATP came to Boston, there I was, a kid from Quincy High, with my first serious girlfriend, sitting smack in the second row of Symphony Hall. We sat rapt, listening to Flip Phillips, Illinois Jacquet, Ray Brown, and Buddy Rich; the list was endless. We sat so close we watched the beads of sweat rolling down their faces, saw the dents and scratches on their instruments, sometimes even heard their witty or sarcastic asides to one another. One by one a spotlight would shine on a different soloist: Dizzy blowing "I Can't Get Started"; Coleman Hawkin's classic rendition of "Body and Soul"; Lester Young, nicknamed the "Prez" by Billie Holiday, strutting out in a little porkpie hat playing "Polka Dots and MoonBeams." I said, "Gerri, look at Lester, you can only see the whites of his eyes." He was playing beautifully, but he was very, very stoned.

Meanwhile, jazz was exploding in clubs all over Boston. The level of artistry was so high I believe I became spoiled forever. Here were the stylists, the homegrown soloists and bands everybody else followed. Missing any of it was inconceivable to me. I had a parallel life going on: school, work, and home were all one thing, and then there was music.

The Jazz Workshop on Boylston Street across from the Prudential Center was run by Fred Taylor, "The Jazz Guy," who had been promoting jazz longer than anyone in the city. Downstairs in this dark, subterranean club, we soaked up the sounds of pianist Bill Evans, Stan Getz, and Thelonius Monk.

Adjacent to this club was Paul's Mall, also headed up by Fred Taylor. He brought in pop acts as well as Jose Greco. Today, he's busy booking acts at Sculler's in the Doubletree Guest Suites in Boston.

The Stables, a club on Huntington Avenue near Symphony Hall, showcased some of the most innovative jazz, especially between 1955 and 1959. That was where I first met Herb Pomeroy, who became one of the America's leading jazz musicians. His Herb Pomeroy big band, one of the most original ensembles of its time, included Charlie Mariano, Joe Gordon, Lenny Johnson, Ray Santisi, and Dick Johnson, among others. When Herb was teaching at the Berklee College of Music, he kept the band going and they performed extensively throughout New England and the United States.

Like many young jazz fans, I was a member of the Teenage Jazz Club, which met at a venue called Storyville on Massachusetts Avenue in Copley Square. George Wein, another great jazz impresario, some say the greatest, and the founder of the Newport Jazz Festival, opened the club in 1950, advertising it as "Boston's Original Home of Jazz." Though it was a humble setup, a long narrow room with the stage jammed into the right hand corner, a pantheon of greats glorified that space. Drinking nothing stronger than cold Coca Cola on a Sunday afternoon, I'd hear Duke Ellington's orchestra, Sarah Vaughan, the Four Freshman, Johnny Mathis, Billie Holiday, Charles Mingus, Lee Wiley, Ella Fitzgerald, Dave Brubeck, Paul Desmond, Erroll Garner, and Pee Wee Russell, among so many others.

So Gerri and I discovered jazz together, both of us just seventeen and out in the real world of smoky dance bars and dives, no longer at home sitting in front of a record player. Instead we were present as living, breathing jazz was being created on stages before us.

For me, it was a time of discovery, of venturing farther and farther out into the world and being astounded by the ever growing beauty I found in it. In his stunning book, *Jazz Is*, Nat Hentoff evokes this feeling of awe when, as a teenager in Boston walking by the Savoy, he recalls:

> *. . . a slow blues curls out into the sunlight and pulls me indoors. Count Basie, hat on, with a half smile, is floating the beat with Jo Jones's brushes whispering behind him. Out on the floor, sitting on a chair, which is leaning back against a table, Coleman Hawkins fills the room with big, deep bursting sounds, conjugating the blues with the rhapsodic sweep and fervor he so loves in the opera singers whose recordings he plays by the hour at home.*
>
> *The blues goes on and on as the players turn it round and round and inside out and back again, showing more of its faces than I had ever thought existed. I stand, just inside the door, careful not to move and break the priceless sound. In a way, I am still standing there.*[1]

■ ■ ■

My decision to attend Boston University in 1958 was determined not only by the fact that they had one of the best communication departments around, but because they had their own student-run radio station, WBUR. One of my earliest assignments was to go out and interview whomever I could get to say, "Sure, kid, I've got fifteen minutes. Shoot." At first I was a bit apprehensive, but I soon learned the first law of the physical universe: people love to talk about themselves.

This proved true for the not-so-famous as well as the famous, people who I was sure would slam down the phone as soon as I stuttered out my request. Stunning me to momentary silence, Howard Hanson, renowned composer and dean of the prestigious Eastman School of Music for over forty years, gave me the thumbs up. So I trucked off to the Copley Plaza Hotel with my LP of his second symphony, the *Romantic*, which he was happy to sign. He told me that one of its themes is performed at the conclusion of all concerts at the Interlochen Center for the Arts in Michigan, and that his opera, *Merry Mount*, received a still unbroken record of fifty curtain calls. Think about that a moment: *fifty curtain calls*. Hours must have gone by. I think my favorite story, however, is the one he shared about proposing to his wife by dedicating *Serenade for Flute, Harp and Strings* to her because he was unable to find the spoken words to form the question.

Among my first interviews was the delightfully salty-tongued American soprano, Eileen Farrell, daughter of the singing O'Farrells. Again, I started off a bit intimidated, but I relaxed a bit more with every frank, irreverent answer she gave to my scripted interview questions. A major opera singer who also created a career in popular music, Eileen called herself the "Queen of Crossover." When we met, she'd been singing at Boston College after a major blowout with Rudolph Bing, general manager of the Metropolitan.

I asked her, "Wasn't it wonderful to sing at the Met?"

She said, "Not if they give you crap you don't want to sing. I want to sing Wagner."

No beating around the bush with her. All I know is, she had her good six years at the Met and went on from there to sing all over the world.

As part of my program at Boston University, I worked at WBUR. I was pretty much handed the position of Director of Public Affairs, which meant I had to manage a department. I soon saw that in radio you were either in sales, management, or talent. Quite quickly I learned that nothing about managing anything appealed to me, and I could stomach sales only if I was primarily talent.

My repeated requests were finally answered with *The Sound of Jazz*, a show I hosted once a week. I was on the air for the very first time. Of course it was the sound of jazz, but it was also the sound of twenty-year-old Ron Della Chiesa, a person and persona I was still trying to get to know and develop. On one hand I was starting to enjoy a bit more confidence about my appearance: learning how to dress, making sure I had the right little Tony Curtis curl on my forehead, and hell, maybe even take the time to glance at myself in the mirror before I left the house in the morning. All this seemed to make a difference with my image around campus, and women, in particular. But who was Ron Della Chiesa, really, as an announcer on the air? With this question in my mind, I did what I do to this day: listened to everything I could get my hands on, from network radio to local talent; from announcers at the end of their game to obscure voices who somehow drew me in. I was open to everything and everyone.

I used to marvel at Jess Cain, an announcer with an acting background who choreographed a unique morning show for over thirty-five years at WHDH in Boston. During three o'clock in the morning commutes from Hingham to Boston, he created skits, riffs, song parodies, and dead-on impressions complete with sound effects that he'd throw on the air the minute he got to the studio. "What you're trying to do," he told the Boston *Herald American* in 1974, "is jolly people into facing the day."[2]

Symphony Sid was another great talent and influence on me. He was famous for his hipster lingo, love of bebop, and knowledge of the black music scene. Sid was the top jazz DJ at the time, broadcasting live nightly from the all-glass studio at the High-Hat, which was on the corner of Columbus and Massachusetts Avenue in Boston. The great tenor saxophonist Lester Young, "The Prez," even wrote a theme song for him called "Jumpin' with Symphony Sid":

*Jumpin' with my boy Sid in the city*
*Mr. President of the DJ Committee*

Many of Sid's commercial sponsors granted him wide artistic license even though he tended to get increasingly stoned as the night went on, at times forgetting where he was. Here's an ad for "Solray Soap," as only Sid could pitch it: "It'll make your skin the grooviest *and* pimple free, baby." He even bebopped his

Boston radio legend Jess Cain and me, 2005.

Author's collection.

way through a commercial for a funeral home: "There comes a time for every hep cat on this earth to cut out from this world. My good friends at Samuel's Funeral Parlor have got you covered. When you split the scene, they'll lay you out in style. So remember, make your arrangements early, 'cause when you're gone, baby, you're *gone*."

Bill Marlowe, who Frankie Dee in his web-tribute named "The Baron of the Airwaves," was another icon of Boston radio I listened to incessantly. His real name was William Moglia but he changed it for a radio career that spanned over fifty-five years, beginning at age nine when he performed a soliloquy from *Hamlet* over WEEI. Bill would personalize every show with dedications, and he was a passionate advocate of the Great American Songbook, passing that torch on to me in many respects. He began the Frank Sinatra Show in Boston, even taking me aside one day and imploring me to continue this tradition.

Though Bill had a weakness for the horses and a barstool—he was no stranger at either Suffolk Downs or Jimmy Mag's in East Boston—he'd get right on the air after spending an entire night doing his thing with no sign of a hangover, a skill I soon learned was no small accomplishment. A tall, good-looking guy, Bill could talk his way out of any situation. He especially loved creating wild commercials for restaurants. In a voice as smooth as hot Velveeta, he could make a second rate restaurant sound like the Waldorf:

*You know, friends, Pasquale's Pasta Palace on Route 1 in Revere is really special to me. Why? It's the pasta, my friend. These people* understand *al dente. It's not just a word to them, this is the pasta your mother made, this is the pasta your grandmother made, this is the pasta from the homeland; it's the pasta you dream about when you're far from home and* dreaming *about pasta . . .*

William Pierce, who hosted more than three thousand Boston Symphony Orchestra broadcasts at WGBH over a span of thirty-eight years, had the following suggestion about being on the air: "Always act as if you're a guest in someone's home; they've invited you to dinner, so behave appropriately." His words made a lot of sense to me then, perhaps even more so these days when so much of radio is angry, mean spirited, and negative, which is not my style. With his warm, calm, assuring delivery, Bill was the original voice of the BSO, creating a style many symphonic music announcers employ to this day.

Another radio giant on the dial during my BU years was Norm Nathan, known for his all-night jazz show *Sounds in the Night,* as well as for his motto: "(my goal is to) try to leave the world a little sillier than I found it."[3] His largest stage came in the mid-1980s, when he joined WBZ and did an all-night talk show that never had a real topic. In fact, he liked to "keep it light" by avoiding politics or social issues of any kind. Nathan's specialty was leading listeners down his own imagination's winding path, convincing one caller that Athol, a town in Western Massachusetts, was originally a town in Africa that was so beloved it was reconstructed, brick by brick, on Route 2. I'm asking you, what's not to love?

Early radio idol Milton Cross on the air at the Met Opera House, early '30s.

I also tuned in to Tony Cennamo, who started *New Morning* at WBUR in the late 1960s, a five-hour show of pure jazz that ran for fifteen years, a one-man revival of jazz in Boston. And, of course, I listened to all I could of the perennial Milton Cross, a class act who still introduced the Metropolitan Opera every Saturday afternoon, bringing opera alive for me and millions of others.

One day I was wrapping up my show at WBUR when I saw a guy with thick, horn-rimmed glasses step into the control room behind the glass. I'd heard of Arnie "Woo Woo" Ginsberg, but I was floored when he pretty much hired me on the spot to be an announcer for a variety of programs at WBOS.

Arnie, a radio engineer turned disc jockey, had one of the biggest followings in Boston, and he was one of the first rock-and-roll DJs in the city. I'll never forget his on-air voice—a continual adolescent crack mixed in with a cacophony of buzzers, Bermuda bells, kazoos, car horns, oogahs, and train whistles. It seemed to capture teen angst with a kind of knowing wink, and even turn it into a sort of victory. Regardless, he was hyper as all get out, a fast-talking mic artist who knew his music backwards and forwards, and I was thrilled to be working my first commercial job for him.

The station was located on Commonwealth Avenue near Boston University. From six o'clock to midnight during my last year of college, I ran the board, opened the microphones, spun the records, and did commercials. Arnie appointed me as announcer and English host for a series of ethnic programs: the *Irish Hour with Tommy*; the *Polish Variety Hour with Karl*; the *Boston Greek Hour with George*; *Music of the Near East with Michelle*; *Italian Melody Hour with Gino*; and *Songtime* with the Reverend John Debrine from the Ruggles Street Baptist Church. Suddenly, I was very *busy*.

On the very first day and the very first time I opened my mic, I introduced Reverend John DeBrine and his show. We enjoyed a brief repartee, and I thought everything was going along amazingly well. It almost made me nervous—how well everything was going—so after a while I checked on what was actually being broadcast and found his sermon was completely muffled, the Word of God under a giant pile of laundry. I realized his mic had not been on for about fifteen minutes. I'd like to say I coolly assessed the situation, calmly found the right button and pushed it, but instead I started to panic. Considering that a ten-second screwup on the radio feels like a century of hellfire, imagine what a quarter of an hour of garbling the Good News of the Gospel of Jesus will do. After a few frantic phone calls—nobody was around—I finally found the right switch and collapsed in my chair, praying for the Reverend's and the Lord's forgiveness.

With the ethnic shows I was really thrown into the fire and tested. At times I felt like Lucy at the chocolate factory, stuffing her face as the assembly line speeds faster and faster and out of control. I just couldn't keep up with who needed what, when, and for how long. There were times my life was actually threatened, but hey, I had fun.

**First day at my first radio gig, WBOS, 1959.**

Author's collection.

Here's the deal. The show hosts would do their commercials in their native language then cue me since I had to repeat the commercials in English. It was all live: no room for error, no seven-second delay, nothing. I controlled both our mics and all the music. Even though my guests were sitting right across from me, keep in mind that they were speaking in another language; so if they would forget to cue me in English instead of their native language, I had no time to react.

The hosts would just stop talking and hand me things to play. Shows ran back to back, and the music and programming would change radically. There was a lot of passion in those shows, not only about pride of content, but about the ticking clock. You had to be very careful not to go over your allotted time, even by a few seconds. These guys bought their own time and sold it. Heated arguments, actual fights, broke out nightly about this: George the Greek would start screaming to Gino, who did the Italian hour, "You bastard! You went over! That's *my* time. You owe me ten seconds, you son of a bitch!" Meanwhile Karl would be saying something to me in Polish as I tried to cue up the first record for the Irish hour.

Karl, who did the Polish hour, looked like Nikita Khrushchev, never smiled, and was always pissed off. He glared at me across the desk as he hammered away in Polish, flinging records at me. It was just a matter of time before I screwed up his show. Finally, it happened. One night after his show wrapped up he walked me into a corner in the studio, spitting, "You put on the wrong record. You made me sound like a fool. So now I am going to wring your neck." Luckily George the Greek hadn't gone home yet and pulled him off me.

George had his own excitement one night during the Greek Hour. For five dollars each he did funeral announcements for families of local Greeks who'd passed away. Somberly and slowly he read the names of the dead while I played a sad dirge in the background. One night I answered the phone as he was doing his thing. It was this hysterical guy speaking half in Greek, half English. I just handed the phone to George and cut in with a commercial to give him a chance to talk. In seconds he grabbed me and said, "Oh my God, the guy's not dead!" I handed him the mic and backed off. It was the guy's brother instead who had died, so George had to correct that on the air. I was kind of thrilled that for once something wasn't my fault.

And then there was Tommy Shields' Irish Hour, sponsored by Irish Airlines. I had the chance to do the commercial for this program because it was all in English. A tailor by day, Tommy was on every night from eight to nine o'clock. All of the Irish pols would come up and be interviewed on the show, and we'd feature the popular Irish performers of the time—John Feeney, Connie Foley, and Ruby Murray.

Irish Airlines wanted us to do more to promote them, so they sent Tommy and me out to visit Cardinal Cushing in Brighton, who had just returned via the airline from a trip to Ireland. I'm sure they had high hopes that the Cardinal would say something positive about his trip. Regardless, Tommy and

I were nervous about the whole thing. I'm not sure what his reservations were, but for me, to extract a commercial from the man who had confirmed me a few years back at St. John's Church in Quincy felt a wee bit off.

Tommy and I stood outside the Cardinal's residence in Brighton on a cold, rainy November night. Shivering, we looked at each other, shrugged, and lifted the huge brass knocker. Both of us jumped when the Cardinal himself answered the door.

He was disarmingly casual. After asking us if we wanted anything to drink—we both passed—he led us into his library, a stunning room with floor to ceiling bookshelves on each wall and the biggest desk I've ever seen.

"So, where are you boys from?"

"The Irish Hour on WBOS, your Eminence," I said. "We called earlier..."

"Well, you've got to give my best to that Woo Woo Ginsberg. What a great guy. He's my engineer for the rosary. Did you boys know that?"

"We listen to you recite the rosary every night," Tommy said. "It's wonderful."

I put the reel-to-reel recorder down on the table and pressed "Record."

"So, how was your trip to Ireland?" I prodded.

He waved away the question and sat on the edge of the desk. "Pure misery. Never seen such rain. The coldest wettest four days of my life—couldn't wait to get home. And the Irish priests!" He paused and shook his head. "What a bunch of lazy do-nothings." He pointed at me, then Tom. I believe I was shaking at that point. "It's not enough, you know, to preach two sermons on Sundays and take the rest of the week off. Ignore the flock completely. Did *God* do that?"

"No, uh, your Eminence."

"No, God put in six and *then* put his feet up."

Beads of sweat broke out on Tommy's brow. "So how was the flight, sir?"

"The what?"

"Your transportation, sir."

"You mean the plane?"

Tommy nodded vigorously.

The Cardinal took on a faraway look. "You know...flight...flight is a miracle. Don't you think so? I don't understand how airplanes actually fly, but there must be some element of the divine at work. But the flight itself? A nightmare. Screaming babies, no air—*zero* leg room—I had to fold my knees up under my chin. Plus, we were two hours late getting in."

He finally glanced over at the tape recorder, watching it for full revolution. "Well, did you get enough?"

After some discussion on the ride home, Tommy and I agreed: some interviews were simply never meant to make it to the airwaves.

At WBOS I was introduced to John Henning—a dear friend since the day we met in 1960, who has since, sadly, passed away. John had his first job in broadcasting at WBOS, going on to become one of the foremost television news broadcasters in the country. Robert MacLean of the *Boston Globe* called him "one of the best street reporters in the history of Boston TV news."[4]

John Henning and me, Copley Plaza Hotel, early '70s.

Author's collection.

When WBOS moved to the top floor of the Somerset Hotel on Commonwealth Avenue in Kenmore Square, there was a whole new glamour to my job. The view of the skyline from the studios on the seventh floor was unbeatable, especially combined with the steaks and martinis you could have sent up from the restaurant below. Between shows, John and I would head down to the Keyboard Lounge on the first floor, hang out, and listen to Bob Winter at the piano bar. A brilliant musician, Bob would put together a medley of themes from old radio shows, and we'd play a trivia game. The waitress at that time, a Boston University theater major, was Faye Dunaway.

Years later Faye and I met again over dinner at the Via Matta in Park Square with my wife Joyce and Faye's former husband, Peter Wolf of the Jay Geils band. She had just appeared as Maria Callas in the play *Master Class,* by Terence McNally. It took a little prompting, but finally she remembered hanging out with us those many years ago at the Keyboard Lounge.

I had kept working nights at WBOS after graduation from BU in 1959, but I also had taken a full time job at the shipyard in Quincy owned at that time by Bethlehem Steel. Being a night owl, the 6:00 AM to 3:00 PM shift took some getting used to, as did the earcracking sounds of steel being cut and shaped by huge machinery into parts of ships.

As a runner, my job was to deliver orders of all kinds of parts from the pipe-fitting shop to the plate yard to the basins where the ships were pieced together. Before my eyes, fantastically huge tankers and passenger ships rose up, looming like giants over us as we ran under their shadows, putting together their inner workings. There seemed to be a bottomless demand for ships at the time.

Though the work was tedious in many ways, I did meet my share of guys I'll never forget. I kept hearing about this one character, the "Butterfly," until finally my boss Red and I stopped him long enough to get his story.

Red, not looking up from pounding a pipe, said, "Hey Butterfly, Ron here wants to know how you got your name."

Butterfly unzipped his pants, and there it was: a monarch tattooed on his Johnson. With no small pride he flipped it back in, saying, "Yah, did that when I was stoned out of my mind in the navy. Didn't hurt till the next day, but then the motherfu—(shriek of ship's whistle)—hurt for a year."

Another guy, Dudley, worked in the plate yard. Red said to me one day, "I don't envy you going to see Dudley."

"Why?"

"You'll find out."

Regardless, I had to deliver orders to Dudley, so off I went. Dudley was bent over a pile of steel, tossing things behind him like he was looking for something. "Beautiful day out today, Mr. Dudley," I said.

He looked up at me with a murderous glint in his eye. "What the hell is good about it? For chrissakes, it's miserable. Give me the damn order and get out of here." Then he picked up a thick section of pipe and threw it at me, missing my ear by a centimeter.

Dudley might have been an extreme example, but there were a lot of disgruntled people working there. After a couple of years at the shipyard it was clear that this was not my career. I wanted full time radio or nothing.

The world, however, had a few other plans for me. At Boston University, I had been in the ROTC (Reserve Officer Training Corps Program.) Once you got out of college, you were vulnerable for the draft and your name got on the list. My choices were either be drafted, join up as an enlisted man for four years, or be an officer. I wasn't anxious to repeat what my dad had been through; in fact we talked it over, and he said if you're going to be in the army, be an officer; it'll be easier for you. In the end, I dropped out of ROTC after graduation and joined the Army Reserves, which was a six-year obligation: six months active duty and six years in the reserve, which is how I ended up a sergeant. I figured I'd stick it out for the six months and pray for the next six years. You had to go to a meeting once a week and Camp Drum for two weeks every summer. Camp: I figured hmmm, I could handle camp, though I suspected roasting marshmallows and singing around the fire were not part of the regimen.

Turned out, I was lucky in a sense; my service fell between the Korean War and the Vietnam War—a relatively quiet period, though Vietnam was escalating. My basic training began at Fort Dix, New Jersey, where most of the guys were a bit older and had been in Korea. I looked up to them and felt comfortable around them, but I was thrilled to run into Paul Antonelli, the crooner friend I grew up with in Quincy. He had gone to the BU School of Music and started up a group called "The Chariots." Just as I was starting basic training, he was getting out, so I got a lot of tips from him.

I know it sounds like a cliché, but basic training did put hair on my chest in a certain sense. I was a guy who loved to be home—my meals cooked for me, my laundry done, the smell of homebaked cookies wafting up to my room. I mean, what's not to love? I liked being home. I liked it then; I like it today. As I lay there in my bunk under my scratchy blankets, clothes folded at the ready for 4:00 AM wake-up call, guys exhausted and snoring all around me, I wondered if I could deal with this. But when someone is screaming in your ear to get moving—*you*, not the guy next to you or someone in a movie—you pretty soon realize it's all up to you now. You want breakfast? Get up before dawn and make the bed. Assemble and disassemble a rifle, then do it again. You want dinner? Go on a twenty-mile march. Dig a foxhole in ten minutes. Put up a tent in five. I missed my home, but I got it done.

In my second week of basic training during early morning formation—we're talking 4:30 AM, dark, raw, and cold—the company commander asked if anybody could play an instrument.

Actually, what he said was, "ANYONE WHO CAN PLAY AN INSTRUMENT, STEP FORWARD."

Some primitive part of my brain, disconnected from any past trauma from playing the trumpet, and perhaps—who knows—connected to the part that knew this might do something for me said, "Yes, Ron, you play an instrument," as my foot took a step forward. A few other half-asleep liars stepped forward too, I was happy to see.

The sergeant stopped in front of me, turned, then lasered into me with his stare. I looked just past his left ear like a good soldier.

"*You* play an instrument?" he barked.

"I do, sir!"

"And what might that be, soldier?" he asked as if he could hear me practicing, off-key in my room.

"The trumpet, sir!"

Again the long stare. As if already disgusted by my playing, he shook his head and resumed pacing in front of us. "All right. Musicians." He spat on the dry ground. "You're going to go home this weekend. Get your instruments and bring them back. We're putting together the company band."

That weekend I got my pass and grabbed the bus home, dug out my book of marches and banged-up old horn, valves all stuck shut, and woodshedded for forty-eight hours straight. No drinking, no girlfriends, just honking on that thing like a crazy man all weekend. My mom brought meals up to me, by ten in the evening begging me to stop so she could sleep. Dad just said, "Son, go do what you gotta do."

Back at Fort Dix the following week we had our first rehearsal. The band leader was this short chubby guy with a chronically pained expression. I honestly couldn't tell if he was laughing his ass off at me or if he was deeply impressed. It didn't matter. That forty-eight-hour cram session turned out to be the best damned move I could have made. That gig in the army band made the rest of my basic training a piece of cake. No more K.P., no digging foxholes, no latrine duty, passes *every* weekend, and the crowning glory, a chance to sleep till eight in the morning on occasion—nirvana for a night owl like me, and unheard of in Uncle Sam's army.

So even though that old horn never got me fame, money, or girls, it got my head out of the toilet, and for that I will be forever grateful.

# Nice Work If You Can Get It

AROUND 1960 MY CAREER TOOK ANOTHER TURN. I'D BEEN SENDING out my tapes with WBOS behind me as a reference if my phone rang with an offer. One day, it did. WBCN successfully lured me over to their offices, then located at 171 Newbury Street in the Back Bay.

WBCN stood for Boston Concert Network, and it was founded by electronics genius T. Mitchell Hastings, who also owned WNCN in New York, WXCN in Providence, and WHCN in Hartford. Mitch was a visionary mentored by

Radio innovator T. Mitchell Hastings and me at WBCN, early '60s.

Author's collection.

Major Edward Armstrong, a pioneer who developed the technology for FM radio. Mitch invented one of the first FM transistor radios, the Hastings Junior, which was named after him. Passionate about keeping classical music alive, he created WBCN in hopes of forming a nationwide system of classical music stations.

Don Otto, program director and another BU alum, was the guy who hired me and put together our playlist. My first day on the job was Christmas morning, and I did my own classical music show. I was on the air five days a week, mornings and late nights, but the station was twenty-four-hour classical music at the time.

It was the perfect gig. WBCN was a powerful station, and we had long leashes to do some great, innovative things. I met Nat Johnson there, and he became another one of my dearest friends who ended up taking a similar path to WGBH in the end.

But at WBCN we bonded over a love of opera. Nat and I would record operas for the legendary Sarah Caldwell, founder of the Opera Company of Boston and the first woman to conduct at the Met. Her office was right across the street at the time, so we found every excuse to be over there and learn all we could from her. Sarah became known for not only putting together complex works under pressure, but for reimagining standard operas with fantastic staging and costumes. These fresh, provocative productions are remembered to this day, and include the American premieres of *Moses und Aron* by Arnold Schoenberg, new interpretations of *Don Giovanni, Otello,* and *A Trip to the Moon,* using some of the most incredible talent of the time who were just as eager to work with her: Renata Tebaldi, Placido Domingo, Beverly Sills, Marilyn Horne, Boris Christoff, and Donald Graham, among others. One quote captures her spirit perfectly: "If you approach an opera as though it were something that always went a certain way, that's what you get. I approach an opera as though I didn't know it."[5]

I'll never forget meeting a young Placido Domingo, fresh from singing *La Boheme* with soprano great Renata Tebaldi, who at the time was approaching the end of her career and retirement. Since the elevator was broken, Domingo climbed the three flights to our attic studio, asking on the way if we had a piano. We didn't, but if we had, I would have had one of the first recordings of Domingo singing.

My kingdom for a piano! Oh, well.

At one point during the 1960s, I decided to do an all-Toscanini show. I wrote to Walter Toscanini, telling him who I was and about my passion for his father Arturo's work and its association with my very young childhood. I was shocked when actual reel-to-reel tapes started showing up in my mailbox at WBCN, steadily, week after week. I pictured this elderly, dignified gentleman standing in line at the post office every Friday, sending me, a complete stranger, live performances of his father conducting the NBC Symphony Orchestra. Many of these tapes had not even been released yet.

I took them home and listened to them over the weekend. The energy in these performances knocked me out! I was able to put together and broadcast two or three weekly shows until the station stopped playing classical, at which point I had to write Walter a letter to let him know the bad news. His reaction was to invite me to the Toscanini homestead in New York for a visit. When I arrived he invited me downstairs where he'd converted the area into a studio which housed all his father's recordings. He asked me what I wanted to hear; I suggested Wagner. In the middle of a rehearsal of *Siegfried*, Arturo lost his temper and started screaming like a wild man at the orchestra: "Vergogna!" (Shame) or "Infamia!" (This is an infamy!) I was quite literally blown back in my seat. What an amazing opportunity to get some insight into the maestro himself. Known for getting the most out of his orchestra, it became clear how much he demanded not only of himself but of his musicians.

Arturo Toscanini lived to age ninety and conducted up until his death in 1957. Years later, Nat Johnson at RCA was instrumental in working to reissue and re-release Toscanini recordings on compact disc (from LPs). In the end, both Nat and I had the pleasure of knowing and interviewing both Walter Toscanini and his son Alfredo.

Since WBCN was the flagship station, we'd tape our programs and send them to the other stations, which could work out fine or, just as often, *not fine*. There could be some monumental screwups. During one scorching hot July day on a drive through Hartford to New York City, I tuned in to our sister station, WHCN, looking forward to hearing one of my shows. Instead I heard "Oh Little Town of Bethlehem," "Silent Night," and other holiday favorites. Nobody was there. The station was on automatic pilot. Unintentional Christmas in July.

At the time we played LP records: you know, those ancient, hat-sized pieces of vinyl. In any case, there were times I might report for duty in the morning after, let's say, a lively night on the town. Head throbbing, four hours' sleep. And I was there alone. The goal on days like these was to hunt down one of the longer recordings, such as one of Beethoven's or Mahler's symphonies that clocked in at well over an hour. This allowed plenty of time for a jaunt (or maybe a crawl) down to Hayes Bickford or the Schrafft's Coffee Shop four stories below our studios. Two over-easy with bacon and links, maybe a small stack with homemade syrup, several cups of coffee enjoyed over the Lifestyle section did wonders for me one morning before I leisurely made my way back to the studio where I heard Beethoven's Fifth, which had never progressed past the first eight beats for the past fifteen minutes:

*BA BA BA BAAAAHHHHH...BA BA BA BAAAAHHHHH....BA BA BA BAAAAHHHHH...*

...and so on. Not my most shining moment in radio, I have to say.

I would like to imply that this happened only once. I really would. But I can't. The call of Schraffts was so strong that over the course of ten years at WBCN

I took the plunge downstairs a few more times. There aren't many ugly expressions in radio, but here's one: "Dead Air." Before we figured out it might be a good idea to bring a portable radio down with us, I had made the miscalculation of overestimating the length of Ravel's *Bolero*: 15 minutes, 50 seconds. My breakfast and scanning the sports section: 21 minutes, 34 seconds. The sound of a record turning in the groove is horrible. People called in thinking the station was off the air or that the show was cancelled. Mr. Hastings certainly had some choice words for me. Sponsors didn't love it either.

On another day I got back in time from a lightning fast lunch to find I was locked out of the building. The words "despair" and "panic" jumped alive for me as I banged on the filthy windows of the snoring super's dank basement apartment, screaming, "Joe, wake up! There's *dead air!!*"

■ ■ ■

Though a man ahead of his time in many ways, Mitch Hastings was not the consummate businessman. He had a lot of trade agreements in place of real advertising revenue, and all of his stations employed the same format, which turned out to be, in the end, not a profitable concept.

He was also a wildly eccentric gentleman, a disciple of the self-described psychic Edgar Cayce, who claimed to be in touch with the citizens of the lost continent of Atlantis. A Harvard grad from an established Boston family, Mitch espoused a variety of philosophies and topics, but as the years went on and the station started to teeter financially, the philosophies and topics just got weirder and weirder.

Keep in mind the world was changing dramatically. I was twenty-five, it was 1963, and the Vietnam War was escalating; American protesters exploded in number, marching in Harvard Square and across the country. And I will always remember, like everyone alive at that time, where I was when I learned about the assassination of John F. Kennedy. My dad and I had plans to see the symphony together that day, and stopped for lunch at a restaurant across the street from Symphony Hall. Above the deep fryer a soap opera was droning on. Walter Cronkite broke in. We watched his shattered face: "Shots were fired in Dallas, Texas..."

We walked across the street to Symphony Hall, where many had not yet learned about the awful news we'd just witnessed. The Boston Symphony opened with their first piece, which was followed by a long intermission. The conductor, Erich Leinsdorf, came out and announced that the President of the United States had been assassinated. A huge gasp, crying. When the crowd quieted, he announced that they would play the funeral march from Beethoven's *Eroica* Symphony.

I had seen President Kennedy just a few weeks before, on the steps of the Copley Plaza hotel during a visit to Boston for the Harvard–Yale game. We made eye contact and smiled at each other; there was a brief sense of recognition. His charisma was such that I could feel it from clear across the street where I stood.

The world was rocking on its axis in so many ways, including musically. The tide of rock 'n' roll was building against the relatively still oceans of symphony, jazz, American Songbook, and other established traditions. WBCN was there: the calm beach sunning itself as the tsunami of rock gathered strength.

As our sponsors started to bail and ratings dipped, Mitch's behavior progressed from odd to simply inexplicable. But he had a good heart and, no matter what happened, always had faith that everything would be fine.

One day I came to work to discover he'd hired a new program director who wanted to implement a crazy system for the playlist involving wheels and charts. The theory being that if we could mix our tempos a bit, maybe this would spice things up. Actually, it's hard to believe the absurdity of these words even as I am relating them, but here are some more: all the records were marked "Slow," "Medium," or "Fast," and were stored in bins labeled as such. No more playing three "Slow" records in a row! Even though symphonies have, ahem, things called *movements,* which often range in tempo, we'd have to follow this bizarre chart that told us when to choose records from the three different bins.

At least *that* strategy didn't last long.

Maybe because of his intimacy with radio waves in general, Mitch came to believe that he could communicate with spirits. Not a problem if all was going well, but it wasn't. Many a day I'd show up at work and find Mitch standing in his office in his underwear, shaving, and conversing with himself in the mirror. Again, not a problem in and of itself, but one morning I noticed the studio clock, a huge expensive piece that showed the time around the world, was missing.

"Um, Mitch...?"

"Ron! Good morning, my friend. Glad you're here. I was just thinking about the lost continent of Atlantis—"

"Mitch, the clock is missing. They must have repossessed it after we left last night—"

"Ahh," he said, carefully shaving his neck. "Time. What a concept! I'm so glad you brought it up. What is the past, really, and who can know the future? It's all about the present, Ronnie. I am a parishioner at the Church of *Right Now;* have you heard of it?"

"But how will we..."

He glanced at my wrist. "What's that on your arm, but humanity's way of keeping time?"

I sadly looked at my watch. Unpaid bills danced before my eyes; the studio's and my own.

The next day I found him again in his underwear, this time eating a doughnut at his desk and tinkering with a radio, the parts scattered all over its surface as if the radio had exploded. Behind us, big grunting bruisers were dismantling our AP and UPI teletype machines which brought us news from around the world. Mitch had skipped paying the bills that month.

"Mr. Hastings, the AP guys are here! They're pulling the wires! They're taking the machines!"

He snapped a tiny part of the radio back in place and took a big powdery bite of doughnut. "That's okay, Ronnie, have faith. These earthly matters always have a way of working themselves out."

"But how will we do the news?"

"Just read from the newspaper, my friend, if you even care to do that."

"The *newspaper?*"

He cocked his head. "You make a good point. Newspapers are absurd. Irrelevant. Tomorrow will be the same thing, different parties involved. But the same old stories: love, war, strife, poverty, suffering, mixed with the occasional bright spot: a new bridge, or perhaps your favorite sports hero makes the touchdown to win the game. I just can't concern myself with it."

The next day, although I was pleased to find Mitch wearing pants and a shirt, I also found a notice taped to the door saying the power would be shut off by nightfall if we didn't pay the bill.

I said, "Mitch, they're gonna cut the juice. We're going to go off the air."

He looked at the notice. "What a bunch of soreheads. A bunch of bureaucratic slaves. Have faith," he said, and left. "I'll go scare us up some funds."

The really crazy thing is, he always did. Yes, we were off the air for twenty-four hours, but somehow he paid that bill. Actually, I'll never know *who* paid that bill.

Inevitably we would go weeks without getting paid. Once we didn't get paid for an entire month—I mean *nobody* got a check. But we hung in there, as dedicated to the classical music as he was. The miracle being that he was able, for years, to scrape up the money to keep us on the air and finally pay us all we were owed. He'd find investors, people who believed in his vision. My friend John Henning even joined the madness there for a few years.

■ ■ ■

In 1961, I was promoted to program director, a position I held until 1968. Meanwhile, our stiffest competition was WCRB, which also ran full-time commercial classical radio. During the fifties and early sixties there was enough cultural interest to support these two stations, but WCRB was really running away with it.

Meanwhile, Harvard Square was jumping with Vietnam War protests, demonstrations, even the occasional riot complete with tear gas. These scruffy kids would come in off the street and visit the station, fresh off the picket line and reeking of pot. They'd flop on the couch and beg us to play The Kinks, The Doors, Janis Joplin, Chicago. Mitch would show up an hour later, take a sniff at the studio, and say, "Now that's a mighty interesting odor we have here. Have our little friends been by again?"

"Yeah, and they brought us some LPs..." I said, staring at a pile of Jefferson Airplane, Sly and the Family Stone, the Beatles.

"Well, isn't that nice."

But Mitch couldn't avoid the writing on the wall. He told us it was time to ask for money—up, down, and sideways—in whatever ways we could think of; at the same time he began to give in to the notion of a mixed format. There was a time when we'd segue from Bach fugues in the evening to Cream blasting in at midnight.

The station began its transition to an underground progressive rock station format on the night of March 15, 1968. WBCN's first rock DJ, "Mississippi Harold Wilson" (Joe Rogers), was the first to use the station's new slogan, "The American Revolution." At first the new format was only heard during the midnight hours, but in May 1968 they went full time. In early June, the station's air staff: Peter Wolf, Tommy Hadges, Jim Parry, Al Perry, and Sam Kopper, who later became Program Director, were joined by Steven "The Seagull" Segal. In December, Peter Wolf left to take the J. Geils Band on the road, and Charles Laquidara was hired to take over the 10:00 PM to 2:00 AM air shift. He later hosted *The Big Mattress*, which became one the most popular radio shows in Boston radio history. The station became a social and political force from the moment it hit the air, both defining and promoting popular culture and politics in Boston for the sixties' boomer generation in a way that nothing had before.

Meanwhile, thank God, I had been working part time at WGBH, getting my feet wet there. Mitch wasn't the only one tripped up by the rock invasion; I mean, here I was, weaned on the American Songbook, Broadway, film, jazz, opera. These guys were playing Vanilla Fudge. The whole scene soon morphed into this raw, cutting edge rock station, from Prokofiev to the Stones, in what felt like a matter of weeks.

In the end though, T. Mitchell Hastings was kept on as a figurehead at the American Revolution. He very much wanted to keep me on the ship, even though I think we both knew my days there were numbered.

"Well, Ronnie, how would you like to be Public Service Director?"

"I'm sorry, I think I'll have to be moving on."

"Well, we're going to miss you." He got a misty look in his eye as he stared out his office overlooking Newbury Street. "Atlantis is really out there, you know, we just have to find it. These things happen. With today's technology and my know-how, it's a matter of time before this magical place is discovered and brought to light..."

Though Mitch never did discover Atlantis (as far as I know), he did sell the station in 1968 for a couple million dollars to Infinity Broadcasting. Rude, irreverent, and wild, it became one of the most listened to and profitable rock stations in American history.

■ ■ ■

I entered a new phase of my life when I joined WGBH on a full-time basis in 1969. I think I'll be forever grateful for the opportunity to be surrounded by the kind of talent and innovation that especially marked my earlier years at the station. The explosion of creativity going on in the studios, then located at 125 Western Avenue in Allston, was remarkable. Full operas were performed and broadcast from the studio. Leonard Bernstein himself conducted his orchestra for live broadcast. Gunther Schuller, Pulitzer Prize–winning composer, would visit and perform regularly. *The Boston Pops, Evening at Pops, Boston Symphony Live, This Old House* (still on today) were new in public television and really set us apart.

Though I started off hosting *Morning Pro Musica*, over time I cycled through an eclectic mix of shows. I became the television booth announcer not only for Channel 2, but part time at Channels 4 and 7 as well. Though now, of course, everything is taped and automated, in those days there was always a live announcer at the station. Every single station break: live. There was no spacing out or missing a cue.

I am also indebted to the good people at WGBH—the best—who had my back every time I was slow to round the technological learning curve. During all my decades at the station, advancements seemed to come fast and furious. To think we used to edit reel-to-reel tape by actually cutting it with a razor and taping it together. Over the years we progressed from vinyl to CD, analog to digital, reel-to-reel to cassettes, and finally to digital audiotapes or DAT. Accessing my playlist from a computer versus literally a pile of tapes still feels like a revelation at times.

During those early years I felt myself growing into my own skin—filling in the person I had sketched out as a bad-at-baseball thirteen-year-old broadcasting into a cardboard microphone in my bedroom. I could feel myself blossoming professionally, like my brain was expanding. I had such a sense of being in the right place at the right time, with the right people, doing the right thing with my life.

Isolated in the studio, I became more intimate with sound than ever before. Bands and the musicians in them, the orchestras, all became the voices of friends. I no longer had to read the notes to know who was in the room with me; there's Count Basie's band, that has to be Renata Tebaldi as Desdemona, that's George Shearing on piano, there's Illinois Jacquet's "Flying Home." Even the BSO had its own signature sound.

I also learned the absolute necessity of being able to turn it on and off. By that I mean: no matter how I was feeling: upset, sad, angry—when that second hand swept past go-time—I was BING!—on the air. In radio as on Broadway: the show must go on.

But if I wanted a few role models for broadcasting, I didn't have to look far; legends surrounded me, both figuratively and literally. The always elegant William Pierce, the original voice of the Boston Symphony Orchestra, had an office a few doors down. I'll always be grateful for his one-on-one lessons on articulation and pronunciation, especially for German, French, and other classical music names.

Photo courtesy of BSO archives.

Bill Cavness was another terrific voice to learn from. An actor, musician, and scholar, Bill knew more about literature than anybody I had ever met. His series "Reading Aloud" brilliantly presaged talking books, bringing Pasternak's *Dr. Zhivago* alive—including voices and dialects of all the characters—in 1958, when the book was not yet widely available. In his thirty-plus years on the air, he made over a thousand recordings for the series. When asked why he devoted his life to radio, as opposed to television or other media, he responded, "Radio, with its appeal to a single sense, forces me to keep my imagination working, while television somehow blocks much of its use. When I listen to a play on the air, I design the sets, costumes and the lighting—even the faces and the characters' movements."[6]

Robert J. Lurtsema hosted *Morning Pro Musica* for close to thirty years with a sonorous voice that was a cross between Orson Welles and Arthur Godfrey. He was known for his long, languorous pauses between phrases, which sometimes stretched out over a minute, an eon in radio time. Audiences either tolerated it or hated the habit, but most people loved the way the show began: birds chirping for a few minutes at seven in the morning...much, much longer if he was running late. On one occasion he was running at least twenty minutes behind because of a blizzard. He called to tell me two things: "start the birds," and to instruct me in no uncertain terms to stay off the air. He said if I went on, people would think something had happened to him. I'll never forget the sight of him standing in the doorway looking like the abominable snowman.

Each morning began with a different theme, as he systematically explored a composer's work such as the Beethoven string quartets or Dvorak symphonies, at times sampling more obscure music, confessing that he learned as much as his audience through his explorations. Respighi's *Ancient Airs and Dances Suite* and Giovanni Gabrieli's triple brass quintets were among his opening themes.

As a compatriot, Robert J. was a bit of an enigma. Here was a man who, on April Fool's Day, made fun of his own pomposity by chirping out the birdsongs himself (surprisingly few people noticed!) and who did a great Gabby Hayes—all off microphone. He was obviously a beloved character: when the station threatened to end his news broadcasts, there was such an uproar that his role was immediately reinstated. People loved his style. On the other hand, there were some unfortunate experiments including the pure, unmitigated strangeness of R. J. Lurtsema wandering around the Museum of Fine Arts, rambling on about paintings and sculptures, quoting ponderous references and dates at great length. Brevity is the soul of wit, especially in radio.

Many of us learned where Robert's sense of humor about himself began and ended. Nat Johnson, a great talent with the triple roles of producer, announcer, and audio engineer, preceded Robert J.'s role as host of *Morning Pro Musica*. When Robert had Nat back as a guest on his show to discuss Nat's program about the organ, "The King of Instruments," Nat made the mistake of calling him "Bob." Lurtsema snapped off the mic the moment the show concluded. "Never, *never* call me Bob on the air...."

Louis Lyons was a great character on Channel 2, a crusty veteran newsman for *The Globe* and the *Christian Science Monitor*, as well as curator of the Nieman Foundation. Blessed with a quick delivery, he was not a man to mince words. One evening, a crew member was signaling him to conclude his broadcast. He said, right on the air, "Young man, I'll get off when I want to. Don't signal me like that. The news director told me I can have as much time as I want." And he went right back to the news. In a neighboring studio, Boston's leading drama critic, Elliot Norton, might be interviewing Richard Burton, Carol Channing, Rogers and Hammerstein, or Anthony Hopkins on *Elliot Norton Reviews*. In our studio kitchen, Julia Child was revolutionizing the world of cooking in her uniquely charming way on *The French Chef*. Any time I had the chance I would watch her do her show. One episode stays with me: while preparing suckling pig she took out the cooked one to showcase the finished product. With her little finger she tucked the tail of the pig right inside, commenting, "And you know, there's always a convenient place to put this little tail."

Speaking of candid moments, nothing keeps you on your toes quite like doing live radio or TV. I was a bit intimidated by WGBH in the early years because the place had such a vaulted reputation. I was so excruciatingly conscious of every word I uttered on the air and its implications that it was only a matter of time before

I royally screwed up and stepped in a big pile of it. But the thing is, you have to keep going. You cannot go back, and more than that, I've learned it's best to pretend that whatever dud has fallen out of your mouth, it's best to feign complete ignorance. Not unlike a belch at a state dinner: let it pass. *It never happened.* Yes, I will try some of that merlot, thank you so much...

It was a beautiful fall day, and I had just finished with *Morning Pro Musica.* I took a few seconds to look over a simple public service announcement: a reminder about Daylight Savings Time.

I said, "Daylight Savings Time is right around the corner. This coming Saturday night, in fact. So don't forget to set your cocks back before you go to sleep. And that's it for our show today..." I don't remember anything else. It's possible I blacked out.

Yes, it went out over the airwaves to thousands of Boston listeners. No, I was not reprimanded. I think the powers that be knew I needed no such beating, that I would be replaying, red-faced, that particular quote in my brain for the rest of my natural life.

Another good rule of thumb in broadcasting is to never assume the mic is off. In fact, it's best to assume it's on while you're in the studio, the hallway, the cafeteria, the men's room, even all the way out to your car where it might be okay to open up the pipes and scream, curse, or whatever you need to do. I know this because once I didn't follow that rule. One morning in the TV broadcast booth, after being assured my mic was off, I went on a rant about the program director and, lo and behold, the mic was on, and *we were live.* I wrapped up with something like, "Well, at least that sonofabitch isn't listening," when in fact, the sonofabitch *was* listening. The only reason I think I wasn't maimed was because there was enough music playing over my comments that the object of my scorn really didn't hear what I said. What did listeners hear? I'll never know.

I do know I went directly to mass after work that day.

On one beautiful spring morning, perhaps inspired by Robert J., I decided to do a little piece on the swallows of Capistrano that return every year after a fifteen-thousand-mile journey—almost a complete flight around the world—to the Mission San Juan Capistrano in California on March 19th. A compelling phenomenon of nature, and one certainly worthy of a few minutes of radio time. After much shlepping around, I found a tape of swallows chirping and cheeping, and I cued it up to play after the Capistrano story.

"So that's the story, folks," I said. "No one knows why, but every year on this very day, thousands of swallows flock back to this beautiful mission by the bay, their cheerful chirping sounds a welcome sign of spring..." cue the engineer to play the bird tape: "Cheep, cheep, chirpy-chirp..." then some guy comes on and says, "...and those are the sounds of swallows from Hudson Bay, New York."

I hadn't listened to the whole tape.

A hundred corrective measures flew through my brain (like sparrows on the wing) but I realized, no: just let it go.

I'm telling you, between that and my advice about setting your cocks back in the fall, it's a good thing I had a sense of humor and forgiving bosses.

I had my rough days in front of the camera too. Occasionally I did the news for Channel 2 on Sunday nights, when the entire crew numbered just three: the announcer (me), the engineer, and the technician in the video room who jockeyed the tapes.

Picture this: I'm sitting at a desk, the one camera locked on me, monitor off to one side. We're live. I'm reading the news and I'm reading the news and I'm reading the news when I hear this little voice in my headset say, "You better slow down, you're reading too fast." I notice that, sure enough, seven minutes remain in the hour while in fact only a half page of news remains to be read.

I slow down, but how slow can you read, anyway? I drew out the weather report as much as humanly possible. "Will it rain tomorrow? Well, you know, we have a forty percent chance of precipitation, but you never really know, do you, especially in Boston! Of course there's a higher chance of rain on the Cape and islands because of the effect of oceans on the weather..."

"Keep going, keep going..." came the engineer's voice through the headset.

"...which can tend to really pack a wallop, especially if a hurricane is expected, which of course, we certainly don't anticipate today..."

"...go on, Ron, go on..."

"...and so why don't we have a look at those headlines again, in case any of you missed anything..." and I reread the headlines, which brought us up to the hour, at which point I said, "And that's the news."

The camera stared at me and I stared back, watching myself watch myself in the monitor.

Nothing happened.

"Keep going, keep reading, we can't find the tape." Panic now in the voice.

"Thank you, and that's the news..."

"Anything, Ron, *just talk...*" the engineer pleaded.

I flipped the AP sheets over, desperate for any other news that might be lying around.

Blank paper.

I actually opened the drawer praying for anything—old news, a flyer, a coupon, a shopping list, something to fill the time.

"And don't forget to stay tuned for the David Suskind show, a favorite here and across the country for ten years now, and we've got a great show lined up for you tonight..."

"Go...go...one more minute!"

I think it's then I realized the engineer's cues were being broadcast as well. I don't know how I maintained bowel control. I imagined tapes flying in the video room as the technician tore through stacks of them looking for the next show.

"And don't forget, we've got Julia Child coming up for you too. Yes, *The French Chef* with Julia Child herself waiting in the wings, so don't touch that dial. That woman can do things with a suckling pig that would blow your mind..."

"Don't stop!"

"...and let me tell you, no one goes home hungry here at the station after one of her shows. So stay tuned, and we'll have that wonderful new show for you coming right up. All kinds of great stuff coming up for you..."

I stood up.

"No, Ron, don't leave—one more minute, come on..."

"Thank you very much, and that's the news."

And I walked off as the camera stayed, broadcasting an empty chair behind a desk for four endless minutes before we found the tape.

■ ■ ■

Though I had begun to grow steadier on my "air legs," I had a ways to go with my personal life at the time. I had married a wonderful woman named Jackie; however by 1970 things were not only extremely shaky between us, but Jackie became pregnant.

Around the same time, my dad suffered a stroke which impacted his ability to do the thing he enjoyed the most: paint. It also meant that my mother had to take on the role of caretaker for him. Within two years, my dad had passed away.

Though we tried counseling, in the end Jackie moved to Phoenix with our almost two-year-old son, Aldo, and I was left in a lonely suburban house staring at an empty crib, wondering how everything could have gone so completely wrong. Working at one radio station and two TV stations and frankly being less available than a husband and father should be had finally taken its toll.

What was important in life? I began trying to figure that out, in big ways and small, and everything in between. By this point, I knew two things: the marriage had ended, and that I must nurture a relationship with my son. Though my schedule was challenging, I had to devise ways of going to Phoenix to spend time with him, as well as bring him here to Boston to do the same.

I had him for the summers, and we squeezed every moment of joy we could out of every day. He loved animals, so I took him to the Trailside Museum in the Blue Hills where we roamed around the woods for hours. Or we'd hit the go-carts (he was obsessed with them!) at amusement parks in the South Shore, or slip off to Faxon Park in Quincy to see the ducks. I took him swimming at Nantasket Beach, where I first learned to swim, and he spent time with my mother who I'm so grateful he got the chance to know.

Every other Christmas was a desert Christmas, when I visited Aldo in Phoenix under the hot summer sun. We'd do the same things out there as in Boston: zoos, go-carts, collect bottle caps for prizes; whatever rocked a five-year-old's world, I wanted to do.

I thanked God for music, and my wonderful, life-affirming job. I was beginning to understand that this is life, now, every minute we're here, that it's not some practice run, and that everything I said or did mattered and had consequences. I also learned to understand that though I certainly had my faults, I was not to blame for everything.

Pain breaks you open in good ways sometimes. I started to think about what I really wanted out of life, both for Aldo and myself. Never mind building a foxhole; having a kid makes you an adult—changing diapers, staring at the face of a human you helped create. Now *that's* frightening. Not to mention all the love I felt, and it was a new kind of love I hadn't experienced before and didn't know existed. I remember listening to a lot of love songs at that time, or songs of love and loss, which is I guess one definition of the music of the Great American Songbook. Tony Bennett, Rosemary Clooney, Bobby Short, Ella Fitzgerald, Nancy Wilson, anything by Gershwin. I would listen and think about what I did right, what I did wrong, what I had hopes of doing right the next time around. In short, I returned to my original therapy: music. Filtered through the newly beat-up lens of my life, each song took on new meaning and richness, and carried me, note for note, through those dark and troubled times.

## CHAPTER 5

# A World of Joyce

IN JANUARY OF 1977, I BEGAN HAVING CONVERSATIONS WITH THEN-station manager John Beck, who'd been turning over a few ideas about my hosting a new show. At the time, WGBH featured a mixed format in the afternoon, and we both agreed the lack of consistency was thinning down and scattering our audience. When I suggested the concept of a show on American music beginning around the turn of the century, he surprised me by jumping on the idea. I asked him why he wanted to take a chance with something so new and untested, and he said, "Ron...*anything's* better than what we have on the air right now!" I still have a good laugh every time I think about that remark. Together we came up with the name *MusicAmerica* as a fitting title for this pilot.

If someone asked you to describe *MusicAmerica* in a word, I doubt you could do it. Imagine visits in your living room with Dizzy Gillespie, Benny Carter, Illinois Jacquet, David Raksin, John Williams, Gunther Schuller, Fred Astaire, Eileen Farrell, and countless others. From noon to five every weekday for what turned into eighteen years, I hosted what I would describe as a variety entertainment show that had at its core the Great American Songbook, showcasing the innovators of American music.

As the years went by, the show became a mix of live interviews with current recording artists, as well as selections from genres such as Broadway, jazz (America's classical music), film, and swing. The show was an opportunity for me to debut something never before attempted in the Boston radio market. Certain stations aired segmented shows such as all folk, all jazz, all blues, all classical, but *MusicAmerica* was a novel concept. Even though I'd been involved in radio for over twenty-five years (thirty if you count my debut at age ten!), I never dreamed I'd see the day when I'd be able to express myself on the air the way I could with this show. I was quite simply a kid in a candy store.

What a privilege for me to develop this program! If someone had asked ten-year-old Ron what he'd like to do as a broadcaster when he grew up, the answer would have been: *MusicAmerica*. At my fingertips were literally thousands of pieces, decades of wonderful music from which I could create a unique show each day. The opportunity to overlap that collection with the work of current artists and modern interpretations of the standards only added to the richness of what I could offer.

■ ■ ■

In the beginning, the show focused on American classical music composed from 1900 to 1960: works by George and Ira Gershwin, Aaron Copland, Ferde Grofe, Walter Piston, and Charles Ives, with a lively, soulful mix of selections by Cole Porter, Irving Berlin, and so many other greats of that era.

On the very first show, which broadcast on September 1, 1977, I wanted to ease into things slowly with classics I felt most people loved and would recognize, though of course the plan was to grow the program organically through listener feedback and my own musical intuition. I opened with the "Three Preludes" by George Gershwin, followed by his *Concerto in F.* From there I played Ellington's "A Drum Is a Woman," then Leonard Sillman's "New Faces of 1952," dropping in favorites by Cleo Laine and jazz violinist Joe Venuti.

As time went on, I became more comfortable introducing changes here and there. Feedback from listeners helped shape the show; in fact, an ongoing, eighteen-year conversation with my audience led to the rich variety that followed. The pure synergy of the thing inspired me every day of the week. Talk about jumping out of bed excited to go to work in the morning! I believe I was creating that show at some level of consciousness twenty-four hours a day and loving every minute of it.

For me, the best thing about *MusicAmerica* was its spontaneity. I could play whatever I wanted and put together shows based on what was happening at the moment. Listeners never knew who would drop by for "delightful conversations," wrote *Boston Herald* columnist Joe Fitzgerald, concluding: "...Dizzy Gillespie, Dave McKenna, Joe Williams, Margaret Whiting, maybe the Four Freshmen, (listeners) just knew (they) were going to love it."[7]

If a musical was in town such as *Ain't Misbehavin'*, I'd invite the entire cast on the show. If Tony Bennett, or Bobby Short, or Rosemary Clooney, or Mel Tormé happened to be in Boston, they wouldn't get back on the plane before stopping by. I might get a call tipping me that Hoagy Carmichael, Jr. was in town, and we'd book him for the following day. And they wouldn't just pop in and chat for a few minutes, they'd stay for half an hour, an hour, sometimes even longer. Most interviews took on a life of their own, but I found the best format was to splice in generous tracks of the artist's classic recordings between frank discussions about their lives, their musical philosophies, and their creative process.

As much as I enjoyed snaring out-of-town talent as they were passing through, I felt just as passionate about inviting local artists on the show, especially those who might be at the beginning of their careers: Rebecca Parris, Scott Hamilton, Deborah Henson-Conant, Gray Sargent, Steve Marvin, and Donna

Byrne became regulars in the studio. As the show gathered steam, *MusicAmerica* became a magnet for this new talent. Artists like Harry Connick Jr., Diana Krall, Wynton Marsalis, and Michael Feinstein also came on the show when they were first starting out and touring through Boston. I loved being in the position to provide another place (besides Scullers or the Regattabar) for budding or local musicians to showcase their talent.

One of the best aspects of the show was the freedom to do impromptu tributes to musical giants who had passed away, or those who had a birthday. Many times, after learning in the morning that an artist had died, I was able to play a wide selection of their work that very afternoon, or certainly the next day. With a library of recordings at my disposal, I could create a tribute that put an artist's entire career in perspective. When Fred Astaire died in 1987, I realized I had one of his only interviews: the tape from the day we called him on his eightieth birthday; so I was able to put together a five-hour show on his career. I did similar tributes for Sarah Vaughan, Ella Fitzgerald, Dizzy Gillespie, Stan Getz, and Mel Tormé after they passed. Though sad to see these giants go—many of whom had become dear personal friends—I felt privileged to have interviewed them.

Half the fun was advancing my own musical education as I put the show together from day to day. As much as I knew about music, I'd inevitably learn something new. Let's say I was playing a whole afternoon of Duke Ellington. I'd put on "Take the 'A' Train" and maybe the soloist was Ray Nance, inspiring me to dig out and play something by Ray. Or perhaps the alto sax player was Johnny Hodges, leading to something by Johnny that might turn me on to other aspects of Ellingtonia. On an artist's birthday, such as Cole Porter, I grabbed the chance to play his lesser known works, most of which I had never heard before. Listeners would call and tell me about musicians I never knew existed and send me their recordings. I branched out even further, interviewing authors with a connection to the musical world such as Neal Gabler who wrote *Winchell: Gossip, Power and the Culture of Celebrity* about Walter Winchell, a journalist immersed in the lives of Broadway celebrities in the twenties and thirties. Putting the show together was like opening a series of musical doors that led to daily discoveries of obscure musicians, composers, and others whose voices I felt should be heard.

To me, there was an everlasting quality about this music. The feeling that, yes, all things must pass, but these songs would forever be deeply appreciated and enjoyed by the American public. I realize now that timeless quality was a huge draw for the show. The peerless music of the 1920s, '30s, '40s, and '50s will never be duplicated. There is only one Leonard Bernstein and *West Side Story*, just one Rodgers and Hammerstein and *Oklahoma!*, one George Gershwin and "Rhapsody in Blue." The songwriters and lyricists—Cole Porter, Harold Arlen, Hoagy Carmichael, Harry Warren, Johnny Mercer—were the innovators of their time, while the musicians—Louis Armstrong, Lester Young, Coleman Hawkins,

Mabel Mercer, Ethel Waters, Lena Horne, Anita O'Day, Sarah Vaughan, and Ella Fitzgerald—all added their unique interpretations to original scores. Just as one never tires of discovering elemental truths in great literature or entering the landscape of a Winslow Homer or John Singer Sargent painting, in my mind it is just as impossible to be untouched by the experience of listening to Gershwin, who drew classic musical portraits with notes on the page.

Perhaps the lead editorial in the August 31, 1995, edition of the *Boston Globe,* which lamented the decision to cancel the show, expressed it best: "We are not talking nostalgia here. Many songs carry memories to some listeners—they are a vital part of the nation's cultural fabric. But this goes beyond history. The songs continue to be sung, to be revived, and reinterpreted, because they are literate and swinging in a way that makes them a monument for the ages."[8]

■ ■ ■

As the years passed, *MusicAmerica* became an integral part of the Boston jazz scene and exploding music culture. Over sixty thousand listeners tuned in to hear who was in town, what I'd seen that week, and what was coming up. And it worked both ways: well known musicians told me the show had no equivalent in other cities they played, and thus sent the talent my way. Musicians have their own network, of course, so word about *MusicAmerica* spread from coast to coast. Eventually, my phone rang off the hook with club owners and promotion managers hoping to snare their clients some air time.

I always thought the show had a New York feel to it. I was very influenced by New York radio hosts such as William B. Williams at WNEW. William B.'s format was also a mixture of big band, swing, jazz, and the American songbook. As host of the William B. William's show, he had the chance to interview top talent such as Lena Horne and Nat King Cole, even befriending Frank Sinatra, Jr. when he recorded an early broadcast at the station. Williams mused that if Benny Goodwin was the "King of Swing" and Duke Ellington was a duke, then surely Sinatra was "Chairman of the Board," a moniker Frankie ended up embracing completely.

I also listened closely to Al Jazzbo Collins' show, broadcast from his studio which he called "The Purple Grotto," referencing not only the purple paint job but the stalactite-like sound-absorbing cylinders suspended from the ceiling that completed the feel of a funky cave. Or, it could have been the weed. In any case, "JazzBeaux" (as he renamed himself) was one of the coolest voices in radio for fifty odd years, a professional hipster who did everything from twisted readings of *Grimm's Fairy Tales:* "78's from Hell: Grimm's Fairy Tales for Hip Kids," to a brief stint in 1957 hosting NBC's *Tonight Show* (then called *Tonight! America After Dark*), to his own TV show in the early sixties where he hosted personal favorites such as Moe Howard from *The Three Stooges.*

Jean Shepherd—the great American iconoclast—kept me laughing and in awe of his prodigious talent. A master of stream of consciousness monologue, writer, narrator, and radio host, Shepherd told hilarious, off-the-cuff tales on his radio show about growing up, working in steel mills, being in the army, pretty much anything that came to mind. We carried his show, taped in New York at WOR, on delayed broadcast and commercial-free.

Edward Grossman in a January 1966 *Harper's Magazine* piece caught his spirit:

> Quite as often, (his) stories are informed by a mass of details that ring true. Shepherd has total recall of the name, rank, and shape of everyone in his barracks; of who played first base for the Chicago White Sox in 1939 (Zeke "Banana-nose" Bonura); of how the Little Orphan Annie *theme song used to go. He remembers sounds, too, and mimics them convincingly. His Heat Lightning over Camp Crowder is distant and ominous; the Gurgling Sink over which his mother struggled forever "in her chenille bathrobe" is enough to give a hardened plumber pause. Together with this realism Shepherd throws in a stiff dose of hyperbole. When his squad is on an overnight hike, for ex- ample, it covers "ninety-seven miles," and the mercury during the day hovers at "a hundred and forty," while at night it plunges to "eighteen below zero."*[9]

His satiric narrative style was considered a precursor to the work of Garrison Keillor, Spalding Gray, and especially Jerry Seinfeld who said in the "Seinfeld Season 6" DVD set, "he formed my entire comedic sensibility—I learned how to do comedy from Jean Shepherd." Seinfeld even named his third child "Shepherd."

I especially admired "Shep" for his uncanny ability to convey the feeling that he was speaking to every listener personally, one on one, through the radio. He inspired me to reach out in my own way to my audience every chance I got. I'd toss out a movie or jazz trivia question or play a piece of music, giv- ing people a chance to win tickets or CDs. What band first recorded "Sing Sing Sing"? What was Tony Bennett's real name? It always amazed me what a knowledgeable audience I had out there; they were so hip I could never stump them. No matter how obscure the question, there was always someone out there who knew the answer.

I also enjoyed programming according to mood. If it was a dark, dreary day, I'd play upbeat music. As *Boston Herald* columnist Joe Fitzgerald recalled in his 1996 piece on *MusicAmerica*, "On the muggiest day of the summer, (listeners) were apt to hear Tony Bennett's 'Winter Wonderland,' or Ella Fitzgerald's 'Santa Claus is Coming to Town,' or maybe Billy Eckstine teaming with Sarah Vaughan on 'I've Got My Love to Keep Me Warm.'"[10]

But at times I did feel like shaking it up behind the mic to make sure people were paying attention. One day I decided to conduct the show as usual, but with one exception: I'd substitute names of Italian dishes for the real last names of the singers or musicians. "And that was Joey Tetrazzini on drums, Rico Manicotti on tenor sax, and Bobby Gnocchi on the horn...next up, Freddie Ravioli and his band with, 'One O'Clock Jump...'" Finally, *finally*, at the end of the day, someone picked up the phone and called to ask, "Hey...are those real names?"

■ ■ ■

Being on the radio is incredibly personal. There is this beautiful intimacy to it. You're in the studio by yourself, yet you're everywhere. Because of the quality of what I programmed for *MusicAmerica*, listeners often told me they felt they should be dressed up while tuning me in. But in reality, people did and of course do very intimate things when they listen to radio. As the voice of the program, I was in their car, their workplace, or their bedroom, the voice they listened to as they danced with their loved ones in their living room. In fact, I often wondered what percentage of the population of Boston was conceived during the hours of twelve to five on weekdays. The guests I brought in—Rosie Clooney, Mel Tormé, Tony Bennett—spoke so frankly and vividly about their careers and even their personal lives. I'd play their music, the kind that many people fell in love to and grew up with, or raised their children with the show in the background. People wrote telling me they planned their day around the show, or that they didn't get out of the car until a piece of music or an interview had ended. I even heard from one of the world's leading orthopedic surgeons, Dr. Michael Goldberg, who told me he scheduled his days in the operating room around *MusicAmerica's* opera programming.

In essence, I never knew who was listening or how what I played would affect them. I felt that, for one gentleman in particular, the show helped provide a validation for a life that revolved around a record collection. Through Larry Katz, I had access to one of the greatest assemblies of recordings I've ever encountered. An avid devotee of big band releases and radio shows, Larry was an endearing sort of eccentric who I first met at a benefit I was doing for a hospital in Lynn. Clutching a stack of LPs, he stepped right up to me and asked for my autograph, then offered to share his entire collection with me. He began to bring his records to the station until finally we had to create a special section of the library just to house his massive donation. I was blown away by what he'd gotten his hands on over the years! This incredible resource made it possible for me to play rare performances of artists no one else had access to, such as Streisand's very first recording, shows from the Golden Age of Radio including those starring Jack Benny, Burns and Allen; big band broadcasts by Benny Goodman, Duke Ellington, Stan Kenton; and so many others. Larry was my secret weapon for programming the show if I was looking for some obscure piece of music. A lot of collectors don't like to share their material, but Larry was bighearted and just the opposite.

If a broadcaster is lucky, there's at least one time he feels humbled by the understanding that his show has made a unique contribution to someone's life. One day I picked up the phone and heard loud wheezing, followed by the voice of a young man who told me his name was John Hersey. He said, "That woman you're playing now, when did she start singing?"

"That's Billie Holiday. In the thirties."

"Why was she called Lady Day? Who gave her that name?"

"Lester Young. But I'm not sure why..."

The next week he'd call and ask me to play anything by Louis Armstrong, peppering me with questions about his life. About the evolution of jazz. Or why Dizzy's horn was bent up that way. Week after week he called, entranced with George Gershwin's "Rhapsody in Blue" and begging for more, or enthralled with Coleman Hawkins' classic recording of "Body and Soul."

Finally, one week his father called and explained what I think I knew intuitively: his son had polio and was in an iron lung. He asked me if I would visit his son someday, speaking so softly I had to ask him to repeat the question. I realized later he never thought I would say yes. Ed Henderson, then president of the Boston Jazz Society, joined me on my trip to meet this young man and his family.

My visit to John Hersey changed me forever. I'll never forget this handsome young man, his head the only thing free of the enormous cylindrical machine that was keeping him alive. His face was full of life, vibrant, lit with excitement about music, thrilled that we were there to answer his questions in person. He'd been in that machine since he was five years old.

I could barely collect myself to do the show that afternoon.

After a few months his calls tapered off, then stopped completely. His father called to tell me that his son was at peace, and that for the first time he was able to lay him down on the couch in the living room, something John had always wanted to do. I told him how grateful I was to have had the chance to spend some time with his son.

One day I got a call from a man whose wife was dying of cancer. He wanted to thank me for playing the music they had fallen in love to. His wife's favorite song was Sinatra singing "Time After Time," which, knowing these were her final days, I played as often as I could. In a couple of weeks he called to tell me that it was the last song she heard.

Everybody felt they knew me, which was usually fine, but there were those who thought they knew me a bit too well, and I had to draw the line. In a scene straight out of *Play Misty For Me*, a woman kept calling and talking to me as if we'd been out together and knew each other well, each call becoming more suggestive as time went on. She'd ask me if I was feeling all right that day; tell me I didn't sound as upbeat as usual. Very creepy. Then the photos started coming: her in bed in a nightie holding a cocktail...anyway, I stopped taking her calls, the mail tapered off and hopefully she's moved on to someone a bit more receptive by now.

■ ■ ■

Just as there were great experiences with listeners as well as not so great, the same held true for interviews, though I have to say most of the time they went extremely well. I certainly did my best to make guests feel at ease, learning as much as I could about them and playing their music as they were on their way to the studio. I was keenly aware of their often grueling schedules: many times musicians arrived jetlagged and exhausted or hungry, or who knows, they could have just come off another interview where the host barely knew who the hell they were. I would watch them physically relax the minute they understood that I knew them and their work. And if things were going well, it was fun to keep things rolling, perhaps ask them to stay longer, mixing in chat about music history, or relevant composers and lyricists.

Though most interviews were a delight, two stand out as particularly difficult. Besides the chronically troubled Anita O'Day, who simply got up and left in the middle of an interview, my all-time worst was with Eartha Kitt.

I certainly knew that the woman was volatile. It was common knowledge that she'd driven Lady Bird Johnson to tears at a 1968 luncheon at the White House when she remarked, "You send the best of this country off to be shot and maimed. No wonder the kids rebel and take pot."[11] But I wasn't quite prepared for the storm that blew into my studio that day. She showed up with her agent who sat scowling as Ms. Kitt arranged herself, wordlessly, in the studio. It was clear she was in a nasty, foul mood. I thought I had my trump card, however, because I'd tracked down a rare recording of one of my favorites of hers, "Lilac Wine," to start the show.

I opened the mic and said, "We are delighted to have the wonderful dancer, actor, and recording star Eartha Kitt with us in the studio today—"

"Why did you play that? That's the worst record I ever made."

"Well, I've always liked that one, and I'm sure our listeners enjoyed—"

"Out of everything I've recorded you had to choose *that?*"

A touch of dead air here as I regrouped... "I just saw you in *Timbuktu*, and I thought you were terrific. The dancing, the singing...I'm so impressed with the diversity of what you can do—"

"You don't know what you're talking about, do you?"

I cut her mic and spliced in a song in record time.

Eartha took a sip of coffee and made a lemon-sour face. "You actually *serve this to guests?* It tastes like it's been sitting around for days. It's weak, it's cold..." Her agent jumped up to remove the offending brew. I told her it was all we had, but she went on attacking me. It was a complete disaster. I let her rant and rave and stuck to music till the show was over, and she stormed out and proceeded to destroy *Say, Brother*, a TV show being recorded next door, where she was scheduled next.

All I knew was I didn't deserve what she was dishing out, and somehow I wanted another crack at her. In fact, when I met her again, twenty years later at a benefit I was hosting, I was mentally locked and loaded to return fire if I caught any. She couldn't have been more delightful and charming; it's possible she remembered nothing of meeting me or our interview.

On that particular day, however, it took me a while to collect myself and finish the show with my dignity intact. I was still processing things at the end of the day even as I was packing up to leave, wondering what sort of response I could expect from listeners, management, and so on. In short, I had a little black cloud over my head.

Just as I was closing the studio door, the phone rang. I stepped back in and picked it up.

"Ron Della Chiesa."

This soft-as-cotton voice said, "Hi Ron, I want to thank you so much for playing my music."

"Who's this?"

"Peggy Lee."

"Peggy Lee..." I nearly dropped the phone. "*The* Peggy Lee?"

"Yes."

"Peggy—Ms. Lee—I'm such an admirer of yours!"

"Well, that's one reason I'm calling, Ron. I heard about you out here on Bellagio Road in Hollywood. The musicians on my records—they say you play my music all the time."

"Every chance I get."

"And you know, I'm so glad you're playing my newest CD. A comeback is never easy, but I'm in my sixties now, so it's even harder. I so appreciate all you're doing."

"It's an honor to play your music, Ms. Lee."

"Please call me Peggy."

And so Peggy Lee and I became friends even as I watched from my studio window as Eartha Kitt's limo took a long, slow left turn out of the lot toward Boston. But that's the way it was at WGBH: challenges every day followed by moments of magic.

■ ■ ■

Though *MusicAmerica* was rolling along, gathering momentum and evolving every day, I was doing my best to have a personal life and not doing too well. After my marriage ended, I'd entered another long term relationship that looked headed toward a real commitment, but in the end it didn't work out.

And though I was a man about town, which sounds fun and rather romantic; unless you're with someone you have an affinity with, it can make for a lonely life. Or at least it did for me. I began to look for other ways of challenging myself.

While my dreams of becoming a great baseball pitcher were on the wane, I still felt the need to do something athletic. Preferably something that didn't involve too much coordination or the letting down of team members. Something like—running! For years I'd thought about running as I watched joggers trotting along the Charles River; finally one day I stopped thinking about it, bought some decent running shoes, and hit the pavement.

On the first day I ran a mile. I thought I was going to die. It was wrong to be that sore. But after babying myself for a few days, I took to the road again and made sure that no matter what, I ran two miles a day. My blisters had blisters and my hamstrings screamed afterwards but no matter how tired I was or how bad the weather, I did my miles.

I can't remember exactly when two miles started to feel like nothing, as natural as breathing, or when I felt I could have kept on running another two miles with ease. But I do remember one gorgeous fall day, the air tasting like apples and smoke, and I never felt so alive. I was forty-three years old, and except for those times lifting weights as a teenager, I'd never before felt this sense of grace and integration with my own physicality.

Running saved me. Problems with women, my worries about my son Aldo, issues with the show—everything fell into perspective as I racked up mile after mile. The rhythm of running was like music to me; I ran listening to Lester Young, Sousa marches, Count Basie, even Wagner. Sometimes I was so lost in the running and music I had a sensation of floating. I began to run the four miles to the studio and back from my South End condo, keeping a change of clothes at the station.

One day I blew into the building after a brisk run from home and made my way to the men's dressing room to shower. Standing in his underwear, shaving, was Vincent Price. It took me a few seconds to reconcile the star of the *The Fly*, and *The Pit and the Pendulum* and at the time the voice of *Mystery Theater* at PBS with the tall man I saw before me, who calmly kept on shaving as I stood there staring.

"I'm sorry, Mr. Price, I can come back..."

"No, no, come in, young man. I'm just finishing up here."

I introduced myself, tossed my bag down, and threw some water on my face. "You have such a wonderful voice, I've enjoyed listening to it all my life."

"My voice is nothing. You want to know who had the most beautiful voice in the world? Mario Lanza. We'll never hear anything like that again on earth, my boy," he said, rinsing his razor.

"Weren't you in that movie with him, *Serenade?*"

"Indeed I was."

"Such a shame he died so young," I said. "At least we have another great tenor today, Pavorotti!"

"You're right. Thank God for Luciano."

We talked for at least fifteen more minutes about his art collection, acting career, cookbooks he'd written, everything under the sun until he was clean shaven and dressed. Later I learned he'd offered to lend his best *House of Wax* voice on outgoing messages for several WGBH employees. He was a real gentleman, as elegant as you might imagine him. I, on the other hand, did my show in sweaty jogging clothes because I had completely forgotten to take a shower.

■ ■ ■

If you had told me back in 1978 that I'd soon meet a woman who would change my definition of what a relationship could be, I'd have laughed in your face, laced up my Nikes, and taken off for a really, really long run.

But that's exactly what happened. And I'll be forever grateful it did.

At the time I'd rescued myself from the suburbs—I am *not* a suburban guy, I learned—and settled into a small rental in Somerville. I started hanging out with my dear friend, jazz pianist Dave McKenna. Whitney Balliett, critic for the *New Yorker*, called him "Super Chops" and "the hardest swinging jazz pianist of all time."[12] Dave had an unending repertoire at the keyboard and a magical left hand that made a bass player seem unnecessary. For a decade he was pianist-in-residence at the Copley Plaza Hotel where everyone came to hear him play, including Tony Bennett, Stan Getz, Zoot Sims, Rosie Clooney, and Joe Venuti. But beyond his brilliant playing, he was a gas to hang out with, always up for fun, and ready to hit the town for a good meal.

Inman Square was our stomping ground. We had our choice of the Inman Square Men's Bar, Ryles, Joe's Place, and the original Legal Sea Food, although mostly we'd hit the Chinese restaurant around the corner for the General Tso's Chicken.

It was around this time that I started hearing about Joyce.

Marty Elkins, a singer friend of Dave's, told me Joyce ran this place called the Turtle Café with her business partner, Nancy Madden. Every time Marty saw me she said the same thing: you have to meet my friend Joyce; you'd like her, I just know it. And every time Dave and I went out, we'd walk right past the Turtle, with its glowing purple neon turtle in the window. I never stopped in at first; I'm not sure why. Later I learned Joyce had spotted me a number of times passing her restaurant on my way to eat Chinese next door. Finally Dave said, "Look, I'm playing at the Turtle next Friday. You're coming with me."

A world of Joyce: Joyce Scardina Della Chiesa.

Author's collection, photo by David Wade.

You know how sometimes you know things even before you know them? That's what that Friday night was like for me. I put a bit more thought into what I wore; I heard Marty's words, "You have to meet this woman" over and over in my brain, and my heart beat a little faster as I walked past the Chinese place, ignoring that moo goo gai pan smell, and opened the heavy mahogany door of the Turtle Café for the very first time.

And, oh dear God, there she was. This woman mixing cocktails behind the bar with supernatural grace. A ringer for Jennifer Jones. I was officially knocked out.

And then she *smiled at me.*

I guess I must have frozen in place because Dave gave me a shove and I snapped out of my stupor.

"Joyce," Dave said, "This is Ron Della Chiesa."

"Oh yes," she said, with a brief glance, "I listen to you sometimes."

And that pretty much wrapped that up. In contrast to my needing life support, she didn't seem particularly blown away.

Then again it was Friday night, and the place was booming. A bar full of boisterous patrons, a jam-packed dining room, and Dave already off and getting set up to play his set. A cutting edge restaurant (which became the East Coast Grill), the Turtle featured contemporary regional cuisine, changing its menu every three or four days not only for variety but to continually showcase whatever was in season, a relatively new concept for its time. On weekends they'd feature jazz greats: Teddy Wilson, Scott Hamilton, Sammy Price, and Gray Sargent, a guitarist who these days works with Tony Bennett. I quickly learned it was a great place to hang out and meet these legends, since they'd usually come in to eat before their gig. I had to admit, it was certainly cooler than the Chinese place.

The next morning I got Marty on the phone and asked her a thousand Joyce-related questions.

"I think she might still be involved with someone, I'm not sure," Marty said. "But I get the feeling that's wrapping up."

"I'm in the same boat. Sort of in, sort of out, but mostly out. You know, the gray area."

Marty laughed. "I hear you. But Joyce is an amazing woman, you know. Did you know that Gordon Hammersley started out at the Turtle as Joyce's protégé? The woman cooks like a dream."

"No kidding," I said, staring at the lonely can of King Oscar Sardines and stale bread on my counter.

"Oh yeah, and she's not so crazy about Chinese food."

By the end of the day I couldn't take it any more. I looked up Joyce's number, picked up the phone, and dialed.

She answered. She sounded a little sleepy.

"Hi, Joyce? It's Ron Della Chiesa. We met last night at your restaurant. I really enjoyed meeting you."

"Oh yes...hi Ron. It was fun to finally meet you; I mean I listen to you all the time."

"Wow, that's great. We should get together sometime."

"I don't know about that..."

"Well, when's your night off?"

"Tonight, actually."

"Do you want to go out tonight, then?"

"I can't...."

"Why not?"

"Well, I'm ironing. I have this big pile of ironing to finish up."

"OK, then maybe some other time."

And we said goodbye. I stared at my dingy bachelor pad a few minutes, threw on my jogging clothes, and went out for a six mile run in the pouring rain.

A few weeks later I was sitting at the bar at Ryles nursing a drink when I caught sight of her sitting at a table with Nancy and Nancy's fiancé, Bill. Though I began to sweat, I decided this was my turn to play it cool. Well known Boston radio personality sipping a martini at the bar; what could be more alluring than that? So I sipped and gazed thoughtfully at the middle distance.

But I couldn't help stealing glances at her. Finally she noticed me, and all she did, all she moved in fact, was one eyebrow. Just lifted it a tiny bit.

I put down my drink and sprinted over to her, leaping over a rail and knocking over several chairs in the process. Once everyone stopped laughing, I was finally able to talk to the woman I'd been dreaming about for weeks.

People have asked me what first attracted me to my wife, and of course, that was her beauty and her smile. I have to say, however, that she was and is my soul mate. How many women would sit through the Wagner's entire Ring Cycle with me, crying at the same moments, or finding the same things hilarious or moving? I remember both of us nearly losing it when, while watching *Tosca* one day at the Met, the heroine jumped off the balcony of the Castel Sant'Angelo screaming, "We'll meet in hell!" and bounced right back up from the mattress beneath the stage.

One of our slogans at WGBH at the time was, "A World of Choice." My personal slogan quickly evolved into "A World of Joyce." In short, I don't think I've ever had so much fun with anyone in my life. Her sense of humor; a similar passion for opera, jazz, the symphony; her love of travel; her fabulous cooking; all there, of course, but mostly it is her heart, that intangible essence of humanity and kindness that defines her that allowed us to bond. Moreover, she embraced my son, Aldo, as if he were her own. She even put him to work at the Turtle Café when he was ten or eleven, trusting him to do the work, which instilled in him some real pride.

Though our courtship mostly took place all around Inman Square, we loved going to Nantucket on weekends. She was living on Beacon Hill with a gay couple when we met, George Ormiston and Vince Scardino; but over time, she moved into my condo, and we became a part of each other's families.

I also appreciated that we not only had our own passionate interests, but that there was a mutual respect about their pursuit. By that I mean, in particular, that she put up with my growing marathon addiction. I'd been training with my old friend Paul Antonelli, now a distance runner who convinced me to jump in the pack (unofficially as a "bandit") at the Boston Marathon, where to my amazement I not only completed the race, but crossed the finish line at just over four hours.

After running two more Boston marathons illegally, I ran the New York Marathon three times with a number, Joyce cheering me on all the way. I think she understood how important this was to me. The third race was especially grueling; a cold rain beat down for most of it, and I almost quit. At the very beginning of the race on the Verrazano Bridge I saw a woman running roped to a blind man, and I thought, my God, if they can do it, I can drag my sorry butt through this. Twenty-two miles later in Harlem, cramping badly in both legs, I had serious doubts about crossing the finish line; then, like a vision, I saw a man under a wide black umbrella who looked like Fats Waller handing me a cup of something. It was Sammy Price, a jazz piano player known as the "King of Boogie Woogie"

Finish line, New York Marathon, 1983.

Author's collection.

and a guest on *MusicAmerica*. He'd actually shown up to support me, as he said he would! He handed me the cup of water saying, "It's gin, buddy. Go, paesan, go!" He was like an angel guiding me through those last excruciating steps. In the end I made my best time of 3:43, not bad for a forty-eight-year-old, and close to my goal of 3:30. The best part was seeing Joyce's smiling face at the finish line and hearing Sinatra's "New York, New York" blaring over the loudspeakers, then celebrating with her family in Brooklyn.

We were married on June 10, 1986, Joyce's birthday. To this day she reminds me that she rates two presents on that day: anniversary and birthday, but hey, it's fewer dates for me to remember. The Turtle Café also closed the same year, and it was an honor to host that event with old friend John Henning and feature Dave McKenna, who on that final night burned up the piano like I'd never seen before or since.

CHAPTER

# I Miss You in the Afternoon

Becoming the voice of the Boston Symphony Orchestra on October 4, 1991, has to be one of the highlights of my life in broadcasting. Henry Becton Jr., President of WGBH at the time, was a major force in developing PBS in America. Under his guidance, WGBH launched some of television and radio's best known shows—a full one-third of PBS's prime-time lineup—including *Nova*, *Masterpiece Theater*, *Mystery!*, *The American Experience*, and *Frontline*; groundbreaking children's shows *Arthur*, *Curious George*, and *Zoom*; and radio series including *The World*, *La Plaza*, *Basic Black*, *Classics in the Morning*, *Jazz With Eric in the Evening*, and a *Celtic Sojourn*.

Henry was so excited about our new presence at the BSO that he arranged for our first broadcast to be videotaped, not only to capture the inner workings of what we did in the booth, but to mark the beginning of a new era. Twenty years later we're still bringing the Boston Symphony Orchestra to thousands of listeners from that very same booth.

By that point in my career, I was rarely nervous on the radio, but being the voice of the BSO took some getting used to. Keenly aware of whose shoes I was filling—the venerable William Pierce with his thirty-eight years and three thousand BSO broadcasts—I at first took a more conservative path than usual with my delivery. In fact, early broadcasts probably came off a bit stilted before I relaxed a bit. I believe that by now we have our own signature sound and identity: a more conversational, casual style; however, I've never lost sight of the huge responsibility I carry in this role.

Being the voice of the symphony both at Symphony Hall in Boston, and in Tanglewood, week after week, year after year, is a profound honor. I feel as if I've come full circle: from being a young boy enthralled by Milton Cross, the voice of the Metropolitan Opera, to actually sitting down in the broadcast booth and lending my own voice to the BSO. The challenge is to properly represent this wealth of talent: the musicians, conductors, and singers who've spent a lifetime preparing for the chance to appear at the hall. I can't help but picture not only the countless hours of practice, but the grueling auditions, in fact, everything it must take to be able to walk across that stage to your seat with your instrument, whether you're Joshua Bell with his violin or a debuting soloist with a magnificent voice. I try not only to present these performers in the best possible light, but also to provide an all-around exciting, challenging, and vibrant listening experience.

Symphony Hall itself carries its own *gravitas*. It is considered to be one of the three most perfect halls in the world: a hallowed, sacred place. I'll never forget one winter evening when Tony Bennett had just wrapped up a rehearsal and sound check. Everyone had left the hall—the technicians, musicians, and a few spectators—leaving only myself, Joyce, and Tony. He walked across the stage and said he wanted to try something out; did we have time to stay? We said of course, and moved to sit closer to the stage. He sang, *a cappella*, just a few measures of Jerome Kerns' "All the Things You Are." It was one of the most beautiful things I've ever heard, just the natural sound of his voice, all alone in that place. It was as if the hall took his sound, caressed it, and gently handed it back to us. Afterwards he leaned down, touched the stage, and said, "It's in the wood, all this music, for all these years."

I love the fact that the show is live, and I never know exactly what's going to happen. Is there chemistry between the conductor and the orchestra, or not? When there is a disconnect the audience picks up on it, just as when they know they're witnessing a spectacular musical synergy between conductor and orchestra. Regardless, nothing is more exciting than watching a conductor tame an orchestra, which feels to me at times like it has some sort of wild energy all its own: as if it could kill you with sound during a Schoenberg crescendo or just as easily curl up in your arms during a Brahms melody.

I often get goosebumps as I sit listening in the booth when I realize this symphony or opera will never be played or sung the same way twice. Beethoven's Fifth, conducted by ten different conductors, will result in ten unique performances. In fact, there are times when the performance is so hair-raisingly brilliant that I want to bolt out of the broadcast booth, sprint right into Huntington Avenue traffic, and shake passersby who are calmly going about their lives and say, "Don't you know what's going on inside this building? You're crazy to be missing this!" and drag them in to listen…

■ ■ ■

When I say you never know what's going to happen in live radio, I mean just that. Many years ago, during the Boston Jazz Festival, (then held at Symphony Hall), we were planning to broadcast Benny Goodman and his orchestra. The problem was, he didn't show up on time. Thank God I had another broadcaster with me, my friend Tony Cennamo. We winged it with no script—nothing, ad-libbing about the history of jazz, the weather, more history of jazz, the history of the hall, until thirty-five minutes later Benny showed up, leaning against the piano and probably half in the bag. Regardless, he put on a great show, definitely worth the wait.

From any vantage point—either in the audience or from the stage—the tiny broadcast booth where I sit, and its equally tiny window, is not easily pointed out. I have to see what's going on while at the same time remain inconspicuous. Unlike my cohorts, producer Brian Bell and audio engineer Jim Donahue who watch the action onstage via TV monitors, I enjoy a perfect bird's eye view of everything that happens on that stage through that small window. And it can get pretty jammed down there, especially with an opera cast, or the Tanglewood Festival Chorus (the BSO's in-house chorus), where as many as eight soloists might take the stage. Just watching the changes unfolding below can be like witnessing a miracle of coordination: getting everyone on and off and on again with the proper music, music stands, and instruments; moving the piano on or off; assembling the risers for the chorus. Every last detail must be planned with precision. Stage manager John Demick and his crew have been doing a masterful job for decades.

Meanwhile, up in the booth, we're on our own tight production schedule. I might arrive at six for a seven o'clock broadcast start; but Jim, one of the best audio engineers in the business, has been in his booth since rehearsal that afternoon, making sure the sound is balanced, bringing the brass down if it's too booming, bringing up the timpani if that's needed. Brian Bell and I sit down with his lead-in script to do the final edit, drop in whatever sound clips are needed, and time it out. This first script, which we call the pre-game show, usually includes a historic BSO recording and precedes the live broadcast. At intermission, Brian might air previews of upcoming performances, present an analysis of that evening's symphony, or interview the soloist or guest conductor. With his vast knowledge of music and music history, as well as an intuitive sense of what the audience is curious about, Brian's contribution to the program is incalculable. After the live broadcast, I tie it all together with a bit of *ad lib* over the applause, acknowledging the conductor and soloist. Lastly, we fill in with another BSO recording until we leave the air at eleven in the evening.

We didn't always have Brian to create fascinating musical commentary or lively interviews to fill the down time during intermission. Twenty years previous, in William Pierce's day and when the BSO program was produced by Jordan Whitelaw, nothing was broadcast during the eighteen- to twenty-minute intermissions except ambience. That is, the sounds of people as they got up from their seats, distant conversations, the rustle of programs, a cough here and there, basically nothing. The feeling back then was that going to the symphony was akin to attending church: everything about it should be treated with a kind of reverence. Nothing should infringe on the audience's experience of being at the symphony, even if in fact they were *not* at the symphony. They were, of course, driving their car, or at work, or at home, ironing, wondering what the hell was wrong with their radio.

During those early years, every week without fail, I'd get a call from some elderly lady. (It actually could have been the same lady every week, who knows?)

"Hello, is this the symphony?"

"Yes, it is. This is Ron Della Chiesa with WGBH radio."

"Well, where did everybody go? Is it over?"

"No, ma'am, it's intermission."

"Is my radio broken?"

"I don't think so. You're hearing sounds from the hall. Your radio's fine."

"I heard someone coughing."

"Yes, you did. The orchestra will be back in about twenty minutes."

"Are you sure you're still on the air? I thought that fund drive would never end. I know it's not much, but I've sent in my check every year for forty years now."

"And we so appreciate that, ma'am."

"And you'd think after forty years I wouldn't have to listen to coughing."

"The orchestra will be back soon, ma'am."

"But I like you, Ron, you're a nice man."

And so on.

One day during some downtime in the broadcast booth we tallied it up. Twenty years of twenty-minute intermissions broadcasting "ambience" totaled over SIX MONTHS of dead air.

■ ■ ■

"I miss you in the afternoon..."

It's amazing how often I heard that comment from friends, from strangers, and even from people walking down the street who recognized me that fall of 1995, after *MusicAmerica* was cancelled. Even today, over sixteen years later, I still hear the same lament. That simple sentiment revealed more to me about the impact of radio on people's lives than five decades of working in the business.

As much as *MusicAmerica* was my creation and my passion, I never would have predicted the impact its cancellation would have on my life, as well as the lives of countless listeners. It's possible I'd been having such a good time those eighteen years—interviewing some of the greatest songwriters, musicians, and composers on earth—that I didn't realize how successful I'd been in conveying my enthusiasm for the American Songbook. I don't think any realization has given me such joy on one hand, and such melancholy on the other.

On August 31, 1995, the *Boston Globe* featured the following editorial:

*"The Day the Music Dies?"*

*Would the Museum of Fine Arts stuff its John Singer Sargents in the base-
ment? Could the Wang Center forget the great American musicals? What if the
Symphony did without Copland or Ives? Or the Pops without Sousa?*

*Impossible? WGBH is doing it.*

*Today's edition of* MusicAmerica *on WGBH-FM is scheduled to be the last,
ending 18 years in which the program has been an ornament to the city.*

*Over those years, Ron Della Chiesa has won the loyal affection of many
thousands of listeners—a fact that apparently tone-deaf potentates at the station may be beginning to realize—for hanging the ornament with such style and
verve.*

*But the ornament itself, as Della Chiesa is the first to say, is the music.
Leading with Gershwin and Ellington and tumbling through decades of fabulous writers and performers,* MusicAmerica *has given soaring voice to what is
unquestionably the golden age of American music.*

*Like the powerful music that sustained black South Africans through bond-
age to freedom, America's great songbook will survive.*[13]

Of course I knew my audience was out there, but I had never pictured the
thousands of people who were listening to me. Instead, alone in my tiny studio,
I'd conjure up a loved one or a friend, a fellow music lover sitting in front of me
who was as excited about what I was playing as I was.

At the time the program was cancelled, *Boston Herald* columnist Joe Fitzgerald
wrote: "Soon the music of Johnny Mercer, George Gershwin and Jerome Kern is
going to be a little harder to find in a world that needs to hear it more than ever,
and the worst part of it all is that it didn't have to be this way."[14]

No one was more surprised by the cancellation than I; however, when I think
back on it, the whole thing wrapped up with very little fanfare. On a hot, humid
July day in 1995, I was called into the Program Director's office and told that
*MusicAmerica's* last broadcast day would be on the thirty-first of August.

One part of my brain was listening to the "why" behind the cancellation:
strategic planning goals dictated programming consistency, in this case, classical music throughout the day. As things stood, the afternoon was a mixed, unpredictable blend of features, music, and interviews...nothing that fit into their
"strategic plan." Another part of me was busy piecing together clues from the past
months, all pointing to the show's demise. In its early years, the show aired from
noon to five, but over time became pre-empted for other programs both at the
beginning of the show, pushing the start time to one or two o'clock, or eroding it
at the other end, thereby shutting me down at four o'clock or earlier. Still, I left
the office in a state of shock.

■ ■ ■

Until we'd made these program changes, I had no idea of the true size of our audience. Numerous articles popped up in the local print media including the *Boston Herald* and the *Boston Globe*, the *Regional Review*, the *Current*, the *Patriot Ledger*, the *Brockton Enterprise*, and so many others.

A letter by Frank Haigh printed in the *Brockton Enterprise* best summarized the feelings of a lot of listeners as they grappled with the loss:

> *Who will be there to present live performances from Studio One and pre-*
> *view the new CDs by local talent like Dick Johnson, Rebecca Parris, Kenney*
> *Hadley, Steve Marvin and Donna Byrne? Who will interview Herb Pomeroy,*
> *Dave McKenna and play recorded interviews with legends such as Sammy*
> *Cahn and Arturo Sandoval? Who will announce daily the increasing num-*
> *ber of local venues featuring live entertainment? Where will stars like Cleo*
> *Lane, Tony Bennett and Mel Tormé go when they're in town, just to chat*
> *and give us some insights into the personality behind the talent? Who will*
> *celebrate the birthdays of Artie Shaw, Frank Sinatra and Nancy Wilson*
> *with tribute programs? Who will educate, promote, enlighten and keep alive*
> *the music that has distinguished America culturally? Others may try, but*
> *no one will do it with the class and style of Boston's mayor of music, Mr.*
> *Ron Della Chiesa. It may be crossed off your broadcast schedule, but replace*
> MusicAmerica? *It can't be done.*

The real impact of the cancellation was on the listener who felt the loss of a close and intimate companion in the afternoon. Through letters and calls, I learned I wasn't the only one who took the sudden loss of the sort of joy and comfort this music provided very hard. People wrote or called telling me they first learned about Frank Sinatra, Jr. from my show, or about Louis Armstrong, Luciano Pavarotti, or Ella Fitzgerald.

I was also surprised by the age range of my audience. Ten-year-olds who regularly had come home from school and listened to the broadcast contacted me to tell me goodbye; nursing home residents called me in utter disbelief and dismay; office workers, students, lab technicians, doctors, mothers at home all reached out to me. Grieving listeners called to tell me that the music I played was music they'd heard in their homes growing up, when they first got a radio, and that the music sustained them in difficult times. Things were made a bit worse by the timing: many were still away at the end of August and didn't hear about the cancellation till weeks later, sending a second shock wave through the listening community.

■ ■ ■

What had snowballed into a cause célèbre finally marshaled itself into the "Save MusicAmerica Committee," an organization that included media and business leaders from throughout the city and beyond. An initial trio, Steve Lowe, John Brady, and Carp Ferrari, were instrumental in giving the group real structure and goals. They demanded a meeting with station management. Entertainers and club managers were also disappointed and saddened by the loss of the show. *MusicAmerica* had given Scullers and the Regatta Bar as well as other venues generous, commercial-free radio time to promote whoever was in town. Carol Sloane, one of the country's great jazz singers as well as a substitute host for me whenever I was away, also lent her strength by becoming a member of the committee. Even the legendary Tony Bennett voiced his support for the cause during a Boston concert, calling WGBH's cancellation of the show a "music emergency." Tony added that he made it a practice to never speak out on issues from the stage, but this was important enough for him to break that rule.

Money donated to the cause of reinstating *MusicAmerica* was placed in the Save MusicAmerica Trust. Even though funds ballooned to over sixty thousand dollars, negotiations with management proved fruitless. The station was as adamant in their decision to drop the show as listeners were who dug in their heels to save it.

That's when the sleepless nights started. I had to walk a narrow line between my management and my audience. *MusicAmerica* was the peak of everything I'd done in radio. That eighteen-year body of work represented not only the biggest part of my career, but also the time during which I had the most impact on people's lives: it was the ultimate experience in sharing the music I loved and believed in. The issue became, for me, that a refusal to budge, to wholeheartedly join my supporters, likely would result in my being let go completely. Eighteen years of tenure could be gone in a flash.

As funds continued to grow in the Save MusicAmerica Trust, pressure mounted to determine just how to use it. Justifiably, people wanted the money either to be used in some way to bring *MusicAmerica* back, or to have the money returned to them. More sleepless nights for me as I tried to negotiate an ever tighter space between that rock and this hard place. I wanted nothing more than to be true to those who were sacrificing so much for me, at the same time I had no wish to sacrifice myself. I understood the evanescent nature of entertainment, with radio being no exception. Even I was surprised at times at how long I'd been able to stay in the game, and I had no wish to throw in my chips. I thought of that old joke:

"Get me Ron Della Chiesa!"

"Who's Ron Della Chiesa?"

*That's* how quickly one can be erased in the entertainment world.

Over time a halfway measure was reached. The Trustees of Save MusicAmerica decided to use the money to sponsor a series of concerts, beginning with an "Ellington Evening" with Carol Sloane and Herb Pomeroy. The idea was to create an even deeper and wider support base to further pressure WGBH to renege on the cancellation. I knew that eventually the station would come to me with some sort of compromise. I just wasn't sure I wanted to hear what it was.

■ ■ ■

Though the *MusicAmerica* circus had turned me on my head, I knew I had to save some of myself for my family and their needs. My mother had turned ninety-five that year and had begun to require some live-in help. Like a gift from the gods, a wonderful woman named Jerri Walsh, in her seventies and in better health than my mom, volunteered to live with her. A registered nurse who (like my mother) said the rosary daily, Jerri was someone I couldn't have conjured in my wildest dreams to step in and be a companion for my mom, but there she was. For four years they did everything together: shopped, read, ate meals, watched television, even met and visited each other's children and their spouses. I really believe this wonderful woman extended my mother's life for years.

Over time, my mother met and got to know Jerri's daughter and her boyfriend, who was illiterate. Even though she'd begun to lose her vision through macular degeneration, Florence took on an amazing project. She was determined to teach this young man how to read. A schoolteacher until age seventy, my mother had a profound understanding of the learning process, as well as how different people approached and assimilated knowledge. She displayed such sensitivity with this man, helping him overcome shyness and embarrassment to finally reach the point, as the light faded from her own eyes, when he could read her the *Patriot Ledger*.

Meanwhile, I'd begun my own project with my mother. Deep in the recesses of mom's basement were boxes and boxes of my father's letters, which he'd written to her while in the war: nearly five years of letters from the very first one he wrote after being called to active duty to the time he was discharged in 1946. She asked me to read them to her one last time, now that she was legally blind. Most of these letters I'd never laid eyes on before.

Every week we'd sit down in the sunny kitchen in the home I grew up in to go through a dozen or so letters or post cards. The letters dealt with all he had gone through in the war; passages could be harrowing or simply tell stories of just how hot it could get in Virginia in the summer. As I read, I occasionally came across some very personal thoughts from my dad, and I'd ask her if she was sure she wanted me to read them. With no hesitation in her voice, she insisted I go on. One of the letters revealed I would have had a brother or sister, had things been different. I had to stop for a moment to take that in, but again, she insisted I continue reading. My father also included clever little cartoons he drew or wonderful sketches of animals he saw in Virginia: cows, pigs, horses, chickens. I did my best to describe them to her. I'm not sure who enjoyed the process more: me as I got to know my dad through his words, or my mother, whose face was a constantly changing map of emotion: delight, joy, sadness, even laughter as she relived those years with me.

I would leave her at these times feeling such a mix of joy and melancholy: joy in sharing these letters with her with such frankness, melancholy because I couldn't truthfully answer all her questions, often about my own son. Without fail she would ask about Aldo, and I would say he was fine, doing well with his life but just too busy with work to visit.

In reality, even I hadn't been able to deal with my son. His teenage years had been marked by the typical rebellion, but he hadn't grown out of it as quickly and neatly as we had all hoped; in fact, I divided my time between worrying about him and worrying about my radio career. Joyce was remarkable during this period: never giving up on Aldo, always reaching out to talk to him, taking care of me as I stumbled in my own attempts to try to fix everything, often revisiting guilt from the divorce so many years ago. Countless times I had to remind myself of what was and was not in my power, a philosophy that on one hand can be depressing, while on the other, freeing. But the fact was, my son was drifting away from me, and I will never forget that feeling of helplessness.

■ ■ ■

At WGBH the pressure mounted. I continued to find myself between the conflicting obligations of supporting the Save MusicAmerica committee and maintaining my longstanding career. After months of meetings and discussions, the station finally issued a press release stating I would be hosting a new program called *Classics in the Morning* along with *The American Songbook* for two hours on Friday nights. I'm sure they saw it as the best of both worlds: appeasing my audience by keeping me there in some fashion, as well as providing the station with classical music all day, the programming consistency they needed to satisfy their so-called "strategic planning."

In the end, I had no problem hosting a classical music show; after all, that's the way I started my career at the station in 1969, so it was almost like going back to my roots in radio. The disappointment was that the show would be pre-taped, thereby lacking the spontaneity of *MusicAmerica*. I missed not only conducting interviews but creating a show that felt timely and alive. I found myself grateful on one hand to still have a presence on WGBH, while on the other I was still hurting from the loss of what had become so much a part of me.

A promotional piece the station created for the show a few years previous really captured the meaning the show had for me:

*Some of its hardcore listeners will try to nail it down as a jazz show. And there is jazz, lots of it. Jazz that goes way back, and jazz musicians who are just blowing through town and thought they'd drop by the show to talk and play for their friend, Ron Della Chiesa. But the heart of Ron's show beats with more than jazz. It rushes with the pulse of the big bands; Glenn Miller, Benny Goodman, Harry James. It taps along to the elegant touch of Astaire and Rogers. It delights in the lyrics of Cole Porter and the style of Sinatra. It lingers on the musical sidestreets of America, listening to forgotten songs. It celebrates Ellington, Bernstein, Gershwin and Copland. It plays Broadway. It runs like a happy kid across an endless playground of 20th century music.*

■ ■ ■

In a pattern perhaps common to all humanity, loss seemed to come in clusters during those years. My mother, a depression baby who never owned a credit card but who made a storybook home and loving life for my father and me, passed away on a cold February night at the New England Medical Center.

Around this time, however, all sorts of positive things started happening in Aldo's life. I sometimes wonder if she was controlling the whole thing from above—if all that praying and saying the rosary was finally kicking in. I still imagine how happy she would be, knowing Aldo as the fine young man he eventually became.

He was terrifically gifted with his hands (probably something he got from my dad), not only great at building things and woodworking, but he could also take apart and put together any car you could think of. When some well deserved work came his way, he found himself able to stabilize financially and to take justifiable pride in that accomplishment.

I'll never forget the day we visited my mother's grave at Mount Wollaston cemetery. It was a brutally cold day, a deep freeze locking the ground under sheets of unbroken whiteness. Hands jammed in the pockets of his too-light coat, Aldo asked if I could give him some time alone. For a good fifteen minutes he kneeled at her grave, crying his eyes out.

Before that moment I'd never realized what she meant to him. Perhaps that was the first time he'd understood that as well. From the cemetery we paid a final visit to her home, to say goodbye to the place he'd spent so much time, including entire summers, as a very young boy. He kept asking on the way over what we'd kept and what we'd given away or thrown out. In every room we walked through he shared a memory of her I had no idea he possessed or even experienced. When we reached the living room he stopped short and gasped.

"You didn't get rid of the chair, did you?" he asked.

"What chair?"

"The green one in the corner, you know, that old beat up recliner."

"That's at Goodwill, probably long gone."

"I wanted that chair. I sat in her lap in that chair while she read me Christmas stories. Out of everything in this house that's the only thing I cared about keeping. I wish you'd have let me know..."

I looked at Aldo like I'd never met him before in my life. Who was this sensitive, sentimental young man with all these warm memories of his grandma? In the coming years he'd talk about things we did together thirty years previous with utter clarity: his first opera, meeting Buddy Rich, the great jazz drummer, or John Williams, the composer. He even remembered things that endeared my mother to me as well: her wonderful muffins, fruitcakes, and pies, the way she made cookies and stored them in old cans so they stayed chewy. He recalled the saxophone I got him as a teenager, which he eventually replaced with a banjo that he still plays, and all the classical music we shared together, the jazz concerts, the bumper cars, everything I'd concluded didn't register with him during the time he had stopped communicating with me.

All this was a revelation. That my son, no matter how much evidence there was to the contrary, had been listening to me, and to Joyce, and to everyone else who loved him the entire time. I realized that you never know how what you say will be tapped for strength by the listener. Children will remember a kindness, remember love, and revisit those odd, sweet gestures or words in their hearts when they need nourishment or strength, most likely long past the time you were trying to reach them. So I guess the lesson is: you must never give up. I feel blessed to have a son who loves his kids and who wants to be a part of their lives.

## CHAPTER 7

# Fly Me to the Moon

Over a year after the cancellation of *MUSICAMERICA*, a deal was struck between WGBH station manager Marita Rivero and Alan Anderson, then manager of WPLM in Plymouth. This new arrangement was unconventional as well as unprecedented in Boston radio history: a sharing of one radio personality between two stations. Via the WGBH frequency I would be able to continue hosting *Classics in the Morning*, Friday evening's *Jazz Songbook*, the BSO broadcasts, *and* go to WPLM and do something called the *Strictly Sinatra* show as well as Sunday night's pre-taped, shortened version of *MusicAmerica*.

Paul Kelly of Kelly Communications, a true *MusicAmerica* addict and a force in sports communications, was the one who put me in touch with Alan and in the end helped broker the deal with WPLM. Paul knew all about the local broadcast industry and where the different formats were than anybody else. At first Paul played a big part in the Save MusicAmerica committee, but as he got to know me and the situation better, he decided the best thing for me would be to not only maintain my present audience but to seek out other horizons as well.

There was a big difference between seeking other horizons and pacifying management at WGBH; however, Paul went to Alan Anderson at WPLM and convinced him I had enough of an audience and enough of a draw to make it all worthwhile. I don't think anyone has a good count on how many meetings it took to bring everything about, but I am forever in Alan and Paul's debt for the patience and tact they brought to bear on not only helping maintain my presence at WGBH, but in reinventing me in a certain way with the new shows.

There even came a time when I had to take a firm stand with the Save MusicAmerica committee and tell them we were pretty much done; to put down their arms and leave the battlefield. It's not that I didn't appreciate everyone's efforts and their passionate commitment to the show (what better compliment could there be?), but it was time to move on. Sometimes I think if I had not been adamant about my decision, they'd still be rallying. They also wanted back all the money they had raised—understandably—but I believe to this day, well over a decade later, it's still tied up in escrow.

So in the end, with some wonderful people behind me, I was able to turn the cancellation of *MusicAmerica* into a positive thing. There was even a small scholarship set up in my name for musically gifted, college-bound students from Quincy High School. Though I missed the ability to do interviews as well as the spontaneity of *MusicAmerica*, the fact was that WPLM's version of the show kept the music alive. In fact, over the years I've developed a whole new audience for the American Songbook show on Sundays and for *Strictly Sinatra* on Saturdays through the WPLM frequency at EASY 99.1, as well as via their website, EASY911. com, where the show streams online.

In all honesty, I was a bit apprehensive about the Sinatra show at first. Five hours of programming is no easy feat, especially when you're showcasing just one artist, week after week. Even though "Old Blue Eyes" is a worldwide favorite, I had to ask myself if any entertainer could sustain all that airtime. I just had to put my best efforts in the show and see what came of it. After programming a few of them, I came to realize once again what a giant and an innovator Sinatra was, what massive influence he had over the American Songbook. One way of looking at the Songbook, in fact, was pre- and post-Sinatra: his interpretations expanded the meaning of every song he touched. Enigmatic, brilliant, tempestuous—I wanted to present the man, warts and all.

Before Frank there was Bing Crosby, who sang things relatively straight, as they were written. Then Sinatra arrived taking great liberties with both the music and the lyrics, particularly when he teamed up with Nelson Riddle, Gordon Jenkins, and Billy May in the fifties. Frank knew instinctively not only where to pause but how to get the most out of a pause. He knew how to shape a line, so that in many cases, a song sung by Sinatra was an adventure in swing. On the other hand, he could get inside a poignant ballad like no one else.

In the end, a lot of people started spending Saturday night with me and Frank, even, to my shock, kids under the age of ten. After a benefit I did one evening, a six-year-old boy came up to me with his grandfather in tow. He grabbed my arm and pulled me down so I could hear him.

"I love that Sinatra show!"

I asked him, "What do you like about Frank?"

He thought for several seconds, taking the question very seriously. "He's a classy guy. I guess that's it."

"What's your favorite Sinatra recording?"

"'Mr. Success.'"

Now *that's* timeless.

■ ■ ■

In 1999, Joyce and I packed up and moved from our small, South End penthouse condo to a ten-room 1880 Victorian on the Dorchester/South Boston line. I was sixty-one, the time in many people's lives when they think about downsizing, but after twenty years in the condo, we both missed the feeling of living in a one-family home.

Living in the South End was terrific, don't get me wrong. We could literally reach out and touch the John Hancock building, and we were within shouting distance of the Cyclorama, the Boston Ballet, and a burgeoning jazz and restaurant scene. But I was tired of keeping so much of my music collection and memorabilia stored in boxes because of space limitations, and Joyce was craving not only a kitchen of her own design but to live closer to the water. I also thought it was high time to create a studio in my home. We were both losing patience with the condo restrictions via the South End Historical Society. At one point, Joyce painted the steps two shades lighter than "allowed" and was subsequently summoned to a hearing. I think our decision to move was sealed when we learned the committee would grant us six months to repaint the steps in their officially approved color. Ironically, the gentleman who conducted the hearing knew me and told me he was a big fan of mine, but that rules were rules and those steps had to be repainted. Long story short, we couldn't wait till we could build a fence and not cower in fear as the neighborhood committee swooped in to make sure it was regulation height.

It was only the second house we looked at, but we were in love with it. Even though it was just two miles from the condo and the bustle of the city, we felt like we'd moved to New Hampshire. To have a driveway and a yard, however small, was a balm for the soul. Still, I could walk straight up Massachusetts Avenue to Symphony Hall in about an hour.

The location couldn't have been better: minutes from the red line and a short distance to Carson Beach and Castle Island. Walking along the promenade where the sunset meets the skyline is an experience second to none on the Eastern seaboard. We both felt great to be back in a real neighborhood again; for Brooklyn-born Joyce it was an especially strong sense of returning to her roots. Even my music sounded better playing in that house, as if we had all finally come home.

Still, the house needed work, and Joyce did a bang-up job during her seven-month stint as general contractor. Her brother Ritchie, a contractor, and his wife Rosie flew in from Los Angeles to help supervise the renovations. Without them I don't how we would have done it. We're also forever indebted to dear friend Carolina Tress-Balsbaugh, one of America's best interior designers, who took the time to really look at the home, room by room, space by space, and offer priceless advice about paint color, wallpaper, even placement of our furniture, but always with our needs and personalities in mind. Freshly painted in beige and putty with dark-green trim and rich in period detail, the house seemed to exude a wonderful warmth and sense of history. We've been so happy over the years adding our own history: visits not only from our children and grandchildren, but wonderful meals

Joyce has prepared for long evenings of great conversation with dear friends. Renovations to the second floor included the transformation of four small bedrooms to a master suite and guest room, while the third floor turned from an attic to a cozy reading room. I cherish especially the winter nights sitting there under the skylight, reading to my grandchildren or listening to old LPs with our cat Giacomo on my lap, snowflakes swirling outside the window.

The foyer, otherwise known as the "Opera Hall of Fame," is hung not only with rare, autographed prints of opera singers and composers, but with another collecting passion of mine: Disney animation cels. These are transparent sheets on which objects were drawn or painted in traditional animation. Stills from *Fantasia*, *What's Opera, Doc?*, *Lady and the Tramp*, and *Jungle Book* line the walls.

But a descent into the basement is like a journey into my heart, soul, and an over sixty-year romance with music. We renovated it to include a sitting area, a soundproof recording studio, and, of course, the music: floor to ceiling shelves of vinyl, cassettes, and CDs including my very first 78 rpm: "Rusty in Orchestraville" sitting next to the Pinocchio album, both Christmas gifts from my parents. Plaques, favorite photographs and awards, vintage "Nippers" (the RCA Victor dog), antique radios, and microphones line the shelves, along with a fair number of Tom Mix glow-in-the-dark-spurs and Disney toys. My collection of vintage magazines including *Life*, *Playboy*, *Mad*, and *Opera News* fills the rest of the space. Joyce even created a subterranean wine cellar, formally a half bathroom. Removing heating pipes resulted in the perfect 62-degree environment.

Above the reel-to-reel recorder that I used to do some of my very first interviews are photos of Bill Marlowe and Milton Cross, a signed photo of Arthur Fiedler, a drawing of Dizzy Gillespie by "Benedetto," the name with which Tony Bennett signs his artwork, and even a photo of the old Boston Opera House that once stood on Huntington Avenue and Opera Place, today the site of a Northeastern University building. Sometimes as I look around at the magnitude of what has come and gone, I can't help but feel melancholy; but that feeling soon passes, and I become energized by the joy of listening to the music again and again.

Even though the Turtle restaurant is gone, the neon sign still hangs in our dining room over the mantel. The pale yellow walls here make a striking backdrop for a collection of black and white illustrations my father did for a series of "noir" mysteries published in the thirties and forties. In the adjoining living room, an exhibit of his work continues with watercolor seascapes, primitive landscapes, and charcoal portraits, which to me recall the style of John Singer Sargent.

I'd always wanted some way of commemorating my dad's art work, and in the spring of 2000, at the Adams Academy Building, home of the Quincy Historical Society, Joyce and I were finally able to make that a reality. A 1926 graduate of what is now the Massachusetts College of Art, my father studied with the famed sculptor Cyrus E. Dallin, known for his sculpture "Appeal to the Great Spirit," which graces the entrance to Boston's Museum of Fine Arts; and with Ernest Major, a pupil of John Singer Sargent himself. My dad taught art in Waterville, Maine, and Lynn, Massachusetts, before returning to the South Shore, where he became director of the art department at Braintree High School. In his native South Quincy, he directed the Quincy Adams Drawing School, giving art lessons to local children.

It was Joyce who had the vision of rescuing my father's paintings from my mother's basement and having them cleaned, re-matted, and re-framed. The result was the first complete retrospective of Aldo Della Chiesa's art work, from his early student days in the 1920s to his last paintings in the 1960s. I often think how wonderful it would have been if he could have attended this event himself; he was never honored in such a way while he was alive. Quite possibly he would have delighted us all by bursting into song with his gorgeous tenor voice. I do know that the sharing of his love of music and the arts was his greatest gift to me.

Tony Bennett and me at exhibition of Aldo Della Chiesa's paintings in Quincy, June 2000.

Photo courtesy of Joyce Scardina Della Chiesa.

■ ■ ■

Paul Schlosberg, my manager, entered my life in 1990. Here is a guy who doesn't stop moving until all the guests are seated, drinking, and happy. He's got the kind of energy and chutzpah and gutsy promotional instinct one simply doesn't find very often. My producer and partner for both *Strictly Sinatra* and *MusicAmerica* on WPLM, Paul has been instrumental in changing my viewpoint about myself and what I could offer beyond the scope of public radio.

Paul had represented Bill Marlowe, which impressed me no end. He kept drilling it into my head that I had an audience, and that they would follow me, and that I could also reach out for a whole different demographic. In other words, why not start selling myself? No one but me owned my identity, after all. The whole idea at least gave me a sense of control and optimism.

I decided it wasn't the end of the world to work at a commercial station; after all, that's where I'd begun my career. I had to look at it as a new bend in the road, one that had the habit of paying the bills just as well as public radio, often even a little bit better. It was a kick to learn that there were people out there who wanted to jump on board as sponsors, such as David Colella, a big Sinatra fan and the manager of the Colonnade Hotel. David was also responsible for one of the city's most successful restaurants, Brasserie Joe, and he's been one of my loyal sponsors from the beginning. Raphael's restaurant, then in Nantasket and now in the South Shore Country Club in Hingham, and Tony Floramo's restaurant in Chelsea signed on as sponsors. Chuck Sozio, who'd been one of Bill Marlowe's sponsors, turned the second floor of his Neponset appliance store into a Sinatra room, completely fitted out with hip fifties furniture, jukeboxes, as well as my Sinatra show piped in to all his stores. I still return the favor by appearing at home shows with him whenever possible.

But still, flipping my brain over to commercial radio took some getting used to! Here I came from this very insulated and dignified public radio station:

"Good afternoon, and welcome to *MusicAmerica*. Thanks so much for joining us today during our fundraiser. We certainly need your pledges to keep bringing you the best of the American Songbook..."

...to...

"Tony Floramo's in Chelsea is the *place to go*! I'm telling you right now, you've *haven't lived* until you've tasted these succulent baby back ribs! The meat falls *right off the bone!*"

There were days I might interview Yo-Yo Ma in the morning at WGBH, race to WPLM in the afternoon to pitch surf 'n' turf, then in the evening, host an event such as a "Strictly Sinatra Dance Party" at Raphael's on Nantasket Beach, where women would swoon over Dean Martin and Frank Sinatra, Jr. impersonators. Hell, they even swooned over me now and then, coming up to meet me, hug me, touch my hair. At events like this, which continue to this day, I get the chance to ham it up and improvise a bit, indulge my actor-ly side. Here was a Vegas crowd with a bone-deep enjoyment of the American Songbook, unafraid to get up on the dance floor and show their love. It's been so refreshing to get out there and meet the people who've listened to me and the music we all cherish.

Paul created our website, www.musicnotnoise.com, arranging sponsors, handling all ticket sales and promotion. Streaming our music 24/7 has inspired joyous listener responses from Los Angeles to Rome, as well as a phone call from Dan and June Weiner at Galaxsea Travel. They convinced me I had enough listeners to do a cruise. Our first was in 2006, and it was a huge success; we've been doing them ever since. These adventures draw people of all age groups, from young couples celebrating their honeymoon to people in their seventies and eighties who love to dance. The unifying factor is a love for this kind of music. So often by the time we pull into port, lifelong friendships have been forged as well as indelible memories. Our February 2010 "Tribute to Frank Sinatra and Dean Martin Cruise" (featuring Michael Dutra as Frank and Steven Palumbo as Dean) sailed to the Eastern Caribbean on the *Atlantica*.

■ ■ ■

Branching out certainly widened horizons for me, but every now and then I would get in a little over my head. Pulitzer Prize winning composer Gunther Schuller, a friend for over thirty years, once asked me to act as narrator for a performance of his composition, "Journey Into Jazz." This twenty-three-minute piece explores a young trumpet player's discovery of jazz, the complexities of which confound him initially. First, he looks for answers through the music of other jazz musicians but soon realizes the real "journey" to jazz is to follow his own instincts and be himself.

Gunther Schuller began his career playing principal French horn with the Metropolitan Opera Orchestra under the baton of legendary conductors such as Fritz Reiner. Later, he collaborated with Miles Davis on the historic "Birth of the Cool" jazz sessions, developing into a prolific composer of classical and jazz music, ultimately inventing the term "Third Stream," which combines the two techniques. He's the author of two major books on jazz, winner of the Pulitzer, MacArthur Genius grant, and several Grammy awards, as well as the recipient of ten honorary degrees. His full resume would require another nine or ten more pages. Being with this man is like being in the same room with Gershwin or Schoenberg, just an extraordinary experience. He really has done it all, and so when he asked me to be the narrator for this piece, I have to say I was thrilled as well as terrified.

At the same time here was a guy who'd been over to our home for dinner many times, and he was no stick in the mud. We knew just how he liked his steak and his gin martinis, and he was great company. After all, I'd interviewed him a number of times, and he never put on airs. Here's what he said when asked about his teen years: "When I was eighteen, my parents told me you have to sleep eight hours every night. I said, 'God, you know, if I sleep eight hours every night, I'm going to piss away one-third of my life just by sleeping. And life is too short.'"

"Journey Into Jazz" is a complex piece for any musician, but because he liked how I sounded as host of the BSO, Gunther entrusted me to narrate it for a performance at the Sanders Theater in Cambridge. I guess he respected my knowledge of jazz, but when I reminded him I didn't read music all that well and that I wasn't a musician he said, "That's okay, you hear with your heart."

Hopeful words from a good friend; how could I let him down? I squirreled myself away with Leonard Bernstein's narration of the piece. I must have listened to it a hundred times, and eventually I thought I had it down.

We had only one rehearsal at the Sanders Theater. It's there I discovered I did *not* have it down. Improvisation mixed with the notes on the page threw me off, I think. Long story short, I was off by a few measures, which is of course a crime when you have an entire orchestra counting on you, and you are performing the conductor's own work. During a break in rehearsal, Gunther took me aside. He was looking awfully pale.

"Man, you can't read music, can you?"

"I told you back at the house, I played trumpet in the army. It got me out of KP—"

He looked grim. "I don't know what we're going to do."

I pictured our friendship evaporating no matter what I did. "What would you like me to do?"

"I don't know, but I don't have time to work with you. This plays next weekend. I'm going to have to bail you out somehow."

The following week I was sitting onstage with my score in front of 1,300 people and a full orchestra. To say I was sweating doesn't do justice to my poor tuxedo. Gunther's solution was to hire a flutist to sit directly in front of me and cue my every move. At first I thought I was pretty much in the groove, that I might actually live through this.

Then, during the performance, Gunther shouted out over the orchestra something like, "Number fifteen!" and after that it seemed like my cues were all over the place. I was a basket case. Later I learned I was off a couple of beats. My performance still is a blur to me.

I'm still not sure what actually happened up there, but people in the audience told me it was fine. Everything sounded "intentional," which is something I'm still trying to figure out. All I know is, Gunther still takes my calls, and that's what matters.

■ ■ ■

I'm often asked how it felt to finally retire, but in fact, I never really have. I'm busier than ever. After thirty-five years with WGBH, I accepted the buyout package with the provision that I could continue my relationship with the station, hosting the Boston Symphony and conducting Learning Tours. I also serve on a number of boards of arts organizations as well as host a number of events around Boston. The AARP has had me all over the country at conventions drawing sometimes upwards of thirty thousand people. It's been an honor sharing program spots with everyone from Bill Clinton to Queen Latifa and Bo Derek, giving talks on the history of jazz, opera, or the Boston culinary scene. In fact, this past decade has been one of the richest of my life, perhaps because everything feels so integrated now both personally and professionally.

Best of all has been the opportunity for Joyce and me to host the WGBH Learning Tours all over the world, including tours in Santa Fe (known for its opera company's fantastic outdoor stage), Los Angeles, Italy, and Tanglewood, as well as a jazz tour in the Caribbean with Oscar Peterson and Dave McKenna. But I think going to Italy both to conduct several Learning Tours and to enjoy subsequent trips on our own has moved us most of all.

Finally going to the country I'd read about and imagined for six decades was like a dream come true for me. It was everything I imagined and more. Just seeing the countryside, driving through picturesque villages, and sampling unparalleled cuisine with Joyce are among the richest experiences of my life. After attending hundreds of operas, listening to thousands more, interviewing opera singers, and studying opera for as long as I have, it was a revelation to finally set eyes on the countryside, meet the people, and visit the places where opera was born.

We visited Roncole, the birthplace of Giuseppe Verdi, and walked from the farmhouse where he was born to the church where he played the organ as a boy; then visited the Verdi Theatre in the town of Busetto, where he grew up. We stayed at "Il Due Foscari," which is the name of a hotel as well as one of Verdi's operas. Owned by the great tenor Carlo Bergonzi, the hotel also houses an opera school that draws young singers from all over the world to study with him.

We stopped at Puccini's home in Lucca, followed by the Puccini museum. What an experience it was to listen to his music while poring over his final letters to his wife, Elvira. A chronic cigar and cigarette smoker, he died of throat cancer in 1924. Even in bronze as a statue in front of the museum, he holds a cigarette in one hand. We also visited his summer villa, Torre Del Lago, where the Puccini Opera Festival is held every summer. It was there we met his granddaughter, Simonetta Puccini, who I had the opportunity to interview (years earlier) for WGBH. We saw the piano he composed on and visited the small chapel inside the lower level of his home where he's buried next to his wife and son, Antonio.

Parma is, of course, known for its Parmesan cheese and prosciutto di parma, but it's also the birthplace of Toscanini, the maestro who first ignited my passion to learn about classical music. Joyce and I delighted in the beautiful opera house, Teatre Regio di Parma, where only the most discerning opera lovers sit at the top of the gallery. If you don't get it right in Parma, they'll make you sing it over and over until you do. The spirit of the place is interactive, to say the least. On one occasion, an American baritone was having an off-night, and the audience booed him. He cursed at them and called them "cretini." By the time the opera had come to a close, he needed a police escort to leave the building.

One evening, a tenor who'd also had a less than stellar performance the previous night took a cab to the airport. The singer paid the cabbie, then waited for some help getting his bags out of the trunk.

The cabbie squinted at him in the rear view. "Were you singing last night?"

The tenor gulped. "Yes."

Glaring at him, the cabbie popped the trunk. "Carry them yourself."

We visited La Scala opera house in Milan where Maria Callas sang, as well as the Teatro Massimo opera house in Palermo, Sicily, where a good deal of *Godfather III* was filmed.

The culinary and wine tours we did in the chianti region and visits to Tuscany were truly life changing. I used to plan on traveling to places all over the world and I still hope to, but as I get older I want to be more thoughtful about where I spend my time, perhaps confine myself to places that hold deeper meaning for us. I think Italy—for both of us—is a place where we simply can't spend too much time. As I've said many times, it's hard to leave Italy.

Being with Joyce has deepened my understanding of a true partnership, and ours is one I cherish and for which I am so grateful. We're a team, even though I'm not always easy to live with. I listen to over forty hours of music a week, and when we're driving, one hand is on the wheel and the other is constantly cruising the dial. There are weeks we're out on the town five nights a week: for a movie preview, a benefit, an opening, the symphony; and every night feels like an adventure. Last year alone we saw twenty-six operas together.

Traveling with Joyce has deepened our relationship in so many ways. I think that's when you truly get to know someone—when you're jetlagged, lost, or stumbling upon wonders new to both of you. Once or twice a year we escape to Cape Santa Maria Bay on Long Island in the Bahamas where Joyce's mother was born. No TV, no phones, just pristine white beaches stretching out for close to a hundred miles. In fact it was Joyce's Bahamian mother, a natural born chef, who first taught her how to cook.

Our travels also included a cruise on the *Queen Elizabeth II,* an ocean crossing from Boston to London where Joyce was a guest chef and I lectured on jazz; a very long trip to Buenos Aires; and several jaunts to Hollywood. Being movie junkies, we visited the cemeteries: the graves of Mario Lanza, Marilyn Monroe, Cecil B. Demille, Bella Lugosi (who's buried in his Dracula costume), Bing Crosby, Tyrone Power, and Sharon Tate. Al Jolson being the winner in the ostentatious category: High up on a hill near Los Angeles International Airport next to a magnificent white marble fountain is a bronze statue of him on one knee, arms out-stretched in his famous "Mammy" pose.

Thanks to our dear friend Dale Pollock, who was a producer at the A&M Studios and who used to manage the Orson Welles Theater in Cambridge, we were able to visit some off-limit sets at Warner Brothers, 20th Century Fox, and Paramount. We explored the original studio lot where *Casablanca* was filmed, the Paramount lot where Marilyn Monroe had her cottage, Culver City where the original MGM studios were located, and the original soundstage for the *Wizard of Oz.* We also looked up another old friend: Bob Genest, who worked with me at WGBH, but who had transitioned from public radio to CEO of Frederick's of Hollywood, which included a Bra Museum among other attractions.

None of these adventures would have happened, however, if I hadn't gotten over my fear of flying. I had it all—the sweaty palms, the racing heart, the sudden deepening of my Catholic faith every time we hit turbulence. One day Joyce said, "Okay, we're going to take care of this." This was a few years ago when we were traveling more than ever before, and I was probably driving her crazy.

I believe the cure came during that brutal, ten-hour flight to Buenos Aires. We're flying, the plane is bouncing, I'm sweating and flipping out, and she pops an Atavan in my mouth. In a few minutes my whole body melted into the seat and I'm sitting there smiling, then I leaned over and French kissed her. She said, "I think your fear of flying is officially over." But we still carry the Atavan, or at least she does, if she wants to get French kissed.

Still, I've got to have a window seat. I just have to see what's going on, even if it's the ground shooting up to meet me; and I still pray when I step on a plane. But I have to say, going first class helps a lot. I finally got what people were talking about the first time we ditched the coach seats.

FLIGHT ATTENDANT: "What can I bring you?"

ME (RECLINING ALL THE WAY): "What do you have?"

FLIGHT ATTENDANT: "Everything."

ME: "I'll start with a Bloody Mary."

FLIGHT ATTENDANT: "How would you like your filet mignon done, Mr. Della Chiesa?"

ME: "Medium, thank you very much."

AN HOUR LATER:

FLIGHT ATTENDANT: "Can I get you an ice cream sundae? Profiteroles? Tiramisu?"

ME: "Yes, thanks, and also a pillow and blankets and earbuds and a martini and...."

And then I wake up over Rome to the smell of bacon and eggs.

■ ■ ■

Though no one would accuse me of being a curmudgeon, even I sometimes forget that this world is a charmed place. Sometimes it took a little magic to remind me, such as the sight of my eight-year-old granddaughter, Tia, walking off the plane and into Logan airport some years ago. My heart skipped a beat as she emerged from the crowd. She had been a toddler the last time I'd seen her, but here she was dressed to the nines in a pink dress and shoes, carrying a Barbie Doll suitcase. She was just so beautiful. Her face lit up when she saw us, and she broke out in a run and jumped into my arms.

A good friend said to me once, and I concur: "When you get to be our age, it's all about the grandchildren." I don't know who benefits more from our relationship, my five grandchildren or me. My intention is to hopefully instill in them a love for beautiful things: the arts, music, theater, painting, as well as a love for the outdoors, nature, traveling, seeing and appreciating different cultures. While there are the less tangible lessons about how to confront the constant obstacles in life with a positive attitude, things that they will no doubt learn through experience, as a grandparent I somehow wish I could shield them from any sort of pain.

While there are all these things I want to teach them, I find myself delighted by everything they have to teach me. I'm fascinated by their wildly different personalities and interests: Gabby who smiles all the time like a little butterfly and most resembles my mother, Florence; Nico, who is a dynamo of a boy, an athletic kid who fights with his sister Gabby like an old married couple. Dominick reminds me of a young Fred Astaire, and young Donovan, who resembles Mary's father Gabe, looks like he could grow up to be a linebacker for the NFL. Tia, the angel I'll always remember running to us in the airport, is now a lovely young woman who I tend to worry about the most. It seems there's no natural, gentle transition to womanhood these days; instead, the pressures and influences of the media throw so many young girls into the fast lane before they're ready to be there. I'm happy to report, however, that she's turned into a level-headed, sophisticated, but fun-loving young woman, and is headed for her freshman year at Arizona State University.

We don't see them enough because they live in Boise and Phoenix and we're in Boston, but we love sharing the joys of New England with them when they're here. I love reading to them, watching movies with them, just observing them relate to each other. As they're playing around me, laughing or fighting over a toy or generally destroying a room, I suddenly feel as if I'm no longer an only child; that I truly have become, finally, part of a large, extended family, a new and wonderful sensation for me.

■ ■ ■

Many times I'll look back and wonder what would have happened if I'd made different choices. The good news is that every time I do, I see why I made the choices I did. I had a chance to go to Los Angeles and program a classical show there, but in the end I'm glad I stayed here and worked to establish my career in Boston. Most of the time I felt like I was having my dream cake and eating it too, so why leave? Besides, where else could I interview James McCracken, one of opera's greatest who sang a few measures from *Otello* for me in my studio?

I could never envision myself doing angry, mean-spirited talk radio. In fact I can't help remembering what Jerry Williams, one of America's most popular talk show hosts, said to a woman who approached him on a cruise he was hosting. She expressed to him how glad she was to be on the ship with him. His response was something to the effect of, "Ma'am, I'm glad you're on this cruise, but that does not mean I have to talk to you." I'm afraid there are a number of DJs who want to get away from their audience at the end of the day, whereas I seek out opportunities to spend time with mine; I want to be with the people who go to the opera, to the BSO, to Scullers, and everywhere else where wonderful music is being played.

That said, I still had a bit of an actor bug in me, so it was fun to dabble in that. For a year I hosted a show called *Cooking Around Town* at WGBH TV. I'd been approached by a young producer named Laurie Donnelly who has since gone on to become a major PBS television producer. Don't get me wrong: I'm a dunce in the kitchen, but I like knowing how things are done—how to *theoretically* cook a chicken, for example. We'd showcase a different restaurant each episode, interview the chef, then sit down for a memorable meal with chefs such as Gordon Hamersly or Jodie Adams of Rialto.

*Amuse Bouche: A Chef's Tale*—named after the chef's complimentary appetizer—is a documentary film about Barbara Lynch, a remarkable young woman who came out of the projects in South Boston to make a name for herself as an award-winning chef. Cat Silirie, a dear friend and one of the leading sommeliers in the world, introduced us to Barbara when she was an up and coming chef at Rocco's restaurant in Park Square. Through Barbara and Maryanne Galvin, a prolific filmmaker as well as a practicing forensic psychologist, I was given the chance to narrate the film.

One of seven children in a hard-scrabble Irish Catholic family, Barbara lost her father when she was quite young and was raised by her mother during one of South Boston's most troubled periods. Through sheer determination as well as generosity of spirit, this young woman overcame the poverty, depression, and violence that marked her childhood and rose through the ranks of the culinary world to establish herself in a male-dominated business with a 60 percent failure rate. She won the James Beard award, began Number 9 Park on Beacon Hill, and has since become a nationally recognized chef. By 2010 she had opened Drink, B&G Oysters, The Butcher Shop, and Sportello, and published a terrific cookbook called *Stir*. A real life tale of true grit, *Amuse Bouche* compiled family photographs, archival and animated footage, and exclusive interviews with some of the top names in Boston's culinary world including Lydia Shire, Todd English, and Jodie Adams, as well as a few clips of Mayor Menino, and running legend Bill Rodgers as he served food alongside Lynch at a fundraising race in South Boston.

What an honor that Barbara sought me out to narrate such an intimate, inspiring story.

■ ■ ■

Another friendship, this time with tenor sax innovator Illinois Jacquet, led me to be a part of his story in the (1992) film *Texas Tenor*. The first time I laid eyes on him was at Symphony Hall: I was sixteen years old and his playing knocked me out. One morning, more than thirty years later, Illinois drove all the way from the Sandy's Jazz Revival in Beverly to pay me a visit during an early morning jazz show I was hosting on WGBH. While I was playing one of his recordings, he stepped into the studio with his entourage.

"What's your name, man?"

"Ron Della Chiesa."

"Della Chiesa? Well, I can't remember that. Sounds like it should be Ron Delicatessen. Mind if I call you that?"

How could I say no?

Over the years, we became very close, so close in fact that during my marathon running years he wrote a tune for me called "Running With Ron." One day I got a call from his manager, Carol Scherick, announcing that a film crew would be joining Illinois during our interview on *MusicAmerica.* That segment became a part of Illinois' film, *Texas Tenor*—no connection to the Lone Star state: Jean Batiste Illinois Jacquet was born in Louisiana. "Texas Tenor" refers to a certain style of playing the sax which brings out its greatest sound and volume—an excellent technique for rooms that at the time were rarely miked. The film followed his life and music starting at age nineteen when Lionel Hampton had him change from alto to tenor sax and gave him the opportunity to record his groundbreaking solo "Flying Home."

It was so gratifying to be a part of Arthur Elgort's documentary, which I felt really captured not only Illinois' music, but life and spirit. The camera follows him everywhere from the cramped backstage of the Blue Note jazz club in Manhattan, all the way through Europe on the band's debut tour. Shot in grainy black and white, we eavesdrop as Jacquet reminisces with some fellow jazz musicians aboard a floating junket, visits a saxophone factory, and gives an impromptu performance at Rayburn Music, a well-known instrument repair shop in Boston run by Emilio Lyons. Interviews with world class musicians Dizzy Gillespie, Sonny Rollins, and Les Paul, among others, shed new light on the man, Gillespie making note that Jacquet used to make more money gambling on the tour bus than playing his horn. Low key offstage, Illinois came to life with his instrument, inventing a rompin' stompin' sound that has become a feature of great R&B ever since. I think Jacquet became, finally, aware of his stature as a legend, but only commented, "I just want to be part of something that will last."

■ ■ ■

Every now and then I used to just get in my car and drive, without any destination in mind. Actually, that's a lie. I may have taken the scenic route, stopping for this or that along the way, but my heart knew where I was headed if my brain didn't.

I would sit in my car, eyes closed, and it was if I'd turned the clock back half a century...

Something about Nantasket Beach in Hull always drew me back. I guess the curse of a great childhood is that you wouldn't mind reliving it; in fact you seek it. There is a photograph I have of my mother and me at the beach. In it, I'm only six months old. Perhaps it's a little movie my mind has constructed from staring at the photograph all these years, but I swear I remember the moment I was dipped in the vast Atlantic, screeching as the frothy waves licked at my toes. Afterwards, my mother put me down on the sand. I crawled a few yards away, turned, but didn't see her. Pure panic! But I looked again and there she was, next to our bright yellow and orange umbrella under blue skies, laughing and waving at me, and all was well.

In the background, behind my mother and me on the sand, Paragon Park's roller coaster snaked up and down, riders screeching with delight at each stomach-dropping twist or hairpin turn. Dusk was always the best time at the park; the heat of the day had lifted while the ocean's cool breezes finally turned toward land. The funhouse seemed more fun, the houses of horrors a little more scary. The smells of hot dogs and salt water taffy wafted through the air; boys chased squealing girls around the carousel with horses that rose and fell, rose and fell. There was the Congo cruise, kid's fantasyland, the Schlitz Beergarten. Up in the hills, classic wooden hotels dotted every corner: Nantasket, Rockland, the Atlantic House.

Even as a teenager, there was nothing like a day at Nantasket. There were even magical ways of getting there: A train used to make stops at the beach; boats would drop you off from Boston Harbor. For my old friends and me, it was the endless summer. Paul Antonelli, Bob Buccini, and I used to hang out in a section called Little Weymouth and chase girls. When that wasn't happening, we'd just go on ride after ride till we made ourselves sick.

Heaven!

■ ■ ■

One bitterly cold winter day a few years ago, I was walking along the main drag of Nantasket, when I saw a sign in a window: "Condo For Sale," with a phone number. Next to it was a picture of a pyramid-like building on top of Atlantic Hill, the site of the original Atlantic Hotel. My heart skipped a beat. Caruso had sung there; and I knew that Sarah Bernhardt and Gloria Swanson had both stayed there.

I called Joyce.

Time for fantasy to intersect reality, if at all possible. Joyce had, after all, been asking me to bring her along on one of these nostalgic journeys, and that moment had come. Problem was, she was from Bayridge; would this place that held such sentimental value for me strike the same chord of yearning in her?

We walked around; I talked a lot. I told her that back in the 1880s and '90s this was considered one of the major resorts on the Eastern seaboard. Sure, the roller coasters were gone, Paragon Park razed, but the carousel remained... I told her to close her eyes and imagine a Coney Island of the mind...only in Massachusetts...

I plied her with drinks and dinner at Raffael's, which used to be the Surf Ballroom, another old haunt. All the great bands had played there in the fifties and sixties: Stan Kenton's, Count Basie, Duke Ellington; even Arnie Woo-Woo Ginsberg used to do his record hops there.

We took a look at the unit for sale. The style was Frank Lloyd Wright, each unit unique, angular, with walls of glass where it faced the sea. In fact the building leaned out and over the rocks, so the ocean actually reached beneath a few of the lower units. I gasped when I looked out the windows: I could see

Nantasket Rat Pack: Tony Del Greco, Kathy "Red Sox" Darling, me, J. L. Sullivan, and Joe Giampaolo.

the very spot my mother dipped me in the surf so many years ago. Because Nantasket is on a peninsula, we could see north to Gloucester and Rockport, to Graves Light and Boston's skyline, and to the west where the sun was setting in streaks of orange and purple. We were only four stories up, but we felt like we were at the bow of some giant cruise ship just setting out to sea. Joyce smiled and took my hand, and I finally relaxed. I knew this wasn't the last sunset we'd watch from these windows.

Little did I know that moving to Nantasket would lead me to a community of like-minded souls. Thomas Wolfe said you can't go home again, but in a way I have. I've been lucky enough to find a group of guys who share my history; in fact we've dubbed ourselves the Nantasket Rat Pack. During long, hot summer days we hang out together and tell tales about days gone by. We've even got official titles: Tony Del Greco is the president; Charlie Lopresti, VP; John L. Sullivan, treasurer. Wearing our official orange jackets (Frank Sinatra's favorite color, of course), soaking up the sun, and leaning on the rail in front of Barefoot Bob's, we bring it all back: the big bands, old movies, old radio shows, old flames. It's a beautiful thing to be able to share these memories of the golden age of Nantasket Beach with such dear friends.

■ ■ ■

I think it was Picasso who once said, "At sixty, you know everything, but it's too late to do anything about it." I agree in a way: It may be too late to do some things, such as join the circus or train for the Olympic pole vault; but the way I see it, the world is as rich in beauty and delights as it has always been. I'm not saying time isn't a cooker: There is radio time which couldn't be more black and white: a commercial is 59 seconds long, not a breath more; your live show starts at seven in the morning; there is no such thing as: I was stuck in traffic...being late is not an option in radio. But time is slippery too: I'll remember a trip Joyce and I took and I'll say to her, Wasn't that trip to Normandy last year fantastic? When we visited the beaches there in honor of my father and the sixtieth anniversary of D-Day? And she'll say, Ron that was five years ago...and we'll get into a discussion: Can you believe we've lived in this house ten years now? Or that my first grandchild, Tia, turns seventeen this year?

As time goes by, I am ever more respectful of it. How can I best use it? Listening to an opera is four hours, five and a half, sometimes six. It's certainly not something done quickly, in a soundbite, like so many other things in life these days. For those few hours, it owns you; so I say, give in, let it envelop you, sweep you off to parts unknown. A symphony takes an hour or two. An afternoon to visit a museum. But it took Wagner thirty years to create the Ring Cycle, Caravaggio countless hours to master chiaroscuro, while Puccini didn't even live to complete *Turandot*—don't I owe these people a few hours of my time to witness their work? And then it comes down to choosing even further: I admit a strong bias here, but I've often spoken of how I never really explored the world of rock music. My reasoning is: If I only have limited time, I choose Beethoven, or Verdi, or Handel. I also believe it takes time for people to develop a sense of appreciation, and just as often it takes a guide of sorts to help them get there. Part of the mission of my life is to encourage people to take the time to listen to not only "beautiful" music but challenging music as well.

Besides, for me there's a spirituality in music, almost a mystical quality; it's a constant source of sustenance. And seriously, if you're listening to Debussy's "Afternoon of a Faun" or "La Mer," it's impossible to experience road rage.

And what about friendship? Of the hundreds of people one meets in a lifetime, it eventually becomes clear that true connections are rare and wonderful, and worth the time to explore. My friendship with Tony Bennett has been lifechanging for me. Of course I'd been a life-long admirer of his incredible singing, but I never dreamed we would have so much in common both musically and aesthetically. We even reacted the same way when we first saw the Sistine Chapel: we were both so moved we broke into tears.

In so many ways he reminds me of my father: he's a true Renaissance man. He not only paints on canvas, he creates masterful song pictures with his voice. When he sings the lyrics, you feel he's not only singing directly to you, but that he's lived every word. There's a tear in his voice, a genuine melancholy. And yet the man can swing! When he's upbeat, the euphoria is palpable. He's over eighty years old and his voice is in incredible shape, hitting the top notes like a much younger man.

But beyond the talent, Tony is a human being who cares deeply about people, and I think that's what has really kept our friendship alive. He's always asked about my family, while over the years Joyce and I have become a sort of surrogate aunt and uncle to Antonia, his youngest child, especially during the years she studied voice at Berklee and delighted us with her visits. Here's an icon, known worldwide, who's the warmest human being I know. He'll walk in the room with his wide smile and light it up and it's not an act, it's just Tony.

So it all comes back to time: time flies; time is precious, decades pass more quickly than years; so at seventy-three, timing is everything. I didn't think about sixty, though fifty gave me pause. In another seven years, I'll be eighty. Now *that's* interesting; I may actually put that in my date book: "Think about this one." But as long as everything goes as well as it has for me, I'm just a zen guy, hosting the BSO on Saturdays, walking Nantasket Beach with Joyce, and drinking my Diet Yahoo. I feel blessed and lucky to be healthy, even as I mourn dear friends who have passed. But they say eighty is the new seventy, which makes seventy-three the new sixty-three...which sounds swinging to me!

After all, many have said our essence doesn't change over time; we only become more intensely who we are. Joyce Curel watched me at work one day and then observed in a 1992 *Post-Gazette* article: "In his small studio, Della Chiesa is the orchestra's conductor—one hand flipping the records, the other answering the phone, one eye on the clock, the other on the promo sheet, a quick smile to the visitor, feet tapping 1-2-3 the beat, thoughts focused on the upcoming 2:01 PM network news tie-in. He is a boy at play. 'Isn't this fun?' he chuckles."[15]

A boy at play—she got that right.

As Louis Armstrong sang, "What a wonderful world!"

# PROFILES

## I. THE WORLD OF OPERA

### 1. LUCIANO PAVAROTTI

# King of the High C's

Luciano Pavarotti

Photo: The Metropolitan Opera Archives.

It was over thirty years ago, driving to work and listening to the Metropolitan Opera broadcast of *L'elisir d'amore*, when I first heard Pavarotti sing. I was so thunderstruck by the power and beauty of his voice that I had to pull off the road. He stopped the show, literally: the ovation after his aria, "Una furtiva lagrima," went on for close to three minutes. His voice swept me right back to those of the classic tenors I'd grown up listening to, huddled around the radio so many years ago.

Each spring the Metropolitan Opera toured and performed several operas in Boston at the Hynes Auditorium, a rather soulless building in my opinion; however, it was the only venue large enough to accommodate an audience of four thousand as well as the company's lavish sets. Unfortunately, our own beautiful Boston Opera House met the fate of the wrecking ball in the 1950s.

In the early 1970s I was hosting a weekly opera program on WGBH radio in which I interviewed many of the singers on the tour. Needless to say, there was a great deal of excitement for the scheduled appearance of this new young tenor who was the talk of the opera world.

He had been scheduled to appear in Boston to sing in Donizetti's *The Daughter of the Regiment*, an opera he had already performed at the Metropolitan in New York. This opera is the ultimate test for any tenor; it contains an aria with nine high C's. Pavarotti had just recorded an LP entitled *King of the High C's*, already a best seller.

My friend Francis Robinson, assistant manager of the Met, was kind enough to set up an interview via telephone with Luciano at his home in New York City. He told me the tenor's English wasn't that good, but I assured him I'd manage with my libretto Italian. The next day I placed the call and the tenor answered sweetly, "Pronto!"

Our conversation touched on many subjects, including his early days growing up in the city of Modena, where he was born in 1935. His father was a baker who also had a lovely voice and sang in the local church choir. Luciano began to listen to some of his father's favorite opera recordings that included the voices of Caruso, Gigli, and Bjoerling.

When the legendary tenor Beniamino Gigli came to sing in Modena, Luciano had the opportunity to meet him after one of his rehearsals. Young Luciano told him he wanted to be a tenor when he grew up. Gigli patted him on the head and said, "Bravo! Ragazzino, but you must work very hard!" Luciano told me that was the best advice anyone ever gave him, since from that day he'd never stopped studying. The meeting with Gigli was pivotal, because it was then, at the age of twelve, that Luciano made the decision to become a singer.

When I brought up the subject of the great Caruso he said: "As for Caruso, there are no comparisons. To me he is the tenor against whom all the rest of us are measured. To try and imitate his voice, you will destroy yours!"

One of the most charming moments in our interview was when we talked about his childhood days with his hometown friend, soprano Mirella Freni. In later years they were to sing all over the world together. Luciano said, "Mirella and I did everything together except sleep together...we studied voice, attended the opera, partied with friends who shared the same interests, and most important of all, we were 'milk babies' together; we shared the same 'wet nurse'! There must have been something very special in that milk!"

We discussed some of his favorite roles, and without hesitation he chose Rodolfo in Puccini's *La Boheme*. He made his debut in that role in April of 1961 in Italy, and it also served as his debut at the Met in 1968. In many of those performances, his Mimi was his dear friend Mirella Freni.

Some twenty years after our interview, I met Luciano again. He was making a movie, *Yes, Giorgio* in Boston and was scheduled to give a concert on the Esplanade on the Charles River, which was to be used in the movie. Thousands of people packed the grounds. Following the concert, he kindly signed his book, *Pavarotti, My Own Story*, for me. At this point in his career, he was the undisputed reigning tenor of the day.

Walter Pierce, at that time the director of the Boston Celebrity Series, was the gentleman largely responsible for bringing Luciano back to Boston's Symphony Hall for recitals on numerous occasions. These events became almost yearly and were always sold out. One evening Walter scheduled a post-concert dinner at the Wang Center with the great tenor. I'll never forget the sight of Pavarotti eating pasta with one hand and signing autographs with the other!

When Luciano died in 2007, Italy and the world mourned for months. The outpouring of love and affection at his funeral was overwhelming. I received dozens of calls from listeners all over the country asking to hear his music. National Public Radio aired a tribute in which they included my 1972 interview with him, one of his first on radio. A PBS special on his life raised more money for the station than any in history.

He was an exuberant, larger-than-life man who succeeded in bringing opera to the masses, not only through "The Three Tenors" phenomenon, which won him millions of fans, but through his appearances on *Johnny Carson* and *Saturday Night Live*, as well as performances with everyone from Barry White to Celine Dion and Vanessa Williams.

## SUGGESTED RECORDINGS

### Pavarotti's Greatest Hits: The Ultimate Collection

This two-CD collection is a must for Pavarotti lovers. He sings most of the famous tenor arias known the world over including "Nessun dorma," which became his signature showpiece during his period of "The Three Tenors."
London/Decca 2LH2458000

### Tutto Pavarotti

Another two-CD set devoted primarily to Italian song by Tosti, Leoncavallo, Lucio Dalla, and others. The great Caruso made many of these songs famous during the early days of recording.
Decca 425681

### Volare with Henry Mancini and His Orchestra

This unique collection presents Pavarotti in a collaboration with Henry Mancini who did the arrangements of popular favorites such as the title song made famous by Dean Martin. It gives us a chance to hear Luciano in a lighthearted mood, and he charms us all!
Decca

# The Incomparable

Placido Domingo

Photo: The Metropolitan Opera Archives.

My first encounter with Placido Domingo was in the mid-1960s when I was working for radio station WBCN, which at that time stood for Boston Concert Network. During those years the station was devoted 24/7 to classical music and opera. The studios were located at 171 Newbury Street in a cozy attic jammed with LPs, a newswire, and a small recording suite. Directly across the street was the headquarters of the Opera Company of Boston and its legendary music director, Sarah Caldwell.

My colleague Nat Johnson and I were given the responsibility of recording all of Sarah's operas for her own personal archives. One of those operas happened to be Puccini's *La Boheme*, which featured the great Renata Tebaldi and the young tenor Placido Domingo. At that time, Domingo had already performed numerous roles with the New York City Opera and was well on his way to becoming an established name in the opera world.

It turned out to be a magnificent performance. I invited the young tenor to join me in our modest studios for an interview. He graciously consented, and we talked about his early days as a singer. Born in Madrid in 1941, he told me both his parents were well known performers of a special kind of operetta known as "Zarzeula," a Spanish lyric-dramatic genre that alternates between spoken and sung scenes. The family later settled in Mexico where they formed their own company. It was there the young tenor started to seriously study voice. He had his major breakthrough in the New York City Opera with his debut as Pinkerton in *Madama Butterfly*. From that moment his career took off, and he began to perform in the world's major opera houses.

At the end of our interview, I congratulated him and thanked him for his time. He said, "I enjoyed it...tell me, do you have a piano somewhere?"

I replied, "Unfortunately no; this is a low budget operation."

"Too bad...I would be happy to record some arias for your show."

To this day I regret that I was unable to record Domingo. What a tape that would have been!

My next meeting with him followed his debut at the Metropolitan Opera. The Met had come to Boston for the annual Spring Tour in 1972, and I attended a performance of *Andrea Chenier* at the cavernous Hynes Auditorium. The title role was being sung by the great tenor Franco Corelli; and the baritone, Anselmo Colzani, was having vocal difficulties. An announcement was made during intermission that despite his indisposition, he would continue with his performance. While in the lobby, I spotted Domingo and reintroduced myself. He remembered our interview and remarked that he was covering for Corelli. I asked if this was one of his favorite roles and without hesitation he replied, "Oh yes...I've sung it several times along with the baritone role, and I can conduct it also!"

From those days to today, Domingo reigns supreme, singing well over a hundred and fifteen roles in Italian, French, German, Russian, Spanish, and English; everything from works by Tchaikovsky to Wagner's *Parsifal*, to even more obscure roles such as Alfano in *Cyrano de Bergerac*. He is also an established conductor, Artistic Director at both the Los Angeles Opera and Washington National Opera, teacher, and a great champion of young singers. As the late Beverly Sills mentioned in the preface to the book *Domingo, My Operatic Roles*, "he has the charisma of a rock star, the charm of a movie star, the voice of a god, and superstar is written all over him."[16]

Another critic once said, "If Pavarotti is the King of the High C's, then Domingo is undoubtedly the King of Opera."[17] I recently heard him sing a performance of *Otello*. Although in his mid-sixties at the time, his voice rang out as clear and strong as if he were still the young tenor I heard in Boston all those many years ago.

## SUGGESTED RECORDINGS

### Bravo Domingo

A two-CD collection of arias and songs by Puccini, Rossini, Verdi, Donizetti, Mascagni, Weber, and Bizet.

DG 459352

### Domingo Songbook

Selections include:

> "Maria"
> "Siboney"
> "Besame Mucho"
> "Autumn Leaves"
> "Time After Time"
> "Annie's Song"
> "Moon River"

Sony MDK 48299

### Domingo: Scenes from Wagner's Ring

This collection features highlights from *Siegfried* and *Gotterdammerung*.

EMI 557242-2

### Placido Domingo Sings Caruso

Domingo in a program of arias made famous by the great tenor.

RCA 09026613562

# Viva Verdi!

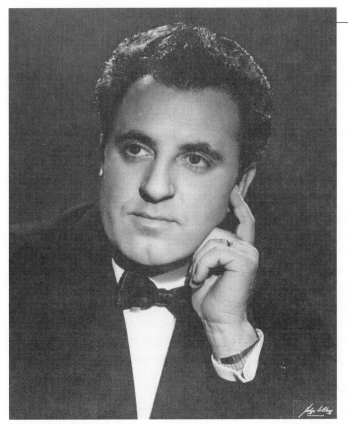

Carlo Bergonzi

Photo: The Metropolitan Opera Archives.

In the fall of 1988, I was invited by Phyllis Curtin, the head of the Boston University opera department (herself a renowned singer), to observe a master class given by Carlo Bergonzi. With the aid of good friend and translator Gino Gemmato, I was also given the opportunity to interview the maestro directly following the class.

Born in 1924, Carlo Bergonzi is a tenor who is mostly associated with the operas of Giuseppe Verdi, including many lesser-known works which he helped revive. Known for his superb diction and elegant phrasing, he is also acclaimed for his attention to style required in Verdi's operas.

The master class itself proved a transforming event to witness. Carlo was coaching a young soprano, and as an exercise had her sing an aria from *La Boheme*. He prompted her to sing the aria measure by measure, and after she vocalized each phrase, he sang it back to her. We all sat mesmerized as we witnessed a kind of metamorphosis. Carlo examined everything—the phrasing, interpretation, meaning, tempo—conducting her as if she were an orchestra unto herself. He stopped her phrase by phrase, word by word, even syllable by syllable to make her think about what exactly she was singing and ultimately delivering to her audience. It was as if we were watching someone learn how to hear, then vocalize, in a way that united meaning, emotion, and sound. I've never witnessed anything like it before or since. By the time she'd sung it the second or third time, she'd created a whole new interpretation of the aria. It was like watching a musical butterfly emerge from a cocoon, the way Carlo drew fresh notes out of her, transforming her and the piece entirely.

Carlo granted us just a few minutes of his time after the master class, but we were still able to glean some wonderful insights into his world before he was off to his next engagement.

On Busseto, Verdi's hometown, Carlo said: "There is something in the air that has produced great voices. The town is steeped in music! From the time they are children, everyone is encouraged to sing, both in church and at home. The spirit of Giuseppe Verdi is everywhere you look: there are posters of him in stores, portraits of him in the town hall, and of course at the Verdi library. Every year, young singers gather to study, to train in Busseto at the school I founded, the Academia du Bussettiana.* For two months a year, we host between eighteen and twenty students chosen from all over the world, and every June there's a contest among the students where only the best of those are chosen to perform."

On vocal longevity: "Never forget the phrasing of the diaphragm. If you don't study technique, you will not sing properly. There is no such thing as a closed sound or an open sound, only a covered sound. What is a covered sound? To place the breath on the diaphragm and turn the sound in the mask, abolishing the closed and open sound. These are the best medicines for longevity.

"You must remember that the life of a singer is a life of great sacrifice. Every single morning of the forty years of my career—summer, winter, during vacations, no matter what my schedule—I never neglected my exercises. I vocalize for forty minutes and do breathing exercises for half an hour. This is the way to preserve the voice."

On the use of modern day settings for operas: "Earlier you mentioned *Rigoletto* set in little Italy. This phenomenon doesn't even exist for me. I am a traditionalist. I believe operas should be performed the way composers intended at the time they were written; however, I do believe greater emphasis should be placed on strong acting ability, rather than the singer just standing there singing. But again, I believe it would be ridiculous to see Rigoletto dressed in contemporary clothes; the opera was written in 1850! Please don't misinterpret me—I'm not against young directors, I know directors who are great talents; but the system should remain the same, the traditions must remain the same. I admire Zeffirelli and Ponnelle because somehow they're successful in keeping the basic spirit of

*Now known as the Academia Verdiana

the composer while bringing them into contemporary times. I loved, for example, singing with Renata Tebaldi for the soundtrack of *Moonstruck*."

On classic Verdi phrasing: "What is lacking today is the ability to sing the true Verdi phrasing, which is one of the reasons these students came to Busetto. There are some very fine voices around, but they don't know the Verdi phrasing."

On his legacy: "I've had the great fortune of singing in the golden age of opera. God has given me the longevity to sing with Renata Tebaldi, Leontyne Price, Joan Sutherland, Birgit Nilsson, Carlo Tagliabue, and Robert Merrill, and to work with conductors such as Bruno Walter, Tullio Serafin, Georg Solti, and Fausto Cleva."

On his connection with Gigli: "I am one of his greatest fans and have had the wonderful fortune of singing baritone to Gigli's tenor: as Marcello in *La Boheme*, Belcore in *L'elisir d'amore*, and Germont in *La Traviata*. He was like a father to me. I'll never forget his generosity, especially with his advice to young people. What good luck I had to know him as an artist and a human being. I'm looking forward to paying tribute to him on the hundredth anniversary of his birth, which is in 1990. I want to celebrate the man who has not only inspired my own life's work but to also help young singers as he has helped me."

Now retired, Bergonzi spends most of his time at I due Foscari, a beautiful hotel in Busseto, which Joyce and I have visited many times.

## Suggested Recordings

**Carlo Bergonzi—The Sublime Voice**

Opera aria and scenes by Puccini, Verdi, Mascagni, Leoncavallo, Cilea, and
Ponchielli.

Decca 467-023-2

# Esultate!

James McCracken

From the Howard Gotlieb Archival Research Center at Boston University.

A contemporary of Placido Domingo, James McCracken began his career singing comprimario roles, or minor roles such as pages. He made his debut at the Met in 1953 as the toy vendor in *La Boheme*, but left in 1957 because Rudolph Bing refused to offer him any major roles. James quickly realized he wasn't getting anywhere with these secondary parts and took to heart that he had more potential as a singer than these roles offered, ultimately deciding to take himself to Europe to build up his repertoire.

Upon his return, James found that Bing had changed his tune. McCracken had come back so strong from overseas, evolving into such a talent that he simply could no longer be ignored. Critics flipped over him. He had built up valuable experience singing big roles including Otello repeatedly in Bonn, then in Zurich, gradually coming to the attention of the larger European houses. He specialized in the biggest Italian parts: Calaf in *Turandot*, the tragic clown, Canio, in *Pagliacci*, and the Verdian heroes of *Aida*, *Il Trovatore*, and *La forza del destino*.

The high point in his career came in the early 1970s when he starred in new productions in five consecutive seasons at the Met: *Otello*, *Carmen*, *Aida*, *Le Prophete*, and *Tannhauser*, his first Wagnerian part, which he accepted after declining many earlier offers to sing Wagner.

In Norfolk, Virginia, in 1954, he sang in a concert performance of *Samson et Dalilah* with mezzo-soprano Sandra Warfield, and the two were married soon after that. In short order, Warfield had a Met contract as well—slightly better than the tenor's, but still modest. They had four roles between them by 1955: they sang Nathanael and the Voice of Antonia's Mother in the matinee of *Les Contes D'Hoffmann*, then the Judge and Ulrica in *Un Ballo in Maschera*.

McCracken was a true dramatic tenor, a larger-than-life character possessed of a dark voice with tremendous power. Yet in person, he was the kind of guy you felt like having a couple of beers with. A big, burly presence with a great white mane of hair, he exuded the gruff aura of a man who might work in a steel mill. As an actor onstage he was completely involved in the role.

I'll never forget his performance of Otello in Providence in May of 1985 when he shook the rafters. I visited him backstage afterward, and found him sitting in his dressing room, his daughter blow-drying his hair off his face to cool him down. I told him it was one of the most powerful performances of Otello I'd ever witnessed and he said, "Thanks, but I think I could have done better," in his oddly high speaking voice.

During our 1983 *MusicAmerica* interview, he asked if he could vocalize along with himself as Otello in the death scene. What an unforgettable show: listening to McCracken on record along with the real live McCracken in the studio.

He also commented on the vagaries of critics: "Twenty-five hundred people could have enjoyed a performance, but then twenty-five hundred more could read a bad review and decide not to come." And on his own battles with stage fright: "I have my own profound inner turmoil going on before each and every performance. Let me tell you, you earn your money before the opera even begins. But then, after your first at bat, you just settle."

Here was a man who had an incredibly charismatic presence even before opening his mouth to sing, yet he sang like there was no tomorrow. His face reddened with concentration each time he vocalized from what sounded like the bottom of his feet. He used that voice to fill enormous auditoriums, and in the end, his technique proved equal to the challenge of sustaining a thirty-year-plus career, most of it at the top of his profession.

# Suggested Recordings

## McCracken Onstage with the Vienna Opera Orchestra

Arias from *Il Travatore, Faust, La forza del destino, Die Meistersinger, Tannhauser,* and *Otello.* James McCracken was at the peak of his vocal powers when he recorded this collection in 1965.

Decca CD 475 6233

## "Gurrelieder" by Arnold Schoenberg

James McCracken teams up with soprano Jessye Norman, the Boston Symphony Orchestra, and the Tanglewood Festival Chorus, conducted by Seija Ozawa in this 1979 Gramophone award-winning recording.

Philips—CD 475-7782

# Hip Heldentenor

Ben Heppner

Author's collection.

Ben Heppner is one of today's leading tenors. Born in Canada, he was a finalist at the 1988 Met Opera auditions and won top honors when he received the first Birgit Nilsson prize in the same year. He has enjoyed singing in the world's great opera houses and concert halls and has recorded with some of the foremost singers and orchestras of our time.

His vast repertoire encompasses such diverse roles as Calaf in *Turandot*, the title roles in *Lohengrin, Tristan and Isolde, Peter Grimes, Otello, Andrea Chenier, Samson et Dalila*, and Florestan in *Fidelio*, to name just a few.

No other tenor today combines the lyric beauty of the voice of Jussi Bjoerling with the stamina and power of Lawrence Melchior. Along with his intelligent musicianship and sparkling dramatic sense, Ben has set new standards in one of the most demanding repertoires in all opera.

He has been honored on numerous occasions for his recordings and performances including being named a member of the prestigious "Order of Canada" in 1999. Ben even has a street named after him in Dawson Creek, British Columbia.

During our first interview in 1999, we spent a good hour discussing his career and his fascination with jazz and the popular songs he heard on the radio as a youngster. He told me that much of his early education was simply listening to the voices of Nelson Eddy, Vera Lynn, Jeanette MacDonald, and his mother's favorite, Mario Lanza. Some of his fondest memories from childhood were of those days surrounded by music at home and in church.

At the time we talked, Ben had just released his first "non operatic" CD entitled *My Secret Heart: Songs of the Parlour, Stage and Silver Screen.* He shared with me that these songs were chosen with his mother (who was born in 1910) in mind: they're all in English and all love songs. These are the songs he imagined she would have heard on the radio growing up, from the first one: "Roses of Picardy," to the last: "Be My Love."

During the course of our interview, we played almost the entire collection from *My Secret Heart.* The phone began ringing off the hook with listeners clamoring to find out where to get this CD. To this day, this collection remains his favorite of his many recordings.

The evening of our interview, my wife Joyce and I attended his recital in Jordan Hall at New England Conservatory. Ben had mentioned that it was one of his favorite halls to sing in. Known for both its stellar acoustics and intimacy, Jordan Hall remains the perfect venue for many of the world's greatest artists.

Among the many attending the sold-out recital that evening was another friend, the great Irish actor Shay Duffin. It turned out that Shay was a big fan of Ben's and was in town performing his one-man show: *Confessions of an Irish Rebel,* based on the life of Brendan Behan.

The recital was a tour de force and included arias, lieder, as well as songs from the new CD, *My Secret Heart.* When Ben sang "Roses of Picardy," I watched as tears streamed down Shay's cheeks. Heppner received a standing ovation. The night was an unquestionable triumph, both critically and emotionally.

Ben has been back in Boston numerous times since, appearing not only at Jordan Hall, but with the Boston Symphony Orchestra and the Berlin Philharmonic in Mahler's *Das Lied Von Der Erde,* singing with the great bass baritone Thomas Quastoff. He continues to add roles to his already vast repertoire including Siegfried, the ultimate test of any Heldentenor.

Some of the qualities that endear Heppner to his many fans all over the world are his unpretentious personality, humor, and good nature. During a performance of *Parsifal* at the Met, Ben (singing the title role in the final scene) raises the chalice high to give communion to the knights of the holy grail. In this pivotal scene, the chalice is supposed to glow bright red in a totally darkened theater, while Wagner's music surges passionately, creating a fantastically dramatic effect. In this instance, the chalice remained completely dark.

Afterwards, Joyce and I went to see him backstage. I said, "Ben, you were so godly as Parsifal, I hope you'll hear my confession."

"But the light in the chalice didn't go on; did you notice that?"

"Yeah, but you pulled it off anyway."

In the end, he had the good humor to laugh it off, and we all went out for a beer.

In a 2004 interview, Ben had some fascinating things to say about James Levine, then Music Director of the Boston Symphony Orchestra.

Regarding the rehearsal process, Ben said: "James Levine is the only one who gets to know you instead of giving pedantic notes such as: faster, you're flat, more crisp, etc.—those are given, but by musical assistants. He looks at the big picture—your personality. He understands you as a person first, *then* leads you through the performance. He always phones before a performance and during intermission, and if things aren't going that well—he may say I can see you're struggling, but he'll say I'm with you all the way. He provides a level of support that's unique in a *way* that's unique.

"James is very concerned about the connective tissue between notes—it's always an issue for him to follow the Italian tradition in this way—and after a while all I would need to do was follow minute movements of his hands to make these vocal transitions. During the final two weeks of rehearsal for *Otello* I requested a few hours of coaching with him and it's funny, we didn't play it, we just talked it through, and it was enormously helpful. He just has this special way of dealing with people."

In the spring of 2010, in Dallas, Texas, Joyce and I had the privilege of hearing Ben in a magnificent performance of an opera based on Melville's *Moby Dick* by Jake Heggie and Gene Scheer.

## SUGGESTED RECORDINGS

**Ben Heppner: Great Tenor Arias**

Selections include some of the most popular and best loved arias from *Aida, Carmen, Turandot, Manon Lescaut, Andrea Chenier,* and other operas.

RCA 090126-62504-2

**Ben Heppner Sings Ricard Strauss with the Toronto Symphony Orchestra**

Conducted by Andrew Davis

A magnificently sung collection of arias from *Daphne, Guntram, Intermezzo, Die Frau ohne Schatten, Der Rosenkavalier,* and other Strauss operas.

CBC SMCD5142

**Ideale: Ben Heppner: Songs of Paolo Tosti**

A must for any lover of Italian song by the so-called "Prince of Melody." Selections include:

"Ideale"
"A vucchella"
"Entra"

"Vorrei morire"
"Seconda mattinata"
"Chitarrata abruzzese"

DGB 0001850-02

**My Secret Heart: Songs of the Parlour, Stage and Silver Screen**

Ben's personal favorite and mine, too.
Selections include:

"Roses of Picardy"
"I'll See You Again"
"Serenade"
"The Desert Song"
"Be My Love"
"I'll Follow My Secret Heart"

RCA 09026-63508-2

# La Dolce Vita

Marcello Giordani

Photo: The Metropolitan Opera Archives.

I had the great pleasure of interviewing tenor Marcello Giordani for the WGBH Forum in January of 2009. Born in 1963 in the small Sicilian town of Augusta, Marcello showed talent at an early age, taking private lessons in his church choir. He finally left a job in a bank and moved to Milan to study voice with Nino Carta.

As he tells it: "My father was a big opera fan. He always wanted one of his children to sing, or at least to love opera. I'm the youngest of four, and the other brothers couldn't care less about it. I was twelve years old when my father played my first opera for me; I heard Jussi Bjoerling and Giuseppi Di Stefano and fell immediately in love. I used to sing along with the recordings—probably a little too loudly, but my father encouraged me. My love for singing grew more powerful every day.

"I moved to Catania when I was eighteen and studied with Maria Gentile, but I won my first competitions in Spoleto in 1996, eventually coming to New York to study technique. Twenty-five years later I'm still here, now with my wife and two sons; and now I consider the Met my second home, and America my second land, or country. The truth is—when I was in my twenties, Europe wasn't interested in me. I'm lucky to be here where the opera community feels more forgiving, more supportive of young singers, indulgent of mistakes developing singers might make. I'm very thankful."

On keeping the voice in shape: "There are four thousand seats at the Met and no mics, so you need to fill that house with your voice. The very last person in the very last row has to hear you, even if you are singing pianissimo.

"Some say it's like sports. We are vocal athletes responsible for training the muscles of the voice. I vocalize every single day, try to be healthy, be careful what I eat, not drink too much, not smoke, of course, or ever be in an environment with smoke. It's kind of a boring life. I'm also mindful of not talking when I don't need to.

"I was always very conscious of choosing my repertoire because it affects my voice. My feeling is that I have a gift and I need to take care of it not just for myself, but for those who come to hear me. There are so many different shades of tenor, but now, after thirty years of singing, at age forty-six I know I'm a lirico spinto. I never pushed myself to be this; perhaps people believe I could have been singing a heavier repertoire, but this is how my voice has developed naturally."

On learning a role: "My approach certainly has evolved over the years. At first I would just learn my part, that's it; then I learned that didn't work! Hey, I was young...of course now I know I need to understand the context—the whole story. Now I have a routine: first, I learn the real history surrounding the story, read everything I can about the composer's own process, the concepts or thinking behind a piece of work, then I go over the libretto. My second step is to look at the orchestral score, if I have it. See which instrument is playing at certain points in the score...sometimes it all makes sense and sometimes not as much, so I ask the conductor: why this instrument at this moment and not another one? It's an interesting, beautiful process. My next step is to study with my vocal coach and my pianist, and after that I listen to whatever other versions of the role I can get my hands on."

On recalling roles, often years later: "It's amazing how the brain retains what it's been exposed to. The music will bring you to the words. Maybe I haven't sung a role for five years, if I hear a certain phrase of music, it comes to me like I sang it yesterday. In fact, if I made a mistake somewhere in the score five years ago, or had vocal trouble of some kind, five years later the music will cue me to make the same mistake. It's maddening, because when I was younger I made so many mistakes with so many operas, used the wrong rhythm or vocal position. I'll catch myself making the same mistakes I made twenty years ago! It takes so much work to retrain the brain and the muscles, it's unbelievable."

The affect of high definition TV on opera: "It's made a big difference in so many ways. Instead of reaching four thousand in one night we can hopefully reach millions. People who would never otherwise know my name approach me now. And it makes *us* more aware of what we're doing. It's a challenge, because now the camera is on us and we can't break our concentration. It's a great way to critique myself as well."

On keys to success: "Even though I have roles I'm comfortable in, I choose to be versatile. I don't want to be thought of as only being able to sing Italian opera. I'm lucky because I'm able to switch roles, vary my repertoire. I'm in a privileged position; at this point I'm able to really choose what I want to sing. This is the best job on earth, and I do it with great joy."

Any opportunity you have to hear this remarkable singer is not to be missed. Giordani is the real thing, a true Italian tenor in the great tradition of Gigli, Del Monaco, and Bergonzi.

## Suggested Recordings

**Marcello Giordani—tenor arias by Bellini, Donizetti, Mascagni, Rossini, Verdi, and Bizet**
NAXOS 8557269

# "Knew Them All!"

Norman Kelley with me wearing Shuisky costume

Author's collection.

Born in Maine in 1911, Norman Kelley was one of the most unforgettable characters I've ever known. I was just leaving a Met opera audition in the early 1990s when he was pointed out to me as someone I should meet. Even though he was in his eighties at the time, I was struck by his fireplug energy and simply his joy in being Norman Kelley. That night we went out for dinner, and both Joyce and I fell in love with this charismatic dynamo of a man who prided himself on knowing everyone from Carmen Miranda to Jose Ferrer to Gigli, even a number of singers who opened the original 1883 Metropolitan Opera House!

He clearly hailed from a different era. Always elegant, always dramatic, from the way he tilted his chin just so, to the way he walked into a room—even the way he posed under the lights—Norman was a joy to be around. When asked about any opera singer alive he would spout, "I knew them all! Every single one!"

The fact is, he did know them all. His career went back to the 1930s where, as a young man, he performed at the Radio City Music Hall. From there he worked his way into opera. His first big break came at the Met when he sang Herod in *Salome.* He went on at the last minute, the third cover after two replacements became unavailable. When he finally walked out onstage, the woman singing the role of Salome had no idea who he was. After this performance he received a note from Rudolph Bing, of which he was quite proud, and which he would often quote: "From now on, all of the Herods will be sung by Mr. Kelley."

Norman also sang the role of Mime (an extremely challenging role) opposite Wolfgang Windgassen, a well-known tenor at the time, in an otherwise all-German cast of *The Ring.* Reviews singled Kelley out as one of the best in the cast—though at the time he didn't speak a word of German.

After thirty-five years in New York, Norman left the state and settled in Easton, Massachusetts, where he stored all of his costumes in a series of trunks in his new home. He worked a great deal with Sarah Caldwell and the Opera Company of Boston on projects including *Good Soldier Schweik* by Robert Kurka. He also learned Russian and French operas by rote since he didn't speak any Russian or French.

One such role was Prince Shuisky in *Boris Godunov*, which he sang in a production directed by Sarah Caldwell in the mid sixties. Singing the title role was Boris Christoff, one of the greatest interpreters of the Czar. During a rehearsal of the final scene in which the tormented Czar falls from his throne and dies, Norman calmly stepped over the prostrate Christoff and ascended the steps to take his seat on the throne. He told me that as he did so he could hear the enraged Christoff muttering, "How dare you upstage me you porco miseria (miserable pig); this is *my* music, Mr. Kelley!" To which Kelley replied, "My dear Mr. Christoff, please tell me where it says that in the score?"

Finally Norman did learn German; in fact, he is perhaps best known for his translation of *Hansel and Gretel* into English from the German, a translation used all over the world to this day. This work generated handsome residuals for him for many years to come. When the Met revived *Hansel and Gretel* in the nineties he proudly invited Joyce and me to the performance for which he rented box seats.

Some of the most memorable times with Norman have to include an Opera tour we took together in 1997, a two-week journey in the Mediterranean. I'll never forget walking around Istanbul with Norman as he instructed passersby who were smoking to quit this ugly habit: "Put that down! It's bad for you, bad, bad, bad! Don't you know it's bad for you?" When he wasn't saving people from themselves, he would be strutting all over the ship, getting to know everyone, inviting them to our table, especially if he suspected they were lonely, and never forgetting to fill them in on his career: "I sang over 600 performances of 'The Consul'! I knew them all—sang them all, every single one of them!"

Norman put on elaborate soirees at his home in Easton, where he'd open up his trunks full of costumes and let everyone wear whatever they wanted. He had an entire barn filled with opera memorabilia: letters, programs, props, costumes, wigs, backdrops. If it had to do with opera, he had it squirreled away in that barn.

These were wonderful evenings, full of laughter and singing. He loved supporting the careers of young singers and encouraged them to perform with him during these parties. I remember him saying, "Someday you'll get old and you'll know what it's like, but you've got to keep going, you've just got to do it! You've got to go forward." To this day I miss his outrageous *joie de vivre*, his flamboyant spirit, and his inspiring love for life and music.

Norman died at the ripe old age of ninety-six, and left over twelve hundred vocal scores and his recordings to the New England Conservatory of Music in Boston. He also donated a magnificent portrait of the American soprano Lillian Nordica (also born in Maine) to the Metropolitan Opera.

# The Opera Couple

Photo: The Metropolitan Opera Archives.

Richard Cassilly

Photo: The Metropolitan Opera Archives.

Patricia Craig

Richard Cassilly trained in Baltimore, where he won the Arthur Godfrey Talent Scout Show in the early 1950s singing "Macushla," an Irish song made famous by John McCormick. He studied voice, married, started a family, and performed for several years at the New York City Opera Company before moving to Europe with his family to kickstart his career, a path similar to that of James McCracken. He landed in Hamburg, where he became a renowned heldentenor, a "heroic" tenor specializing in Wagner and other rigorous Germanic roles.

By the mid-1960s he had sung leading roles with almost every major opera house in Europe, including La Scala, the Opera National de Paris, the Vienna State Opera, and the Bavarian State Opera. Cassilly also forged a strong collaborative partnership with the Royal Opera in London, appearing in that house almost every year from 1968–1982. In 1978 he joined the roster of principal tenors at the Metropolitan in New York, where he spent the majority of his time until his retirement in 1990. His repertoire at the Met was incredibly wide, including Herod in *Salome*, as well as extremely demanding roles in operas by Berg, Schoenberg, Benjamin Britten, and Dalla Piccola. The *New York Times* described him as "a burly tenor with a bright ping on the top notes who had a supple lyric quality [to his voice]," and "was known to bring a musical intelligence and uncommonly clear diction to his work."[18] His portrayal in the title role in Benjamin Britten's opera *Peter Grimes* was praised by the composer as one of the finest he had ever seen.

Richard's life changed forever one evening when he played Don Jose opposite a very beautiful Micaela. Patricia Craig was a brown-eyed stunner, a soprano beloved for her Puccini and Verdi heroines, including *Madame Butterfly* and *La Boheme*. Very soon after their meeting, Richard's first marriage ended and he married Patricia. He had found his soul mate.

I'll never forget the evening Pat saved a performance of *The Queen of Spades* for Seiji Ozawa when the soprano Mirella Freni lost her voice. Ozawa was hours from canceling the performance when he enlisted Pat to take over the role for the evening. Everyone else who knew the role could sing it in Russian; Pat could sing the role, but only in Italian, which she did. Pat always sang with incredible passion; she reminded me and many others especially of Renata Tebaldi.

As their careers began to wind down, Richard and Pat moved to Brookline, Massachusetts, where they both became teachers: Richard taught at Boston University, while Pat taught at the New England Conservatory.

I first met Richard during a master class taught by Carlo Bergonzi at Boston University. It wasn't long before Richard, Pat, Joyce, and I bonded and began sharing good food, restaurants, and opera stories. They were like your average couple next door, except they just happened to sing roles like Peter Grimes, Mimi, Tannhauser, Otello, Billy Budd, and Tosca!

A big, burly Paul Bunyan type, Richard loved beer and oysters. For a time, he and I would haunt a restaurant in Brookline, where every Wednesday, oysters were a quarter each from five to seven in the evening. We showed up hungry week after week and kept on ordering dozens upon dozens of oysters until finally they canceled the special.

Richard also loved to cook, and he and Joyce were constantly sharing recipes. Every year he and Pat would have a "Crab Bash" at their home. Richard flew in boxes and boxes of crabs packed in ice, and we would sit around his kitchen table with mallets smashing open big fat Maryland crabs. Inevitably Norman Kelley would be with us, under good lighting, of course.

Toward the end of his career, Richard tried a variety of new roles, including a controversial production of *Tannhauser*, updated by the infamous Peter Sellers at the Chicago Lyric Opera. Instead of a minstrel knight, Peter Sellers made Tannhauser a Jimmy Swaggart bible belt hellion who steps off a 747 with a flask and ends up with a hooker in a hotel room.

I asked Richard how he felt about performing in this production and he admitted that his heart wasn't into it. In the end, he just wasn't one to be thrown into contemporary revisions of classic operas.

In January of 1998, Richard died of a cerebral hemorrhage at his home in Brookline. The world lost a great artist and we lost a dear friend.

Joyce and I feel fortunate to have seen him in his last performance, which was as Herod in *Salome*. Both Richard and Pat turned out great students, and Patricia continues to do so. Many are singing with major opera companies all over the world.

## SUGGESTED RECORDINGS

### *Tannhauser* by Richard Wagner

A DVD of a Metropolitan Opera performance from 1982 with Richard Casilly in one of his most famous portrayals of Tannhauser. The cast also includes Eva Marton, Tatiana Troyanos, and Bernard Weikl, conducted by James Levine.

### Patricia Craig Live!

A double-CD collection of soprano arias from *Madama Butterfly, Tosca, Otello, La Rondine, Carmen, Pagliacci, Faust, La Traviata*, and many other operas. Available at Casamagda@aol.com.

# The Singing Actor

Jose Van Dam

Photo: The Metropolitan Opera Archives.

A Belgian bass-baritone born in 1940, Jose Van Dam entered the Royal Brussels Conservatory at the age of seventeen where he studied with Frederic Anspach. He graduated just a year later with diplomas and first prizes in voice and opera performance, debuting as the music teacher "Don Basilio" in Rossini's *Il Barbiere di Siviglia* at the Paris Opera. It is there that Van Dam sang his first major role, Escamillo in *Carmen*. He remained with the Paris Opera until 1965.

But as he shared with me in the studio in our *Classics in the Morning* interview, his career started much earlier than seventeen: "I was eleven years old when I started singing in a boys' choir. One day a friend of my parents heard me and took me to sing in his church where I became a soloist. When I was thirteen my voice broke, and I went to my teacher with a lot of questions about my future. But then it hit me, I knew I would be an opera singer, and nothing and no one could convince me otherwise. I just knew this would become my life, it was strange: it's as if I didn't choose it; it chose me."

On his work in film: "I loved working on *The Music Teacher*. It was a low budget film, but a very poetic one. There was this osmosis in that film that should be in every film: the music was wedded to the image, so that the connection became almost an unconscious one for the viewer."

On acting onstage: "When you are onstage and you play a role, the minute you're conscious of acting, it's simply too much. It must be natural. When I go onstage I become Mephisto, and I react as if I am the character onstage. I am no longer Jose Van Dam."

On playing the devil: "You must remember one thing: the devil in the street looks like you and me. I try to play it with lots of humor and irony. When I play Mephisto, I play it as if he knows the end of the story, because he does! He has this confidence, almost arrogance, because he has the power of knowing what's going to happen. The devil isn't stressed, he doesn't have to seem dangerous, he's the master of the situation. It's all a joke for him."

On preparing for a performance: "I don't do anything special, really. I'm like a battery. I feel myself charging up during the day, and by evening I'm ready to go—just very ready to enter into this personage, whoever it is, by the time the curtain rises. I can be myself one minute and the next minute I'm on stage and bang—I'm the devil! I'm Leporello, Mephistopheles, whoever I need to be. I have great concentration onstage. Perhaps I'm lucky that way. I'm not worried or nervous the day of a performance. It's funny—I don't need to be the devil all day to be the devil at night; I simply don't think about the role until I'm there, facing the audience."

On his favorite roles: "I've loved all my roles, so I have no favorites, really. I just sing and play every role I can. I've been singing for thirty-five years, yet there is such a legion of work to choose from, even though I have an insatiable musical curiosity. Think about it: Schubert alone wrote more than six hundred Lieder, and you have Schumann and you have Wolf; you have Strauss. In the French there's Faure, Chausson, Ravel, Debussy, Poulenc—it's endless!—and if I tried to sing all of this I would need fifty more years to begin to get through it all."

His advice to young singers: "It's simple. Sing, learn, listen. Remember, you are a singer but you are musician first. Your voice is your instrument, but don't just listen to opera, go take it all in: the world of music is your school, and the world is at your feet."

## Suggested Recordings

### The Very Best of Jose Van Dam

A program of arias and songs by Gounod, Offenbach, Hahn, Bizet, Delibes, Massenet, Berlioz, and Verdi.

EMI CD #2005

# Star Spangled Baritone

Robert Merrill

Photo: The Metropolitan Opera Archives.

Born in Brooklyn on June 4, 1917, Robert Merrill became one of the most popular American singers of the twentieth century. As a youngster he wanted to become a baseball player, but blessed with a natural voice, he decided to study singing. His early days were spent as a crooner at weddings, at Borscht Belt venues in the Catskills, and on a local radio station. He also appeared as a straight man to comedians Danny Kaye and Red Skelton.

He won the Metropolitan Opera auditions in April of 1945, debuting on December 15 of that year as Germont in a performance of *La Traviata* that featured Licia Albanese and Richard Tucker. The great Toscanini chose him to sing this role in his historic broadcast with the NBC Symphony Orchestra.

On top of all his other talents, Robert had a killer sense of humor. As Germont in *La Traviata*, while Violetta lay dying, then perking up, then dying again...and dying a little bit more...he whispered to Richard Tucker (as Alfredo): "If this broad doesn't kick off in a few minutes, we're definitely going to miss the 6:20 to New Rochelle."

Richard Tucker and Merrill enjoyed playing tricks on each other. The tenor got his revenge when during a performance of *La forza del destino*, Robert Merrill as Carlo opened a locket that is supposed to contain a picture of his sister Leonora, pitting Carlo and Alvaro against each other. Instead, Tucker enjoyed watching his costar's face redden as he opened the locket, since Tucker had replaced the photo of Leonora with a rowdy photo of a couple making love.

On an episode of *Candid Camera*, Alan Funt placed Merrill, his face covered with a towel, in a barber chair, while an unsuspecting customer strolled in and sat down next to him for a shave. After a few quiet moments, Merrill burst into song with a fabulous rendition of "Largo al factotum," with its, "Figaro here, Figaro there, Figaro up, Figaro down." His face frothy with shaving cream, the customer jumped to his feet, exclaiming, "My God, what a voice! You should sing opera!"

But I think my most personal brush with his humor was standing next to him many years ago, in, of all places, the men's room at the Friar's Club. We'd been standing next to each other for, let's say, more than the usual amount of time, when he said, ever so casually, "How's the old prostate, kiddo?"

The first time I heard him sing was during a Met tour in Boston in the mid 1960s. The opera was *Rigoletto*, and Merrill's interpretation left an unforgettable impression on me. His sound was not only ringing and beautiful, but also big and powerful, recalling the great Italian baritone Titta Ruffo.

Later I got to hear him in many of his other signature roles: Amonasro in *Aida*, Escamillo in *Carmen*, Figaro in *The Barber of Seville*, Marcello in *La Boheme*, and Count de Luna in *Il Trovatore*. We shared many wonderful moments during our radio interviews talking about not only the world of opera, but also another of his favorite subjects: baseball. Bob, along with his wife Marion, were die-hard Yankee fans and even had the uniforms to prove it. No one ever sang the "National Anthem" at Yankee Stadium like Merrill. Being a member of the Red Sox Nation, I always delighted in discussing the greatest rivalry in sports with him.

I thought it fitting that when he passed away, he was watching the Red Sox in the first game of the 2004 World Series. Several years ago, along with my friend Nat Johnson, we produced a tribute to Merrill for Public Radio called: "Star Spangled Baritone." The show received favorable response from all over the country, and we shared many wonderful memories of interviewing him at the Friar's Club in New York.

Robert considered tenor Jussi Bjoerling to have one of the most beautiful voices of all time, and they shared a great friendship both on and off stage. The culmination of this relationship was on November 30, 1950, when they recorded what is considered to be the definitive version of the famous duet from the *Pearl Fishers* by Bizet. Beloved by so many, Robert was also a confidante of Frank Sinatra, who consulted with Merrill many times over the years whenever he had vocal problems.

The Robert Merrill Scholarship was instituted at the Julliard School of Music in New York City to assist young singers in their studies. Also, Bob's wife Marion informed me that the local post office in New Rochelle would be named after Robert Merrill. In Brooklyn where Bob lived, a street in his childhood neighborhood of Bensonhurst is also named after him. What a fitting tribute to one of the greatest singers of all time and a dear friend.

## SUGGESTED RECORDINGS

### Jussi Bjoerling and Robert Merrill: The Pearl Fishers Duet

A must for lovers of beautiful singing! Remastered by my good friend Nat Johnson, this collection not only includes the definitive version of the *Pearl Fishers* duet, it also includes duets from *Don Carlo, LaBoheme,* and *La forza del destino.* I can assure you this is a collection you'll return to again and again.

RCA7799-2 RG

### Robert Merrill: America's Greatest Baritone: Live Performances 1944–1990

Produced by Ed Rosen and taken from live radio broadcasts of operas and recitals, this collection offers the opportunity to hear Merrill in some of his most exciting performances before an audience. Most of these performances have never been heard before and are real collectors' items.

Selections include:

"Figaro's Aria" from the *Barber of Seville* (1944)
"Di Provenza" from *La Traviata* (1950)
"Il Balen" from *Il Trovatore* (1969)
"The Prologue" from *Pagliacci* (1954)
"Some Enchanted Evening" (1953)
"Old Man River" from *Show Boat* (1988)

Biographies in Music BIM 710-1

### *La Traviata* by Giuseppe Verdi: Complete Opera

Recorded in 1969

This excellent cast includes Anna Moffa as Violetta, Richard Tucker as Alfredo, and Robert Merrill in perhaps his most famous role, Germont. He was never in better voice and turns in a career-defining performance. Fernando Previtali is the conductor in a performance recorded in Rome.

RCA 09026 68885-2

# Devils and Demons

Jerome Hines

Photo: The Metropolitan Opera Archives.

Hines was born and raised in Hollywood, California, where his family was involved in the film industry. He studied mathematics and chemistry at the University of California while also taking vocal lessons. Imposing and regal at 6'6", Jerome Hines seemed born to be a basso—as he put it—"specializing in devils, demons, and despots."

As he tells it: "I started my career at eighteen on the West Coast tour of the Civic Light Opera, where I did *Pinafore* with John Charles Thomas. The year after, I debuted at the San Francisco Opera as Monterone in *Rigoletto* and Biterolf in *Tannhauser*, which is when I began singing with all the Met singers. At twenty I did my own concert in the Hollywood Bowl with the Los Angeles Philharmonic. About two years later I began singing Mephisto and Faust around the country. Then the war began slowing me up, so I went back to teaching chemistry and math. But the minute that time was over, I called the Met and they said come on over. My audition aria was the Boris monologue and Faust serenade. I was offered a contract right away. Did my first Mephisto at the Met just after my twenty-fourth birthday.

"The next season I sang eleven different operas, and little by little I just took over the repertoire; and that's when they let some of the older basses go because they had me. There were so many legends still around.

"In the end I did more Gurnemanzes than anyone in Met history, every bass role in the Wagner repertoire: Wotan, Fasolt, King Mark, King Henry, you name it! But like all American singers, I think, I didn't want to be known as a Wagnerian; we all wanted to do the Italian and French repertoire as well as the Russian."

On teaching and coaching: "I began the Opera Music Theatre Institute in 1987 and have really enjoyed coaching young singers. Mark Delavan—my pride and joy—when he came to me, he was so terrible my manager said, 'Look, don't bother, don't waste your time with the guy.' Mark sang for me and it was just awful—he sang right through his nose—and I yelled at him about it. When he sang back at me it was amazing; I mean it was as if he got it, right then and there. I asked him what clicked and he said, 'When Jerry Hines yells at me I yell right back at him.' Within a short time he had made it—he became just brilliant musically, and as an actor.

"We also had Franco Corelli at the school. If you ask me he was *the* tenor of the twenty-first century. Hell, even if you don't ask me, he just was. Franco taught for me at the school, and I think it really helped pull him out of his depression at the time: he was very broken hearted because he was losing his voice. We were great friends for the two and half years I knew him.

"Another great story: when I told one my students, actually one of my pet students with a gorgeous voice—Craig Hart—that I was going to be singing the role of King Phillip II of Spain in *Don Carlos* in Boston, he burst out with: 'I want to do the grand inquisitor with you!' I had to say to him, 'Craig, I'm warning you now, I'll be out there to blow you off the stage and I'll be expecting the same from you!' That's what it's all about, that scene. It's one of Verdi's greatest duets: the competition between church and state, and he pits two leading basses, I mean they are *fighting* each other, and of course, any two basses are going to want to wipe up the stage with the other. It was a thing of genius."

Hines spoke about his opera on the life of Jesus, *I Am the Way*: "I began fooling around with the idea for this opera in the early 1950s. I have to say it's through the writing of this work that I really became a Christian. I first produced it in 1956 with the Salvation Army in New York, where they have a beautiful auditorium. Nine of the disciples were not singing parts—they were just walk-ons—but we had trouble finding people, so we ended up using pan handlers from Skid Row; and it actually worked out great. Hey, what they lacked in talent they made up with enthusiasm! In 1963 the *Voice of Firestone* played a twelve-minute clip; and in 1966, we did it at the Lincoln Center with the help of the Lily Foundation. In 1968 we did a worldwide broadcast on the *Voice of America*, put it on at the Met, where it sold out two weeks in advance, and finally, [we] put it on in Moscow at the Bolshoi in 1993."

We were never fortunate enough to have *I Am the Way* come to Boston; however, Jerry continued performing virtually till the end of his life. In fact his last performance was as Philip II with Boston Bel Canto Opera at the age of seventy-nine. He also penned three books: a memoir called *This is My Story, This is My Song*, and two books on singing, *The Four Voices of Man* and *Great Singers on Great Singing*.

## SUGGESTED RECORDINGS

### *Don Carlo* by Verdi

A 1955 live Metropolitan Opera recording with Jerome Hines singing one of his greatest roles, Philip II of Spain.

Andromeda CD #5018

# West Virginia Diva

Eleanor Steber

Photo: The Metropolitan Opera Archives.

I interviewed Eleanor Steber for *MusicAmerica* in September of 1988, start-ing off the show with a recording of "Pace, Pace, Mio Dio" from *La forza del destino.*

As the piece ended, she laughed and commented, "Fantastic that you chose that to play! That was the first aria I could remember in my ears, because my mother sang it; she studied with Rosa Ponselle and had such a lovely voice. She used to really go to town singing 'Pace,' and we kids—all three of us—used to learn it sliding down the banister."

At that, Eleanor sang the first few measures of "Pace," and it was easy to picture three kids swooping down a banister to that gorgeous decrescendo! Even at seventy-four, her voice remained supple and silvery.

Born in Wheeling, West Virginia, or "as we called it: 'west by God Virginia,'" Eleanor won the 1940 Met auditions, debuting in *Der Rosenkavalier.*

Eleanor spoke to me about opera in the 1940s: "A whole new school of opera opened up between 1940 and 1950, especially as a result of the Met Auditions of the Air. A new generation of singers became real contemporaries of mine: Richard Crooks, Lawrence Tibbett, Gladys Swarthout, and Robert Merrill. And we really had the best of it. Not only did we have the advantage of getting right on the stage, as opposed to cutting our teeth in Europe, but we were doing repertoire—there were a variety of roles we could do.

"We also had the best conductors and coaches at that time, though maybe not for the best reasons. They had to flee Europe because of Hitler. It was my great fortune to have been brought to the attention of Bruno Walter, who I called my godfather. He got me into Mozart because of the way I sang, a style developed wholly and completely via my years with William L. Whitney at the New England Conservatory of Music in Boston. It was that style Walter recognized—in fact when he was told I was American he didn't think it was possible; he said there is no American that can sing like that today. Together we all made a magnificent company, which carried through to the closing of the old Met.

"We also had great conductors like George Szell, and I was lucky enough to study all my German roles with Felix Wolfes, all my Italian roles with Renato Bellini, a direct descendant of Bellini himself, and my French with Jean Morel."

On the role of radio in opera: "The radio helped a great deal in developing a true listening audience for opera, which in turn helped with getting funding for the Met. *The Voice of Firestone* and the *Bell Telephone Hour* brought young singers to the attention of the American public, just fabulous exposure for us. That's when a whole new operatic style happened; they really pulled out all the stops with these glorious productions. We all came out of that period of the musicals starring people like Fred Astaire, Ginger Rogers, and Jeanette MacDonald, so of course we all felt we had to be glamorous and beautiful as well as great singers!"

On the operatic voice: "At that time there weren't a lot of 'big voices' per se; that is, you had to have projection, had to get to the back of the Met, yes, but that was not what defined the era. I think there's too much emphasis on having a big voice rather than having a beautiful voice and a well placed voice. It's more important to have a quality voice and sound."

On taking care of one's voice: "I'm asked about that a lot. Honestly, I was never conscious of being fanatical about taking care of my voice. I guess we *were* careful about yelling out at baseball games. The way I see it is: we lived, and as a result we performed, and we produced."

Eleanor continued, "It takes a special kind of person to be a great opera singer, just like it takes a special person to be a star baseball player. You have to have all the ingredients. You might have a beautiful voice, and you're missing the brains to go with it. You might have a beautiful voice, and an ungainly body. You might have charisma and not have a beautiful voice, but the charisma won't get you there. [At one time] you had to know how to be a prima donna. Today it's kind of iffy, the prima donna thing! The big hats, the big gestures, the big moments, there's still some of that left but not as much as in my time."

Eleanor was an operatic soprano star who, as she put it, could "not be slotted." Her art ranged from Berg to Puccinni, Mozart, Strauss, Barber, art songs, Lieder, and beyond. Her energy was boundless: in one year between concerts, radio, and opera, she did 233 performances. When she retired, she went on to teach voice at the Cleveland Institute of Music.

## SUGGESTED RECORDINGS

**Eleanor Steber: Her First Recordings (1940)**

Arias and scenes from *Madama Butterfly, La Boheme, Faust,* and other operas.
VAI Audio #1023

# The Feisty Soprano

Eileen Farrell

Photo: The Metropolitan Opera Archives.

My relationship with Eileen Farrell goes back to my student days at Boston University. While working at BU's student-run radio station WBUR, I was hosting an opera program and learned that Farrell would be appearing at Boston College singing in a performance of Poulenc's *Gloria*.

I scouted out where she was staying and called the hotel to inquire about a possible interview. The voice on the other end was unmistakably the lady herself. When I asked if she could join me on the show, she said, "Sure kid, come over after the performance."

I was taken aback by the casual and easy manner of this legendary diva. Later on, when we became good friends, I realized Eileen was like that with everyone: unpretentious, extremely funny, and always just herself. At our meeting I asked her about her relationship with the Met. "I really don't want to sing the roles they're offering me, so I can do without them." Maria Callas herself was said to have quipped: "Who needs the Met, anyway, they don't even have Farrell!"

Such was the admiration for a singer who started out performing with her family as a member of "The Singing O'Farrells" from Woonsocket, Rhode Island. While still a teenager, Eileen landed her own radio show in the early 1940s in New York City, where she belted out opera, musical comedy, and popular songs of the day. Her career as a concert and opera singer took off at age twenty-six when she sang with conductors Charles Munch, Leonard Bernstein, and Dimitri Mitropoulos. She performed everything from Wozzeck to Wagner and made her Met debut in 1960 singing the title role in Gluck's *Alceste*. From 1963–64 she sang forty-four performances in six roles. Other roles at the Met included Leonora in *La forza del destino* and the title role in Ponchielli's *La Gioconda*.

She shared a wonderful story of her performance in that opera. At the time, the great tenor Franco Correlli was singing opposite her as Chenier. In fact Farrell and Correlli were known for their fierce battles onstage. Eileen was in great voice singing Maddalena and brought the house down with her big aria "La Mamma Morta," taking the opportunity to nearly blow Corelli offstage with the power of her voice. Backstage between acts, Robert Merrill, her friend and colleague who was singing the part of Gerard, knocked on her dressing room door and said: "Eileen, keep it up, you've got Correlli sweating!"

She was, in fact, one salty lady who swore like a longshoreman and took no b.s. from *anyone.*

One well known story involved a rehearsal with Fritz Reiner, who was known as not only a great artist, but more infamously, one of the toughest, most tyrannical conductors around. So tough that some musicians were reputed to have taken their lives while under his direction.

No such thing happened with Eileen Farrell. She was rehearsing her part, when Reiner sputtered: "Miss Farrell, you are singing too loudly. Watch your markings! More pianissimo!"

She kept on singing, just as loudly as before.

"Miss Farrell. Your markings! Pay attention! You're too loud!"

She sang on.

He tapped his baton on the music stand. "Miss Farrell, this is unacceptable!"

She kept at it, singing the part as she felt it should be sung.

"Miss Farrell, can't you read? This is piano! Piano. Means *softly.*"

She said, "If you want (expletive deleted) piano, get Dinah Shore." And she walked out.

During her later years, we kept in touch and enjoyed frequent lunches at "Arches," one of her favorite restaurants near her home in Fort Lee, New Jersey. The staff all knew and adored her and upon arrival always brought Eileen her favorite drink, a Manhattan with one rock. She spoke of her love for the great American standards and the album she made with Luther Henderson in 1959, called *I Gotta Right to Sing the Blues,* that proved she was indeed the very first "crossover diva." She loved hanging out in jazz clubs, joining her favorites such as Bobby Short. She had a ball belting out classics from the Great American Songbook, singing cabaret at the Algonquin in New York.

One of the highlights of her career came in 1979 when her friend Frank Sinatra asked her to record with him. Eileen always loved Frank, and their relationship dated to the early days when she appeared on his radio programs.

They got together on August 21 in New York City and made one of the great duo recordings of "For the Good Times" by Kris Kristofferson in a beautiful arrangement by Don Costa. It's included as part of the boxed set of the complete Reprise studio recordings of Frank Sinatra, consisting of twenty CDs.

Highly recommended is Eileen's biography by Brian Kellow, entitled *Can't Help Singing*. She said she had more fun and laughs telling her story to Brian than just about anything she did. It's a great read from beginning to end about the life of a woman who did it all, and as Sinatra would say, "her way"!

## SUGGESTED RECORDINGS

### The Eileen Farrell Album

This is one of the very first of the so-called crossover albums. Eileen shows she is equally at home swinging with the great American standards as she is singing Bach, Beethoven, Puccini, or Wagner. With wonderful arrangements by Luther Henderson, this collection recorded in 1959 has become a true classic. Eileen shows her great admiration for singers like Ella Fitzgerald, Mabel Mercer, and Sarah Vaughan in a style that is uniquely her own.
Selections include:

"Blues in the Night"
"Ten Cents a Dance"
"A Foggy Day"
"Old Devil Moon"
"The Man I Love"
"I'm Old Fashioned"
"The Second Time Around"
"My Funny Valentine"

Sony MDK 47255

### Eileen Farrell: Opera Arias and Songs

This collection features the best of Farrell's opera arias and songs and proves that she was one of the great sopranos of our time. The power and beauty of her voice is showcased to perfection in this CD featuring arias from *La Gioconda*, *Ernani*, *Oberon*, and *Alceste* and a song recital featuring traditional favorites "Summertime," "Danny Boy," and "Though the Years."
Testament SBT 1073

### Eileen Farrell: My Very Best

Between 1988 and 1991, Eileen recorded seven albums of great American popular songs that turned out to be some of the best recordings she made during her final years. Collaborating with arrangers Robert Farnon, Manny Albam, and Loonis McGlohon, often referred to as the "Super Soprano" who once said she loved to sing jazz as much as opera, Eileen proved that she is one of the most celebrated and versatile artists of the century. This collection features her own favorites from this period.

Selections include:

"Stormy Weather"
"My Foolish Heart"
"Little Girl Blue"
"Laura"
"Alone Together"
"Lush Life"
"Happiness is a Thing Called Joe"

Reference Recordings RR 60 CD

# After the Golden Age

Aprile Millo

Photo: The Metropolitan Opera Archives.

After being welcomed into the Met's Young Artist Program at the age of twenty-two, Aprile quickly became one of the most celebrated sopranos of the late twentieth century, known for her spirited and nuanced interpretations of Verdi. On April 4, 1986, Donal Henahan wrote in the *New York Times* of Millo's performance in *Don Carlo*: "Miss Millo sounds more and more like the Verdi soprano we've been waiting for."[19]

While at the Met program she was mentored by such greats as Renata Tebaldi, Zinka Milanov, and Licia Albanese; before that time she had been trained solely by her parents, tenor Giovanni Millo and soprano Margherita Girosi.

I met Aprile for the first time when she came to Boston for her first recital at Symphony Hall in April of 1987. The event was billed as "Aprile in April." At that time she was the talk of the opera world; even Pavarotti praised her as one of his favorites. I was delighted to be asked to introduce her from the stage, as well as interview her the day before the recital.

She told me the following regarding her parents, also opera singers: "My parents were not a great commercial success; it's unfortunate life wasn't kinder to them. But they had the single greatest voices I've ever heard in my life, and I swear I'm saying this without bias! My father sang like a Gigli with Del Monaco's power. Sadly, he had heart problems and it sidelined him. My mother sounded like Muzio: she had this histrionic quality, on the stage you thought she would die before she would let loose of the moment because it meant that much to her. I have a long way to go before I can ever be that good.

"They presented those qualities to me, so I went elsewhere in search of other components: the Oliveros, Callas.... I wanted to adopt her qualities as a singing actress. Some people think of that style as melodramatic or not really true in some sense, but I like this quality; you don't want to go to Kabuki theater and see the men in sneakers, you know?

"There is a way of presenting an art form and keeping it true, and not selling out. When I'm out there singing Verdi I feel like I'm at the end of some long evolutionary line."

On growing up in Hollywood: "I'll never forget picking up the cover of Andrea Chenier's album with the young Mario Del Monaco dressed as a magnificent seventeenth-century French aristocrat with a walking stick! I was in awe. I thought: I must be—I have to be!—where people look this good. It was a wonderful inspiration.

"My parents and I were constantly flying around the world to opera houses. It was a vagabond existence; but when we did settle down, it was in Los Angeles and I went to Hollywood High. I had my time of dating the quarterback, doing normal high school things, but there was something different about me. I was always slightly strange. They called me the duchess or the queen because I always knew what I wanted to do. We are a different breed, you know! Pavarotti compared singing opera to walking a tightrope, you might fall and sometimes you do fall—in front of everyone—and there are the same kind of gasps, and it's just as horrific when it happens, it really is."

On the human voice: "Music without voice is symphony. With the voice [a piece of music] becomes a direct line to the human being who is listening...there are no curves in the road, it goes directly to you because you are hearing a human sound. What I crave in a world that is so coffee-comes-immediately, dinner's-in-the-microwave, [is that] intangible thing, which music can be, and it becomes so much more present and powerful if it becomes expressive, and not just notes on a piece of paper."

On the influence of the golden age: "I am a true admirer of the golden age, an artist imbued with the passion of the old school. I listen religiously to Tebaldi, Milanov, Ponselle. I think what I appreciate about the old school, and what I'm very honored to be grouped in with, is that they sought to express the ineffable—that which would remind people of something greater than themselves. Young students, ones who are serious I think, should know Caterina Mancini; they should know Muzio, Rosa Raisa, Gina Cigna—these names will not fade away. They weren't so much on the beaten track but their contribution to the art form was enormous. I'm just glad, for our future, that the next generation is being exposed to them."

## SUGGESTED RECORDINGS

**Presenting Aprile Millo**

A collection of Verdi arias from *Ernani, Il Trovatore, Otello, Aida, MacBeth, Un Ballo in Maschera*, and *La forza del destino*.

Angel CD #47396

# II. ALL THAT JAZZ

## 1. DIZZY GILLESPIE

# The Diz

Dizzy Gillespie

© Ken Franckling.

Dizzy Gillespie was born John Birks Gillespie in 1917, in Cheraw, South Carolina, to a family of ten. His father, a local bandleader, encouraged his musical side and made instruments available to him from a young age. At four, John was playing the piano; by age twelve he had taught himself to play the trombone but soon switched to the trumpet. He received a scholarship at the Laurinburg Institute but left the school in 1935 to pursue his career, following his idol Roy "Little Jazz" Eldridge, who pioneered black musicianship in a white band. John soon earned the name "Dizzy" for his comical stage antics.

Dizzy Gillespie and Charlie Parker created a form of jazz called "BeBop," in which they broke all the rules. Up until that time, jazz was confined to big band and swing. Diz and Charlie, known as "the Bird," invented a whole other form of jazz musically, rhythmically, and artistically. Its center was 52nd Street in New York City in the mid-forties. At the time it was "hip" to be part of the bebop movement, which also sparked a fashion trend that included suede shoes, berets, and dark glasses worn at night.

Many people were turned off by the layers of harmonic complexity in this new sound...after all you couldn't dance to it; it was more like jazz concert music. Bebop was known as the first modern jazz style, seen at first as an outgrowth of swing, not a revolution. Thelonius Monk, Bud Powell, Kenny Clarke, and Oscar Pettiford were just a few of the musicians known for bebop, jamming at such renowned clubs as Minton's Playhouse and Monroe's Uptown House. Dizzy, now blooming as a trumpet virtuoso and gifted improviser, taught and influenced countless other musicians including trumpeters Miles Davis, Fats Navarro, Clifford Brown, Arturo Sandoval, Lee Morgan, and Jon Faddis.

In addition to changing the music world forever with bebop, Dizzy was a pioneer in founding Afro-Cuban jazz, which mixed Latin rhythms and African elements together with jazz and even pop music, particularly salsa. One of his most famous songs, "A Night in Tunisia," has a distinctly Afro-Cuban feel, and one of Dizzy's landmark travel events was his visit to Cuba, where he jammed with local musicians and met Fidel Castro, who was a big fan.

I first heard Dizzy in Boston in the late 1950s at Symphony Hall. He was performing as part of Norman Granz's "Jazz at the Philharmonic" series. From my seat near the front of the stage, I had a great view of Dizzy's iconic trumpet tilted up at a forty-five-degree angle. His breath came hot out of the bell of his horn, cooling into a gauzy smoke as it hit the air. I couldn't help but think that this cat's playing was on fire! Word was that his horn was damaged at a gig and he just kept it that way. Others said he saw and copied a horn like it in England: a trumpeter had it bent up like that because his vision was poor and it made reading music easier. Whatever the reason, Dizzy's inflated cheeks, horn-rimmed specs, beret, scat singing, and unique sound blew me away.

My dear friend Charlie Lake, "the Whale," was Dizzy's road manager for almost twenty years. Charlie handled all of Gillespie's business on the road, which was worldwide. He'd make sure he got the right room at the hotel, that he was on time for gigs, handled all of Dizzy's personal requests, set up the stage, and took care of his horn. Dizzy did have a habit of being absent-minded. As a matter of course, Charlie would visit his hotel room while Dizzy would be downstairs checking out. Charlie would leave the room with a huge trash bag full of forgotten items. The Diz had a way of living so much in the present that the immediate past was ancient history. In other words, he just left everything behind. Even money! Charlie swears today that Dizzy has funds abandoned in banks all over the world because he always insisted on getting paid in cash. This was a habit that persisted right up to the end of his career and wasn't just a function of being absent minded. Dizzy had dealt with one too many unscrupulous managers who stiffed him.

Dizzy had a way of charming the world around him by pretty much doing what he felt like doing at the time. According to the Whale, Dizzy once stepped on a plane carrying a watermelon that he was eating pieces of. He said something like, "hey, this is good s—t!" handed it to the flight attendant and suggested she cut up the rest of it and pass it around to the other passengers. In other airport shenanigans, Dizzy lore reports that he'd been caught a number of times going through security with a controlled substance in his pocket, but was let go when he professed complete ignorance and blew his cheeks out to full sail.

The first thing Dizzy would say to the Whale, his manager, when he arrived in Boston was "take me to see Ron." And each time he stopped by it was a celebration: it would take him half an hour to get from the front door to my studio because everybody wanted to talk to him and have their picture taken with him; all the way from the corridor, to the cafeteria, to the studio the Diz would sign autographs, albums, complying with all requests to blow out those famous balloon cheeks and a crack a huge smile.

During a 1990 interview, he shared a story that I think illustrates how the world loved him. "I was in Sicily, the morning of a gig. My cordless mic is hooked to me and my trumpet, and I just kept forgetting that, so I kept dropping it. Well, this one time, I *really* dropped it. The first valve went down and wouldn't come back up. So we're wandering around and we find a repair shop. This young Sicilian cat closed his shop just to fix my horn. We don't speak Italian; cat speaks not one word of English. He spent four or five hours on the horn. Then he had to go by boat to bring it to me, got there a minute or two before the show started."

Dizzy's sense of humor took everyone off guard. He would get onstage and make up what seemed like a real story, then spin it off into ridiculous-land, making the audience howl with laughter, then launch into these soul-searing sounds. He appeared on the *Muppet Show* once with Kermit the Frog. In 1964 he put himself forward as a presidential candidate, saying that if he were elected, the White House would become "The Blues House." His running mate would be Phyllis Diller, and his cabinet composed of Duke Ellington (Secretary of State); Miles Davis (Director of the CIA); Max Roach (Secretary of Defense); with Charles Mingus, Louis Armstrong, Thelonius Monk, and Malcolm X all holding vaulted positions.

I think to hear Dizzy at his best would be to listen to the *Jazz at Massey Hall* album that he recorded in the 1950s with Charlie Parker, when they were really into the peak of the bebop movement. Another favorite of mine is *Diz and Roy*, where Dizzy and Roy Eldrige embarked on a fierce trumpet battle. He made a wonderful album of ballads with strings called *Dizzy Gillespie and Strings* with Johnny Richards where he does a tune called "Swing Low Sweet Cadillac."

As Dizzy got older, time took a toll on his "chops," so he relied more on dancing and clowning around. But when he did play you could hear the old Dizzy in there, and he played right up until the end of his life. My son Aldo, who heard one of his last performances a few weeks before he died, said he was as kind and gracious as he always was to everyone. Dizzy seemed driven by a spiritual power that propelled him even from his earliest days of performing, at times having to confront some of the worst forms of racism. It was a very tough time for black musicians to perform, particularly in the deep South.

Drive and passion took Dizzy all over the world, until the world recognized him with countless awards including no fewer than fourteen Doctorates of Music. He became a world ambassador, sent by the State Department on tours to take jazz to Africa, Europe, and Havana. Jazz is a universal language that conveyed a freedom many were not used to hearing in their own countries. Dizzy embodied that freedom; there was nobody else like him. He was so loose and free, always unaffected, always in touch with joy.

## SUGGESTED RECORDINGS

### Dizzy Gillespie: Dizzier and Dizzier

The best of Dizzy's early recordings from 1946 through 1949. This collection is a wonderful introduction to Gillespie's days recording with other pioneers during the bebop era, including Charlie Parker, Milt Jackson, Don Byas, Ray Brown, and James Moody. We hear Dizzy in a small group setting as well as fronting his big band.

Some selections:

"Night in Tunisia"
"Jumpin' With Symphony Sid"
"Anthropology"
"Kool Breeze"
"Two Bass Hit"
"52nd Street Theme"
"Good Bait"

BMG Music 09026-68517-2

### Dizzy: The Music of John Birks Gillespie

This collection surveys the full range of Dizzy's recordings from 1950 to 1963 and features classic jam sessions with jazz greats Oscar Peterson, Stan Getz, and Roy Eldridge. It also serves as a companion to the biography of Dizzy by Donald L. Maggin entitled *Dizzy, The Life and Times of John Birks Gillespie.* Maggin also provided the excellent liner notes that accompany this collection.

Selections include:

"Bloomdido"
"Africana"
"Caravan"
"Blue Moon"
"BeBop"
"I Remember Clifford"
"Exactly Like You"
"Leap Frog"

Verve B 0004133-02

I highly recommend the book *Dizzy, the Life and Times of John Burks Gillespie* by Donald Maggin.

# Gentleman Ben

Benny Carter

© Ken Franckling.

Duke Ellington once wrote: "The problem of expressing the contributions that Benny Carter has made to popular music is so tremendous it completely fazes me, so extraordinary a musician is he."[20]

Benny Carter, whose seven decades of accomplishments included professional arranger, composer, alto saxophonist, and trumpeter, was about to visit me for an interview. So it was with a minor case of nerves that I welcomed the erudite, sophisticated, crisply dressed gentleman into my studio back in March of 1988. I was struck not only by his elegant yet laid back style, but his humility when speaking of his enormous contributions to music. Along with Johnny Hodges, Carter was the model for the 1930s swing era alto saxophonists, an innovative stylist with his own signature sound not only on the sax, but the trumpet, clarinet, piano, and trombone. A much admired arranger and composer, he was known for not only setting the stage for big band jazz, but for his major contributions to the world of film and television music. His personality and musicianship—the rare ability to collaborate, juggle all kinds of music and musicians with professional grace, which was not a hallmark of jazz musicians at the time—set him apart and made him in great demand.

Plus, how can I put this? The man was just...cool. One of the coolest jazz musicians who ever sat before me.

Born in 1907, Carter was primarily self-taught except for a few piano lessons from his mother. As a young man he befriended Bubber Miley who lived around the corner from him in Harlem, and who happened to be Duke Ellington's star trumpeter. By age fifteen, Benny was sitting in at several Harlem night spots.

Again, with no formal training, he taught himself how to arrange music as a member of Charlie Johnson's Orchestra. Soon he joined Fletcher Henderson's Orchestra, then in 1932 took off for Detroit and McKinney's Cotton Pickers where he became a force not only on alto sax but on his first love: the trumpet. Compositions he wrote during this time, "When Lights Are Low" and "Blues in My Heart," have since become jazz standards.

Within a year he was putting together his own band, which included swing greats Chu Berry, Teddy Wilson, Sid Catlett, and Dicky Wells, and he and his band were often featured at the Savoy Ballroom. By the mid-thirties he was traveling through Europe, playing an inestimable role in spreading the word of jazz abroad; recording and gigging with the top French, Scandinavian, and British bands, especially as staff arranger for Henry Hall's BBC house radio band. Swarms of delirious fans would greet him at every train station. During this time he also led the first international, interracial band.

He returned home in the late thirties, jumping into the swing era in the United States. He not only found work as a soloist, composer, and arranger with Lionel Hampton, but ultimately joined forces with Benny Goodman, Count Basie, Duke Ellington, Glenn Miller, Gene Krupa, and Tommy Dorsey. In a few years he'd pared down to a sextet, which included bebop pioneers Kenny Clarke and Dizzy Gillespie. Always open to new styles of music, the band soon welcomed modernists Miles Davis, J. J. Johnson, Max Roach, and Art Pepper.

By this time Carter had settled in California, and he was steadily taking on more studio work. I asked him about his transition from arranger to composer. He said, "You know, so many things that I've done I've never considered a goal before I did them. Then they fell into my lap and I took advantage...I was lucky, I just happened to be in Hollywood playing in a club with my orchestra, and they were doing a picture called *Stormy Weather*. I was called in to do some arranging, along with a bit of playing and from then on I was stuck in Hollywood for a couple of years."

Carter was also known for securing equal opportunity and equal pay for black musicians in the entertainment business. Before Carter, black artists were restricted to playing for movie soundtracks; Carter was hired not only to perform, but to direct studio orchestras as well as write the music for dozens of films and television programs. He even waged successful legal battles to obtain housing in then-exclusive areas of Los Angeles.

Beyond scoring a number of feature films including *An American in Paris* and *The Guns of Navarone*, he has provided arrangements for nearly every popular singer including Ella Fitzgerald, Billie Holiday, Peggy Lee, Ray Charles, Lou Rawls, Sarah Vaughan, Pearl Bailey, and Mel Tormé. Though Carter gave up full-time leadership of big bands in 1946, he became even more active in the fifties and sixties as a soloist, especially with such renowned groups as Norman Granz's *Jazz at the Philharmonic*. Especially beloved in Japan, he was received there like royalty.

In the seventies he turned his talents in a new direction—education. Invited to participate in classes and seminars offered by Morroe Berger, a sociologist at Princeton University, who eventually wrote Carter's biography *Benny Carter—A Life in American Music,* Carter was inspired to share his wealth of knowledge with eager music students. Conducting workshops and seminars at universities such as Princeton, Harvard, and Rutgers earned him honorary doctorates wherever he set foot. Nominated for seven Grammys and winner of two (for "Harlem Renaissance" and "Elegy in Blue)," Benny will always be known to his fellow musicians as the "King."

Our interview took place when the man was eighty-one, and he still had his chops. *A Gentleman and His Music*, an LP featuring an all-out jam session with Scott Hamilton, Ed Bickert, Gene Harris, John Clayton, Joe Wilder, and drummer Jimmie Smith, was cut when he was seventy-eight years old. At eighty he had reunited with old friend Dizzy Gillespie for the release "I'm in the Mood for Swing."

I asked him if it was true that he needed to be cajoled to play his trumpet: "Not at all! What they have to do is wrestle it away from me. I just love to play it—I prefer it to the sax—but I just don't get to play it much, so I have to watch out. I don't think I should be assailing rather than regaling the ears of my listeners, you know?"

During his travels in Europe, Carter spent much of his time visiting good friend Coleman Hawkins. Two recordings that best reflect their synergy are "Honeysuckle Rose," recorded with Django Reinhardt, and the album *Further Definitions*, now considered a masterpiece. I asked him who might have been an influence on Coleman Hawkins: "You know, that really makes me think...today we have so much to listen to, so many choices; but back then, in 1922 when Hawkins was playing with the Mamie Smith Jazz Hounds, what tenor sax players were around? What influences could he have had? I'm telling you, what came out of him, just came out of him! He was a real giant and innovator."

Carter remarked on how much he enjoyed speaking at colleges, but lamented the utter lack of resources available to him as a teenager: "I guess the advantage was that you learned how to swing, first hand. You had to feel it; there were no books to teach you. Sure, there were books with scales and so on, but no scores with solos like a Charlie Parker solo—you couldn't put your hands on that anywhere. I would've given anything at the time to find a Frankie Trumbauer solo book! Sure I could listen to his record; I could copy "Singing the Blues" or "I'm Comin' Virginia" that I loved so much, but that was it.

"But we learned from each other. We put on these endless, no-holds-barred jam sessions. Spur of the moment, for hours right on till morning sometimes. Then they were called 'cutting contests,' but they were as much about collaboration as they were competitive. We had so much fun playing, listening and playing off of each other."

Finally I asked if there was anything in his record-breaking career that he regretted not accomplishing. He said, "Ron, there's really only one. I've been in several movies, but never have I realized my dream of being leading man to Lena Horne.'"

## SUGGESTED RECORDINGS

### Best of Benny Carter

This Grammy winner features Benny in performances with jazz greats Dizzy Gillespie, Phil Woods, Clark Terry, Harry Edison, and many others.
Selections include:

> "Lover Man"
> "Wonderland"
> "Prelude to a Kiss"
> "Remember"

Musicmasters CD 65133-2

### Benny Carter: New York Nights

A 1995 release with some of Carter's favorite younger musicians including Chris Neville, piano; Steve LaSpina, bass; and Sherman Ferguson, drums.
Musicmasters 65154

### Sax a la Carter!

A classic recording from Capitol in 1960. Benny at his very best with Jimmy Rowles, piano; Leroy Vinnegar, bass; and Mel Lewis, drums.

# Jacquet's Got It

Illinois Jacquet

Courtesy of Pamela Jacquet Davis.

Born Jean Baptiste Jacquet in 1922, in Broussard, Louisiana, Illinois Jacquet was one of the true jazz giants of the tenor saxophone. His full-bodied, bluesy, rich sound was instrumental in developing the so-called "Texas Tenor Style," a way of playing the sax to bring out its big sound and biggest volume. Jacquet was known for introducing the screeching technique, as well as an original rompin' stompin' sound that's been a part of R&B ever since. His true jazz and swing fans, however, appreciate the warm, sensitive tone apparent in countless jazz ballads that communicates a sensitive side of one of the last big-toned swing tenor saxophonists.

Though he began singing and dancing to his dad's band at age three, Illinois went on to play drums as a teenager, then switched to soprano, alto, and finally the tenor sax under the tutelage of Lionel Hampton in the late 1930s. It was in 1942 with the Hampton band that he made jazz history with one of the most famous solos ever: "Flyin' Home" became his signature tune, and he included it in every performance that followed. The solo became so familiar that other saxophonists played it note for note.

In one of our interviews, he talked more about that incredible night: May 26, 1942: "I guess I was getting good at being everybody else: I could imitate Lester Young, Coleman Hawkins, Flip Phillips, lots of gifted players, and I did it out of love for their style. But just before I went on, Lionel took me aside and said, 'man, I've heard you be everybody else. Now you go on up there and play like Illinois Jacquet. Go be yourself.' And that was the moment, you know, when I heard my own sound. It was some high-spirited blowing! Anyway, you know the rest; but everywhere I go, *everywhere*, even the end of the world where I was last month: New Zealand, they know that solo. That was a stroke of luck; the grace of God gave me the originality to come up with a solo like that."

After leaving Hampton, Illinois joined Cab Calloway's Band in 1943, finally teaming up with the Count Basie Orchestra in 1945. The early 1950s began his long association with Norman Granz's *Jazz at the Philharmonic.* These concerts, also showcasing Coleman Hawkins, Lester Young, Charlie Parker, Dizzy Gillespie, Roy Eldridge, Oscar Peterson, Ray Brown, Louie Bellson, and Buddy Rich introduced jazz to millions around the world.

It was during this period that he made another one of his famous recordings, this time teaming up with fellow tenor saxophonist Flip Phillips. By this point, Jacquet's wildly swinging improvisational forays were working countless crowds into a frenzy. Their saxophone battle on *Perdido* became a classic, and was performed at every JATP concert from that time on.

I always enjoyed Illinois' visits to my show. He had an infectious sense of humor and told wonderful stories about the legends he played with. One involved Charlie Parker. Stoned and without his horn or any bread, Charlie rang his doorbell one evening. "Bird" was strung out and looking to crash for a day or two. Those few days turned out to be several weeks. Illinois said when he returned from one of his many worldwide engagements, Bird had not moved from the couch that he'd last seen him on!

My first interview with Illinois took place back in the 1970s when he appeared at Sandy's Jazz Revival in Beverly, Massachusetts. At that time I was hosting an all night jazz show called *GBH After Hours.* Sandy Berman, the late owner of the club, called and told me he was bringing Illinois down for an interview after the gig that evening. Needless to say, I knew this would be tiring for Illinois because he would not be arriving until two o'clock in the morning after his performance at the club. I mentioned this to Sandy, but he said, "Don't worry about it—everything's cool."

It wasn't until two-thirty in the morning that Illinois arrived at my studio. I could see he was tired and dragged, but I made sure the first thing he heard when he stepped into the studio was one of his most famous recordings, the original 10" LP of "Black Velvet." His eyes lit up and from that moment on, the interview went just fine. He stayed for more than an hour, told great stories, and dubbed me "Ron Delicatessen"! To top it off, I gave him the LP, which delighted him no end. From then on we became great friends; it seemed he wouldn't leave a city in the world without sending me a postcard. He also never missed a chance to visit me whenever he played in Boston. I was most flattered when he wrote a jazz chart for me called "Runnin' With Ron" (referring to my days as a marathoner) and recorded it with his big band. When my marathon days were over, I started walking for a change. Jacquet said, no problem I'll just have to write a new one for you: "Walking With Ron."

On worldwide reactions to his music: "Once we were in Perugia in Italy, which is a new city built on top of an ancient one, an incredible place. Anyway, we did a gig there, and they wouldn't let us leave the stage. People just couldn't get enough of the music. They have a way of feeling the music there, not just hearing it. I know I'm in another state of mind when I'm playing the ballads, a very soulful state of mind. Anyway, they wouldn't let me leave the stage for quite some time. Even after—I can't remember how many encores. It was getting late, you know. This happened in Spain, too: people would cry, they came up to us crying and touching us, asking for an autograph. It's fascinating how people from different countries express their love for the music."

Illinois championed the careers of many young musicians including the guitarist Gray Sargent who now works with Tony Bennett. A true jazz original, Illinois was the consummate showman who always delighted his audience with his clowning and dancing as well as his fantastic blowing. Credited with more than three hundred original compositions, he even jammed with President Bill Clinton, an amateur saxophonist, on the White House lawn during Clinton's inaugural ball in 1993.

His manager and life partner, Carol Scherick, called me with the sad news one July day in 2004 that he had died after one of his concerts with his big band at Lincoln Center in New York. It turned out to be not only his final performance, but also one of his greatest. I was privileged and honored to be one of the eulogists at his funeral at Riverside Cathedral in New York City. It was a glorious and fitting send-off to an artist revered and loved worldwide, who was finally flying home.

## SUGGESTED RECORDINGS

### Illinois Jacquet and His Big Band: Jacquet's Got IT!

One of the finest big band recordings ever. Illinois put together some of the best musicians for this session that took place in New York City in 1987. His tenor sax solos on all the selections are inspired and soulful, defining what jazz is all about.

Selections include:

"Tickletoe"
"Smooth Sailin'"
"Flyin' Home"
"More Than You Know"
"Blues From Louisiana"
"Port of Rico"
"Running With Ron"

Atlantic Jazz 781816-2

### Illinois Jacquet: The Black Velvet Band

In this collection we hear one of the earliest and best of Illinois' big band sessions, performing with jazz greats Joe Newman, his brother Russell Jacquet, J. J. Johnson, Milt Buckner, and Jo Jones.

Selections include:

"Black Velvet"
"King Jacquet"
"Big Foot"
"Mutton Leg"
"My Ol' Gal"
"Blue Satin"
"Adam's Alley"

RCA/Bluebird 6571-2-RB

### Illinois Jacquet Plays Cole Porter

Illinois once told me the real test of a jazz musician was playing ballads. In this all-ballad collection, backed up by a nineteen-piece orchestra of strings, woodwinds, harp, French horn, and rhythm section, Illinois proves his point with masterful performances of Cole Porter classics. I am indebted to my good friend and Illinois' #1 fan, Dan Frank, for getting this hard-to-find CD. It was Dan who brought Illinois and his great band year after year to the Barry Price Center Benefit concerts.

Selections include:

"It's Alright With Me"
"I Got You Under My Skin"
"I Concentrate On You"
"Get Out of Town"
"So In Love"
"Every Time I Say Goodbye"
"Begin the Beguine"

Argo LP-746 (reissued as a CD with original number)

# Hot, Cool, and Swinging

Stan Getz, Joyce Della Chiesa, and me

Courtesy of Charlie Lake.

If Mozart could play the saxophone, he would sound like Stan Getz. It was in the early 1960s when a friend and fellow jazz enthusiast, Bob Buccini, asked me if I had heard a tune called "Desafinado." He'd heard it on a jukebox and told me to check it out; it was like no other sound he had ever heard. As one of Stan's college students Larry Grenadier remarked, "His sound was striking; it hit you over the head. His timing and rhythm were so strong; he was a master of space and silence."[21]

It turned out to be one of the biggest hit recordings of the time and popularized the Bossa nova, a brilliant blend of jazz and Brazilian rhythms. As it turned out, Stan had been playing in this style for some time. "The Girl from Ipanema," recorded with Joao Gilberto and his wife Astrud, became one of the most well known Latin jazz cuts of all time, winning two Grammys (Best Album and Best Single), sales surpassing even The Beatles' *A Hard Day's Night*. In the last few weeks of 1962, there was a national Bossa nova craze, and Stan found himself on top of the Downbeat Poll for the year.

His nearly fifty-year career began in the 1940s in New York City, when he worked with Stan Kenton, Benny Goodman, and Woody Herman's Second Herd.

While working with Woody's band, his solo on the 1948 recording of "Early Autumn" became a classic. Based on this one solo, Stan had become a star, and everyone wanted to hear him play.

Even as a child, Getz felt the need to play every instrument in sight, but gravitated toward the saxophone. His father bought him one when he was thirteen years old. Stan said, "In my neighborhood, the choice was: be a bum or escape. So I became a music kid, practicing eight hours a day. I'd practice the sax in the bathroom, and the tenements were so close together that someone across the alleyway would yell, 'Shut that kid up!' and my mother would shout back, 'Play louder, Stan, play louder!'"[22]

That unique and cool "Getz sound" was influenced by the great tenor saxophonist Lester Young, "the Prez." It was out of that sound that Stan developed his own style that could be hot, cool, and swinging at the same time.

It was the drummer Jimmy Falzone who first introduced me to Stan. Jimmy had worked with Stan when he was a teenager in the Kenton band. One afternoon in the summer of 1985, I picked up the phone at WGBH and the voice on the other end said: "Ron...this is Stan Getz...Jimmy told me to call you." Needless to say, I was delighted to invite him on my show, and that began our friendship that continued until he passed away in 1991.

He shared some thoughts on music: "It's like a language. You learn the alphabet, which are the scales. You learn the sentences, which are the chords, and then you talk extemporaneously with the horn. It's a wonderful thing to be able to speak extemporaneously, which is something I've never gotten the hang of. But musically, I love to talk off the top of my head. And that's what jazz music is all about. Nothing gives me the same satisfaction of spontaneous interaction."

Stan was known, however, for some remarkably spontaneous behavior. One night in London in 1969, while hanging out and drinking wine with Peter Sellers and Mike Milligan, the conversation turned to Stan's swimming prowess, and eventually they bet him he couldn't swim across the Thames that night. In no time Stan was walking through the lobby with his swim trunks under his terrycloth robe. Up to the very last minute the size of the bets increased, until Stan dove into the dark, cold river water. Panicking, Milligan and Sellers ran to the police to report it, but they were told to go home and sleep it off. They jumped in a cab and raced over the nearest bridge to look for Stan on the other side, where they found him calmly sitting on a bench, dripping wet but enjoying the night air. Getz turned to them and said, "What took you guys so long?"

By the time we met in person, Stan had overcome his life-long battles with drugs and alcohol and was on the road to better health that rejuvenated his playing. During one of our many interviews, he told me he met a lovely woman named Samantha in a health food store near his home in Malibu. They began to date and soon after planned to marry. At the time they met, Samantha had no idea who he was. She soon found out! She bought every Getz recording she could find and only then did she understand who this great man was.

He very much wanted my wife Joyce and me to come to his wedding. Alas, it was not to be. Stan was now battling his new demon: cancer. Though he made great strides in his fight against this disease, it finally overcame him with Samantha at his side to the very end.

Later, when my wife and I visited Los Angeles, we met Samantha for lunch at the Beverly Hilton Hotel. She told us about Stan's great stamina and will to live. After wonderful remembrances, we said our goodbyes. Wanting to see Stan's home, we took a ride north on the Pacific Coast Highway. As we approached the beautiful Malibu area, with the top down on our convertible, from the radio came the unmistakable sound of Stan's music greeting us! It was like he was saying, "Thanks for stopping by!"

## SUGGESTED RECORDINGS

### Getz/Gilberto

No Stan Getz collection would be complete without this one that includes the original hit recordings of the Bossa nova years from March 1963. Stan's musical colleagues include Antonio Carlos Jobim, Joao and Astrud Gilberto, Tony Williams, and Milton Banana.
Selections include:

"Girl From Ipamena"
"Desafinado:
"Corcovado"
"O Grande Amor"
"So Danco Samba"

Verve 810048-2

### Stan Getz: A Life in Jazz

This CD is a companion to the excellent biography by Donald L. Maggin entitled *Stan Getz: A Life In Jazz*. The CD includes many highlights of Getz' distinguished career and collaborations with Gary Burton, Ella Fitzgerald, Chick Correa, J. J. Johnson, and Jimmy Raney. As Maggin points out in his notes, "Nature provided Stan with abundant talents for music that included perfect pitch, a photographic memory, and the ability to create fresh and beautiful melodies in his improvisations."
Selections include:

"Night Rider" (from the 1961 masterpiece *Focus*, an album featuring Eddie Sauter's brilliant compositions and arrangements)
"Summertime"
"Billies Bounce" (with J. J. Johnson and Oscar Petterson)
"You're Blasé" (with Ella Fitzgerald)

"Litha" (with Chick Correa)
"I'm in Love" (with Abbey Lincoln)

Verve 314535119-2

## Stan Getz and Kenny Barron: People Time

Recorded live in Copenhagen in 1991 with pianist Kenny Barron, this duo
collection is one of Stan's last and most brilliant sessions. The two-CD
set features originals by Charlie Haden, Mal Waldron, Benny Golson, and
Eddie del Barrio along with standards by Cole Porter, Benny Carter, and
Jimmy Van Heusen.

Selections include:

"Night and Day"
"First Song"
"East of the Sun"
"I'm OK"
"Like Someone in Love"
"Soul Eyes"
"Gone With the Wind"
"I Remember Clifford"

Verve 314510823-2

# Super Drummer

Buddy Rich
and me

Courtesy of Charlie Lake.

There's a famous story about Buddy Rich that made the rounds shortly after he died in 1987. Buddy had been in and out of the hospital for several weeks undergoing tests for a variety of ailments. During his final visit, just before surgery he was asked if he was allergic to anything. "Yeah...country western music!"

Buddy was born in Brooklyn in 1917 to parents who lived and breathed Vaudeville. Buddy's father Robert noticed that his son could keep a steady beat with spoons at the age of one. At the tender age of eighteen months, Buddy made his first appearance on stage in his parent's act, "Wilson and Rich." At four years old, he presented an act as a drummer and tap dancer called, "Traps, the Drum Wonder," performing regularly on Broadway. At eleven, he led his first band, turning into the second highest paid child star in the world (after Jackie Coogan.)

In the late thirties, Buddy's career took off in earnest when he played at the Hickory House in New York, jamming with Joe Marsala, Bunny Berigan, Harry James, and the great Artie Shaw.

Buddy became known around the world for his virtuosic power, dexterity, groove, and speed, despite the fact that he never received a formal lesson and never practiced outside of his performances. Though he typically held his stick using a "traditional grip," he was a master at the "match grip." He could pull off the one-handed-roll on both hands, do gymnastic crossover riffs, and was once clocked at twenty-strokes-per-second on a single-stroke roll. Aside from his explosive displays, he could slide easily into quieter passages and was masterful at the brush technique. Influences included: Chick Webb, Gene Krupa, Dan Tough, and Jo Jones, among others.

By 1939 he had joined Tommy Dorsey's famous orchestra before enlisting with the marines. During his years with Dorsey, his roommate was a young singer named Frank Sinatra. A volatile combination to say the least! To taunt Frank, Buddy would play as loud as he could when Frank was singing a quiet ballad. One night at the Paramount Theatre, Frank had had it. He picked up a music stand and threw it at Buddy, who luckily ducked in time. It crashed into the wall and to this day, sixty years later, the dents are still there.

During the 1950s, Rich toured with Norman Granz' *Jazz at the Philharmonic*, teaming up with greats like Oscar Peterson, Charlie Parker, Illinois Jacquet, Roy Eldridge, Thelonius Monk, Al Haig, Dexter Gordon, Dizzy Gillespie, and Ella Fitzgerald. These concerts took place at Symphony Hall in Boston, and that's where I saw him for the first time. I couldn't believe his sense of swing and his spectacular drum solos that went on sometimes for fifteen minutes, always bringing down the house.

Buddy had his own successful big band from 1966 to 1974. During that period he appeared quite often at Lenny's on the Turnpike in Peabody, Massachusetts, which is where I met him for the first time. Lenny was kind enough to set up the interview between sets in his dressing room. Stories about Buddy's temper were legendary...firing members of the band for no reason, his tremendous ego, his impatience with boring interviews. Even Dusty Springfield had allegedly slapped him after a rude remark. Needless to say, I was intimidated when I knocked on his door. I walked in and there was Buddy, in his underwear, wrapped in a bath sheet minutes after playing a dynamic drum solo that closed out the set to a standing ovation.

I was relieved to find him quite the opposite of what I expected after all the buildup. He invited me in and graciously consented to a relaxed and revealing interview. He spoke of the days when he was billed as "Traps, the Drum Wonder" and also with great affection regarding his parents who started him on the road to his career. When I mentioned the Sinatra stories, he said, "We were just kids then, and sometimes we liked to bait each other just for kicks. Actually, we're the best of friends, and there's nobody like Frank anywhere. He's the greatest singer of all time."

Later on, I learned it was Sinatra who had backed Buddy's band and was instrumental in keeping him active during an era when rock was the big thing. The story goes that Rich mentioned to Sinatra after a gig that he was interested in starting his own band. Sinatra promptly wrote him a check for $40,000, saying, "Good luck. This'll get you started.'"

At that time, Buddy's band was made up mostly of graduates from Berklee College of Music. He said that Berklee produced the best young jazz musicians of any school in the world and he was always assured of fresh, vibrant talent. It appeared that anyone who played in his band had to deal with the perfection that the leader demanded from each and every one of them. When I asked what made his band so unique, he replied without hesitation, "I do!"

Some years later, Buddy returned for an appearance at a club on the North Shore where I was asked to introduce him. As usual, it was a tour de force, again with standing ovations.

I told Buddy I would drop back to see him in the band bus where he used to ride up front with the musicians. As I entered the bus, I asked him if he'd like to meet my wife Joyce. He said of course. I brought her in and there he was again, clad only in his underwear and signature towel. I said, "Buddy, meet my wife Joyce; she's from Brooklyn like you." He smiled and she befriended him with a meatball sandwich on fresh Italian bread. He said, "Wow! You've made my day!"

Buddy's caustic humor made him a bit of a hit on several TV talk shows including *The Tonight Show with Johnny Carson*, the *Mike Douglas Show*, the *Dick Cavett Show,* and the *Merv Griffin Show*, entertaining audiences with his constant sparring with the hosts and his cracks about various pop singers of the day.

Ross Konikoff once remarked, "Never in my life have I met or heard about another human being who was more born to one task than Buddy. It was this single-mindedness that necessitated his brutal honesty. The only thing that prevented him from playing before the age of three was his inability to enunciate the words, 'Stay with me, goddammit.'"[23]

After Buddy died, I got a call from his daughter Cathy, who as a teenager had sung with her dad's band at Lenny's. I hadn't seen her since then, and she had grown into a tall, beautiful woman. She was coming to Boston to promote some recently discovered live video performances by her father. She told me that unlike all the negative stories about her dad's temper and nasty disposition, as a father he couldn't have been nicer to her or her mother. He called every night when on the road from all over the world to make sure things were going well on the home front. Cathy makes her home in Las Vegas with her husband and works tirelessly to keep the Buddy Rich legacy alive for a whole new generation.

There's a great story the late Whitney Balliett mentions in his book, *Super Drummer, a Profile of Buddy Rich*.

"I'm told I'm not humble, but who is? I remember being interviewed by a college kid once who said, 'Mr. Rich, who is the greatest drummer in the world?' and I said, 'I am.' He laughed and said, 'No, really, Mr. Rich, who do you consider the greatest drummer alive?' I said, 'Me. It's a fact.' He couldn't get over it. But why go through that humble bit? Look at Ted Williams—straight ahead, no tipping of his cap when he belted one out of the park. He knew the name of the game: Do your job. That's all I do. I play my drums."[24]

# SUGGESTED RECORDINGS

### Big Swing Face Buddy Rich

This is a live recording of the first edition of Buddy's 1966 band: arguably his greatest band ever. This is the band that also appeared at Lenny's on the Turnpike during the late 1960s and early '70s.

Selections include:

"Norwegian Wood" arranged by Bill Holman
"Love For Sale" arranged by Pete Meyers
"The Beat Goes On" arranged by Shorty Rogers, vocals Cathy Rich
"Chicago" arranged by Don Rader
"Willowcrest" arranged by Bob Florence

Pacific Jazz CDP 7243 8379 8926

### Buddy Rich Live at Ronnie Scotts

This is a later edition of Buddy's band dating from 1980 with some outstanding soloists like Steve Marcus on tenor sax, Bob Doll on trumpet, and Bob Mintzer on baritone sax.

Selections include:

"Slow Funk"
"Good News"
"Saturday Night"
"Ernie's Blues"
"Beulah Witch"

DRG 91247

### Buddy Rich: The Lost West Side Story Tapes

This DVD is every Buddy Rich fan's dream come true. He was sixty-seven at the time and playing better than ever. These master tapes were thought to have been lost in a fire in 1985 but were discovered and restored in 2000, thanks to his daughter Kathy. The concert includes Buddy Rich standards such as:

"Cottontail"
"Mexicali Nose"
"West Side Story"

In addition, there are interview segments, behind-the-scenes footage of Buddy and rare photos from the Rich Family achieves. For more information about Buddy, go to www.BuddyRich.com

DVD Hudson Music HDZWS01

# The Irascible

Joe Venuti

Author's collection.

Joe Venuti claimed he was born on a ship as his parents emigrated from Italy in 1906; however, many believe and in fact records show he was simply born in Philly. Considered the father of jazz violin, Joe was almost as well known for his groundbreaking technique as he was for being one of the great jokesters and storytellers of all time. Put it this way: anyone who knew Joe has a story.

Joe studied classical violin as a child, the fruits of which he combined with his natural inventiveness on the fiddle to produce his own signature sound, including a technique that allowed him to play four note chords. Childhood friend Salvatore Massaro, better known as jazz guitarist Eddie Lang, became a cherished musical partner until Lang's untimely death. In the late 1920s, Joe and Eddie moved to New York where they became so well known for their "hot" violin and guitar solos that they were commissioned to liven up otherwise stock dance recordings with original twelve- or twenty-four-bar solos.

Many early recordings turned to classics when they teamed up with the likes of Bix Beiderbecke, the Dorsey Brothers, Frankie Trumbauer, Jack Teagarden, the Boswell Sisters, and the young Benny Goodman. Though Venuti and Lang recorded some milestone jazz records during the 1920s, Venuti's career began to lose steam after Eddie's death in 1933. His fortunes improved somewhat in the mid-thirties when he joined the Paul Whiteman Orchestra.

In the mid-fifties, Joe met Bing Crosby and they developed a friendship that continued for many years. Bing gave Joe the opportunity to appear as a regular guest on the *Bing Crosby Radio Show*. This relationship, in fact, seemed to help pull Venuti out of a rough couple of decades following Eddie's passing when he formed his own unsuccessful band and did some anonymous Hollywood studio work. On Bing's show he was able to show off his quick wit, outrageous stories, and gruff charm in the best light.

Venuti's playing was inseparable from his personality, both of which were aggressive, inventive, playful, and punchy. He became known for his "violin capo" technique and his extended swinging pizzicato solos, but his most well known invention, rarely copied because it's both very difficult and extremely wacky, was to unfasten the hairs of the bow and wrap them around the top of his fiddle, with the bow underneath. This arrangement let him play all four strings simultaneously, producing lush four-part harmonies.

Joe virtually disappeared from the music scene during the 1960s, and many people wondered if he was still alive. He put those rumors to rest when in the mid-seventies he resurrected himself and, coming back stronger than ever, appeared at the Newport Jazz Festival and recorded with a slew of pop and jazz stars including Bucky Pizzarelli, Curly Walker, Eldon Shamblin, Jethro Burns, swinging tenor saxophonist Zoot Sims; even concocting and recording "Venupelli Blues" with Stephane Grappelli.

It was around this time that I first met Joe. He was appearing in Boston at the Merry-Go-Round Room in the Copley Plaza Hotel with pianist Dave McKenna. We hit it off right away, and I invited him to be a guest on my show. I'd already heard some classic Venuti stories, such as the time a pianist bugged Joe by tapping his foot...so he nailed the guy's shoe to the floor! According to more than one source, every Christmas Joe would send Wingy Manone, a one-armed trumpet player, a single cufflink.

But my all-time favorite Venuti story reportedly happened at the Hollywood Bowl. Joe was backstage waiting to perform when he realized he was about to share the stage with Roy Rogers' famous horse, Trigger. Being Venuti, while standing next to Trigger, he took his bow and began to saw away at the horse's privates! Needless to say, the horse was turned on and exhibited his unmistakable pleasure and masculinity. You can imagine the audience's reaction when Roy called out to his famous horse to come on stage.

Other tales include Joe pushing a piano out of a fifth-floor window to see what key it would play in when it hit the sidewalk; giving a musician directions to a gig that involved a two-hundred-mile odyssey, which landed him a block from where he started; calling up forty-six tuba players and sending them to a faux gig in Hollywood; chewing up (onstage) a violin he borrowed from bandleader Paul Whiteman.

In our 1990 interview, Joe shared this story: "I was in Italy when this aficionado of jazz heard me play and said, 'Maestro, I think you should give a concert.' Problem was, he was low on cash, but he did have a cheese shop. I told him to forget it; pay me in cheese. I did four concerts for cheese. Man, I was swimming in bel paesa, parmesan, gorgonzola, mozzarella! We ate as much as we could, and I sold the rest."

I asked how he named his songs. He told me: "We made these names up at the session. 'Bullfrog Moan,' 'Jet Black Blues,' 'Going Places,' 'Doing Things,' 'Add a Little Wiggle.' Right there on the spot. One side was called 'Nothing,' the other side was called 'Something.' One side was called 'Flip,' the other was 'Flop.' And I got news for you, that one lived up to its name." Backing instruments named on these recordings include the comb, hot fountain pen, kazoo, and something called the goofus, for which I never learned the translation. He also mentioned the unceremonious way musicians were treated in the thirties: "No rehearsal, just record. Four tunes in four hours, twenty-five bucks and see ya later."

But there was a serious side to Joe that came out in our interview. He told me he started out as a classical musician and played for the great Toscanini. He had a deep love for Italian opera, especially Puccini, and could sing in that gravelly voice of his all the well known arias, as well as play them on the violin. In fact he shared with me that he'd penned an opera in the Italian buffa style, laughingly admitting that it had never been performed.

I'll never forget a lunch I enjoyed with Joe and Dave McKenna at Giro's, a once-famous Italian restaurant in Boston's North End at the corner of Hanover and Atlantic Avenue. This place had an old world sort of dignity; all the waiters wore black tie. Joe insisted on ordering "family style"...that is, everything on the menu! In the middle of this fabulous meal, who should walk in but another legend, Dizzy Gillespie with his manager Charlie "The Whale" Lake. What an unbelievable occasion that turned into when these two started telling stories! I only wish I had my tape recorder when Dizzy took out his famous upright trumpet, puffed out those cheeks, and played a duo with Joe. Everybody in the place stopped to enjoy this once-in-a-lifetime impromptu jam session.

In 1978 Venuti was scheduled to play a jazz date in a club in Chicago. The band arrived a day ahead and received the sad news that Joe had passed away the night before in Seattle, where he made his home.

On opening night, a violin was placed on a chair in front of the band: a fitting tribute to a great man. Joe wowed music fans for over sixty years, playing with breathtaking speed, but always with absolute neatness and precision. This dazzling technique as well as his irreverent, light hearted humor put jazz violin on the musical map.

## SUGGESTED RECORDINGS

### Joe Venuti–Eddie Lang: Great Original Performances 1926–1933

This CD includes some of their finest and rarely heard performances and is a lasting testimony to two of the greatest innovators in jazz. The vintage recordings have been digitally restored by the jazz historian Robert Parker, resulting in excellent sound.

Selections include:

"Bugle Call Rag"
"Four String Joe"
"Krazy Kat"
"The Wild Dog"
"Hot Heels"
"Running Ragged"
"Sensation"

BBC 644

## The Fabulous Joe Venuti

This wonderful collection was recorded in Italy during a Venuti tour in 1971. He was over eighty and still had the swing and verve of a man decades younger. In this session, he was able to record with many of Italy's finest musicians including Lino Patruno and Giorgio Vanni.

Selections include:

"Sweet Georgia Brown"
"After You've Gone"
"Jazz Me Blues"
"Margie"
"Some of These Days"
"Sweet Sue"
"Clementine"

Omega 3019

## Joe Venuti and Zoot Sims

In 1975 Venuti was in New York City and teamed up with Zoot Sims on saxophone, John Bunch on piano, Bucky Pizzarelli on guitar, and Bobby Rosengarden on drums. What a swinging session!

Selections include:

"I Got Rhythm"
"Avalon"
"Russian Lullaby"
"Where OR When"
"Shine"
"I'll See You in My Dreams"

Chiaroscuro CRD 142

# A Lifetime of Swing

Lionel Hampton

154

© Ken Franckling.

Lionel Hampton was perhaps one of the greatest showmen in jazz: he was a vibraphonist, pianist, percussionist, bandleader, and actor. Born in Kentucky in 1908, Lionel was raised by his grandmother. After a move to Chicago in his teens, he took xylophone lessons from Jimmy Bertrand and started playing drums, jumpstarting his career as a percussionist with a group called The Chicago Newsboys.

"Hamp," as he was known by his musician friends, was a styling sort of guy. He always bought the best of everything, including the latest silk shirts and the finest suits. In the 1930s he bought his first marimba and set of drums and became accomplished on both instruments. But it was the vibraphone that brought him lasting fame. At that time, Louis Armstrong had come to California and hired the Les Hite band, enlisting Lionel to play the vibes on two songs. This performance not only galvanized his career as a vibraphonist, but it popularized the use of the instrument ever since.

In 1936 he met dancer Gladys Riddle, who became his wife and personal manager. She developed a reputation as a brilliant businesswoman, responsible for raising enough money for Lionel to start his own band. Their marriage lasted well over sixty years and was one of the most endearing relationships in the jazz world. His beloved Gladys also encouraged him to study music theory at the University of Southern California.

In November of the same year, the Benny Goodman orchestra came to the Palomar Ballroom in Los Angeles, where John Hammond brought Goodman to hear Hampton play. Goodman asked Hamp to join the Benny Goodman Trio, made up of Goodman, Teddy Wilson, and drummer Gene Krupa, expanding it into the Benny Goodman Quartet. This was the first time two black musicians integrated the all-white Goodman Orchestra.

After Lionel left Benny in 1940, he formed his own band, and what a band it was! The musicians who played with Hamp in his Lionel Hampton Orchestra were a who's who in jazz; they included Charlie Mingus, Joe Newman, Art Farmer, Clark Terry, Dexter Gordon, Illinois Jacquet, and singers Joe Williams and Dinah Washington.

In 1942 Hamp recorded "Flying Home," Illinois Jacquet's famous solo that drove audiences into a frenzy and paved the way for rhythm and blues. This recording put the band over the top, and Hamp featured it in every one of his concerts. Illinois' solo was so popular that saxophonists who followed him played his solo note for note.

Also known for his generous role as an educator, Lionel worked with the University of Idaho in the 1980s to establish accessible music education, and in 1986 the University's music school was renamed the Hampton School of Music, ensuring that Lionel's vision will live on. The sixty-million-dollar-project provided a "home for jazz," housing the university's Jazz Festival and its International Jazz Collection designed to help and teach presenters of jazz.

In a 1990 interview with Lionel, he was effusive about the quality of the program: "We've educated over twenty-five hundred students so far; can you imagine that? We present concerts all year long by professional artists. We've had Dizzy, we've had Ella, and they'd perform, then do a lecture, and the kids go gaga over it! And the professors—I've never seen anyone break down music like this—they're so thorough with these lessons, they could teach a tree. And they honor me, they do. When I was there last, an eighty-six piece choir sang 'Midnight Sun,' and you never heard anything so beautiful as the way they did it. Not just two or three part harmony, this was five, six parts—gorgeous."

When I met Hamp in the 1970s, he was appearing with his band on the north shore of Boston. I called him and he said, "Yeah, Gate...come up to my dressing room after the first set." I must say after hearing his opening performance I was a bit disappointed; I guess I expected the band to catch fire...and it didn't, and I worried that Hamp, in his late seventies, might be past his prime.

When I entered his dressing room he was sitting, relaxed and watching TV. His eyes never left the TV during our entire interview. Finally he said, "Hope you gonna stick 'round, Gate...because that first set wasn't happen'n and I'm gonna kick some butt in a few minutes." And did he ever! The second set lasted over two hours, and Hamp seemed to never tire, outplaying all the other younger musicians in the band.

After the performance I went back to his dressing room. With a big grin, he looked at me and said, "Did you dig that, Gate?" This time his wasn't watching TV.

When he died a few years ago, I recalled that evening and thought about the millions of people he had entertained all over the world. Like Louis Armstrong, Hamp was one of our great ambassadors of jazz and good will.

## SUGGESTED RECORDINGS

### Lionel Hampton and the Just Jazz All Stars

Recorded in 1947 by Crescendo

This CD features Hampton and an all star jazz lineup including Charlie Shavers on trumpet, Willie Smith on alto, and Slam Stewart on bass. Among the selections is my all time favorite instrumental performance of "Stardust."

### Lionel Hampton Hot Mallets Volume 1

Lionel himself selected the best of his small band recordings from 1937 to 1939.

RCA 6458-2-RB

### Lionel Hampton The All Star Groups

RCA 2433-2-RB

# Boston's Own

Ruby Braff
with Gray
Sargent

© Ken Franckling.

In 1988 I was privileged to write the liner notes for a CD entitled *Me, Myself and I*, by The Ruby Braff Trio.

Here's an excerpt: "I hereby declare Ruby Braff to be a National Treasure. What other musician has such an extraordinary range of ideas, emotion, and sustained playing? Ruby is unique, not only as a performer, but as a superb interpreter of our country's musical heritage. He has recorded everything from 'America the Beautiful' to 'White Christmas.' Like his idol Louis Armstrong, Braff has great respect for the melody, embellishing it in ways others envy. Ruby has played and recorded with a multitude of jazz greats including Benny Goodman, Roy Eldridge, Coleman Hawkins, Buck Clayton, and Pee Wee Russell among many others. Very few musicians share his eclectic repertoire. He is equally comfortable talking about the wit and wisdom of Dorothy Parker as he is discussing the music of Tchaikovsky and Puccini. His iconoclasm is well known. You can't fake it with Ruby. He tells it like it is, and his artistry is a reflection of this."

Since his passing in 2003, his stature as one of the all-time great innovators in jazz has only grown. I'm still amazed at the output of classic recordings he made during the last decade of his life. Thanks to the efforts of Mat Domber of Arbors Records, Ruby was able to record prolifically, choosing his own material along with his favorite musicians.

Interviewing Ruby, however, was always an experience that kept me walking on eggshells. Here was a guy who fired his own music teacher; he was eight at the time. Ask him the wrong question, and the result could be an outburst of invectives that made you wish you could disappear!

Here are some classic Ruby quotes that still ring in my ears:

"It took civilization two thousand years to give us Louis Armstrong, and fifteen minutes for Bruce Springstein."

"Rock music is built on three chords and two of them are wrong!"

"The state of music today is so bad, even Sammy Kaye sounds good to me these days."

Stories about Ruby Braff are legendary. Here's one that has made the rounds:

While traveling in Japan, Ruby approached the front desk at his hotel to ask if there were any messages for him. Politely he was told, "No, Mr. Braff, no message today." Ruby then asked if there was any mail for him; again the polite man bowed, and replied, "No, Mr. Braff, no mail today." Fuming by now, an agitated Ruby said, "Well for Christ's sake, do you have the time?!"

A New York musician related another classic Braff story to me. After a gig, Ruby was heading home in a cab with some fellow musicians. As each musician was dropped off, Ruby bad mouthed the guy to the remaining cats with caustic remarks like, "That guy's a moron and can't play for s--t." After another musician was dropped off, Ruby said, "His chops are gone, I wish he'd take up bowling." Finally, as the last musician said good night to Ruby, he thought to himself, "I'm sure with no one left, Ruby will bad mouth me to the cab driver!"

More memorable quotes from the man himself:

"I've always hated the trumpet. I didn't choose it. I wanted the B-flat tenor saxophone. (My parents) brought home this peculiar thing with valves on it, which I hated forever. Never did care for it."

"Unfortunately, I'm mostly self-taught. I hope to fix that one of these days."

Referring to famous musicians such as Tommy Dorsey and Artie Shaw: "They looked like they were having such a marvellous, glamorous life, living in hotels, so well-dressed. It seemed like the epitome of luxury. I had no idea that they were all miserable!

"My first records were made in Boston, for a label called Storyville, and for Savoy Records, with Edmond Hall and Vic Dickenson. But they were terrible recordings—off broadcasts, mainly. Very sad things. I couldn't play, either. The one made in a club where you could hear the audience more than the music was one of the better records.

"Sure, they've made statements about my supposedly combining a modern approach with a feeling for traditional forms. Well, people say all sorts of things, because they want to categorize and label. I've only ever had two labels: Either it's good or it stinks.

"So it's silly. Is (a musician) playing good or isn't he? That's the only thing that counts. But I know a lot of people don't agree with me, particularly the critics. They must put labels on music, so they can have it like canned goods on their shelf.

"Talent is something that very few people have, really. And there are no geniuses. Maybe Louis and Duke are something in jazz. But they keep throwing these words around."[25]

All this from a musician who could be sensitive and caring when he wanted to be. He carried on a phone relationship with my mother, when she was in her nineties, until she died. She always looked forward to her extended conversations with Ruby on a wide variety of subjects from Al Jolson to the price of milk. She said he always was a delight to talk to and called her mostly when it was raining!

As Dan Morgenstern, director of the Institute of Jazz Studies at Rutgers University, has said, "Ruby was able to reach heights of artistic achievement granted only an exalted few in the final decade of his life. He put everything he could muster into his horn despite his emphysema. The strength he corralled to override his condition was truly super-human coming from such a small, and increasing frail body."

Braff's overriding love for music came through in one of our last interviews: "Every kid in school should know the names of Harold Arlen and Duke Ellington and George Gershwin, and they don't; and I think it's an awful thing. The 1930s was a unique time for our country: it was a gentler time, a gentler country; people were more innocent in a more innocent land, and with it came great beauty. That, *that* is the chunk of American music that should be cared for and loved and protected until the end of time."

## SUGGESTED RECORDINGS

### Ruby Braff: In the Wee Small Hours in London and New York

Recorded in London and New York in 1999, Ruby performs a collection
of standards showcasing his unique sound backed up by a lush
string section arranged and conducted by Neil Richardson and
Tommy Newsom.

Selections include:

"In the Wee Small Hours of the Morning"
"April in Paris"
"Pennies from Heaven"
"Love Walked In"
"White Christmas"
"Old Folks"

"All Alone"
"You're Sensational"

Arbors ARCD 19219

### Ruby Braff: Very Sinatra

This 1993 recording is devoted to songs made famous by Frank Sinatra. Ruby
was a great admirer of Frank's artistry both as a musician and singer.
Braff often said that he'd like to live exactly as Sinatra does, except for
two roadblocks... "I can't afford it, and I couldn't pass the physical!"
The tasteful arrangements are by Dick Hyman who plays both piano
and organ along with Bucky Pizzarelli on guitar, Michael Moore on
bass, and Mel Lewis on drums.

Selections include:

"All the Way"
"The Second Time Around"
"My Kind of Town"
"Nancy"
"Come Fly With Me"
"Lady is a Tramp"
And an original by Ruby entitled "Perfectly Frank"

Red Barron JK 53749

### Ruby Braff and His Buddies

This collection won the best jazz record of 1995 by the combined critics of Jazz
Journal International and is one of the most swinging recordings that
Ruby ever made. His jazz buddies included the great Dave McKenna on
piano; Scott Hamilton, tenor sax; Gray Sargent, guitar; Marshall Wood,
bass; and Chuck Riggs, drums.

Selections include:

"It's Alright With Me"
"Ain't Misbehaving"
"Swinging on a Star"
"Them There Eyes"
"On the Sunny Side of the Street"

Arbors ARCD 19134

## 9. GEORGE SHEARING

# It's Shearing You're Hearing!

George Shearing

Courtesy of George Shearing.

It was in the early 1950s that I first heard the singular sound of the George Shearing Quintet on the radio. George was making history with his Capitol recordings that featured piano, vibraphone, guitar, bass, and drums. Later on he added Latin rhythms to shake up the group's style. His improvisations were unapologetically romantic, but I always picked up a hint of whimsy in his music, reflecting the warmth and offbeat humor of the man I had the pleasure of getting to know.

Born in London in 1919, George was the youngest of nine children. Though congenitally blind, George began learning how to play the piano at three years of age. His father delivered coal, and his mother cleaned trains at night after caring for the family during the day. With only four years of formal training at the Linden Lodge School for the Blind, George found his inventive, orchestrated jazz voice through not only listening constantly to recorded jazz musicians, such as Teddy Wilson and Fats Waller, but also by jumping into performance, joining an all-blind band while still in his teens for the sum of five dollars a week.

In 1947 he emigrated to America where he put together the George Shearing Quintet, known for its shifting personnel over the years, which has included Margie Hyams, Chuck Wayne, John Levy, and Denzil Best. Shearing became known for the lock-hands technique, which is when the pianist plays parallel melodies with the two hands, creating a full, rich sound. He was accompanied by a low-key rhythm section including guitar, bass, drums, and vibraphone. The Shearing sound commanded national attention when he gathered his quintet to record "September in the Rain," which sold 900,000 copies. Among other songs recorded by his Quintet include: "Mambo Inn," "Let's Call the Whole Thing Off" and "I'll Never Smile Again."

George's piano style was heavily influenced by the "bebop" sound created by Charlie Parker and Dizzy Gillespie in the 1940s. During his early years in America, George performed on 52nd Street in New York, known as "Swing Street," with jazz greats Oscar Pettiford and Buddy DeFranco. His U.S. reputation was permanently established when he was booked into Birdland, the celebrated jazz spot in New York, eventually inspiring "Lullaby of Birdland," composed in 1952. He acknowledged creating the piece in just ten minutes, but as he told the *Christian Science Monitor* in 1980, "I always tell people it took me 10 minutes and 35 years in the business, just in case anybody thinks there are totally free rides, there are none!"[26]

George shared some of his thoughts on his early years in New York: "Here I am this young English cat in shades, and they didn't know what to make of me. There was an obvious language barrier. I think it was George Bernard Shaw who said that the Americans haven't spoken the English language in over a hundred years. My wife reminds me of the significance of the 4th of July and I say, yes, but remember you still owe us for the tea.

"Fifty-second Street is gone now, but it was everything back then. Before the quintet, I did several months at the Three Deuces, with Oscar Pettiford on bass and J. C. Heard on drums. I had the best possible training just being opposite these giants. No better apprenticeship has ever been offered. Once we had Ray Brown on bass, Charlie Smith on drums, and one fine day, Ella Fitzgerald. I got someone to take over my intermission piano for a few nights, anything to play with Ella. We did 'Flying Home,' and we were still going at 2:30 in the morning. I'm telling you, it was like I died and gone to heaven."

In the late 1950s he began performing classical concerts with symphony orchestras all over the world with arrangements by Robert Farnon, a masterful pianist with great technique and his own sound. But with George, as many have said, you always knew, "It's Shearing you're hearing."

This was especially true because Shearing's harmonically complex style mixed swing, bop, and modern classical influences. In fact his fascination with classical music resulted in guest performances with concert orchestras in the 1950s and '60s; his solos frequently borrowing musical patterns or phrases from Debussy, Delius, Schumann, Rachmaninov, Vaughn-Williams, Elgar, and particularly, Erik Satie. Shearing's delicate touch and fanciful nature made him an ideal interpreter of Satie's work.

He commented in our 1996 interview on his tendency to incorporate classical phrasing: "I try to be diversified both in my music and my investments." His famous "Shearing sound" continued until 1978 when he disbanded his quintet and began to work in other musical contexts.

In the same interview, George commented on his relationship with Mel Tormé: "We are the best musical marriage that I know, Mel and I. We like the same composers, we feel the same way about music, and unlike many singers, Mel is also an arranger. He's not my favorite pianist, but hey, I'm not his favorite singer. He's a consummate musician. If I play something and I make a change, *within that bar*, Mel will be on it a split second after I am, real time. Therefore, he doesn't worry about me trying to memorize accompaniments so I can play the same way every night to cover up his insecurity. In other words, *I can improvise*, and isn't that what jazz is all about? Mel would never, ever ask me to play something the same way. If you can't be free to improvise, tell me, what's the point?"

George's dry wit always delighted me. In 1990 I interviewed him after an appearance at the Regatta Bar in Cambridge, where he announced the release of a solo piano CD, one of several on the Concord label. Among other topics, we discussed his countless trips performing concerts all over the world, making him one of the most frequent flyers of all time. He said, "You know, Ron, they've asked me to form 'Shearing Airlines'...of course I'll be flying blind but plan on sticking my white cane out the window to determine our wind velocity and direction."

His thoughts on jazz from our 1996 interview: "When I let the George Shearing Quintet go in 1978, it was time. But absence really does make the heart grow fonder, because by '94 I'd become recharged somehow. I'd get out of the shower every day geared up with a new arrangement for the same combination that I'd been so sick and tired of in 1978.

"But I have to say, jazz has changed: it's become so complex. I came over to this country knowing about Glenn Miller, Teddy Wilson, Art Tatum, Benny Goodman. Now I go in these clubs and hear these complex lines that Charlie and Dizzy were playing and I think, *what?* Good God, do I have to go back to school for another ten years? It's a strange feeling."

Granted an honorary doctorate of music from three universities, knighted by the Queen of England, composer of over three hundred songs, invited by presidents Ford, Carter, and Reagan to perform at the White House, and, last but not least, creator of countless terrible puns, George Shearing and his music will always be near and dear to my heart and to the hearts of millions of fans worldwide. Of his passing in early 2011, Dave Brubeck said, "George paved the way for me and (the Modern Jazz Quartet) and even today, jazz players, especially pianists, are indebted to him."[27]

## SUGGESTED RECORDINGS

### How Beautiful Is Night

George Shearing with the Robert Farnon Orchestra

This collection brings together the magnificent piano artistry of George Shearing with one of the greatest arrangers of our time, Robert Farnon, whose lovely arrangements provide the perfect setting for Shearing and his quintet along with Farnon's orchestra. These wonderful interpretations of great standards make for a sublime listening experience. Some selections on this CD include:

"Dancing in the Dark"
"Heather on the Hill"
"Oh, Lady Be Good!"
"Our Waltz"
"Days Gone By"
"Put On a Happy Face"

Telarc CD-8332

### That Shearing Sound—The New George Shearing Quintet

George Shearing revisits his famous quintet years with a newly formed edition of the group. It's a joy to hear that famous sound once again in new versions of the classic quintet selections popularized in the late 1940s and early 1950s that made this group one of the most popular and listened to in the history of jazz.
Selections include:

"East of the Sun"
"I'll Never Smile Again"
"I Hear Music"
"Girl Talk"
"Autumn Serenade"
and Shearing's most famous composition, "Lullaby of Birdland"

Telarc CD-83347

### George Shearing—Favorite Things Solo Piano

Starting in the 1970s, George Shearing started to record a series of solo performances that showcased his remarkable talents that he explored as a much younger man in his native England. In this collection we hear echoes of his love for classical composers like Tchaikovsky, Brahms, and Satie. This shows Shearing to be the true master of interpolating quotes from classical music into his solo piano performances.
Some selections from this CD:

"Angel Eyes"
"Taking a Chance On Love"
"Summer Song"
"Moonray"
"P.S. I Love You"
"It Amazes Me" (also featuring George's vocalizing)

Telarc CD-83398

# Three Handed Swing

Gray Sargent and Dave McKenna

© Ken Franckling.

Pianist Dave McKenna was born in Woonsocket, Rhode Island, to a musical family. His mother played violin and piano; his father, a postman, played drums part-time; and both of his sisters were singers. As a youngster, Dave was largely self-taught and began playing weddings and bridal showers by the time he was fifteen. One of his early champions was "Boots" Mussulli who had a band in nearby Milford, Massachusetts. Boots later became a member of the Stan Kenton Orchestra and appeared frequently at the Crystal Room in Milford.

As a teen, Dave joined Charlie Ventura's band that featured jazz greats Conte Condoli (who also worked with Stan Kenton), saxophonists Al Cohen, and bassist Red Mitchell. In 1950 he played with the Woody Herman "Herd" until the army tapped him for a stint in Korea where, of all things, he became a cook! He said he learned how to make pancakes for five hundred guys but couldn't figure out how to make pancakes for one.

After the service, Dave teamed up with many other jazz greats including Gene Krupa, Stan Getz, and Buddy Rich. In the mid-1960s he settled on Cape Cod with his wife Frankie and their sons Stephen and Douglas. He made his home there for many years, becoming pianist in residence at "the Columns" in West Dennis.

I was introduced to Dave by our mutual good friend Dick Johnson, who was playing with him at that time on the Cape. He insisted I go out and hear Dave, telling me he was simply the greatest. What an understatement that turned out to be! Up until that time, no pianist I had ever heard got more out of the piano than Dave. He played swinging bass lines with his left hand while his right had the ability to play ballads, swing, or bop. Though he started out as a big-band side-man, he became best known for his distinctive solo playing, with that powerful left hand making a bass player seem unnecessary.

This technique, rooted in the jazz piano tradition of an earlier era, was built around these powerful bass lines, elegantly voiced chords, and a loving approach to melodies, especially those of the Tin Pan Alley standards that were the foundation of his playing. Unlike many of his contemporaries, he was more likely to ornament a tune with graceful embellishments than to spin off wild riffs, abandoning the melody. This "three handed swing" style brilliantly evoked the rhythmic structure normally provided by a three- to four-person band. To hear Dave play solo piano was to hear a whole symphony orchestra.

After leaving the Cape, Dave became pianist in residence at the elegant Plaza Bar in the Copley Plaza Hotel for almost ten years. He became so popular there that many of the icons of the music world such as Tony Bennett, George Shearing, and Rosie Clooney would drop by to listen whenever they were in town. In the world of classical music, André Previn, John Williams, and Kurt Musur were also great admirers. It was during this period, the early 1970s, that my wife Joyce and I made a trip to Italy with Dave. It was a long desired dream of his to sample the cuisine he loved so much.

Florence turned out to be one of the highlights of that trip. While walking past a disco we noticed an announcement in the window that jazz pianist Romano Mussolini would be playing that night. Dave had heard of Romano and mentioned he wouldn't mind seeing him. That evening we arrived to the sounds of loud, abrasive disco music. The manager informed us that Romano was running late but was on his way. Soon after that, several gentlemen showed up who appeared to be bodyguards with Romano in tow. No mistaking that this man was the son of "Il Duce"; he looked just like his father.

After the first set, I introduced Dave to Romano, who invited him to play. Dave graciously declined, but Romano insisted and Dave finally said sure, as soon as I finish my dinner. He played a heartfelt and swinging Duke Ellington medley which had the place in an uproar, as well as Romano, watching in disbelief as Dave's hands flew across the keyboard. Even the cooks came out of the kitchen when they heard his music. Afterwards, more plates of beautiful pasta arrived at our table...now Dave was in heaven!

McKenna couldn't have been happier during his later days with the success of his beloved Red Sox and Patriots. Years ago he even penned and recorded a musical tribute to his idol Ted Williams, called "The Splendid Splinter."

McKenna's fans are worldwide and his theme medleys from the Great American Songbook rival those of any other musical artist. I rarely saw him with written music at the keyboard: he'd committed thousands of songs to memory. As modest as he was, he once described his playing to me with these words, "I'm just a saloon piano player who's killing time." But what better way to kill time than listening to Dave play! His humility and laid-back style always seemed to contrast with his masterful playing. He was truly one of the all-time keyboard greats in a class with Art Tatum, Oscar Peterson, Teddy Wilson, and Nat Cole.

## Suggested Recordings

### An Intimate Evening with Dave McKenna

This collection of live solo piano performances from 1999 recorded at the Sarasota Opera House is a good example of how Dave rewrites the song without losing the melody and includes several of those famous McKenna "theme medleys":

The Thought Medley includes: "The Very Thought of You," "I Had the Craziest Dream," and "Thinking of You."

The Street Medley includes: "Easy Street," "Broadway," "42nd Street," "Beal Street Blues," and "On Green Dolphin Street."

The Change Medley includes: "There'll Be Some Changes Made," "Change Partners," and "You've Changed."

The Letter Medley includes: "I'm Gonna Sit Right Down and Write Myself a Letter" and "Love Letters."

The Time Medley includes: "Time After Time" and "Time On My Hands."

Volume 10 Arbors 19264

### Dave McKenna Giant Strides

One of the best of McKenna's saloon piano recordings showcasing his great sensitivity and signature approach to standards as a master improviser and piano innovator.

Selections include:

"Yardbird Suite"
"I Got the World on a String"
"Lulu's Back in Town"
"Walking My Baby Back Home"
"Windsong" (a beautiful waltz by Bob Wilbur)
"Dave's Blues" (an original by Dave and my personal favorite)

Concord CCD 4099

### Dave McKenna and Grey Sargent Live at Maybeck Recital Hall

Dave McKenna once said, "I really dig the way Grey Sargent plays. He's right up there with the best of them." McKenna and Sargent played for years around the New England area and something extra special happened when they got together as you will hear in these selections.

Selections include:

"Sheik of Araby"
"Girl of My Dreams"
"Deed I Do"

Concord CCD 4552

# Born to Play

Dick "Spider" Johnson

Courtesy of Pamela Johnson Sargent.

Very few musicians I've been privileged to know were as versatile as Dick Johnson, Brockton's gift to the jazz world. In the notes for Dick's first Concord CD in 1979 I wrote, "Dick has the musical talent of three or four great players rolled up into one, and can swing mightily on alto, tenor, soprano saxophone, flute, and clarinet." Though he was affiliated with big bands for over fifty years, touring with Charlie Spivak, Benny Goodman, Neil Hefti, Buddy Morrow, and Buddy Rich, for the past twenty-five he was director of the Artie Shaw Orchestra, hand-picked by the late "King of Clarinet" himself.

His dexterity and speed on all the reed instruments earned him the nickname "Spider," and it was an album he cut for Concord Records called *Spider's Blues* that opened a new chapter in his life. His manager, Bill Curtis, sent the record to Dick's idol Artie Shaw. The response from Artie was ecstatic: not only did he say it was the best clarinet playing he'd heard, but he asked if Dick was available to lead Shaw's Band. Making his first road trip in years, Artie came to Boston to personally rehearse the band at the WGBH studios. Audiences young and old came to hear the exciting charts that had made the Shaw Band famous. The orchestra still performs regularly worldwide, bringing big band music at its best to a whole new generation.

I first heard Dick play alto saxophone at a roadhouse in Brockton called the Roma Cafe. It was back in the early sixties and he'd just come off the road with the Buddy Morrow Band. I couldn't believe my ears! On the lower registers his tones were rich, warm, and vibrant, or soaring high above the band when needed. He sounded like a combination of Charlie Parker, Lee Konitz, and Johnny Hodges. I introduced myself and we became good friends.

Dick's wit and boundless energy were an inspiration. Comfortable in any setting, Dick was always a great guest on my show. We talked about everything from favorite movies, his love of food and restaurants, to back-stage gossip, and priceless stories of life on the road. He would break me up as well as anyone else who happened to be in the studio.

Ruby Braff, always one of the most difficult musicians to please, once told me that Dick was one of his favorite players. When I told Dick, he said, "Wow... I guess I'm one of the few cats who made the cut!"

During one of our interviews, Dick shared a story of a failed attempt to snare an applause break: "We were on with Dave McKenna. We thought, how are we going to get half the applause Dave gets. Gray (Sargent) and I just pulled out some bebop tune like 'Crazyology,' hoping to dupe Dave, but then of course *we* had to hang on for dear life. I got an okay round of applause, and then Gray took off and whaled and got the same thing. Then Dave got ramped up. *While he was still playing*, he got a standing ovation. I looked at Gray and we both nodded: forget it; let's just go along for the ride."

When not performing, Dick made his home in his native city of Brockton with his lovely wife Rose. His son Gary carries on the family tradition in music and is one of the finest drummers on the scene. His daughter Pam is married to the brilliant guitarist Gray Sargent, a member of the Tony Bennett quartet.

I was honored when Dick dedicated a big band arrangement called "Spread that Gospel, Ron." I have it framed in my studio and consider it one of my greatest treasures. Thanks, Dick, for all you've done in spreading your gospel worldwide, and may you rest in peace, my friend.

## Suggested Recordings

**Dick Johnson Plays Alto Sax, Flute, Soprano Sax and Clarinet**
Dave McKenna on Piano; Bob Maize, Bass; and Jake Hannah, Drums
A collection of jazz standards showcasing the versatility of Dick Johnson along with the brilliant artistry of Dave McKenna.
Concord CCD4107

**Artie's Choice! And the Naturals**
Featuring Dick Johnson, Lou Columbo, Trumpet; and Gray Sargent, Guitar

This privately produced two-CD collection includes marvelous Dick Johnson arrangements of "Stardust," "Waltz for Debbie," "Indian Summer," "Young and Foolish," "Emily," and other standards.

**Star Dust and Beyond: A Tribute to Artie Shaw featuring Dick Johnson**

Dick's loving tribute to his idol and mentor Artie Shaw that includes such classics as:

"I Concentrate on You"
"My Funny Valentine"
"My Romance"
"Gone With the Wind"

# Boss of the Blues

Joe Williams

© Ken Franckling.

Joe Williams was a renowned jazz vocalist, a baritone who sang a mixture of blues, popular songs, ballads, and jazz standards. He was born Joseph Goreed in 1918, in Cordele, Georgia, to his eighteen-year-old mother Anne. Determined to make a new life for her and her son, Anne moved to Chicago and worked as a cook for four years until she could afford to send for him to join her and his grandmother and aunt.

Joe was a teenager when he began singing in nightclubs around the windy city. He reveled in the then-rebellious sounds of Louis Armstrong, Duke Ellington, Ethel Waters, Cab Calloway, and Big Joe Turner, among others, even starting up his own gospel vocal quartet called "The Jubilee Boys," as well as teaching himself to play piano. His first job was at a club called Kitty Davis's, where he cleaned latrines and sang for tips. Even though the most he took home was five dollars a night, both Joe and his family were convinced he could make a living with his voice, so he dropped out of school at age sixteen to pursue his dreams.

Word of this young man's heartfelt tone and impeccable timing spread fast, and he was soon tapped to apprentice with jazz legends Coleman Hawkins and Lionel Hampton in the early 1930s. His first real break was in 1938, when Jimmy Noone invited him to sing with his band, which soon created a reputation for Williams not only at Chicago dance halls but from coast to coast on national radio stations. Soon he was touring with the Les Hite band, which accompanied such greats as Fats Waller and Louis Armstrong.

Lionel Hampton hired Joe to fill in for his regular vocalist in 1942. By the time the band's former singer returned, Williams was in great demand. He made his first recording with Andy Kirk and His Twelve Clouds of Joy in the late 1940s. Following stints with Albert Ammons, Red Saunders, and trumpeter "Hot Lips Page," he had his first big hit with "Every Day I Have the Blues," which became his signature song.

After teaming up with the Count Basie Orchestra in 1954, he became an international star and one of the most important elements in the Count Basie Band. He stayed with Basie for seven years, recording "Every Day I Have the Blues" among countless others. Joe shared some details with me during our 1990 interview: "I needed to get out of Chicago, I'd been there all my life. Basie told me, 'Look, I can't give you what you want or what you deserve, but why don't you come with me around the country and see what people think?' Within a year's time, we did our first tour of Europe. First stop was this tiny town outside of Stockholm, where there were ten thousand people jammed into this park. When we finished, ten thousand people were standing on their chairs and cheering. I said, 'What are we going to do, Base?' And he said, 'For once, you're going to quit while you're ahead.'"

Later on, Joe toured with his own group and made wonderful recordings with other jazz giants including George Shearing, Cannonball Adderley, the Thad Jones–Mel Louis Orchestra, and the Capp/Pierce Juggernaut Band. With his good looks and charismatic presence, he became a familiar face on television, appearing on Johnny Carson's *The Tonight Show*, as well as the Joey Bishop, Steve Allen, Merv Griffin, and Mike Douglas shows; even taking a turn as father-in-law "Grandpa Al" on the *Bill Cosby Show*. Joe also spent much of his time performing at jazz festivals, appearing in movies, and playing various jazz cruises. He told me he had been on so many cruises that when in port, he never bothered to leave the ship.

Joe possessed a powerful, rich baritone capable of singing ballads, blues, and standards in his own swinging style. I'll always remember my radio conversations with him in which he expressed his great love for classical music, especially opera. He recalled to me his early days in Chicago where he never failed to miss the Saturday afternoon broadcast of the Metropolitan Opera hosted by Milton Cross. At that time, black artists were nonexistent at the Met, and I always felt that Joe would have had a great career as an opera singer if the times had been different.

Once he blew me away in the studio when he sang a few measures of King Phillip's aria "Ella giammai m'amo" from Verdi's *Don Carlo*. After hearing that, there was no doubt in my mind that he would have made a great Verdi basso!

Duke Ellington wrote about Joe Williams in his autobiography, *Music Is My Mistress*, "He sang real soul blues on which his perfect enunciation of the words gave the blues a new dimension...all the accents were in the right places and on the right words."[28]

Joe passed away in Las Vegas where he made his home. He was singing right up until the end and still in magnificent voice. His wife Jillian said he always looked forward to his visits to Boston where he had so many friends who packed the house whenever he performed here. I feel privileged to have known such a remarkable man and to be able to share his treasured legacy.

## SUGGESTED RECORDINGS

**Joe Williams with the Count Basic Orchestra, directed by Frank Foster**

In this exciting 1993 recording, made during a live performance at Orchestra Hall in Detroit, Joe revisits some of the big hits he had with Count Basie.

Selections include:

> "Hurry On Down"
> "Honeysuckle Rose"
> "Sometimes I'm Happy"
> "The Comeback"
> "Roll 'em Pete"
> "Sugar"
> "There Never Be Another You"

Telarc CD 83329

**Jazz 'Round Midnight**

A collection showcasing the matchless style of one of the great singers of blues and ballads. Joe Williams is in great company in this superb CD with vocalist Shirley Horn, the Count Basie Orchestra, pianist Norman Simmons, and "SuperSax" conducted by Med Flory.

Selections include:

> "Embraceable You"
> "Come Rain or Come Shine"
> "Never the Less"
> "Teach Me Tonight"
> "When Sunny Gets Blue"
> "I'm Beginning to See the Light"

Verve 314 527 034-2

**Here's To Life: Joe Williams with the Robert Farnon Orchestra**

Joe told me that this is one of his all-time favorite recordings, and it's one of mine as well! Recorded in London in 1993 with a full symphony orchestra and arrangements by Robert Farnon, this CD is difficult to surpass. The beauty and range of Joe Williams' voice in his interpretation of these ballads is astounding.

Selections include:

"Here's To Life"
"What a Wonderful World"
"If I Had You"
"Young and Foolish"
"A Time For Love"
"Maybe September"
"When I Fall In Love"

Telarc CD 83357

# III. THE GREAT AMERICAN SONGBOOK

## The John Singer Sargent of Singers

Tony Bennett, Joyce Della Chiesa, and me

© Roger Farrington.

After an appearance with the Count Basie Band at the Berklee Performance Center one snowy evening in 1982, Tony Bennett made his way to the Copley Plaza Hotel to visit his old friend, jazz pianist Dave McKenna. There he sat in with Dave and sang a few songs to the delight of the lucky people who just happened to be in the audience that evening... which included Joyce and me.

After his set, Tony joined us for dinner. This was the beginning of a long and beautiful friendship with one of the great icons of the American Songbook. During that first meal together, we touched on topics ranging from the Italian Renaissance, Frank Sinatra, John Singer Sargent, David Hockney, Dr. Martin Luther King, American and Italian film; to Martin Scorsese, Robert DeNiro, Federico Fellini, and Victorio DeSica.

All this before the gelato.

Born Anthony Dominick Benedetto on August 13, 1926, in Astoria, New York, Tony Bennett can handily be called one of the greatest friends the American Songbook has ever known. With a poet's imagination and an artist's passion, his work has endeared him to millions around the world for over fifty years.

Even after decades of worldwide acclaim, Tony has never forgotten his humble origins or his dear mother Anna, a dressmaker who was widowed at an early age and brought up Mary, John, and Tony on her own. During one of our interviews, Tony related a story that turned out to have a profound impact on his thinking and ultimately his career. One day a woman approached his mother with inferior cloth and said, "I want you to make me a beautiful dress with this." Anna gave the material back to the woman and said, "To make a beautiful dress, you need beautiful material."

Tony learned a real lesson that day! Never compromise your artistry by singing inferior songs; always choose quality and you'll never go wrong. Sinatra gave him the same advice many years later.

Bennett's father came from Calabria, Italy, and opened a grocery store in New York, but it was Tony's uncle, a tap dancer in vaudeville, who gave him an early window into show business. By age ten Tony was already singing wherever he could, soaking up the sounds of Al Jolson, Eddie Cantor, Bing Crosby, Louis Armstrong, and Jimmy Durante. At New York's High School of Industrial Art he studied music and painting (he was especially taken at the time with caricatures), but dropped out at sixteen to help support his family. Though he worked as a copy boy and runner for the Associated Press in Manhattan, he also landed several gigs as a singing waiter in Italian restaurants in Queens.

Even though he worked hard to put food on his family's table, New York's burgeoning jazz scene was an inspiration to him, and his home life felt relatively secure. He told me: "I loved my early days growing up in Astoria. It's where Woody Allen makes his films! Back then it was half Italian, half Greek, just a nice middle-income neighborhood which had a real country feeling—lots of trees—but was just fifteen minutes from the city."

Drafted into the army in 1944 during the final stages of World War II, Tony's experiences as an enlisted man read like an adventure novel. At basic training, Benedetto tangled with a sergeant from the South who lacked tolerance for Italians from New York City, assigning him backbreaking doses of KP and latrine cleaning. Assigned as a replacement infantryman for a unit that suffered heavy casualties in the Battle of the Bulge, Tony moved across France into Germany, ultimately fighting on the front line which he described as a "front row seat in hell."

In Germany, Tony and his company fought in bitterly cold winter conditions, hunkered down in foxholes as Germans fired on them. After crossing the Rhine, he and his company engaged in brutally dangerous house-to-house, town-after-town combat to smoke out the German soldiers, narrowly escaping death several times. At the war's end, he helped liberate a Nazi concentration camp near Landsberg.

Tony's dining with a black friend from high school—when the army was still segregated—led to a demotion and reassignment to Graves Registration Service duties. Subsequently, he sang with the Army military band under the stage name Joe Bari, which was a partial anagram of Calabria.

The entire experience made him not only a patriot but a pacifist, and he would later write, "Anybody who thinks war is romantic obviously hasn't gone through one."[29]

Upon his discharge, Tony learned the discipline of singing bel canto, which would, along with other healthy habits of living, keep his voice in good shape even to the present day. As he said during our 1985 interview, "I joined the American Theater wing when I got out under the GI Bill of Rights, learned popular phrasing under Myriam Speare who taught Peggy Lee and Helen O'Connell. She was right on 52nd Street, which was the land of geniuses, I'm telling you. The golden era of singing. It was the age of miracles when we saw the highest level of American music: Stan Getz, Nat King Cole, Stan Kenton, Woody Herman, Duke Ellington, Count Basie, Jimmy Lunceford."

He continued to perform while waiting tables, incorporating the style and phrasing of other musicians such as Stan Getz's saxophone technique and Art Tatum's piano, a brilliant way of helping him improvise and interpret songs. In 1949 Pearl Bailey lit up when she heard him sing and asked him to open for her in Greenwich Village where her friend Bob Hope sat in the crowd. Hope immediately took him on the road, billing him as Tony Bennett, where Mitch Miller caught his act and signed him with Columbia Records.

His first big hit, "Because of You," a ballad with lush orchestration, caught fire via jukeboxes, reached number one, and stayed there for ten weeks in 1951, selling over a million copies. His recording of "Blue Velvet" attracted screaming teenaged fans at the Paramount Theater in New York, where Bennett did seven shows a day, starting at 10:30 in the morning.

Along with the hits came Bennett's marriage to Ohio art student Patricia Beech. Two thousand female fans dressed in black gathered outside New York's St. Patrick's cathedral in mock mourning. Bennett and Beech would have two sons: Danny in 1954 and Dae in 1955.

"Rags to Riches," an uptempo big band number, was his third number one, followed by "Strangers in Paradise," which became a hit in the United Kingdom, launching his career as an international artist. Even in the fifties when rock and roll was taking hold and pop singers beginning to struggle, Bennett placed eight songs in the top 40.

In the late fifties, Bennett's pianist and musical director Ralph Sharon suggested a change in strategy. Noting a short shelf life of bubble gum material, he advised Tony to go with his instinct and record a jazz album. Tony's first long-playing album, *Cloud 7*, showcased these new jazz elements while 1957's *The Beat of My Heart* used jazz musicians Herbie Mann and Nat Adderly and Latin stars Candido Camero and Chico Hamilton. The first male pop vocalist to sing with Basie's band, Bennett followed up with *Basie Swings, Bennett Sings*. This was also a time when Tony was building up the quality of his nightclub act, staging a highly successful performance at Carnegie Hall and finally appearing on *Johnny Carson's Tonight Show*. He was, in fact, Johnny's first guest ever.

The next few years saw the release of "I Left My Heart in San Francisco," which would become Bennett's signature song, and "The Good Life" on his album *I Wanna Be Around.* But as the sixties wore on, there was no escaping the Beatles and the British invasion and with them a dimming focus on pop, standards, and jazz. Beyond some minor hits mostly based on show tunes, Bennett's star seemed destined to fade, perhaps even go out.

Bennett's marriage ended soon after an attempt to break into acting in a 1966 film, *The Oscar,* an experience he neither enjoyed nor sought to repeat. To round out the misery, Bennett cut the album *Tony Sings the Great Hits of Today!* featuring Beatles tunes and a psychedelic art cover. The fact was, Bennett became physically ill at the thought of stepping into the studio to record these songs. Years later he would equate this experience to that of his mother being forced to produce quality from cheap raw material.

Bennett married Sandra Grant, an actress he'd met filming *The Oscar,* and with her had two daughters, Joanna in 1969 and Antonia in 1974. Deciding to take his career into his own hands, Bennett started his own record company, Improv. Even though under this label he recorded classics such as "A Tribute to Irving Berlin" and "What is This Thing Called Love?" as well as two historic albums with pianist Bill Evans, Improv lacked distribution and by 1977 was out of business. Bennett hit bottom with no recording contract, no manager, a second marriage in failure mode, and the IRS trying to seize his Los Angeles home.

In 1979 he called his sons Danny and Dae and said, "Look, I'm lost here. It seems people don't want to hear the music I make."[30] Danny not only listened, but came to terms with the fact that what he lacked in musical gifts, he more than made up for in business sense, the opposite constellation of abilities his father possessed. Danny moved his dad back to New York, began booking him in colleges and small theaters to cleanse him of a "Vegas" image, and by 1986 Tony was re-assigned to Columbia Records, this time with creative control. The release of *The Art of Excellence* brought him to the charts for the first time since 1972.

Danny Bennett felt that younger audiences would connect with his dad's music if only they were given a chance to hear it. By the mid-eighties, even rock stars such as Linda Ronstadt began recording albums of standards and songs from the Great American Songbook. A younger, hipper audience caught on to Tony Bennett's magic with the help of appearances on *Late Night with David Letterman, Late Night with Conan O'Brien,* even a series of benefit concerts organized by alternative rock radio stations like WBCN in Boston. In 1990 he released *Astoria: Portrait of the Artist, Perfectly Frank* in '92, and *Steppin'Out* in '93; the latter two earning gold status and Grammys. At the MTV Music Awards he stood shoulder to shoulder with the Red Hot Chili Peppers and Flavor Flav. As the *New York Times* put it, "Tony Bennett has not just bridged the generation gap, he has demolished it."[31] At age 68, after winning the top Grammy prize of Album of the Year, it was clear that Tony Bennett had come all the way back.

I can't help but think of Tony as the John Singer Sargent of singers: the man paints portraits in song with his voice. Like a magnificent oil or watercolor, his voice evokes delicate shades of light and color, capturing the essence of the lyric and melody. That signature catch or tear in his voice seems to spring naturally from every experience the man has ever had.

The bel canto element in his singing conjures an Italian operatic style that makes each song like a mini aria. Sometimes I hear the sound of Giuseppi DiStefano; at other times, Claudio Villa. Occasionally Tony will simply put the mic down and sing naturally, often "Fly Me to the Moon," partly to demonstrate to the younger members of the audience the dying art of vocal projection. He shared with me that his Calabrian grandfather would stand up high on a mountain and sing, his voice carrying throughout the village.

I'll wait as you imagine *that* for a moment....

There's a song called "When I Lost You" that Irving Berlin wrote when his first wife passed away. Tony recorded one of the verses a cappella. If you aren't moved hearing him paint that picture of loss I might question your status as human. A cut on the album, *The Playground*, called "My Mom" by Walter Donaldson, a song about mothers, also never fails to get to me each time I hear it.

In a 1985 interview I asked Tony how he keeps things fresh, after having sung these tunes hundreds and hundreds of times: "It's the combination of elements: the crowd, the atmosphere of the place, the performers I'm with." We talked about his comfortable rapport with the jazz musicians he worked with at the time: Ruby Braff, Zoot Sims, Bobby Hackett, and Dave McKenna. He called Bobby Hackett: "one of the great melody players of all time. In fact, it was Bobby who introduced me to Louis Armstrong who said, 'I'm the coffee, but man, Bobby's the cream.'"

The artistry of Bennett and Sinatra have often been compared. In the early fifties when a young Bennett was first signed with Columbia, he was warned against trying to become another "Old Blue Eyes," encouraged to develop a style of his own. Both are poets, but Sinatra's sound seems more robust, almost brash at times, while Tony, who works more closely with his arrangers, is smoky, mellow, and intimate. In the end, however, the admiration between the two was mutual. Sinatra said, "For my money, Tony Bennett is the best singer in the business. He excites me when I watch him. He moves me. He's the singer who gets across what the composer has in mind, and probably a little bit more."[32]

The same kind of commitment Bennett brings to the music and lyrics of Jerome Kern, Cole Porter, George Gershwin, Johnny Mandel, Johnny Mercer, and others he brings to his more introspective passion: drawing and painting. The small boy who drew chalk pictures on the sidewalks of Astoria is alive and well in the man who picks up a paintbrush every day in quiet contemplation of the things he loves most: the rolling hills of Tuscany, the time-worn faces of jazz masters Duke Ellington and Dizzy Gillespie.

"Like music," Tony said, "art is a lifetime of study, which is the whole adventure, but it's a terrific amount of work if you are serious about it." A self-confessed "museum freak," Tony visits museums and galleries all over the world, learning the subtleties of various styles and studying the great masters, always making time to sketch the view from the window of his hotel suite, or as he told me, "photograph whatever strikes me, if I don't have time to sketch or paint. I especially love landscape painting, and there is no better day for me than one without plans, when I can wander in the countryside." We shared a laugh over an exhibit of the impressionists at Boston's Museum of Fine Arts; reviewers were still questioning whether Monet was "good" or not, since as Tony put it, "no one likes him but the public." One of Tony's favorite artists is good friend David Hockney, for whom he painted "Homage to Hockney," on permanent display at the Butler Institute of American Art in Youngstown, Ohio, where Tony was recently named one of America's greatest artists.

Tony has often said that he sees colors when he sings; that his voice is affected by the weather: even the rain or sun influence his sound. I can certainly close my eyes and see not only colors but images and often entire scenes in my mind. "Because of You," "When Joanna Loves Me," and "Strangers in Paradise" animate entire worlds inside my head. I believe his intuitive sense of the relationship between sound and image is enhanced by his daily involvement with both of those worlds.

The United Nations has commissioned his work (under his family name of Benedetto) on two occasions including their 50th anniversary, and his work, including lithographs, is part of permanent collections at the National Arts Club in Manhattan, the Smithsonian, as well as numerous galleries around the world. Among the many owners of original Benedettos are the late Cary Grant, Carol Burnett, Whoopie Goldberg, the late Frank Sinatra, Donald Trump, Oprah Winfrey, Mickey Rooney, Katie Couric, and contemporary artist Robert Rauschenberg. I am also the proud owner of a pencil sketch he did of me while at dinner at the Four Seasons with Joyce and his wife Susan. Tony carries his sketchbook and uses it everywhere he goes: airports, airplanes, restaurants, concert halls, and cafes, saying "Airplanes are best. People aren't moving too much and they're pretty much stuck there!" Many of his works are included in *Tony Bennett: What My Heart Has Seen* in 1996, as well as the best-seller *Tony Bennett in the Studio: A Life of Art & Music*.

I confess that a lot of what I admire in Tony I admired in my own father; in fact some of the similarities are striking. Both served in World War II, both stood up for African American friends. In my father's case, he vouched for a black enlisted man who was being court-martialed for setting up his tent incorrectly (using the butt of his rifle to pound down a tent stake as opposed to a rock or similar item.) The black soldiers under his command thanked my father with a shaving kit, something he treasured all his life. Shortly afterward, he was assigned to Graves Registration in Hawaii (now known as "Mortuary Affairs"), which involved everything from making identification via bodies or body parts, to recovery of bodies, to ensuring proper burial. As noted earlier, Tony was assigned to these same duties (though not sent to Hawaii) because he shared a meal with a black friend. Like Tony, my father also painted not only with his voice, but on canvas; the two men share this wonderful Renaissance quality. Both men stood up for what they believed when actions like these were certainly met with more resistance than today.

A staunch believer in the Civil Rights movement, Bennett took part in the Selma to Montgomery marches in 1965, and he consistently refused to perform in apartheid South Africa. Tony donates so much time to charitable causes he is sometimes nicknamed "Tony Benefit."

On a purely personal note, he is just a very good friend in a world where true friendship simply isn't as common as you might think, certainly not as common as it should be. A certain vulnerability based in real strength and a sense of self seems necessary, both for friendships between men and, from what I've heard, women as well. My birthday simply doesn't pass without some good wishes from him, and he asks after Joyce and our family with genuine interest. Over the years Joyce and I have become a surrogate aunt and uncle to Tony's daughter Antonia, a wonderful young singer who stayed with us a great deal, especially during her years at the Berklee College of Music.

In 2007 Tony married his long time partner Susan Crow, and together they founded Exploring the Arts, a charitable organization dedicated to creating, promoting, and supporting arts education. They simultaneously founded the Frank Sinatra School of the Arts in Queens, a public high school focused on teaching the performing arts.

Now in his mid-eighties, Tony tours steadily, doing one hundred to two hundred shows a year. He has sold fifty million records worldwide, won fifteen Grammy Awards, and released more than seventy albums. But it doesn't matter if he's singing to thousands or just a few, his voice finds you and holds you.

His persona has crossed over to the youth market like no other artist before or since, introducing millions of young people to the American Songbook not only through his appearances on MTV and elsewhere, but through the efforts of his son and manager, Danny Bennett. His encouragement of fellow singers shines through in the 2001 album *Playin' With My Friends: Bennett Sings the Blues*, which also features Stevie Wonder, k.d. lang, Bonnie Rait, B.B. King, and Ray Charles. All in all, one of the cleverest, classiest comebacks I've ever seen.

A focus on diet, exercise—especially tennis—and proper rest keeps his voice preternaturally young and able to reach the high registers, a feat that becomes more difficult with age. As he told me, "The theory is that after thirty-five your voice goes down. My idols are Ellington, Fred Astaire, Bing Crosby, and Maurice Chevalier. They sustained what they had as they got older; they hung in there. That's what I'm hoping to do; in fact I have an ambition to get better as I get older." He added that he had no intention of retiring: "If you study the masters—Picasso, Jack

Benny, Fred Astaire—right up to the day they died, they were performing. If you are creative, you only get busier as you get older. In fact, I need two lifetimes. I'll never get it finished. I have that many ideas about what I'd like to do and what I'd like to learn."

## Suggested Recordings

### Tony Bennett "The Playground"

No better way for kids and adults to come together. Some selections include:

> "Bein' Green"
> "Put on a Happy Face"
> "When You Wish Upon a Star"
> "My Mom"

Columbia CSK 41457

### Tony Bennett "The Art of Romance"

When it comes to love songs, there's no better singer than Tony. This CD, conducted by composer/arranger Johnny Mandel, includes such favorites as:

> "I Remember You"
> "Where Do You Start"
> "Don't Like Goodbyes"
> "Close Enough for Love"

RPM Records/Columbia CK 92820

### Forty Years: The Artistry of Tony Bennett

This four-CD boxed set, together with an informative booklet and anthology, is a must for any Tony Bennett collector. It includes just about every hit he recorded along with some lesser-known songs.
Selections include:

> "Boulevard of Broken Dreams"
> "I Wanna Be Around"
> "Maybe This Time"
> "Because of You"
> "Stranger In Paradise"
> "I Left My Heart In San Francisco"

Columbia/Legacy 46843

# The Embodiment of Elegance

Courtesy of Charlie Lake.

James Gavin sums up Bobby Short's impeccable artistry in his excellent book, *Intimate Nights—The Golden Age of New York Cabaret:* "By the early seventies, Bobby Short had become a national symbol of style, sophistication, elegance and good taste. The influences of his idols, the vivacity of Fats Waller and Cab Calloway, Duke Ellington's suave refined manner, the supreme dignity of Ellington's vocalist Ivy Anderson, combined to give Short a polish that none of his colleagues could equal."[33]

I couldn't agree more! New York's premiere cabaret singer and pianist, Bobby Short owned a charisma and presence unrivaled for the time. With his beatific smile and charm he could light up a room just by walking in it, even before singing a note. What other entertainer could get away with wearing a tuxedo with a bare chest and bow tie and look absolutely dashing? Many times interviewing Bobby in my studio, I felt I should have been wearing my tux as well.

His extraordinary career began in Danville, Illinois, in 1926, where one of his classmates was Dick Van Dyke. The ninth of ten children in a family of modest means, he began playing piano as soon as he could reach the keys. He told me that both his parents played piano, in fact "most households back then had a piano and at least one person who could play it." A child prodigy, Bobby learned to sight read stunningly early, though he would insist to host Marian McPartland, "I played by ear, Marian, I still play by ear."[34]

Though young Bobby was playing the piano and singing in roadhouses by the time he was nine, he also lived in the traditional world of family, church, and school, gaining a great deal of musical experience from glee clubs. As he said in his memoir, *Black and White Baby*, "It was all right (performing in clubs) with my mother because I was in the care of a man whose aunt was a church friend of hers."[35] On weekend nights a family friend would take him from taverns to vaudeville stages where a hat was passed as he played and sang. Two years later he left for Chicago and was managed by agents. "I had no idea the image I projected,"[36] he later recalled, but he did remember singing "Sophisticated Lady" in a high, squeaky voice.

Another formative influence was the radio. He explained: "It was very elegant. We were left alone a lot at night as kids, but there was always the radio. My great loves were Ella Fitzgerald, Art Tatum on the piano—who completely blew my mind and changed me forever—Fats Waller, and Earl Hines. I also adored Mildred Bailey, Walter Fuller, Arthur Lee Simpson, the Andrews Sisters, Fred Astaire, and Bing Crosby. They were my world."

New York came just before he turned thirteen. Though arrangements were made for a tutor, he was more excited about his first theatrical photographs that involved custom-made white tails and ankle-length camel's hair coats. Long before graduating from high school, Bobby was performing at Harlem's Apollo Theatre where he met his idols Duke Ellington, Fats Waller, Benny Carter, and the legendary pianist Art Tatum.

In our 1988 interview we talked about his first impressions of New York: "It was 1937. I was just a kid and 52nd Street *was* 52nd Street. One could hear Joe Marsala, and Adele Girard, Art Tatum, Billie Holiday, Francis Faye, Roy Eldridge. It was a jazz street. When I came back in 1945 there was still a lot of jazz, but it was starting to fall apart. Of course it's still called jazz street but it's not like it was. After gigs we musicians would leave and hang out or jam between sets at Mintons, or at the White Rose on 6th Avenue. There were smart nightclubs popping up in the 1940s: Leon and Eddies, Tony's, where Mabel Mercer sang; then across the street, Bonds and Cook, the piano team would perform." Bobby was always generous with his impromptu late-night performances at the various cafes and restaurants up and down 52nd Street.

In 1968 he was offered a two-week stint at the Café Carlyle in Manhattan to fill in for George Feyer. Accompanied by Beverly Peer on bass and Dick Sheridan on drums, Short parlayed this visit into a nearly four-decade run, becoming the pianist in residence at the Carlyle, much like Feyer had before him. Sitting an arm's reach from his audience, Bobby combined an effortless elegance, intelligent vocal phrasing (perfected at the feet of Mabel Mercer and Ethel Waters), a talent for showcasing little known songs while reinvigorating standbys, infectious high spirits, and unflagging professionalism to produce an ever growing and enamored audience.

Many have said that Short was a living library of America's popular songs, knowing more songs than even he has counted, but also who wrote them, who sang them, and when and where they were first performed. Stephen Holden of the *New York Times* wrote, "(Bobby) infuses traditional society piano with the rollicking animation of Harlem vaudeville...even sadder songs convey a high style *joie de vivre*."[37] Night after night, the rich and famous of New York society came to be entertained in what turned out to be one of the longest engagements of any performer in the history of show business. He became a symbol of Manhattan sophistication, attracting royalty, movie stars, socialites, sports figures, captains of industry, and jazz aficionados, and was a guest performer at the White House when President and Mrs. Nixon entertained the Duke and Duchess of Windsor. Woody Allen included him as a must-see for his characters' tour of his beloved city, granting him a part in *Hannah and Her Sisters*.

Our interview marked his twenty-first year at the Carlyle: "It's more than where I make my bread and butter. It's the nicest room of its kind in New York, thus in the country, so perhaps the world. There are no other rooms where one can go and be that comfortable I think, but it wasn't always like that. It took me a number of years to make it what I dreamed it could be, to make it the kind of room where people would actually come in and sit and listen to what I was doing."

He reflected on his audience: "Great saloon singers had firm blues and jazz roots. Cabaret had to get the audience's attention in a hurry, then it was always a struggle to keep their attention. You need that sense of urgency, and with a jazz background you came out prepared to entertain your audience. They didn't want to hear 'Softly As In a Morning Sunrise,' they wanted to hear something peppy. They were out there drinking, out in a bar for some kicks and to have a good time."

For me, a trip to "Gotham" was not complete without a visit to the Carlyle. Every performance Bobby gave was a master class. His dedication to the "Great American Song" made him equally brilliant at capturing the lighthearted spirit of Bessie Smith's "Gimme a Pigfoot" as he was at pouring out the honey in Gershwin and Duke's "I Can't Get Started With You." I can honestly say I learned more about a song from him that anyone else.

During our interview we often discussed his favorite composers and songwriters; he named Rodgers and Hart, Jerome Kern, Harold Arlen, Vernon Duke, Noel Coward, and George and Ira Gershwin; although Cole Porter always topped the list. The sophistication of Porter's lyrics seemed tailor made for him, and Short's playful sense of humor was in perfect sync with Porter's loaded double-entendres. He told me he was once asked what these lyrics meant from a Porter song: "I will be your broom if you'll be my dustpan." He said, "What's this song *about*? Sex! I mean, is it me? I don't think so."

He had a talent for insinuating the sensuality behind a song. As Whitney Balliett put it in a 1970 *New Yorker* profile, Short "stripped (a song) to its essentials—words lifted and carried by the curves of melody," also noting that his baritone, often tinged with laryngitis, "lends his voice a searching down sound, and his uncertain notes enhance the cheerfulness and abandon he projects."[38]

Bobby also dedicated himself to championing African American composers who contributed to New York's musical theatre, including Eubie Blake, James P. Johnson, Andy Razaf, Fats Waller, Duke Ellington, and Billy Straythorn, showcasing their work not in a didactic way, but simply as equals to their white contemporaries.

I once shared an unforgettable lunch with him and his good friend Charlie Davidson at the Ritz in Boston. Bobby would frequent Charlie's legendary "Andover Shop" in Harvard Square to augment his natty wardrobe (his name appeared on numerous "best dressed" lists). Our lunch lasted several hours and was constantly interrupted by excited fans. He was always gracious and accommodating to each and every one who stopped by.

When I reflect on Bobby, I envision a kind of intimate live performance and unabashed romance that is sadly missing today. Though he called himself a saloon singer, the world will remember him as one of the greatest cabaret performers of the twentieth century. What Fred Astaire was to dancing, Bobby Short was to American Song.

## SUGGESTED RECORDINGS

### The Café Carlyle Presents Bobby Short: You're the Top Love Songs of Cole Porter

Robert Kimball who wrote the notes for this CD said: "Seeing and hearing Bobby Short perform Cole Porter songs at New York City's Café Carlyle is one of THE quintessential experiences of Gotham Nightlife." When you listen to this CD, you'll know why. It was recorded live in 1998 with Bobby's marvelous band and wonderful arrangements. It's the next best thing to being there!

Selections include:

"I Concentrate On You"
"You Do Something To Me"
"In the Still of the Night"
"I've Got You Under My Skin"
"You're the Top"
"You're Sensational"
"Can-Can"

Telarc CD 83463

### Bobby Short: Late Night at the Carlyle

This collection captures Bobby in a more intimate setting with his trio. It was recorded in 1991, live at the Carlyle with Beverly Peer, bass and Robert Scott on drums. In addition to being a great singer, Bobby gives us a chance to hear his expertise on the keyboard playing some familiar and less familiar selections from the American Songbook.

Selections include:

"Do I Hear You Saying I Love You?" (A lesser know Rogers and Hart tune from 1928)
"Body and Soul"
"Street of Dreams"
"Paradise"
"Satin Doll"
"Love is Here to Stay"
"After You, Who?" (A rare Cole Porter tune dropped from the 1934 film version of the *Gay Divorcee*)

Telarc CD 83311

### Bobby Short: Swing That Music

This recording came about when Bobby rediscovered favorite music from his early days; it reminds me of the classic Teddy Wilson sessions with Billie Holiday. All the arrangements are by trombonist Dan Barrett who teams up with Howard Alden, guitar; Chuck Wilson, reeds; Frank Tate, bass; and Jackie Williams, drums. The music is irresistible, infectious, and a delight from beginning to end.

Selections include:

The Title Song: "Swing That Music"
"Tenderly"
"Drop Me Off in Harlem"
"If Dreams Come True"
"Gone With the Wind"
"Sleep Baby Don't Cry" (A haunting lullaby by William Archibald and my personal favorite)

Telarc CD 83317

## 3. FRANK SINATRA, JR.

# His Own Way

Photo courtesy of Frank Sinatra, Jr.

As the only son of larger-than-life icon Frank Sinatra, Frank Sinatra, Jr. was dealt a complex hand that he has played out with dignity, his own native talent, and a kind of learned elegance for over fifty years.

Born in Jersey City, New Jersey, in 1944 to Frank and Nancy Sinatra, young Frank rarely saw his famous father who was either on the road nonstop or making films. Keenly interested in music as far back as he can recall, Frank Jr. began performing in local clubs by his early teens. By age nineteen he'd been enlisted as the vocalist for Sam Donahue's band (Frank called Sam a "musician's musician") and spent a good deal of time learning the business from Duke Ellington who had taken the young man under his wing.

Around nine o'clock one evening in 1963 at Harrah's Tahoe Lounge in Nevada, nineteen-year-old Frank was finishing up a room-service chicken dinner with John Foss, 26, a trumpet player with the Dorsey band. There was a knock at the door; Frank invited the visitors in. After announcing themselves as delivery men, Barry Keenan with his friend Joe Amsler (both 23 and greenhorn criminals) whipped out a gun, tied up Foss, and forced young Frank, blindfolded, into their beat up Chevy. In minutes they were navigating their way through a blizzard in the Sierra Nevadas toward a hideout eight hours away. Two days later, a $240,000 ransom was paid just before the two young men were arrested and ultimately sentenced to short prison terms. Though it was soon shot down, a rumor circulated that the whole thing had been a publicity stunt to juice up Frank Sinatra, Jr.'s career.

By his early twenties, Frank Jr. began life on the road. In fact by age twenty-four he'd performed in 47 states, 30 countries, and been a guest on several TV shows: everything from *Family Guy* to a role on *The Sopranos* where it was unclear if he was riffing on or confirming all the lore involving his father and the mob, and where Paulie Walnuts calls him "Chairboy of the Board." Despite being offered the role of Vic Fontaine on *Star Trek: Deep Space Nine*, a show he admired, he declined, saying he only would accept a role as an alien.

He'd also advanced to becoming the opening act for several bigger names at various casinos. All the while his reputation preceded him: a consummate performer with exacting musical standards for his musicians, including nonstop rehearsals till they got it right.

In 1988 Frank Jr. put his own career on the back burner to act as his father's musical director and conductor. Poet/vocalist Rod McKuen said, "As the senior Sinatra outlived one by one all of his conductors and nearly every arranger, and began to grow frail himself, his son knew he needed someone he trusted near him. (Frank Jr.) was also savvy enough to know that performing was everything to his dad, and the longer he kept that connection with his audience, the longer he would stay vital and alive."[39]

Frank Jr. recalled the life-changing moment this way: he was socializing with friends in Atlantic City when his phone rang. It was his father, asking him to conduct for him. "After my friends revived me with smelling salts, I said, 'Why do you need me?'"[40] The answer was simple: he couldn't get a conductor to understand what he wanted done; he thought another singer might understand it best. Skeptics have cried nepotism, but nobody knew Frank Sr.'s music like Frank Jr. So for the last seven years of Frank Sr.'s career, his son was there as collaborator and confidante. "It (also) meant," Frank Jr. said, "that for the first time in my life, I spent time with him. And it was a gift from heaven."[41]

A year before his last performance, Frank Sr. told his son he wanted to put together an album of ballads with real swing, featuring the best soloists. Young Frank agreed and asked if there was anything else. The answer was "No. Get outta here."[42] One of the songs he brought his father was "The People You Never Get to Love." The elder Sinatra was delighted with the song, but age caught up with him and he never recorded it. So Frank Jr. sings the Nelson Riddle arrangement with the big orchestra every chance he gets.

No doubt there have been endless hours of barstool philosophizing over whether there is room for two Sinatras in this world: a worshipped legend and the phenomenal living talent. One was born in a world where people flocked to big band and swing; the other strives to continue a legacy against a tide of rock, pop,

R&B, and Lady Gaga. In the end, I really believe each Sinatra will find his own place in musical history.

But the fact remains, even though he passed in 1998, Frank Sr. is still everywhere you look, or listen: "weddings, cocktail soirees, movie soundtracks, TV commercials, elevators..."[43] It can't be easy to establish your own identity in the shadow of this mountain, especially when you've inherited such similar abilities. Regarding the doors that have swung open for Frank Sinatra, Jr. all his life, he has remarked, "A famous father means that in order to prove yourself you have to work three times harder than the guy off the street."[44]

It hasn't been easy. He's heard his own show referred to as "Jurassic Park." In a 2006 *Washington Post* article, Frank Jr. himself says, "There is no demand for Frank Sinatra, Jr. records. There never has been. Rod Stewart's now doing the Great American Songbook. So is Harry Connick Jr. and Michael Buble. The truth is, Frank Sinatra, Jr. has been doing it for forty-four years."[45] He said all this without a shred of self pity or irony, then turned on his heels to warm up his band for a two-night gig at the Hilton on the Boardwalk.

The "strip" in Vegas will certainly never be the same. It's about making the most amount of money with the smallest investment possible, and entertainment, per se, is just not the primary draw any longer. Frank commented: "You know what the big draw is in Vegas today? The shopping centers in the malls."[46] These days, a New Year's gig for Frank is more likely to happen in Spokane than anywhere else.

But he has persisted in being one of the few big bands still touring, even though there aren't many clubs still booking. They don't have the room or often the money it takes to make the numbers work all around, yet the man is still out there, still filling the 1,450 seats at the Atlantic City Hilton where New Jersey is still "Sinatra country," as he put it. At a 2006 show the line formed a couple of hours before showtime, filled with, among others, the aging but devoted members of the "Sinatra Social Society."

In my interviews with Frank Jr., it's clear to me that a number of factors keep him going despite what he may interpret as a lack of success: his access to hundreds of his father's arrangements and his entire musical library, his decency and generosity to all his sidemen—a well known fact in the music business. Saxophonist Terry Anthony, who has played with both Frank Sr. and Jr., has only the highest praise for not only Frank Jr. the man, but for his treatment of band members. "He's a great musician and bandleader," Frank's guitarist Jim Fox said, "He takes us to dinner...the best restaurants in town. He says it's the best thing for him—to get us all out on the road, get us all together. Now, how sweet is that?"[47] And the checks always arrive, whether a gig was missed or not.

But what impresses me when I speak with Frank is his intensity and passion for music, igniting a lifelong flame of curiosity that has probably carried him through some tough times as well as turned him into a wealth of knowledge. He not only knows every third trombone part, every cello part of every arrangement, but he studies jazz history with the ferocity of a scholar.

Frank is an aficionado of the Eddie Sauter and Bill Finnegan orchestra of the early fifties; in fact that post war period seems to interest him the most. He told me he worked with Sauter and was a student of Bill Finnegan, who "taught me more in about six lessons than three and half years of music school." As he described the orchestra in our 1997 interview, "They took the big band, trumpets, trombones, sax, plus the rhythm section, and they luxuriated that sound with a French horn, percussion, and just a few symphonic touches. This was the Sauter–Finnegan sound."

We also talked about Tommy Dorsey crashing the color barrier with his recognition and selection of Sy Oliver for his band: "In 1961 my father was very sentimental so he made an album called *I Remember Tommy*. Back in 1937 black musicians only played with blacks and whites with whites. Dorsey said, What difference could this make? If I want somebody from the black bands, I don't particularly give a damn. So he hired Sy Oliver who was the man who'd written almost the entire book for a man named Jimmy Lunceford, who took all the credit for it. Later Dorsey used Ernie Wilkins and others, but Sy Oliver broke the color barrier for white jazz bands over seventy years ago."

In 1997 I had an unparalleled opportunity to speak with Frank about George Gershwin, for whom "there has never been a bigger fan." He told me his father was also fascinated with the man, but Gershwin died just as Sinatra senior was getting into the business.

Along with his orchestra, Frank Sinatra, Jr. traveled to ten cities to celebrate the centennial of the Gershwin brothers: Ira, born in 1896, and George, born in 1898. Bringing his own core players, Sinatra Jr. engaged local musicians in each town to perform *Porgy and Bess* as well as a number of Billy May, Don Costa, and Nelson Riddle arrangements.

Frank called Gershwin "the man who took jazz out of the gutter. He did it in 1916 with a little ditty he called, 'When You Want 'em You Can't Get 'em; When You Got 'em You Don't Want 'em,' which was a failure till his next song, 'Swanee' that Al Jolson made into a monster hit in 1920. Soon after that, Gershwin wrote *Rhapsody in Blue*, and at that moment jazz began being taken more seriously."

Many great composers of the time, including Rachmaninov, came to hear this music, and they were enraptured. Frank said, "Ravel wrote his *G-Major Piano Concerto* after meeting Gershwin. With the trombone slurring and everything, that was *his* jazz! And so it went on—Paul Hindemeth, Francis Poulenc—all of them so taken by this new music. George Gershwin made an honest woman of jazz—to quote many sources!"

I pointed out that *Porgy and Bess* debuted in Boston at the Colonial Theater in 1935, forging a good connection to the city, then traveled on to the Alvin in New York for its debut in November of that year. Frank reminded me that the reception was anything but smooth: "Gershwin paid through the nose for the blasphemy of creating something like *Porgy and Bess*! The things they said about him: that he was a sinner trying to become a saint, that he gave up writing good songs to write bad concert pieces; they called him a vulgar composer. Then again they slammed Beethoven, Wagner, Stravinsky—who nowadays is called the greatest musical mind of the twentieth century. The late George Gershwin is in excellent company.

"Irving Berlin, Gershwin's friend, said he was the only songwriter who became a composer. Here was a man who wrote piano preludes, concertos, rhapsodies, symphonic tone poems—then suddenly out of nowhere—here comes an opera."

Before his tragic death at age thirty-nine, Gershwin often rhapsodized about the music that surrounded him, saying that he could never get enough of it and that he wanted to write a ballet and a string quartet among other things. Frank said, "So many times I've asked myself what we did *not* get because Gershwin didn't even live to be sixty-five. I can only speculate on the masterpieces we will never hear.

"We do have the arrangements of Gershwin songs written for my father by Nelson Riddle, Billy May, and Don Costa. Ella Fitzgerald, who we lost in 1996, gave her music to the Library of Congress, and the Gershwin curator there got us the Gershwin arrangements from her album, the one she made with Nelson Riddle, so from that we've scavenged together our program."

Though Frank named "Fascinating Rhythm" the quintessential Gershwin song, he had a real story about "The Man I Love," the Gershwin song torch singer Helen Morgan crooned one sultry evening in New York. Evidently the song was a hit in England, but they couldn't give it away in the States until the cabaret song-stress began singing it in clubs across the country.

Frank said, "I had the incredible opportunity of meeting Ira Gershwin, who signed his book to me: *Lyrics on Several Occasions*. People who knew both Ira and George said they were a hundred and eighty degrees out of phase. George was very gregarious: if he wasn't playing tennis he was taking boxing lessons to stay in shape. He loved going to people's homes and playing piano. He'd go to his friends' homes and play the songs, and by the time the show hit Broadway every-one thought it was a revival; they already knew the music.

"Ira was very quiet; he'd sit in his chair, observing, making notes. The opposite of his brother. But Ira was very kind to me when I met him—I actually spent an evening with him. I was singing in a dance band with Helen Forest who at the time sang with Harry James, Benny Goodman, and Tommy Dorsey.

"I was just twenty years old. Helen got up in her sequined gown and sang the torch song 'The Man I Love' as only she could. In the audience a nice-looking middle-aged man in a well tailored grey suit took his glasses off and started to cry.

"After the show I introduced myself to him and learned who he was. I brought him backstage to meet Helen. She took his hand and said, 'You're that man who wept when I sang 'The Man I Love.' I hope I didn't bring up any bad memories.' I told her that she was speaking to Mr. Ira Gershwin.

"'Oh,' was all she managed to say, as all the color drained from her face. After a moment he embraced her and said, 'Young lady, I've never heard it sung any better since George and I wrote it over forty years ago.'"

The great jazz historian Nat Hentoff basically ignored Frank Sinatra, Jr., assuming he'd be a pale shadow of his father, until he heard his most recent CD, *That Face!,* released on Rhino in 2006. "Backed by an invigorating swing band, his singing made me feel good with his personal, signature sound, infectious jazz timing and conversational phrasing."[48] Frank's fifteen minute song and monologue *Over the Land*, inspired by the nation's bicentennial in 1976, describes "how that flag grew in impact, where it went and the troubles it survived during its travels'"[49] since the war of 1812. After the piece was performed by the U.S. Air Force Symphony Orchestra, a U.S Marine officer visited Frank Sinatra, Jr. in his office with a warrant that commandeered the piece for placement in the National Archives. I have to agree with Hentoff who remarked "that beats a Grammy."[50]

Frank Sinatra, Jr. endures. A passionate devotion to the Great American Songbook—to music—has drawn him to stages worldwide for decades. "I think the dirtiest word in the English language is retirement," he said. "This is my job, and I've been doing it for 47 years."[51]

In our interviews, through his dry wit, charm, and direct manner, Frank Sinatra, Jr. has simultaneously addressed and diffused comparisons with his father. He names the gorilla in the room and we all can relax, but there are times this deference comes at his own expense—or at the expense of his own talent. So many times I've wondered what better, more respectful ambassador his father could have asked for. During one interview, Frank told me of a performance in Canada that changed his mind about performing "My Way." Out of respect for his father, he had never before sung it in concert. It made me wonder if his father would have had the same consideration.

Frank said, "I had never performed 'My Way.' I thought it was presumptuous of me to do it. But by the end of this concert, the audience was chanting 'My Way, My Way!' and they weren't stopping. My guys asked what to do, and I said send out the arrangement and let's see what happens."

At that moment Frank paused and seemed to light up from within. "They loved it. So now we do 'My Way' everywhere we go. You know, Ron, I just never believed that they would accept it from me."

## SUGGESTED RECORDINGS

**That Face! Frank Sinatra, Jr.**

Selections include:

> "Spice"
> "Cry Me A River"
> "You'll Never Know"
> "Feeling Good"
> "Girl Talk" (Duet with Steve Tyrell)
> "Walking Happy" "What a Difference a Day Makes"

Orchestra conducted by Terry Woodson.
Reprise CD R2 70017

# Style's Back in Style

John Pizzarelli

© Ken Franckling.

Born in Paterson, New Jersey, on April 6, 1960, jazz guitarist, singer, and bandleader John Pizzarelli is one of the hottest acts in jazz today. Internationally known for classic standards, he combines his hip, swinging style with contemporary sophistication to create his own unique sound. A veteran radio personality, Pizzarelli hosted *New York Tonight* on WNEW from 1984–88, recently launching *Radio Deluxe with John Pizzarelli,* a nationally syndicated program co-hosted with his wife, Broadway star Jessica Molaskey. Josh Getlin of the *L.A Times* says that John and Jessica's relaxed, off-the-cuff show "combines the retro feel of a 1940s living room broadcast with a boomer's passion for the Great American Songbook,"[52] bringing warmth, humor, and that long-lost "live radio" feel back.

Born into music, John is the son of jazz guitarist "Bucky" Pizzarelli and the nephew of Peter and Bobby Pizzarelli, both virtuoso banjo players. John told me that some of the great musicians of the day visited his home, recalling especially a Christmastime jam session with Zoot Sims, Joe Venuti, and Les Paul playing the old standard "Out of Nowhere." When John was just six, his dad began taking him to recording sessions where he met jazz greats Dave McKenna, Slam Stuart, Dizzy Gillespie, and Erroll Garner; he even sat in with Marian McParland during her decade-long reign at the Hickory House, a New York city jazz spot.

All this great music had a profound impact on young John. His eclectic record collection included not only the music of his generation: the Beatles, James Taylor, and Billy Joel, but also *Sinatra and Strings*, Billie Holiday's *Lady in Satin*, and Clifford Brown's *At Basin Street*. These artists were an inspiration to him, and in turn he integrated elements of their style into his music. He told me, "I'm very lucky to be able to perform the music I loved and learned as a child to express myself not only as a singer and player, but also as a composer."

Along with Ray Kennedy on piano, John's brother Martin Pizzarelli on bass, and drummer Tony Tedesco, this current edition of the John Pizzarelli Trio is one of the finest since the late Nat Cole, entertaining people from Berlin to Istanbul. In fact, two albums: *Dear Mr. Cole* and *P.S Mr. Cole* are musical tributes to a man John credits as the one who inspired him to pursue jazz. The trio swings hard and plays ballads beautifully.

John shared with me that he's recorded twenty-three albums of his own, as well as other joint recordings with his father, and appeared on more than forty albums by other artists including James Taylor, Natalie Cole, Rickie Lee Jones, Dave Brubeck, George Shearing, Rosemary Clooney, his wife Jessica Molaskey, even the Boston Pops with Keith Lockhart. John has also appeared on Broadway, numerous TV shows including *The Late Show With David Letterman*, and a well known Fox Woods Casino commercial, "The Wonder of it All."

Another passion the Pizzarelli family shares is their love of Italian cuisine. At home, the combination of good times, homemade pasta, particularly gnocchi, and fine wine was a tradition the family shared with all their guests from the music world.

I looked forward to John and his father visiting me at the studio, not only for the music and the company, but because they always brought a calzone or pepperoni and cheese pizza to a hungry and appreciative DJ! Our conversations always turned to food: where to get it, which cities had the best restaurants, what they served, and especially, where the best pizza could be found. With a name like Pizzarelli, they should know! Top on their list is Pepe's Pizza in New Haven, Connecticut, best known for their white pizza. In Boston it's the original Regina's in the North End, or Santarpio's in East Boston which they were partial to since that was Sinatra's favorite when he was in town.

John and his trio love to entertain not only musically but with their witty banter and infectious humor. One of their most requested numbers is John's own loving tribute to New Jersey, his home state. With his clever composition "I Like Jersey Best," John gives us a long overdue tribute to the much-maligned state. During one of our interviews he said, "When you consider what New Jersey has given the country—Bucky Pizzarelli, Bill Evans, George Van Epps, Fort Dix, Wildwood, Atlantic City, Route 22, the Polaski Skyway, Abbott and Costello—this was a song that had to be written!"

He shared with me how the song came to be: "Actually, it was written in 1981 by Joe Cosgriff, on a whim, in about forty minutes at a diner. He gave it to a mutual buddy, Phil Bernardi. We recorded it in 1983, then played it before the New Jersey state assembly in the state house in 1985. The goal was to make it the official state jingle.

"That day it was the first order of business. They slammed the gavel down and there we were, my dad watching from the balcony. Joe Peterro, the guy who was sponsoring the bill to make it the official state jingle, stood up and said, 'Before I start, John's father is a very famous guitar player named Rocky Pizzarelli, and we'd like him to stand up and take a bow.' Anyway, I sang and they were very receptive, they stood up and applauded. And now it's *still* the unofficial state jingle. But it's funny, crowds get on their feet for it here. It's twenty times more popular in Boston than in New Jersey. In New Jersey, they don't get it."

A huge Red Sox fan, John has played baseball most of his life. In April of 2007, his trio played at a BLOHARDS luncheon at the Yale Club. John said, "It's the Red Sox Nation chapter in New York, stands for the Beloved Loyal Order of Red Sox Diehards, something along those lines. Every time the Red Sox play in New York these fans meet at the Time Life building, and they sure know their stuff. I want to wear my Sox hat, but Jessica won't let me wear it in New York City."

A lighter-than-air jazz guitarist, John teases easy swing and Bossa nova pulses from his guitar creating a sound that melds the virtuosity of George Benson and Les Paul with the soul of Django Reinhardt. Any time you have the opportunity to catch a Pizzarelli performance, do it! A consummate entertainer, he is quite simply one of the best of his generation.

## SUGGESTED RECORDINGS

**John Pizzarelli Trio . . . Live at Birdland**

This 2-CD set, recorded at Birdland in New York City, captures a live performance of the trio at its very best. It includes not only outstanding interpretations of standards and originals, but also John's delightful commentary on a wide range of topics from Dizzy Gillespie to Frank Sinatra.

Selections include:

"Three Little Words"
"I Like Jersey Best"
"Just You, Just Me"
"Manhattan"
"Isn't It a Pity"
"Only a Paper Moon"
A James Taylor Medley

Telarc 2CD-83577

**John Pizzarelli: Dear Mr. Cole**

John and his trio in a loving tribute to Nat Cole.

Selections include such Cole favorites as:

"Nature Boy"
"Route 66"

"Sweet Lorraine"
"Straighten Up and Fly Right"
"Little Girl"
"Unforgettable"

Novus 63182-2

### The Rare Delight of You...John Pizzarelli with the George Shearing Quintet

This collection offers the opportunity to hear the musical magic of Pizzarelli's
voice and guitar combined with the legendary sound of the George Shearing
Quintet...a match made in heaven!

Selections include:

"September in the Rain"
"Indian Summer"
"Everything Happens to Me"
"Be Careful, It's My Heart"
"Lulu's Back in Town"
"Shine On Your Shoes"

Telarc CD 83546

# "Don't Call Me the Velvet Fog!"

Mel Tormé

Courtesy of the Tormé Estate.

In my mind, Mel Tormé defines multi-talented more than any other artist in the history of American music. Born in Chicago in 1925 to a career that spanned some six decades, he has been respected as a jazz singer, actor, composer, arranger, pianist, drummer, author, child radio performer, and raconteur.

From childhood, his hobbies and interests were numerous: he was an aficionado of World War 1 aircraft, American movies, and the so-called "Little Books." A child prodigy, he sang for his supper at four years old in Chicago's Black Hawk restaurant with the Coon-Saunders Orchestra, and was working the vaudeville circuit soon after that.

In his early teens he was already writing original songs and acting on radio shows such as *Little Orphan Annie* and *Jack Armstrong*. His first big break came in 1942 at seventeen when he joined the Chico Marx orchestra as a singer, arranger, and eventually as the band's drummer. He made his first film, *Higher and Higher*, with Frank Sinatra in 1944.

His recording career ramped up in 1946 when he took over a vocal group which he named, "Mel Tormé and his Mel-Tones," modeled on Frank Sinatra and "The Pied Pipers," and started recording for Decca and Music-Craft. Hits included Cole Porter's "What Is This Thing Called Love?" but his best-known song, which became a classic, written with partner Robert Wells, was "The Christmas Song." Inspired to write the tune in order to cool off on a hot July day, Mel and Robert dashed off the song in a mere forty minutes.

I first interviewed Mel on the radio on July 2, 1985. He commented on his early influences as a singer: "Listening very early on to Ella Fitzgerald, Duke Ellington, Connie Boswell and the Boswell Sisters, the Mills brothers; a little later on to the Andrew Sisters, all influenced my intonation and phrasing. These were the groups that helped me formulate ideas for the Mel-Tones, the ones who got me to the place I am today, from how I interpret a song or add my own filigree to a phrase. And I have to say, (alto saxophonist) Phil Woods is the greatest, and Artie Shaw blew my mind as a bandleader."

We also talked about his favorite composer: "Johnny Mercer was my number one lyric writer of all time. The great thing is the admiration went both ways: he'd always wanted to compose something just for me. Unfortunately, we talked about it but never did it, which is a big regret of mine."

I've also been struck by Mel's self-deprecating sense of humor. Though he never loved his nickname, "The Velvet Fog" (which often morphed into "The Velvet Frog"), he had a wonderful openness to reinvention. Mel took advantage of opportunities to appear on TV even when the roles had a tongue-in-cheek reference to what was perhaps his fading star. The irony is, these many appearances led to an even greater following among Gen-Xers who learned about him in a series of Mountain Dew commercials and on an episode of *Seinfeld* where Mel dedicates a song to Kramer. He even made nine appearances as himself on the 1980s situation comedy *Night Court*, whose main character Judge Harry Stone (played by Harry Anderson) was painted as an unabashed Tormé fan, which in fact was true in real life as well.

Like Tony Bennett, Mel never sold out in the 1960s when rock music started to take over; in fact he referred to rock and roll as "three chord manure." This period was perhaps his roughest time, so challenging in fact that he considered changing professions and becoming an airline pilot. But slowly, through a series of club performances, he began to reconstruct his following. He stayed with the classics from the Great American Songbook, commenting during our interview, "As time went on, I literally recorded the entire American songbook. I grew up during the depression—I learned these songs when they came out, which was kind of nice. I *lived* them, I grew up with them. I didn't turn thirty and say hey, time to learn these songs. They are part of me."

In 1982 Mel began a long association with Concord Records, helping to pioneer jazz. While at Concord, he worked with jazz greats Marty Paich, Rob McConnell and the Boss Brass, the Frank Wess/Harry Edison Orchestra, and perhaps most fruitfully with jazz pianist George Shearing.

The excellent albums *An Evening With George Shearing and Mel Tormé* and *Top Drawer* earned Mel the Grammy for Best Male Jazz Vocalist in 1982 and 1983. In the course of our interview, Mel had this to say about his special musical relationship with George Shearing: "George Wein—the entrepreneur who put together the Newport Jazz Festival—he was the one who got the bright idea of putting us together, and George Shearing and I have been friends since the fifties. And yes, it was life-changing to finally win two grammys after thirteen nominations, but I am upset that George was never recognized in this way. It took working with George to make this happen, period, and he wasn't even nominated. I share those two awards with Shearing; he is, at minimum, half the reason I won them. As George says, we breathe together, we literally think identically on a musical plane, and that's why I think the records are so successful."

Though a stickler for maintaining and protecting his voice—Tormé insisted on eight hours of sleep before every performance, never smoked, avoided drafts, and rarely drank—he was obligated to perform for the London Sessions after an eight-hour flight during which, tormented by a painful divorce, he didn't sleep. I'm so glad he relented and recorded anyway. His performance of "All in Love Is Fair," set off by Phil Woods on alto sax solo and accompaniment, never sounded so emotive, so heartbreakingly beautiful, as if he truly was confessing his own personal story in that song.

The author of five books, including a biography of drummer Buddy Rich, a remembrance of Judy Garland's controversial television series, an autobiography, as well as a loving tribute to singers who influenced him: *My Singing Teachers,* Mel was nothing if not terrifically gifted. But I've always felt that Tormé was unrivaled singing ballads and "scatting"—he was simply one of the best jazz singers in the business. His 1962 hard-driving R&B song "Comin' Home, Baby" prompted gospel singer Ethel Waters to say, "Tormé is the only white man who sings with the soul of a black man."[53] Composer and lyricist of over three hundred songs, he was blessed with a lush sound along with perfect pitch and brilliant intonation...simply put, Mel never sang a bad note!

Mel performed worldwide in concerts and jazz festivals throughout the world until the time of his unfortunate stroke in 1996 that ultimately took his life. I treasure the many times he came to visit me in the studio. His broad range of musical knowledge was always refreshing, and I knew of no one else who could carry on conversations about so many other topics relating to the history of entertainment in America.

## SUGGESTED RECORDINGS

### The Mel Tormé Collection: 1944–85

This four-CD collection represents Mel Tormé's collaboration with the greatest songwriters, arrangers, and musicians in this four-decade history. Also included is an extensive photo biography of Mel together with a complete discography and recording history of all of the songs included in this collection.

Among the selections:

> "Night and Day"
> "Blue Moon"
> "All of You"
> "It's Delovely"
> "Lullaby of Birdland"
> "Cheek to Cheek"
> "Sunday in New York"
> And a knockout Gershwin medley that includes twenty songs!

Rhino R271589

## Mel Tormé: The Best of the Concord Years

This two-CD collection picks up from 1985 and presents some of the finest recordings that Mel made during the last two decades of his phenomenal career. During this period, Tormé was in his vocal prime. This collection includes performances with George Shearing, Cleo Laine, the Marty Paich Dek-tette, Rob McConnell and the Boss Brass, and the Frank Wess/Harry Edison Orchestra.

Some of the selections included:

> "Stardust"
> "The Carioca"
> "Born to be Blue"
> "These Foolish Things"
> "Pennies from Heaven"
> A Duke Ellington medley
> And Mel's version of his own "The Christmas Song"

Concord CCD2-4871-2

## Mel Tormé–George Shearing: A Vintage Year

Recorded live in concert.

Selections include: "The Midnight Sun," "When Sunny Gets Blue," "Someday I'll Find You," "The Way You Look Tonight," "Bittersweet," "New York, New York" (medley), and "Little Man You've Had a Busy Day."

Concord CCD 4341

# The Thrush of Columbus

Nancy Wilson

© Photofest, Inc.

Nancy Wilson began her career singing in church choirs and dance bands as a teenager, though she'd made the decision to be a singer by the age of four. Some of the greats who influenced her during those formative years included: Carmen McRae, Sarah Vaughan, Dinah Washington, LaVerne Baker, Ruth Brown, Nat King Cole, Louis Jordan, and Lionel Hampton's Little Jimmy Scott. Every spare moment was spent either listening to the radio at home or at "the juke joint down the block."

Over the years her repertoire has included jazz and blues, show tunes, standards, even gospel and R&B. She's been described as a "storyteller," "a professor emeritus of body language," a "consummate actress," and "the complete entertainer."[54] She prefers the term "song stylist," and explained why in our 1991 interview: "I never sing standards the same way twice because each time it's a separate performance. 'Song stylist' is more me than anything else because I'm interpreting the lyrics as well as acting. I like to put myself into a song so that it will touch you, not to impress you with my voice. I want it to mean something to you ten years from now. Each song is a vignette, a little play. For me, it starts with the lyric."

After touring with Rusty Bryant's Carolyn Club Big Band in her late teens and early twenties, she made her first recordings in 1956 under Dot Records, joining the ranks of Capitol Records in 1959. There was no question that this was the label to be with at that time. Capitol had become one of the most popular and best selling record companies in America; their roster of singers included Nat Cole, Peggy Lee, Dean Martin, Kay Starr, and Frank Sinatra who joined the label in 1953.

Touring with Rusty's band brought her to New York where she met and became great friends with saxophonist Cannonball Adderley. Nancy allotted herself six months to achieve her goal of having Cannonball's manager John Levy sign her, and Capitol Records as her label. When she got the call to fill in for Irene Reid at the Blue Morocco, she called Levy to let him know, and he showed up for the gig. John called the very next day, recorded a demo with her, and within five days Capitol was on the phone.

About her early days she told me, "I made it a point to be on the spot when anyone ever needed a singer. I finally quit college and starting singing full time all over the west." She had everything going for her: she was young, beautiful, and gifted with her own unique sound, but at the time of her Capitol contract, she was up against the rising tide of rock and roll. This did not stop Ms. Wilson from turning out albums that became classics. In all she recorded over seventy albums—virtually the entire American Songbook—and worked with the best arrangers in the business including Nelson Riddle, Billy May, and Oliver Nelson.

Her move to New York gave her access to regular work in jazz clubs around the city where she came to the attention of George Shearing, among others. They hit it off right away when she sat in with his group at New York's Basin Street East. The result was one of the most successful Capitol albums ever, called *The Swingin's Mutual*. It was a great success commercially, artistically, and musically. The jazz critic Leonard Feather called it one of the most logical and successful collaborations of the year.

Her debut single "Guess Who I Saw Today?" was so successful that between 1960 and 1962 Capitol released five Nancy Wilson albums.

She shared with me the special significance of "Guess Who I Saw Today?" during one of our interviews: "The first time I heard the song was in 1952, in Ohio, when I was fifteen. Carman McCrae sang it, then Eydie Gorme. I ended up singing that song every night I performed for as long as I can remember. I think it struck a chord as the ultimate song about betrayal. It was the song John Levy heard when I came to New York, and I know he signed me because of it."

In the late seventies, Nancy acted in a variety of television shows including the *Andy Williams Show*, the *Carol Burnett Show*, and the *Flip Wilson Show*, while sales of her albums with Capitol were second only to the Beatles, surpassing even Frank Sinatra, Peggy Lee, the Beach Boys, and early idol Nat Cole.

Nancy never compromised her talent with inferior material or arrangements. She exuded class, making everything sound smooth and effortless and choosing to work with the best musicians of the day: Benny Carter, Harry Edison, Pete Candoli, Dick Nash, and Shelley Manne, to name a few.

When I introduced Nancy from the stage of Symphony Hall in the mid-eighties, she turned in one of her finest performances singing and swinging as only she could. She sang many of her signature songs, including "Guess Who I Saw Today?," "Now I'm a Woman," "Peace of Mind," and "How Glad I Am," which won her a Grammy in 1964. My wife Joyce has always been one of Nancy's biggest fans and was very excited about meeting her backstage before the concert. Joyce's big thrill came when Nancy asked her to zip up her dress, and they enjoyed a little repartee. She was on cloud nine all night!

Recently a friend gave me a DVD of a live performance of Nancy in the 1970s when she sang in a concert with the Count Basie Band and the great Joe Williams. I was knocked out by her dynamic stage presence and her ability to relay the lyrics so purely as to make you swing, swoon, or cry. She not only inspired the musicians, but the audience as well. *Time* magazine said after her career-launching turn at the Coconut Grove, "She is, all at once, both cool and sweet, both singer and storyteller."[55]

After an appearance in Las Vegas, she confessed to me how privileged she felt to work with so many legends in the music business. Nancy singled out pianist/arranger Jimmy Jones who she described as one of the greatest and most underappreciated arrangers and conductors she had every known. We also talked about her successful radio stint when she hosted *Jazz Profiles* for National Public Radio in 1995. She said how much she enjoyed presenting the music and the artists she loved so dearly. Her casual conversational style with the musicians made the series great listening and a highly informed history of jazz.

I asked what she thought made the greats great: "They had energy and wonderful voices of course, but there was a freedom about Ella Fitzgerald that you will never hear today. It's never going to be that good again. There's not enough room; we don't give them the space. We don't give young singers the opportunity to be Lena Horne. But the music will be heard on a radio dial somewhere. In the air someplace, the music is still there."

However she's known, as "Sweet Nancy," "Fancy Nancy," or "The Girl With the Honey-Coated Voice," Nancy Wilson continues to delight and captivate us. She's still active today as a performer, educator, and simply a hard worker for all sorts of charities, including the National Minority AIDS Council. Of her 2005 award where she was inducted into the International Civil Rights Walk of Fame at the Martin Luther King Jr. National Historic Site, she says, "This award means more to me than anything else I have ever received."[56]

Though she's the winner of three Grammys and countless other honors, I think it takes a certain kind of courage to look inside the lyrics of a song and bring out its heart, and for that kind of joy I'd like to offer her my personal thanks.

## SUGGESTED RECORDINGS

### The Best of Nancy Wilson: Ballads, Blues and Big Bands

Quite simply, this three-CD set is the definitive collection of Nancy's Capitol
recording years. It includes an excellent discography by Pete Welding and
is a must for lovers of her music.

Selections include:

"When the World was Young"
"Nearness of You"
"Midnight Sun"
"Satin Doll"
"Willow Weep For Me"
"Fly Me to the Moon"
"Angel Eyes"

Capitol CDP 834886

### Nancy Wilson Turned to Blue

www.mcgjazz.org

One of Nancy's most recent recordings, this collection features a chamber music
quartet with pianist Billy Taylor as well as an all-star big band. The lovely
arrangements are by Jay Ashby, John Wilson, and Llew Matthews.

Selections include:

"This Is All I Ask"
"Be My Love"
"Taking A Chance On Love"
"Old Folks"
"I'll Be Seeing You"

MegJazz MCG J1022

### Nancy Wilson's Greatest Hits

This CD is a wonderful collection of the best of her Columbia songbook
recordings.

Selections include:

"Guess Who I Saw Today?"
"The Two of Us" with pianist Ramsey Lewis
"Loving You" duet with Peabo Bryson
"Hello Like Before"
"When October Goes"

Columbia CK 65542

# Blue Rose

Rosemary Clooney

© Photofest, Inc.

I had the opportunity to interview the woman Tony Bennett called "the most beloved singer in America" over thirty years ago during one of her many appearances in Boston. Though her life was marked with both personal and professional struggle, her voice was a source of joy and comfort for countless fans, and her fortitude an inspiration to all of us.

Her decades-long mastery of America's popular song was preceded by a life of poverty in Maysville, Kentucky. As she describes in her first memoir, *This For Remembrance*, Rosemary's earliest years were rocked by an emotionally unstable and alcoholic father, and a mother who took off for California taking her brother Nick but leaving Rose and Betty behind. One night after celebrating the end of World War I, their father took off—household cash in hand—never to return. To eat, Rose and her sister collected soda bottles. When the girls left home one evening to compete at an open audition at a Cincinnati radio station, their rent was past due, their utilities about to be cut off, and the phone disconnected. "The Clooney Sisters" won the audition, jump-starting their career in 1945 at WLW in Cincinnati with a twenty-dollar-a-week late-night gig.

Though Betty left the act in 1949, their work had caught the attention of bandleader Tony Pastor. At the time, Tony was handled by Joe Schribman, who was Charlie Schribman's nephew, booker for Glenn Miller as well as all the major big bands around New England. Summers she'd perform at "The Surf" at Revere Beach and "The Pier" in Old Orchard Beach, while in the winter, there were enough good ballrooms in the area to keep her singing.

In the early fifties, Rosemary left the area to try her luck in New York City. Her arrival there was perfectly timed: World War II had depleted personnel for many bands, and audiences were coming out in droves to listen to the likes of Bing Crosby, Doris Day, Frank Sinatra, Peggy Lee, Ella Fitzgerald, and Dinah Washington.

Rose was signed immediately with Columbia Records, and it was there that Mitch Miller, her agent, convinced her (she actually had no choice—do it or she was gone) to record "Come On-A My House." She resisted cutting the song with every fiber of her being, citing everything from how uncomfortable she was with putting on an Italian accent, to calling the double entendres in the song a cheap lyrical device. Richard Harrington of the *Washington Post* said what bothered her most was the "deliberately disordered phrasing, so antithetical to her instinctive devotion to craftsmanship."[57] Just like "A-Tisket, A-Tasket" proved for Ella Fitzgerald, this "pop fluff" (as Michael Feinstein put it) she hated morphed her into a star, topping the charts with gold, making "Rosie" a household name and landing her on the cover of *Time* magazine—the first female singer ever to do so. Rosemary followed up with other hits: "Hey There," "You'll Never Know," and "This Ole House," as well as a popular television show.

These milestones set off a successful twenty-year cycle, made more so by her beloved friend Bing Crosby who signed her to co-host a songfest radio show on weekday mornings on CBS. Film roles abounded, including *White Christmas*, a white-hot hit that featured Bing, Danny Kaye, and music by Irving Berlin, and where Clooney was lauded for her sultry performance of "Love, You Didn't Do Right By Me." She also toured for six years with her own group, "Four Girls Four," which included singing stars Margaret Whiting, Rose Marie, and Helen O'Connell.

Rosemary shared with me her great admiration and love for Bing Crosby. "I learned so much from Bing about being a true professional, as an actor and singer. He was the most relaxed and casual performer I've ever known. Nobody worked harder to get things absolutely correct, which I think enabled him to be relaxed and casual onstage." She also named Sinatra as one of her biggest influences; his ability to reach out and touch his audience seemed to leave its mark on every singer who came after him.

To the amazement of her family and friends, Rosemary eloped with actor Jose Ferrer, sixteen years her senior, in the summer of 1953. The marriage thrilled the tabloids—Ferrer was quite the ladies man—but was not easy for Rosemary. They moved into a glamorous Beverly Hills home once owned by George Gershwin (and as she reminded me—next door to Ira) where she and Jose entertained the toast of Hollywood: Jack Benny, Nat King Cole, Billie Holiday, and others with opulent poolside galas. Their first child was born in 1955; by 1960 they had five children. They divorced in '61, then married and divorced again in '67.

In the midst of all this turmoil, Rosemary made her signature album called *Blue Rose*, a collaboration with Duke Ellington and Billy Strayhorn. Not well received at the time, it has since become a jazz classic, proving she could cross over into the world of jazz as a singer with popular hits. She told me a bit about how she prepared to sing the title song: "Duke wrote that for me; Billy produced it. He said to sing it as if I was listening to Duke on the radio and getting ready for a date; putting on makeup, fixing my hair; and it was such a wonderful word picture for me. That's the rendition you hear on the album."

Meanwhile, pressures mounted for Rosie that would have toppled anyone. The stress of raising five children on her own while pursuing careers as a TV, movie, radio, and recording star fueled an addiction to tranquilizers and sleeping pills. She recalled in her autobiography feeling prey to the "fifties myth of family and career."[58]

Her house of cards came crashing down in 1968. When close friend Bobby Kennedy was assassinated in Los Angeles, she was standing only yards away. As she said in 2002, "I had my nervous breakdown the same time the country did."[59] This tragedy, compounded with her split from Ferrer, resulted in a mental collapse that took her years to work through and eventually emerge from. Later she would touchingly revisit and possibly expunge this pain through her recurring role as an Alzheimer patient on *ER* opposite nephew George Clooney, her moving performance winning her an Emmy nomination.

She launched her comeback at a performance at Tivoli gardens in Copenhagen in 1972; but in essence she had to start all over again, taking humble gigs and counting on help from old friends. In 1975 Bing invited her to appear with him in concert at the Los Angeles Music Centre. The duo took it on the road to Chicago, New York, London, even to Ireland, and finally Rosie's career was reborn. Playing "How Are Things in Glocca Morra" on *MusicAmerica* inspired Rose to share this story with me: "Bing asked me to go, and it was the perfect time. He supported an orphanage in Dublin—they had a band—and the children came out to the tarmac when we arrived. These little tiny kids played, 'Where the Blue of the Night Meets the Gold of the Day'—the oldest was just sixteen. When we left Ireland and they met us at our plane in the morning with the fog rolling in and they played, 'Come Back to Erin,' there wasn't a dry eye on the tarmac, let me tell you."

Each year for the next two decades she made a new album on the Concord Jazz Label run by Carl Jefferson, the sweetness of her voice mellowing to a more golden tone, with a touch of swing. Her popularity also surged in supper clubs across America. She had a chance to get back in the studio and work with musicians like Scott Hamilton, Dave McKenna, and Warren Vache as well as the bands of Woody Herman and Count Basie. In fact, her second professional life was built around making some of her finest recordings by master composers Irving Berlin, Cole Porter, Harold Arlen, Johnny Mercer, Ira Gershwin, and arranger Nelson Riddle, all who happened to be friends of hers. As she said in an interview with *Lear's* magazine, "I can even pick the songs. The arranger says to me, 'How do you want it? How do you see it?' Nobody ever asked me that before."[60]

I always felt that Rosie had the ability to sing the lyrics of a song as if she had lived the experience. Her voice had humor, honesty, and heartbreak in it, caressing you completely. Stephen Holden of the *New York Times* said: "An audience became her extended family, encouraged to rest its collective head in her lap as she poured out musical bedtime stories for grownups. Though written by others, those stories, told in her voice, came across as nuggets of personal experience."[61]

John Pizzarelli said: "She was a great person to eat and drink with, and she was a wonderful presence when she held court."

Barry Manilow: "She comes from that great time when singers respected the songwriter and music arranger more than anyone, and their interpretation showed that!"

Michael Feinstein: "Humor was a Clooney trademark, and she laughed abundantly. She also cried abundantly and had sadness...I wish I could have eased (some of that) for her. But no one could. If she only knew how much she gave to others...how we treasured the expression of her personality through her voice."

Diana Krall: "The best piece of advice she gave me was, 'Just sing the damn song, honey, it's all right there!'"[62]

As for myself, she was constantly thanking me for supporting her career, but the truth was she was just as supportive of mine. I suspect she felt in some sense like a lot of people's favorite aunt, the one who loved you no matter what. Always self-deprecating, she could tell a funny story about meeting the pope as well as tales of her stay in the Mayo clinic. I also felt that a big part of her legacy was perseverance: her return to stardom in the fickle world of show business was a triumph.

In 1997 she married long time love, dancer Dante DiPaolo, who said he fell for her on the set of the 1953 film, *Here Come the Girls*. *Girl Singer*, her second autobiography, was published in 1999 and like her singing voice, it's warm, free of bitterness, and full of love for her family and life. Recording steadily until her death of lung cancer in 2002, she made a total of twenty-five albums for Concord. The inscription on her award in 1995 for the ASCAP Pied Piper Award called her: "an American treasure and one of the best friends a song ever had."[63]

## SUGGESTED RECORDINGS

### Rosemary Clooney: 16 Biggest Hits

In a career marked by an incredible number of hits, Rosie was tops! During this time, she led the charts with the selections included in this CD. Among them:

"Tenderly"
"Hey There"
"Blues In the Night"
"Memories of You"

"Botch-A-Me"
"From This Moment On"
"Mambo Italiano"

Columbia/Legacy CK 63553

## Blue Rose: Rosemary Clooney and Duke Ellington and His Orchestra

This reissue of a late 1950s session came to me from my good friend, Sal Ingeme. For many years Sal was the gentleman responsible for introducing many of the artists who recorded for Columbia Records, including Tony Bennett, Johnny Mathis, Barbara Streisand, and Rosie Clooney. This historic 1956 collaboration of Rosie and Duke Ellington showcases the talents of two musical giants. The original LP was unavailable for many years until it was restored and remastered on this CD in 1999.

Selections include:

"Blue Rose" (Title song)
"Mood Indigo"
"I Got It Bad, and That Ain't Good"
"Don't Mean A Thing If It Ain't Got That Swing"
"I'm Checkin' Out, Goombye"
"Sophisticated Lady"
"Hey Baby"

Columbia/Legacy CK 65506

## Rosemary Sings Rodgers, Hart and Hammerstein

Backed up by a swinging group, including Scott Hamilton, tenor sax, John Oddo on piano, Jack Sheldon on trumpet, and vocals, Rosie takes us on a journey featuring marvelous interpretations of classics by Richard Rodgers, Lorenz Hart, and Oscar Hammerstein.

Selections include:

"Oh What a Beautiful Morning"
"The Lady Is a Tramp"
"My Romance"
"I Could Write a Book"
"The Sweetest Sounds"
My personal favorite, "People Will Say We're in Love"

In the above selection, Rosie shows off her great sense of humor teaming up with Jack Sheldon. Their banter is infectious!

Concord CCD 4405

# A Song's Best Friend

Frank Sinatra,
Sammy Cahn,
and Paul Weston

© Photofest, Inc.

Sammy Cahn always had a way with words. Even as a scrawny kid in wire rims it kept bullies off his back and got him out of countless scrapes with his parents. His gift turned him into one of the most prolific and beloved lyricists of all time, responsible for putting more words in Frank Sinatra's mouth than anyone else.

Born Samuel Cohen in 1913 in New York City's East Side, Sammy described his childhood to me as having an idyllic quality: "those images from the movies—kids playing in the streets, the pushcarts—that's really the way it was.

"I came out of the lowest part of the lower East Side. If you take one step back you're in New Jersey. I was born on 10 Cannon Street. I've been asked if they ever put a plaque on the building. Not only did they not do that, they removed the building, the street, *and* the neighborhood. Gives you an idea how fleeting fame is.

"When I think of the East Side," Sammy continued, "I can't help laughing because it reminds me of the day I met Cole Porter. I was at a party and someone said Cole Porter wanted to meet me. Next thing I know Cole Porter's coming at me, this man on crutches, but instead of me rushing to meet him, I go immobile. I'm so stunned I can't move. He said, 'Sammy Cahn, I've always wanted to meet you.' All I can say is, '*You* wanted to meet *me?*' He said, 'Yes, because I've always envied you.' Echoing everything he says like an idiot, I say, '*You've* always envied *me?* Why would you envy me?' He said, 'Because you were born on the East Side.' So I asked him, 'What's the big deal with that?' and he said, 'If I'd been born on the East Side I'd be a true genius.'"

Not the most diligent of students, young Sammy (always with a good story for his parents) would cut classes to see movies or go to vaudeville shows. He played violin as a hobby but realized it was possible to make money at it when he saw his mother pay the musicians at his bar mitzvah, so at just fourteen he took off with the Pals of Harmony, the very band his mother had hired. He did local gigs, then took off with the band, playing shows from Atlantic City to the Catskills.

He wrote his first song at age sixteen, describing the experience in his autobiography, *I Should Care*: "It was actually Jackie Osterman at the Academy of Music on 14th Street who inspired my song-writing career...In the middle of the act, [Osterman] took a change of pace and said he'd like to sing a song he'd written. It was a fascinating thing for me to be actually looking at a songwriter—in person...walking home...I began to frame a song in my head. By the time I reached home I had actually written a lyric...The song was a piece of idiocy called 'Like Niagara Falls, I'm Falling for You—Baby!' But if, as...somebody said, a journey of a thousand miles starts with the first step, that was the first step."[64] It was Saul Caplan, the pianist from the Pals of Harmony, who joined forces with Sammy and formed his first songwriting team.

Vaudeville embraced them, but their songs languished. One day in 1935 their luck changed. During our 1985 interview, Sammy explained: "We wrote the song 'Rhythm is Our Business' for Jimmy Lunceford as an opening song, and through the popularity of this one song, we became known as band songwriters. We started writing for all the big bands: Glen Gray, Glenn Miller, Jimmy Dorsey. That was the turning point, the catalyst." At this time the duo was also welcomed into ASCAP, the American Society of Composers, Authors and Publishers.

But it was the song "Bei Mir Bist Du Schon" (means that you're grand) which sold a million copies, that made Cahn wealthy enough to buy his parents a new house. At first he had trouble selling the idea, but a then-obscure sister act from the Midwest happened to cross his path: "One day Lou (Levy) brought the Andrews Sisters: Patty, Maxene, and LaVerne up to our apartment. On the piano was this copy of a song in Yiddish. Patty asked...'How does it go?' I played it for them, and they started to sing right along and to rock with it. 'Gee,' said Patty, 'can we have it?'"[65]

Twice in his life Sammy Cahn changed his name: first to Kahn to avoid confusion with actor Sammy Cohen, and from Kahn to Cahn, some say to avoid confusion with lyricist Gus Kahn. In our interview, he explained another reason for the second name change: "At the time I was working with Saul Caplan. I said: 'Caplan/Kahn,' it just doesn't work. It sounds like a dress firm—you're gonna have to change your name. He said if I change my name you're gonna change your name. So he went to Chaplan and I went to Cahn."

In the thirties Cahn and Chaplan wrote for New York's Vitaphone Studios, a subsidiary of Warner Brothers. Artists such as Betty Hutton, Bob Hope, and Edgar Bergen sang lyrics they'd written, but by 1940 Vitaphone was shuttered. The duo moved to Hollywood still under contract, but the work dried up and they parted ways. Just as desperation began to kick in, Cahn was asked to write a song with composer Jule Styne. "From the beginning it was fun," he remembered. "Jule went to the piano and played a complete melody. I listened and said 'Would you play it again, just a bit slower?' He played and I listened...then I said, 'I've heard that song before'—to which he said, bristling, 'What the hell are you, a tune detective?' 'No,' I said, 'that wasn't a criticism, it was a title: 'I've Heard That Song Before.'"[66]

Styne's collaboration with Cahn proved incredibly fruitful; together they composed songs for nineteen films between 1942 and 1951. Year in and year out, these songs captured top ten spots. *Time* magazine compared their synergy to that of Rodgers and Hammerstein.

Sammy explained to me how "Let it Snow!" came to be: "It was one of the hottest days of the year, even sitting in our studio on Hollywood and Vine to be exact. And I said to Jules, Why don't we get out of here; everybody else has. Let's hit the beach and cool off. And he said, Why don't we stay here and write a winter song. Seemed like a reasonable idea. I went to the typewriter and typed 'oh the weather outside is frightful, but the fire is so delightful, and since we've no place to go, let it snow, let it snow, let it snow.' Just like that.

"After we came down to earth from that hit, Styne looked at me and said, Why don't we write a summer song? And we wrote 'The Things We Did Last Summer.'"

Cahn's recounting of the scoring for the title song of the film *Three Coins in a Fountain* painted a quintessential picture of 1950s Hollywood:

"We were working on a film called *Pink Tights*. Styne wrote a marvelous score, but you're never going to hear it. The film got dumped because Marilyn Monroe ran off to Japan with Joe Dimaggio.

"So there we were at the studio with nothing to do. One afternoon, the producer breezed in and said can you write a song called 'Three Coins in a Fountain'? I looked at him and said I could write a song called 'Eh' and he said don't be funny, we need a song called 'Three Coins in a Fountain' and we need it *yesterday*. I said can you at least tell us what it's about? He said three girls go to Rome and hope to fall in love so they throw three coins in a fountain. And then he was gone.

"So we had a title and a clue. I went to the typewriter, Styne went to the piano. Coming up with the lyrics took five minutes. I handed them to Styne who came up with a theme in half an hour. Still, it was hell coming up with the bridge, but we got it done, not before getting a little salty with each other. In an hour the producer comes back to check on us. Actually, I think he said, 'Okay, let's hear it.' I sang it, he loved it, then he rushed down and sang it for Zanuck, who loved it, then zipped back to me and asked me for a demo. *Me? A demo?* I said. Look, Sinatra's wandering around the halls getting paid millions for not making *Pink Tights*, why not ask him? Long story short, Frank said yes. In the end, it was that very demo record that became the one used for the film."

A collaboration with Nicholas Brodsky for "Be My Love" brought about a remarkable afternoon with Mario Lanza. Cahn said, "The music came before the words for this one, and even though he really was the Irving Berlin of Budapest, Brodsky was a waterfall pianist: I could never hear the melody under all that sound. So I had to take it apart and put it back together. I ended up singing it for Mario Lanza, if you can imagine, who stared at me the whole time with this bemused expression. Then Lanza sang it back, not to me, but *at me*. It was one of the most stunning experiences of my life; he nearly cracked my glasses. Listening to him electronically could never match being in the same room with him."

Though Cahn had met Sinatra back in Frank's early days with Tommy Dorsey, their collaboration intensified when Sinatra introduced Sammy to composer Jimmy Van Heusen, and another songwriting duo was born. Cahn said, "Through Tommy came the enduring and perhaps most satisfying relationship of my lyric writing career—Frank Sinatra."[67] Over time, the men were practically considered Sinatra's personal songwriters. Frank recorded eighty-nine of Cahn's songs including "Love and Marriage," "Come Fly With Me," "My Kind of Town," and "The Tender Trap."

Since Cahn had been Sinatra's roommate at one time, it's possible he had more than the usual insight into the man's quirks and character in general, not to mention his phrasing and sense of timing. Cahn explained, "My wife asked me all the time: Sam, tell me the truth; who came first, you or Sinatra? And I always told her: the song came first.

"The first time Sinatra heard (a song I wrote)," Sammy continued, "was when I sang it to him. It's great fun to sing to him because he's such a marvelous audience; he sits and listens and he takes the back of his thumb and rolls it over his lower lip, and when you finish he just looks at you. No comment. But he's never turned a song down. Crosby's the same way."

Other collaborators included Sammy Fain, Arthur Schwartz, Sylvia Fine, Vernon Duke, Gene de Paul, Paul Weston, and Axel Stordahl. Cahn had especially kind words for these last two: "One of my favorite arrangers in all the world is Axel Stordahl. Sy Oliver was a marvel—what a great, great talent—Paul Weston too. Arrangers are truly the unsung heroes; they paint with sound."

Cahn also proved to be a survivor, nimbly adapting to the changing tastes of a nation as he moved from the musical tunes of the 1940s to the ballads of the '50s. In 1992 he told *Pulse!* magazine that he would love to write songs for contemporary singers like Michael Bolton or Madonna: "My opinion of the music of today is simply put: Whatever the number-one song in the world is at this moment, I wish my name were on it."[68]

At one time, Sammy Cahn was considered the nation's highest-paid songwriter, often earning more than a thousand dollars a word. Over a span of thirty-three years, twenty-six songs with Cahn's lyrics were nominated for Academy Awards for best songs of the year, winning four times. He also won an Emmy for

"Love and Marriage." He even starred in a one-man show on Broadway called *Words and Music*, which critics adored.

I would place him up in the stratosphere with '40s and '50s songwriters Johnny Mercer and Harry Warren, a prolific contributor to the Great American Songbook. His delight in his lifelong involvement in songwriting was palpable during our interview: "Songwriting is one of the great joys of my life. And I've been blessed: if you're a man who puts words in people's mouths, the trick is to get them in the mouths of those who are extremely talented. And that's the kind of incredible luck I've had."

## SUGGESTED RECORDINGS

### An Evening with Sammy Cahn

### Sinatra Sings Select Sammy Cahn

An absolute must for lovers.
Selections include:

> "Come Fly With Me"
> "Time After Time"
> "The Tender Trap"
> "Love and Marriage"
> "All the Way"
> "Three Coins in the Fountain"
> "High Hopes"
> "All My Tomorrows"
> "Five Minutes More"

Capital CD# CDP8380942

### Ralph Sharon Swings the Sammy Cahn Songbook with special guest Gerry Mulligan

Wonderful jazz interpretations of Cahn classics including:

> "Be My Love"
> "My Kind of Town"
> "Teach Me Tonight"
> "Call Me Irresponsible"
> "The Things We Did Last Summer"
> "The Second Time Around"
> "I Should Care"
> "It's Magic"
> "Guess I'll Hang Out My Tears to Dry"

DRG CD 5232

# IV. CONDUCTORS AND COMPOSERS

## 1. ARTURO TOSCANINI

# The Maestro

Arturo Toscanini

© Robert Hupka. Courtesy Estate of Robert Hupka, Arthur Fierro, Executor.

Through serendipity, WBCN, and WGBH, respectively, my life has intersected with three generations of Toscaninis: the world-famous conductor Arturo, his son Walter, and Maestro's grandson Walfredo.

One Saturday in 1948, when I was ten years old, I dragged my father to an RCA Victor trade show at the old Mechanics Hall on Huntington Avenue in Boston. Why? One word: television. I'd never seen one but wanted one with every cell of my being, so I finally convinced my dad to at least have a look at these magical new inventions. We wandered around the showroom, utterly wowed by these cumbersome boxes glowing with small black and white screens. Repeated a hundred times around the hall was a single image: the white-maned conductor himself climbing to the podium and lifting his arms. The first glorious notes of the overtrue to Verdi's *La forza del destino* thundered to life. You could feel the man's intensity and passion blasting through the cathode rays as he chopped his baton through the thick air surrounding his spellbound musicians. They knew what was expected of them, as well as the consequences for not providing it: the wrath of Arturo Toscanini.

Already steeped in opera through not only my father who sang arias while shaving, but through Saturday afternoon Metropolitan opera broadcasts, I could already quote you the liner notes off all the Puccini LPs at home. But this sound— this was a *new* sound! Violins as sweet as voices, crashing timpani, a tremendous energy just under the surface that this wild man on the podium seemed to have tamed with just a stick. I not only had to have the TV, I had to hear every piece of classical music on earth.

More than two decades later, in the early sixties, I found myself at WBCN, when those call letters meant "Boston's Concert Network." In tribute to the impact this man had on my life, I decided to put together an all-Toscanini series. I wrote to Arturo's son Walter about the concept. His response was to immediately start mailing me a different set of reel-to-reel tapes each week. I knew what these tapes meant to Walter; he'd worked with his father during the last years of the conductor's life helping edit tapes and transcriptions of performances with the NBC Symphony. The energy that blasted out of these tapes was just as alive and powerful as what I'd heard and seen in 1948 as the awestruck kid in front of my first television.

It wasn't easy letting Walter know about WBCN's decision to stop playing classical music and become an all-rock station, something that happened just a few months after I'd started the series. Walter's reaction to the disappointment was to invite me to the Toscanini homestead in Riverdale, New York. In the basement of his home, which he'd converted to a recording studio, I was treated to rarely heard reels of Toscanini in rehearsal as he chastised members of his symphony, indulging his constant reach for perfection.

A good thirty years passed. In October of 1993, many years after Walter's passing in 1971, I reached out to his son, Walfredo, after learning he'd spent some time accompanying Arturo during one of his last tours in the states when Walfredo was twenty-one and the maestro in his early eighties. Walfredo granted me the opportunity of a series of interviews for *The Classical Hour* on WGBH, lending myself and our listeners fantastic insights into his famous grandfather.

Born on March 25, 1867, in Parma, Italy, Arturo Toscanini is widely considered to be the greatest conductor of the late nineteenth/early twentieth century. A combination of elements set him apart from all who came before: his celebrated interpretations of Beethoven, Brahms, Wagner, and Verdi, his photographic memory, his phenomenal ear for orchestral detail, his passionate swings from rapture to despair, and his sexual voraciousness are just a few. Radio and TV broadcasts as music director of the NBC Symphony Orchestra from 1938 to 1954 made him a household name.

At age nineteen, Arturo played cello and acted as Chorus Master with an opera company on tour in South America. After several lackluster performances of *Aida* in Rio de Janeiro, the musicians completely lost patience with their conductor who'd apparently been doing such a terrible job that they went on strike, forcing the company to find a new conductor. After no fewer than two replacements were booed offstage, the company turned to their Chorus Master, who knew the whole opera cold. With zero conducting experience, Arturo took up the baton, tossed the score to one side, and led an astounding performance, going on to conduct eighteen more operas with the company.

Back in Italy, he continued to conduct but also returned to the cello, playing in the world premiere of Verdi's *Otello* in 1887 under the composer's supervision. Impressed by Toscanini's faithfulness to the score, Verdi knew he was dealing with no ordinary cello player when Arturo suggested an *allargando* for the Te Deum where it was not in the score; Verdi responded that he'd omitted it for fear "certain interpreters would have exaggerated the markings."[69]

Word of Arturo's brilliance spread. The next decade saw Toscanini not only entrusted with the world premieres of Puccini's *La Boheme* as well as Leoncavallo's *Pagliacci*, but conducting symphonies worldwide. He became the resident conductor at the La Scala Orchestra, leaving for the Metropolitan Opera in 1908, but returning in 1920 to tour with La Scala in the states. In 1930 he toured Europe with the New York Philharmonic to fantastic acclaim from both critics and audiences.

Toscanini initially supported Mussolini in 1919, along with many other bourgeois Italians at the time. By 1923 he had come around to a complete opposite point of view, at one point refusing to play their national hymn at concerts. He became the first non-German conductor to appear at Bayreuth in 1930, while the New York Philharmonic was the first non-German orchestra to play there. In 1931 he was assaulted by Fascist thugs in Bologna, and thereafter refused to conduct in Italy until a new regime was in place. He cursed other musicians, such as Richard Strauss, who continued to perform in Italy after 1933: "To Strauss the composer I take off my hat; to Strauss the man I put it back on again."[70] To a friend he said, "If I were capable of killing a man, I would kill Mussolini."[71]

Exiled from not only from Bayreuth, Salzburg, and now his own country, Arturo conducted with the Vienna Philharmonic in Vienna and the BBC Symphony in London, and what is now known as the Israel Philharmonic in Tel-Aviv, Jerusalem, and Haifa, even traveling with this company to perform in Alexandra and Cairo.

That said, the maestro was limited in where he could go and was at a crossroads of sorts in his career. In 1937 the NBC Symphony Orchestra was created for him, and he returned to the states to embrace this opportunity. The acoustics of the custom-built studio were a bit dry; however, remodeling in 1939 added reverberation to the space now used for NBC's *Saturday Night Live*. During our 1993 interview, Walfredo recalled, "NBC had the idea of promoting classical music as well as sales of radios and recordings. So it wasn't about giving concerts for audiences in the hall, but rather for the millions of people listening at home...no matter how long (Toscanini's) career had been up until that point, in one day more people would hear his concerts than in the past fifty years of his career. So I think that really appealed to him."

Though criticized for ignoring American music, in 1945 Arturo led the orchestra in recording sessions of Gershwin's *An American in Paris* and *Rhapsody in Blue* with soloists Earl Wild and Benny Goodman, as well as Copland's *El Salon Mexico*; he even penned his own orchestral arrangement of "The Star Spangled Banner," which was later incorporated into NBC performances of Verdi's *Hymn of the Nations.*

Toscanini the man has been called incredibly private for such a well known figure; he left no memoir and refused most interviews, and it wasn't until Harvey Sachs' stunner *The Letters of Arturo Toscanini* hit the stands in 2002 that much of the man was revealed. These letters showcased his fantastic energy and complexity: his perfectionism, his quick temper, and boundless sexuality. Some of the most eye-popping letters, both in content and number, are to Ada Mainadi, wife of a well-known cellist. This was when he was in his seventies and she was half his age, and counts as one of untold numbers of liaisons. Keep in mind he was married with four children (though his son Giorgio passed at age five.) His grandson Walfredo put it this way: "he cast his nets wide." Though apparently not a Maestro in the love letter department, he still wrote thousands of them, a fair number to the singer Jane Lawrence Smith, a woman he had never met, but had only spoken to over the telephone.

His conquests read like a punch list of late nineteenth, early twentieth century sopranos including Lucretia Bori, Rosina Storchio, and Geraldine Farrar. Though his wife Carla intercepted a few of these steaming missives, after a while she gave up doing anything about the situation, though the letters to Mainardi take up more than two hundred pages of Sach's book. In them he begs for clippings of her pubic hair ("tiny flowers")[72] and a handkerchief stained with her menstrual blood. In his early seventies the letters devolved mostly into speculation about amorous rendezvous instead of the meetings themselves. He would dash them off after performances as if they were excess creative energy he needed to release. In notes to various lovers he often compared sex to music: "Music has the same effect on me that you have."[73]

But there is no summarizing the man. Did drama come to him or did he create it in life or on the podium for his own enjoyment? "My God, what a life!" he wrote in 1936. "And to think that people envy me! They see nothing but the exterior, which glitters in appearance, but a person's interior, soul, heart? What unknown, unexplored things they are!'"[74] As Michael Kimmelman intuited in his review of *The Letters of Arturo Toscanini*: "You sense him playing a role, the Great Maestro. Life seems to have been a Puccini opera for Toscanini...his ego clearly thrived on excess."[75]

At the same time, he was known for his generosity: he left instructions with Clara, to aid, financially, any member of the La Scala orchestra who approached her, no explanations needed. In 1936 he paid his own expenses for a trip to Palestine to conduct an inaugural concert. The musicians consisted of Jewish refugees from Central Europe, a group that ultimately became the Israel Philharmonic.

Walfredo's recollections in our interview revealed yet another side to the man: "I knew he was (famous) because he had a big Cadillac and a chauffeur, and he had a big house with a marble staircase in Milan, but I didn't know what he did until I went to Salzburg and heard him perform there. So, to me, a seven-year-old kid, he was just this nice gentleman who had these beautiful villas and was very interested in me. The funny part is that no one had seen that side of him. Here he was: this stern and forbidding figure, and suddenly there was this little boy in his arms; and he was just my grandfather, and we were having fun. He had a very relaxed side to him not many people saw."

Though even at family events, his controlling nature came to the fore at odd times: "We'd have these big meals (after a performance) in Riverdale: sometimes eighteen to twenty-four people, sitting down to a very sumptuous meal. But Grandfather himself ate very little: just some soup and a tiny piece of veal, something like that. He'd look at everyone else's plates and say 'How can you eat all that?' At first the women thought it was a great honor to sit next to him, until they discovered they couldn't enjoy their meal because of the constant criticism."

Like many of his contemporaries, Toscanini never took to Schoenberg or Mahler; however, he was ahead of his time in admiring Debussy. He also instigated changes in the physical presentation of symphonic music that would be taken for granted today: he insisted on dimming the house lights during performances, and that audiences refrain from eating, talking, playing cards, and wandering around. Theater in the early 1900s was more a casual gathering place than anything else. He also insisted that musicians adhere to the score, which sounds rather obvious, but his letters tell of a rampant sort of sloppiness that was the rule rather than the exception at the time. His critics have said he pushed things too far the other way: that his performances felt metronomic and too brisk at times.

Another huge innovation Toscanini championed was the creation of the orchestra pit; in fact he installed the first one at La Scala in 1907. Before this time the orchestra played at the main floor level. His biographer Harvey Sachs wrote, "He believed that a performance could not be artistically successful unless unity of intention was first established among all the components: singers, orchestra, chorus, staging, sets and costumes."[76]

Walfredo shared his thoughts about the lowering of the pit: "It was Wagner's idea that there should be a mystic gulf between the audience and the stage and not hundreds of people sawing away or blowing at instruments, and a man waving his hands. So, he was happiest when the orchestra and the conductor didn't exist, but the sound and music and spectacle did. When (my grandfather) caused the pit at La Scala to be lowered, he was concerned that he would be perceived as a prima donna, or that people would think this was a foolish expense. In fact there was a cartoon showing members of the orchestra being lowered into a pit, because Toscanini wanted it down, as far as it could go."

Much has also been said about the differences between Arturo at rehearsal screaming profanities at his musicians in a perhaps consciously theatrical way of extracting "perfection," in contrast to his still passionate but controlled presence during performances. Long stretches of Toscanini on film via NBC cameras captured the piano-wire tight tension between conductor and musician. Walfredo noted that during rehearsal, his grandfather would "make swoops down almost to his feet with his baton...but in concert he limited his motion because (he thought) the audience should be listening to the music, not watching the conductor." Part of what made Toscanini so modern, though, was his constant propulsion even through slower passages, while never neglecting the more subtle layers of the score. His melodramatic flourishes were evident when he left the stage in 1926 during the premiere of *Turandot* precisely at the point Puccini left off due to his death. The maestro simply stated, "Here, Death triumphed over art,"[77] and left the pit. In moments the lights came up and the hushed audience filed from the hall.

Walfredo recalled a few highlights of one of his grandfather's final tours through the states: "After playing something like four concerts a week—this is after full rehearsals—the next morning he'd be awake and ready to tour whatever city we were in: New Orleans, Houston, San Francisco...and I'm telling you we were dead tired but he was always energized by his ability to lead the orchestra and make music for people, many of whom had never seen him in action. It was a triumphal tour for him...and an astonishing thing for me to experience first hand not only his intellectual and musical talent, but his physical resistance and strength as well. It's a little like Nolan Ryan pitching for twenty-seven years. Grandfather pitched for sixty-eight years on the podium, and his intensity never diminished."

I asked Walfredo about his grandfather's pet canaries, a story I'd heard only bits and pieces of before: "He had this enormous cage of canaries, maybe eighteen or twenty of them in there, and he'd set it near the radio and turn it on. In a few minutes they'd perk up and start to sing in sympathy with the music. He loved to see that: these birds just chirping and cooing, flapping their wings with excitement. He told me that he felt that if people heard the best music—the best compositions of his beloved composers played in the best way, such as with the NBC orchestra—then maybe they would feel it too, and they'd be moved to sing along as well."

## SUGGESTED RECORDINGS

**Toscanini Conducts Favorite Overtures Including Works by Mozart, Weber, Verdi, Rossini, and Other Composers**
RCA CD 09026 60310-2

**La Boheme by Puccini**

Licia Albanese, Jan Peerce, and Francesco Valentino are featured in the cast of this historic 1946 broadcast with the NBC Symphony Orchestra. Toscanini conducted the world premiere at Teatro Regio in Turin 50 years earlier on February 1, 1896.
RCA CD 60288-2-RG

**Tchaikovsky: Symphony No. 6 "Pathetique"**

Toscanini conducts The Philadelphia Orchestra in a performance recorded in 1942.
RCA 60312-2-RG

**The Pines of Rome, The Fountains of Rome, and Roman Festivals by Ottorinio Respighi**

Arturo Toscanini conducts the NBC Symphony Orchestra
RCA CD 60262-2-RG

**Symphony No. 4 " Italian" and Symphony No. 5 "Reformation" by Felix Mendelssohn**

Arturo Toscanini from a broadcast with the NBC Symphony Orchestra in 1953 and 1954.
RCA CD 60284-2-RG

# "Where the Word Ends"

Author's collection.

Gunther Schuller, Joan Kennedy, and me

A man of endless musical pursuits, Gunther Schuller is also a joy to not only interview but to spend an evening with over wine and good food. I'm either laughing or learning with him, many times both at once.

Born in New York City in 1925, the son of a violinist with the New York Philharmonic, Gunther is the sort of wunderkind you might imagine composing a score in diapers. The fact is, except for an episode his dad recalls where five-year-old Gunther sat in his bathtub splashing his rubber ducky while humming *Tannhauser*, the musical side of the young composer's brain didn't really light up until a bit later. At age eleven, he became obsessed with a toy xylophone his younger brother Edgar was given for Christmas and composed his first score: thirty bars for xylophone, flute, violin, and piano. He still has it in a notebook somewhere. Some years later he was, as he put it: "bowled over by the *Rite of Spring* in *Fantasia*. For two or three weeks after seeing that I wasn't myself. That [movie] transformed my life, turned me into a musician."

Once the floodgates opened, Gunther worked at nothing but music for up to nineteen hours a day, many times forgetting to eat. He devoured music by Beethoven, Schoenberg, Ravel, and Stravinsky. But when he discovered Duke Ellington and told his father that jazz was just as great as the classical music he'd been listening to, his dad nearly fell off his chair. Statements like these at the time were practically heretical, and presaged Gunther's rebellious, independent, ultimately groundbreaking style of thinking about music.

He began his professional life as a sixteen-year-old principal horn player in the Cincinnati Symphony in 1943, later holding that position with the Metropolitan Opera Orchestra. Of that era, he said: "I feel so fortunate to have been in on that whole decade from 1945 to the beginnings of Bebop, and into the modern jazz period. Those fifteen years were spectacular. Not only were Mario Del Monaco and Lily Pons at the Met, but all I had to do was walk five minutes up Broadway to find at least twelve Jazz clubs, starting with the Aquarium on 52nd Street...all the way up to Birdland, Bop City, and Basin Street East. I met Dizzy in 1948 at the Aquarium, then Ray Brown, John Lewis, and J. J. Johnson. In fact that's how I got to play with Miles Davis—they ran out of horn players."

At this point, equally at ease working with Miles as with Toscanini, the young Gunther was invited by Davis to record sessions that evolved into the classic album, *The Birth of the Cool*. Meanwhile, back at the Metropolitan Opera Orchestra (where Gunther stayed until 1959,) his enthusiasm for jazz was met with a bit less excitement. He told me: "Bob Boyd used to play jazz during intermission at the Met, and it used to infuriate the players who thought jazz was disgusting. We were the young Turks."

When the director of the New York Philharmonic broadcast Schuller's "Music for Brass" in 1956, the floodgates opened. Letters arrived from Aaron Copland, Samuel Barber, and others, all welcoming him into an elite musical society. In the fifties he began conducting a range of contemporary works—including his own—at most of the major symphony orchestras in the world. His term "Third Stream," coined while lecturing at Brandeis in 1957, refers to the "totally logical" stylistic marriage of jazz and classical music, which he developed further while working with the Modern Jazz Quartet and John Lewis. To the critics he said, "'My God, [here are] these two great musics and they are in separate camps—they don't talk to each other, they hate each other, they vilify each other. We've got to get these two musics together.'"[78] Ran Blake, an improvisation-based pianist who became Gunther's student, pointed out that though the term "Third Stream" was new, musical examples such as *Porgy and Bess* were not uncommon. Together, Schuller and Blake created the Third Stream Department at the New England Conservatory. Original jazz compositions "Teardrop" and "Jumpin' in the Future" epitomize the Third Stream approach, blending classical structural sophistication with the swing and fluidity of jazz.

By the early sixties, Schuller had turned away from performance to devote himself to teaching, writing, and composition. An influential and elegant educator, Gunther has been on the faculties of the Manhattan School of Music and Yale University where he succeeded Aaron Copland; Director of the Tanglewood Music Center; and President of the New England Conservatory. In the late seventies he started GunMar and Margun music publishing companies (now part of G. Schirmer/Music Sales/AMP) and the GM Recordings label. The author of more than 180 works ranging from solos to concertos, symphonies, opera, and other works that simply don't fall neatly into any genre, Gunther continues to keep a grueling schedule even well into his eighties.

"Spectra," an orchestral work commissioned in 1958, rearranged musical space by dividing the larger orchestra into smaller chamber groups; at the same time drawing out each instrument's character in such a way that hadn't been considered before. "An Arc Ascending" (1996) was inspired by the photographs of Alice Weston. Other compositions drew on Impressionist and late Romantic tone poems of Debussy and Schoenberg. "Of Reminiscences and Reflections," one of Schuller's two memorials to Marjorie Black, his wife of forty-nine years, won him the Pulitzer Prize in Music in 1994.

Gunther also composed a stunning variety of concertos for solo or small ensemble with orchestra, bringing to the forefront less commonly championed instruments such as contrabassoon, organ, double bass, and alto saxophone. His "Grand Concerto for Percussion and Keyboards" uses more than one hundred percussion instruments. Schuller's two operas include *The Visitation* (1966) based on a Kafka story and an hour-long 1970 children's opera, *The Fisherman and His Wife,* that features a libretto by John Updike, drawn from a Grimm fairy tale.

Schuller's chamber music also embraces traditional and non-traditional trends, from four string quintets, brass and woodwind quintets, to works for solo instrument or voice with piano, and mixed ensemble pieces.

In February of 2009, the Boston Symphony Orchestra and James Levine premiered the very well received "Where the Word Ends," a single movement, twenty-five minute piece, which was also heard on one of our live BSO broadcasts. For someone with little patience for lofty titles, Gunther was especially proud of this one. He said, "[Composers] write a piece and slap some title on it and it's usually something like 'The Moon is Drifting Under the Bridge' or 'The Trees are Falling to the Left'...I'm just not into that. Regardless of whether my music is any good, I think the title is really one of the most beautiful definitions of music itself.'"[79] He shared with me the fact that the composition came to him in a rush—that he couldn't have spent more than twenty or thirty hours on it. "It was almost eerie...Maybe it's ancient age, but for 'Where the Word Ends' I almost didn't have to think about what the next note would be. It was like an improvisation.'"[80] Considered by many to be his best work, *Globe* jazz writer Bob Blumenthal called the piece a "time capsule" of his career, noting echoes of Stravinsky as well as Miles Davis.

Schuller's brilliance exists on many planes, but most strikingly, in my opinion, in his utter openness to the ultimate harmonic marriage of musical styles, philosophies, even instruments. His arrangements of classic jazz, standards, and ragtime music by Paul Whiteman, Duke Ellington, Jelly Roll Morton, and Dizzy Gillespie mesh with his realizations of music by Charles Ives and Thomas Tallis. The author of two books on jazz, Gunther is also the recipient of the MacArthur Foundation's genius award in 1991, the William Schuman Award in 1988 and 1993, Downbeat's Lifetime achievement award for his contribution to jazz, and two Grammys.

We spoke about the process behind writing *The Swing Era*, just how much research was involved and a few of the secrets revealed in the process: "It became a humungous book. Well over a thousand pages, dealing with that era from 1930 to about 1948. The index alone has four thousand words, and something like five hundred musical examples, all of which I transcribed from recordings. All in all I listened to thirty-five thousand records writing this book. I made it my philosophy that if there was anybody I was going to mention in the book, I was going to listen to every recording (they made) that I could lay my hands on. My son George ferreted out all these recordings, and, of course, I begged, borrowed, stole, bought, and copied the rest, so now we have a pretty darn good collection."

Gunther also discovered, through painstaking cataloging of the Ellington manuscripts at the Smithsonian, that a lot of the solos by Johnny Hodges, Harry Carney, and others were actually compositions written by Ellington himself. Gunther explained, "We always marveled about the high integrity of the composition of these solos, but assumed they were improvised out of plain cloth. They were not! Of course they were embellished in a free or personal way, but there they were, right there in Duke's own handwriting." Other surprises included a new appreciation for John Nesbitt, a jazz trumpet player who influenced, ultimately, Benny Goodman and the swing era. Only by actually listening to these countless recordings could he have heard this influence and been able to posit a different place in history for Nesbitt. Cab Calloway was "treated kind of negligibly by jazz historians because he was a comedian...but when you really listen, his singing is so exquisite...he could sing anything from bass to soprano; he had something like five voices in him."

These days Gunther's life is almost exclusively focused on music, though he used to travel the world with his wife Marjorie Black. He still relishes spending time with his two sons: jazz percussionist George Schuller and bassist Ed Schuller. Two young women he calls his secretaries take care of quotidian concerns like groceries and trips to the bank.

Gunther spoke to me a bit about music then and now: "Anything we wrote in those early days, no matter how far out it was, it still had to be danceable. That's all changed now of course. So, it's a different feel, but I love that old feeling. I mean, I'm certainly an avant-gardist in many ways, but there's something to preserving that sense of swing and that style. I don't think we should ever lose that. And I know that's something you agree with, Ron."...and of course he's right about that!

"Where are we headed? Change is exploding all around us, but certain principles remain. Synthesizers are just instruments and not inherently creative; though they're a whole world of sound, they don't guarantee quality. The older I get the more I realize how much there is to do, to learn, how much more music there is to appreciate and understand, especially the ethnic, world music, with record companies catching on. I love the cross fertilization between world musics and jazz and popular music. It's all seething and bubbling...really, who knows what will happen next?"

## SUGGESTED RECORDINGS

### Happy Feet: A Tribute to Paul Whiteman

The New England Jazz Repertory Orchestra conducted by Gunther Schuller, GM
3048

I love this recording! I wore out my LP copy, so I'm delighted to have it released
on CD. Whiteman was one of the great figures in American Music, and
this collection lets us hear why. There's even a cameo appearance by the
legendary jazz violinist Joe Venuti.

### Gunther Schuller: Orchestra Works

A collection of some of the best orchestral writing of the twentieth century
including Schuller's most popular work, "Seven Sketches on Themes of
Paul Klee."

GM 2059

### Jazz Compositions and Arrangements by Gunther Schuller

A must for jazz lovers that showcases masterpieces from the "Cool Jazz Era."

GM 3010

# Wild About Harry!

Harry Ellis Dickson, Leslie Warshaw, and me

Author's collection.

One of the most joyful occasions during my years at WGBH was in 1998, when we celebrated Boston's beloved conductor Harry Ellis Dickson's ninetieth birthday. After sharing some of his wonderful stories during a guest appearance on *Classics in the Morning,* we enjoyed a huge birthday cake, provided by my then-producer Leslie Warshaw. The greeting on the cake expressed the feelings of everyone who ever came in contact with this charismatic and dynamic gentleman of music: "We're All Wild About Harry!"

Born in Cambridge, Massachusetts, in 1908 to Russian immigrant parents, Harry wasn't at first considered the gifted child in the family. He told me: "It was my sister who got the lessons. She had no interest in the violin, but I couldn't get enough of it. I'd sit in during lessons and when the teacher asked a question, I'd answer. Finally he told my mother he'd give me lessons for nothing, so she took the fiddle from my sister and handed it to me. I was six years old."

From then on young Harry barely put the instrument down: "I had a recording of Mischa Elman playing Massenet's *Elegie,* and I played it over and over and over again on our Victrola. I thought if I could ever make a sound like that, my life would be complete. That's when I fell in love with the violin. It was a glorious era of violin playing: I remember especially Jascha Heifitz and Fritz Kriester. Isaac Stern and Itzhak Perlman weren't yet on the scene."

Dickson went on to study at the New England Conservatory, later living in Berlin from 1931 to 1934 where he studied at the Hochschule fuer Musik. By the age of fourteen, he was a professional musician and soon went on to play under such storied conductors as Serge Koussevitzky, Pierre Monteux, Charles Munch, William Steinberg, Eric Leinsdorf, and Seiji Ozawa, as well as meet and study with Bartok and Sibelius.

Beginning in 1938, he played violin with the Boston Symphony Orchestra for forty-nine years, regretfully retiring just before the fifty-year mark. Over the years he took part in several historic events in Symphony Hall, such as the world premiere of Bartok's "Concerto for Orchestra" in 1944 and some of the first years at Tanglewood, the summer home of the orchestra.

Though few can boast of such a superior musical resume, Harry the man was also charming, self-deprecating, and a lot of fun to be around. I always considered him "Maestro" and introduced him as such on the air, but he winced a bit each time, saying, "I'm just Harry, okay?" In 1999 after being named music director laureate of the BSO, he said of his time there: "I've been around so long that all the statues in Symphony Hall were little boys when I started."[81] Harry once told me of a musician he played with in his early years who himself was as old as Methuselah and had actually gone out on a bar crawl with Brahms!

In 1955 Harry began another decades-long musical relationship: this time with the Boston Pops, where he stepped in as assistant conductor for Arthur Fielder. This partnership lasted forty-four years and inspired Dickson's 1984 book, *Arthur Fiedler and the Boston Pops: An Irreverent Memoir*. In one interview Harry said, "I was as close to Arthur Fiedler as anybody ever got, and that wasn't very close."[82]

One evening, Harry observed an interaction between Fiedler and Dizzy Gillespie, probably the most polar opposite musicians in the world. After handing Gillespie the score, Fiedler began to conduct as Dizzy started to float away from the notes on the page, as usual. Fiedler said, "What are you *doing?*" Dizzy said, "Man, you just keep going, never mind about me." So they started again, and Dizzy spun off somewhere into the stratosphere while Fiedler gave him one dirty look after the other, but Dizzy was in his own world.

Harry said, "Later I wondered if any of Dizzy's band read music. I asked his bass player, who was studying the score very intently, if he did. He said, 'Not enough for it to bother me.'"

The story I remember best was when Dizzy put his funky, used handkerchief on Fiedler's podium. Eyeing it as if it were a dead rat, Fiedler knocked it off with his baton. Dizzy put it back. Fiedler knocked it off. Finally Dizzy gave in and held on to it for the duration of his performance.

Harry elaborated about life with Fielder: "Over the years, I wondered how the public could so ceaselessly love Arthur. I realize now it was more than his showmanship, more than the music. Arthur's image was lovable; he had a wolfishness about him, but it was a vulnerable wolfishness. Beyond that, little else mattered to the public about his private life. His eccentricities merely added to the fun."

Speaking of eccentricities, there was always the fire-chasing thing. When not involved in music, Fiedler spent most of his spare time as a "spark," one who learns about fires and races off to witness them. At the time of his death, he had one of the largest collections of honorary fire chief hats of anyone in the country.

Several years ago I attended the dedication of the "Harry Ellis Dickson Park" located on Westland Avenue and not far from Symphony Hall's stage door. It was a gala occasion with Harry's son-in-law, former Governor Michael Dukakis, his daughters Kitty and Ginny in attendance, along with many other dignitaries. When it came time for Harry to speak, the wailing sirens of fire engines speeding by drowned him out. Harry commented, "Just like Arthur to try to upstage me again...but that's OK, he's got a statue on the Esplanade, and now I've got a park next to Symphony Hall."

Appearing with Harry on numerous occasions when he was guest conductor of the Boston Pops afforded me the pleasure of getting to know him even better. Some of my best recollections come from our New Year's Eve celebrations. In addition to hosting the evening, one of my jobs was to count down the last minute to midnight with the audience. Once, during these last dramatic seconds of the year, the huge clock above the stage stopped, cold. Somebody had accidently pulled the plug with seven seconds to go. Harry covered up the goof and we kept on counting down. Afterwards, he had a bit of a meltdown backstage. "This guy has one job to do all year and he screws it up! Imagine...no clock for New Year's Eve...what's *wrong* with these people!" But Harry was not one to hold a grudge; soon we were toasting each other with champagne and welcoming in the new year with our guests.

In 1959 Harry founded the Boston Symphony Youth Concerts. He talked a little about the earliest days: "In the late fifties we were doing some young people's concerts in Brookline with about twenty-five or thirty musicians from the BSO, and they'd been going well. My wife and I tossed around the idea of starting a few in the hall itself; pretty soon they just took it over. We used to have auditions from high schools—just a few would apply—soon we had hundreds of applicants. Ninety-five percent of those we've chosen have gone on to become professional musicians.

"Before our youth concerts there were sporadic ones put on by the BSO, even Koussevitzsky did a few. Wallace Goodrich of the New England Conservatory narrated some, but they were pretty stultifying. Wheeler Beckett's were too long. I don't believe there's special music for young people—you can play anything for them—in fact what their parents may feel is too far out they take in stride. The only concession I make is a listening span; they'll just tune out if it's too long."

Even by this 1989 interview, Harry estimated these youth concerts had reached more than three-quarters of a million youngsters. His hopes were not so much that they become professional musicians, but that they become, as Aaron Copland put it, "talented listeners."

In addition to his consummate musicianship both as violinist and conductor, he authored three very irreverent, entertaining books. His witty anecdotes—some hilarious, some serious—illustrate his days working with everyone from Danny

Kaye to Fiedler, Serge Koussevitzky, Seiji Ozawa, John Williams, and Igor Stravinsky. There are also moving reflections on his wife Jane and daughters Kitty and Ginny, and the effects on his family of the failed presidential campaign of his son-in-law Michael Dukakis. Of Dickson, Dukakis once said, "No one on this planet could wish for a better father-in-law."[83]

In *Gentlemen, More Dolce Please*, he described the era of Koussevitzky as an exciting but turbulent one. Dickson couldn't remember a concert under him when in the end each player wasn't soaking wet and as emotionally spent as the conductor. "Gentlemen," Koussevitzky once explained, "You play all the time the wrong notes not in time! And please, made important, you play like it is something nothing!"[84] To a string player he raged, "How can you play with died fingers?"[85] Another story involved an older woman who said to him after a concert, "'Dr. Koussevitzky, to us you are God,' and he answered, 'I know my responsibility.'"[86] Although Koussevitzky murdered the English language, he became, according to Harry, the conductor by whom all others would be judged.

Of all the many honors bestowed on him, Harry was especially proud of being awarded the title of Chevalier of Arts and Letters by the French government. His longevity also amused him; when asked his age he would say: "I've been through four ages in my life: childhood, youth, middle age, and *you look marvelous!*" His youth concert series has been the pattern for many others and continues to change the lives of millions of children and young adults. Other honors include the Harry Ellis Dickson Center of Fine Arts and Humanities in the Winter Hill Community School, and the Harry Ellis Dickson Orchestral Suite at Madison Park High School in 1983, as well as a scholarship under his name at the Boston Arts Academy.

These days when I think of Harry, I remember a Fourth of July many years ago. I'd left the studio late that afternoon and treated myself to a long walk along the Charles River toward my home in the South End. It was a perfect summer afternoon, hot but dry, and the sun seemed to backlight everything with an orange glow. Soon I was joined by hundreds of people also walking along the banks of the river, young and old all making their way with blankets and picnic baskets as we listened to the oddly pleasing sounds of the Pops warming up. I couldn't believe how many boats were in the river; it looked like you could walk across it just by jumping from boat to boat. As I reached the Esplanade, I could just make out the tiny figure of my friend lifting his arms in welcome as the crowd, now stretching for miles, broke into joyous applause.

## SUGGESTED RECORDINGS

Harry Ellis Dickson, a member of the Boston Symphony for some fifty years, participated in just about every recording session with the BSO and the Boston Pops. He conducted hundreds of "Pops" concerts at Symphony Hall, the Esplanade, and on tour.

## SUGGESTED READINGS

*Gentlemen, More Dolce Please: An Irreverent Memoir of Thirty-five Years in the Boston Symphony Orchestra*

> Published in 1969 by Beacon Press
> Drawings by Olga Koussevitzsky

*Arthur Fielder and the Boston Pops*

> Published by Houghton Mifflin Company Boston, 1981
> Introduction by Ellen Bottomley Fielder
> Epilogue by John Williams

*Harry Ellis Dickson ... Beating Time, A Musician's Memoir*

> Published by Northeastern University Press, 1995
> Forward by John Williams

# Grandfather of Film Music

David Raksin

© Photofest, Inc.

When asked to name his favorite song, Frank Sinatra said, "Laura," without hesitation. A gentleman known as the "Grandfather of Film Music," David Raksin, composed the song for the 1944 movie of the same name.

Born in Philadelphia in 1912, David was raised in a musical family. His father, Isador, played clarinet in the Philadelphia Orchestra and composed and conducted music for silent films and vaudeville. By age six David was at the piano; soon after his dad gave him a clarinet and instructed him on wind instruments. Twelve-year-old Raksin not only had his own dance band, but played in all the dance bands in his high school. He worked his way through UPenn, teaching himself composition and playing in radio orchestras, then moved to New York City where he discovered Broadway and began arranging for record companies.

In the thirties David headed to Hollywood where he studied composition with Arnold Schoenberg, a man of genius but not known for patience. Once Raksin made the error of asking how to compose music for an airplane sequence. Schoenberg replied, "Like music for *big bees*, only *louder*."[87] David also shared with me the infamous Raksin/Hitchcock exchange: Hitchcock didn't want music for lost-at-sea drama *Lifeboat* because he worried audiences would wonder where the music was coming from in the middle of the ocean. Raksin said to pass this on: "Ask Hitch where the cameras are coming from."[88]

He summed up his early days in our 1984 interview: "I got involved with Howard Lanin and his band, but for a while things were slim. I nearly starved to death. Finally, Roger Wolfe Kahn and Al Goodman heard my arrangements. We had Artie Shaw on sax, Tommy Dorsey on second trombone. Gershwin heard me and put me on staff. I think I was nineteen at the time."

Impressed by the ingenuity of Raksin's arrangement of "I Got Rhythm" for Jay Savitt's band, composer Oscar Levant contacted his friend George Gershwin and suggested they meet. Not at all thrown by the gravity of having lunch with Gershwin, Raksin said, "He recognized that I was a colleague who knew what I was doing. He played 'Love Walked In' for me. Didn't sing, just played the melody."

Gershwin recommended Raksin to Harms Music, who drafted him to be Charlie Chaplin's assistant on the landmark 1935 film, *Modern Times*. Notoriously tough on musical partners, Chaplin never had the same assistant two movies in a row. He was also a "hummer," that is, he was unable to transcribe his scores by himself. Chaplin hummed, tapped, and whistled his musical ideas, including the three-note blast of the factory. Unfazed by Chaplin, twenty-three-year old Raksin voiced his own musical opinions. Chaplin fired him, then rehired him after Alfred Newman, a renowned film scorer, told him he was nuts to give him the axe. The score took four months to complete and included the song "Smile" for which Chaplin received total credit. As Raksin said, "There wasn't any point in making a big deal out of it; that's just the way things were back then."

In fact for his work on forty-eight films through the 1930s at Universal Studios, Columbia Pictures, and Twentieth Century Fox, Raksin received not a single credit, since that was reserved at that time for the studio's music department head. Because of his association with Schoenberg, Raksin's music was considered avant-garde by the studio system, and he was commonly relegated to low budget horror flicks such as *The Undying Monster*.

One brief collaboration was with Igor Stravinsky, whose "Circus Polka" Raksin arranged for the elephant ballet. Choreographed by Ballanchine, it was staged by Ringling Brothers and Barnum & Bailey. The troupe bellowed and stampeded on hearing Raksin's work, which he and Stravinsky interpreted as a big thumbs up.

By the time I interviewed him in September 1984, David had well over a hundred film scores and three hundred television scores to his credit. He told the story of how he landed the job composing the music for *Laura*, one of the all-time great film classics. Studio scuttlebutt at the time was that the movie was troubled, "and everyone wants to stay away from a picture with problems because you become a part of those problems." So when director Otto Preminger asked both Alfred Newman and Bernard Herrmann to compose the score, they turned him down, and Raksin was next in line. The very next day Preminger announced he wanted to use Duke Ellington's "Sophisticated Lady" for the film. Raksin told me, "I saw the picture and fell in love with it. I was horrified by the 'Sophisticated Lady' choice. When I told him it wasn't right for the picture, Otto and I had a big blowout. You don't need a tune that already has a past. You need a tune special for your picture." The argument was finally settled when Preminger said, 'Well alright, it's Friday. Show up Monday with something else or we use 'Sophisticated Lady.'"

Raksin continued, "I went home, and it's sort of a corny story, but as a result of receiving a letter from a lady who told me to get lost, I was suddenly improvising at the piano and out came this tune. I'd spent the whole weekend sweating, nothing coming out of me until I read that letter. Lose a lady, gain a tune, was one way of looking at it."

I asked David if he realized what sort of worldwide impact and reception "Laura" would receive even as he was writing it. He said, "I'm ashamed to say I predicted something like it. When you hit something that's really you, you feel it, you know it. Of course that's common, that delusion, for songwriters and composers. But hell, we stayed on the hit parade for twelve weeks."

Raksin told me that when he played it for Preminger, Otto loved it right away. David got to choose his own lyricist, so he chose the best: Johnny Mercer. In the first week after the movie's release he received over seventeen hundred letters.

I listen to "Laura" and I hear a sexy, eerie, beautiful tune. Raksin told me he wrote the entire score around a single haunting melody that he never used in its entirety, creating a sense of yearning; as he said, "the ephemeral girl and the interrupted melody."[89] The idea was to highlight the lingering impact a murdered woman had on those lives she touched while living and breathing. Even Cole Porter said it was the tune he would have most liked to have written. Hedy Lamarr, among those who turned down the script for *Laura*, remarked, "They sent me the script, not the score."[90]

With the composer sitting in the studio in front of me, I played three back-to-back versions of "Laura" (out of four hundred which exist!) first with the composer conducting the New Philharmonic Orchestra, followed by Vic Damone in a vocal version with lyrics by Johnny Mercer, and finally a jazz version by Charlie Parker with strings. Each arrangement evoked a slightly different mood.

David said "Laura" was his biggest breakthrough as a Hollywood composer because it freed him from being typecast for "B" movies. On the other hand, he said, "I actually loved doing those movies because I was never under the gun to write pretty music, and I could compose things that were at the time rather avant-garde."

His score for the movie *The Bad and The Beautiful* almost never happened. At the time he was at MGM with a great production team and a marvelous cast that included Kirk Douglas, Lana Turner, Dick Powell, and Walter Pigeon. Raksin told me Producer John Houseman "kept saying, 'we need a siren song!' But it was another weekend thing, so I was down for the count. I knew what I needed to do: create music that would persuade people that this ruthless studio bigwig was the real McCoy to people he subsequently betrayed—this music needed *charisma*. Anyway, I made a demo of the theme, played it for Vincente Minelli and John. A shrug. Lukewarm reception, to say the least. The two people who saved that tune, championed it, kept it from getting thrown out were Bette Comden and Adolf Green (screenwriters for *Singing in the Rain*) who loved it. They kept saying play it again, and again, until finally, the whole mood changed in that room."

When I asked him if he harbored nostalgia for this period in film making, he said, "This was a great time for the movies, but I don't feel nostalgic about it. Nostalgia is a word that has fallen into evil hands; it's a word exploited by marketers. I prefer to think that if something had merit, and if there are people around who have the ears and eyes to understand it, it will be understood and appreciated.

"The thing I love best about that period is the wonderful time we had. Sounds crazy but we had a ball working under such duress. We'd spend all night working, come in the next day and there was something marvelous to show for it."

He shared his thoughts on altering original scores: "Jerome Kern never liked tunes to be messed with, neither did Dick Rogers. I don't mind, especially if they improve on it. A tune is a vehicle, a mode of self-expression. Actually, I'm amazed Charlie Parker stuck so close to the tune, very unusual for him. For me, as long as tune is in there somewhere, I'm okay with it."

David never won an Oscar, though he did receive two Academy Awards nominations, for *Forever Amber* in 1947 and *Separate Tables* in 1958. He also composed film scores for *Carrie, The Force of Evil, The Secret Life of Walter Mitty, Too Late Blues, The Redeemer,* and *The Day After* in 1984. His three-year radio series, "The Subject Is Film Music," was hosted by NPR and is held by the Library of Congress, which has proclaimed the series "the finest oral history of the (film composing) profession."[91]

Themes for television included *Wagon Train, Ben Casey,* and *Medical Center.* David served from 1962-70 as president of the Composers and Lyricists Guild of America, but his later years were spent teaching composition at the University of Southern California and composing several concert pieces ultimately performed by the New York Philharmonic, the Boston Pops, and the London Symphony. He died on August 8, 2004, in Los Angeles at the age of 92. I was glad to learn that he had completed his memoirs, *If I Say So Myself,* before his passing.

Having David Raksin as a guest certainly rates as one of the highlights in my radio career. He was a warm, witty, and humble man who made an immense contribution to the world of film music.

I cherish his final comments from our interview: "Jerome Kern originally turned me on to composing. I was just a kid hanging out with my dad and they played 'Look for the Silver Lining.' And that was my inspiration. I thought if someday I could compose a piece that does for people what that song does, I will have lived a life worth living. The only other thing I learned later was necessary was to be in love and have children, so when you've done all three you've made it."

## SUGGESTED RECORDINGS

### David Raksin Conducts the National Philharmonic Orchestra in His Film Music

This is the CD I played during my David Raksin interview from *Laura, Forever Amber,* and *The Bad and the Beautiful* RCA disc

The entire film score for *The Bad and the Beautiful* has been released on a Rhino CD. Hard to find, but well worth having.

# Musical Wunderkind

André Previn

Photo by A.P. Mutter. Reprinted courtesy of IMG Artists.

André Previn is a musical polymath: the rare artist who has "successfully made the transition from the podium to the pen."[92] In over sixty years of his professional life he's been a pianist, conductor, composer, and pedagogue. He's written for Broadway, concert and opera houses, led most of the world's major orchestras, won Oscars for film scores, and been knighted by the Queen of England. I had the privilege of only one interview with him; it seemed I could have done twenty more and just scratched the surface of his life and accomplishments.

Most sources state that Previn was born in Berlin, Germany, to a Jewish Russian family in 1929; however, he states the year was 1930. Birth records were lost when his family left Germany in 1938 to escape Nazi persecution. He moved to Hollywood with his family, where as a ten-year-old he steeped himself in film via ten-cent tickets at Graumann's Chinese Theatre down the street.

After an upbringing of classical music study, teenaged Previn fell in love with jazz, emulating and eventually playing with Dizzy Gillespie and Charlie Parker. As a sixteen-year-old he was already composing and arranging music for MGM, bringing to life some of the great musicals: *Kiss Me Kate*, *Gigi*, and *Silk Stockings*. Stage-to-film adaptations included *Kismet*, *Porgy and Bess*, *Paint Your Wagon*, and *My Fair Lady*.

André shared the story behind the creation of the jazz version of the broadway score for *My Fair Lady*. One night in 1956, along with drummer Shelley Manne and bass player Leroy Vinegar, Previn assembled in a little warehouse where Contemporary Records (a small jazz label) was housed. Previn told me, "Shelley was tossing around this idea of recording the score for *My Fair Lady*. I reminded him the only tunes we knew were 'The Street Where You Live' and 'I've Grown Accustomed to Your Face,' so I had my doubts. We had to send somebody to an all-night record shop and buy the cast album. We played one tune at a time—figuring out what each one was all about, changing it around if we wanted to—then recorded it. It took all night, but we started after dinner and by breakfast we'd finished the whole album. I thought it was the most elaborate private album ever made; I never thought anyone would be interested in the jazz meanderings of that kind of show music. It sold a million copies, which was a big deal at the time, especially considering we made that *before* all the songs were a big hit."

Though André confessed fond memories of working in the film industry, he told me it was nothing he wanted to do again. We talked about the impeccable quality of so many of the films made in the thirties and forties: "The reason these movies hold up so well is because technically they're so remarkable; they're perfect. I don't know about the other elements of them as movies, but the sound and music and arrangements—sets, camerawork—they're fun to look at if it's even for that. It's interesting how people dissect things, though; there are these long PhD dissertations on Esther Williams...to us it was just a job."

Part of the reason for the quality of sound was, according to Previn: "these musicians came from the greatest symphony orchestras and dance bands in the world! They just opted for a bit of an easier life, sitting around playing in soft shirts, but it was a great time."

Previn spent decades both conducting symphony orchestras and composing classical music including vocal, chamber, and orchestral music, holding chief artistic posts at the Houston, London, Los Angeles, and Pittsburgh orchestras, and guest conducting with the Boston Symphony Orchestra, the New York Philharmonic, and Vienna Philharmonic among others. During his Los Angeles Orchestra tenure he appeared on televison in the program: *André Previn's Music Night*; in turn he had a series named *Previn and the Pittsburgh* during his time at the Pittsburgh Symphony Orchestra.

Of the Boston Symphony Orchestra he commented, "To make music with the BSO is easier than in most places because they're so wonderful, such a joy, and of course you have the best hall in America. That's not even an opinion, that's a fact."

We discussed the joys and terrors of live performance: I asked him if when a piece was going well, there was any settling into a kind of groove. He said, "That's an extremely dangerous thought! The moment you go there something terrible will happen...you really don't have time to think except for the moment at hand." At the same time he lamented that recordings will of course always sound the same, while "every performance of every piece is different. Anna-Sophie (Mutter) played Tchaikovsky's *Violin Concerto* for a week—the structure was the same, but details were different every night—even the tempo changed slightly. But I think if you 'fix' something to the point of inevitability you're going lose a lot. I like gambling with the unknown during a performance. Things can go either very, very well or quite dreadful, and there's nothing you can do."

Quite a lot can go wrong on the podium, too. He shared some advice from Bruno Walter: "He said no matter how long a piece is, it has only one climax, and your job as a conductor is to figure out where it is and don't arrive at it too quickly. That is just very, very good advice. With some repertoire, it's easy to explode the whole thing one minute in; then there's no place left to go."

Previn knew Benjamin Britten very well, another musical chameleon: composer, conductor, pianist, and violist. When Britten was ill, and had trouble concentrating on books, André told me he offered to send him some records, asking him "what he wanted to hear, and he said what about some Tchaikovsky ballets. I thought he was kidding—you don't really think of Benjamin Britten sitting around listening to the *Nutcracker*, but he thought it was an immortal masterpiece."

Several critics have attempted to sum up Previn's "sound" or at least his musical leanings, even though he's not a man easily summarized. The majority acknowledge his distaste for the dissonant along with his affection for the gorgeous sounds of the French and late Romantic composers: "He's like Richard Strauss with palm trees."[93] Underneath his musical choices and expression lives a cell-deep understanding of how orchestras work, something he learned through coming at music through every possible direction. He's like an anemone that keeps sprouting beautiful growths here and there: he recorded Gershwin standards with bassist David Finck; composed two operas: *A Streetcar Named Desire* in 1998 with Renee Fleming as Blanche Dubois and *Brief Encounter*; a cello sonata for Yo-Yo Ma, a music drama called *Every Good Boy Deserves a Favor* for the London Symphony in collaboration with playwright Tom Stoppard, and in 2009 composed and recorded the jazz album *Alone: Ballads for Solo Piano*. And he's not done yet.

André Previn was also one of the wittiest musicians I've ever had the pleasure of interviewing. His one appearance as the conductor "Mr. Andrew Preview" who struggles with a comically inept soloist on a 1971 British Christmas show was referred to for years. His 1991 book (one of four) *No Minor Chords: My Days in Hollywood* is full of candid stories of his MGM days and called "hilarious" by *Publisher's Weekly*. A memo from Irving Thalberg explains the title: "No music in an MGM film is to contain a minor chord."[94] Previn also had the presence of mind to make note of such priceless quotes as this one from Sam Goldwyn: "My wife's hands are so beautiful I think I'll have a bust made out of them."[95] On Previn's website he answers a series of short questions:

> My motto: "A day without music is a wasted day."
> What I would like to be: "I'm working on it."
> Favorite bird: "Charlie Parker."
> Military deed I most admire: "I don't admire anything military."
> How I would like to die: "Later."[96]

Previn's been to the altar five times: with Betty Bennett, Dory Langdon, Mia Farrow (he is the adoptive father of Soon-Yi Previn), Heather Sneddon, and finally, with the German violinist Anne-Sophie Mutter, who he divorced in 2006, but with whom he maintains an amicable relationship. Known as someone who likes to compose for specific musicians, he composed a Harp Concerto for Pittsburgh Symphony Orchestra's harpist Gretchen Von Hoeson, among many others. In a 2004 performance, Previn's conducting style for his violin concerto *Anne-Sophie,* the rhapsodic piece he named after her, seemed to personify an artistic tension between them. Mutter embraced the sentimental spirit of the piece, which was "surprising, given her fondness for abrasive musical scores....Love, these two experienced musicians seemed to be saying, is a complicated thing."[97]

Previn is the winner of four Academy Awards and several Grammys, as well as countless honors for his stunning musical accomplishments. Just a few include the Austrian and German Cross of Merit, the Kennedy Center's Lifetime Achievement Award, and the Glenn Gould Prize.

But more than his awards and accomplishments, Previn is defined by his love of music every waking moment of the day. He even confessed to being a bit phobic about not having his own music in the car: "Who knows," he told me, "you could be assaulted by a country western station and plough into a truck."

I shared with him a story I'd heard about Aaron Copland: while shopping in a supermarket one day he heard his own music piped in overhead. Someone approached him saying, Isn't it wonderful that your music is playing? His answer was: not really; I have to be in the mood to hear that sort of thing. Previn said, "I agree! It's everywhere: elevators, waiting rooms, shopping malls. It's as if everyone is afraid of silence, whereas I think occasional silence makes the music that much more wonderful."

## SUGGESTED RECORDINGS

**Symphonies Nos. 1–3 by Rachmaninov**
The London Symphony Orchestra conducted by André Previn.
EMI CD 7 64530-2

**Symphonies Nos. 1–9 by Vaughan Williams**
The London Symphony Orchestra conducted by André Previn.
RCA CD 82876 55708-2

**Shelly Manne and His Friends Play Music from My Fair Lady**
Shelly Manne, drums; André Previn, piano; and Leroy Vinnegar, bass.
(This is the recording I discussed with Previn during our interview.)
Contemporary CD 7527

**Uptown with André Previn, piano; Mundell Lowe, guitar; and Ray Brown, bass**
Songs of Harold Arlen, Duke Ellington, and others.
Telarc CD 83303

## 1. JEAN SHEPHERD

# A Voice in the Night

Jean Shepherd

"Okay, gang, are you ready to play radio? Are you ready to shuffle off the mortal coil of mediocrity? I am if you are. Yes, you fatheads out there in the darkness, you losers in the Sargasso Sea of existence, take heart, because WOR, in its never-ending crusade of public service, is once again proud to bring you...*The Jean Shepherd Program!*"[98]

Like thousands of other fatheads out in the darkness, I went under the spell of Jean Shepherd in the early fifties. To this day I remain enchanted. When I first heard his voice on the radio—the first time I erupted in helpless laughter at one of his stories—I knew this was a different kind of talent: one-of-a-kind, brilliant, American, endlessly creative, and entertaining. And like so many have said before: I felt as if he were speaking to *me*, talking about *my* past history with girls, the time *I* bought the ten-cent X-ray vision tube to see through my crush's blouse, the time *I* was invited to the rich girl's house and ate snails and sipped martinis for the very first time. His meandering fables, laced with sardonic wit, filled New York's air waves, permanently shaking radio free of its news/sports/music/call-in straitjacket. A master story teller in the league of Mark Twain, S. J. Perlman, and P. G. Wodehouse, Jean Shepherd took pieces of his own life, endlessly spinning and weaving them into stories that proved pain is often found in humor, with truth not so very far behind.

As a boy, Jean Shepherd or "Shep" as he was later known, would lie in his bed in industrial Hammond, Indiana, listening to the trains rumbling by in the night, "longing to be anywhere else."[99] In our 1985 interview he talked about the trilogy he wrote for PBS and the American Playhouse, how the setting and characters were rooted in his past: "This is urban America, like Detroit or Newark, or a place where everybody works at the plant, lots of railroads, refineries in the distance; there's not much written about places like this and I don't know why, because most of America is this industrial world."

His "old man"—never named in Shep's radio life—was a rabid White Sox fan, beer drinker, and disenchanted political cartoonist for the *Chicago Tribune* before he abandoned his family and fled to Florida. "Profanity was his medium," Jean said, "he hated to hear amateurs swear."[100] Jean described the father in the PBS series to me as follows: "he works at a place he just calls 'thedamnedoffice.' One word. Mother says, how'd it go today? And he always said, whaddya mean, 'how'd it go'? It's 'thedamnedoffice!'...My father always generically referred to it that way. His boss was named Fogarty; he'd simply say: 'thatdamnedFogarty!'"

Shepherd's mom, a perennial presence in his radio dramas, was a "'real mother' who wore a rump-sprung, Chinese red chenille bathrobe (with the petrified egg on the lapel), kept her hair in rollers, and ground out endless meals of meatloaf, red cabbage, and Jell-O."[101] Other characters—in real life and spun into radio tales—included younger brother Randy, eternally whining in the background, as well as assorted bullies, crushes, and neighborhood kids including Flick, Schwartz, and Brunner. When asked if he missed Hammond, Jean said, "That's like asking, 'Do you miss the cold sores you had last week?'"[102]

Ham radio became an obsession with Jean, and in his teens he played the role of Billy Fairfield, Jack Armstrong's sidekick on Chicago's radio show of the same name. He was constantly in trouble during his early stints as a DJ at WCKY, WKRC, and WSAI because he never stopped talking long enough to play records.

In our interview he shared a telling story about growing up in Chicago's south side: "Actually, we were right across the line in Indiana, and we'd go on these field trips. I was seven or eight going to the Warren G. Harding School, so of course I had no idea I was going to a school named after the worst president in history—'course there've been a few contenders since. Other kids were going to the George Washington School, the Lincoln School, but nope, it was the Harding school for us! Mrs. Robinette took us on a field trip to the first national bank, we drove for hours and hours to this bank...we get there and she shows us all these holes in the building; it was shot by John Dillinger. And that was it, that was our field trip, and to this day this is a big sightseer thing. When you grow up with this kind of historical perspective you can see that the star-crossed romance of Josephene Coznowsky is a Wagnerian tragedy."

Shep served two years in the Army Signal Corps where he developed a rollicking distaste for authority, honed his fine sense of the absurd, and stowed away a rich source of material for years to come. He attended Northwestern, the University of Chicago, and Indiana University but never graduated. After a stint at Goodman Theater drama school and selling cars, he broke in to radio on WSAI-FM with his own show hosting a hillbilly jamboree and interviewing wild animal acts. He also hosted a nightly comedy show called *Rear Bumper*. After a brief move to Philadelphia, he finally landed in New York in 1956, where he began his stormy, twenty-one-year career at WOR with an all-night show involving lots of commercials, jazz, and jug music.

Quite a bit later, in 1971, we started rerunning his WOR radio shows along with the groundbreaking PBS television series *Jean Shepherd's America*, produced by my colleague and friend at WGBH, Fred Barzyk. Those radio shows could be quite a lot of editing for me. I had to cut out as many as twenty-two commercials from a show that ran forty-five-minutes, but what a pleasure to finally work so intimately with Jean Shepherd! That voice of his, cozy but full of mischief: an Indiana twang crossed with a New York toughness. His meandering style had often been compared to jazz; his verbal riffs turning and following the muse wherever it went, whether it be his tenure in the Army, druids, the inner workings of a steel mill, King Tut, his boyhood in Hammond's industrial wasteland, or a break to perform a kazoo solo; perhaps 'Yellow Dog Blues.' In between tales, he'd sing along to bad old records or play the Jew's harp. He also liked to break things up by asking his "night people," his "gang," to put their radios in the window at two o'clock in the morning, then crank the volume to ten while he played incredibly loud recordings of train whistles, or yell things like, "You filthy pragmatists, I'm going to get you!"[103]

As Donald Fagen put it, listening to Shep, "I learned about social observation and human types; how to parse modern rituals (like dating and sports); the omnipresence of hierarchy; joy in struggle; 'slobism'; 'creeping meatballism'; 19th-century panoramic painting; the primitive, violent nature of man; Nelson Algren; Brecht, Beckett, the fable of George Ade; the nature of the soul; the codes inherent in 'trivia'; bliss in art; fishing for crappies; and the transience of desire."[104]

A typical Shepherd story began with a throwaway remark about what a "slob" he was as an adolescent. He goes on to describe the endless meatloaf dinners he had growing up: "could have been: meatloaf with carrots and mashed potatoes, or meatloaf with red cabbage and mashed potatoes, or meatloaf with peas....we thought everybody lived this way..."[105] Sure, he'd read about clam dinners—even raw oysters—and collectively gag along with his family: did people actually *live* this way? No *meatloaf*? Unthinkable...until one day in college a girl invites him to dinner, saying, "don't bother to dress..."[106] Not sure what she meant by this, he still dons his best: Penney's sport coat and Montgomery Ward slacks. As he approaches her home the lawns get bigger and bigger, until the houses are completely out of sight down gently curving drives. At a brass-knockered door a butler greets him and the girl, Nancy, rushes up to welcome him, planting a kiss on his cheek—an act never quite as casual in his circles—he takes a long stemmed glass from a tray and promptly breaks it. He learns these are martinis...he says, "My old man would only say, 'How about some booze?' We didn't have any actual names for it—it was just called booze."[107]

Along with the other guests, Jean and Nancy move to the linen and china bedecked table, where she queries him: "Have you had any fresh escargots this season?" Shepherd: "...my meatloaf insides were churning...I couldn't chicken out...so with this little fork, I fished it out...and, Oh my God! It was fantastic! I made a total pig of myself and went slurp, slurp, slurp...and then it hit me...what other things did I think are awful? Late that night...lying in the dormitory room and I can feel them snails, there's an aftertaste, and I begin to suspect...there was a fantastic unbelievable world out there. And I was just beginning to taste it, and God knows where it would lead!"[108]

Through these tales he warned about what life had in store for you: endless discovery, but also loss and betrayal, and somehow it was a comfort. I could, and did, listen for hours, in awe of what came out of him with no script, no preparation of any kind.

I actually watched him do his show a couple of times. In some interviews, Shep claimed he prepared for hours on end, but that's not what I witnessed. In front of him were just a few notes, handwritten, about how he would open and where he was going to land, but there was no script whatsoever. Nothing close to what someone without his sort of mind might have needed to spin off a three-and-a-half-hour show, five nights a week! A show with no guests, and whatever spots he did included well-aimed barbs at the sponsors. He had his stories framed in his mind as well as a sketchy outline to light his way on a path that forked and switchbacked—yet he'd always end the piece exactly on time—a feat of radio engineering that still floors me. He liked to be alone in the studio, and he never liked being interrupted for any reason. Herb Squire, Shepherd's longtime engineer at WOR, learned very quickly he'd better be able to work the board while being Shep's audience for the duration of the show. He said, "You had to react and become very animated in your reactions...because like any performer, if he didn't get a response, he'd feel up the creek without a paddle."[109]

Shepherd hated the word "nostalgia" and balked each time his work was described with that term. He didn't even like to be called a "radio personality." His stories were fetched from his past, but they were no Norman Rockwell renderings; they were more like jumping-off places for hilarious spins on "the absolute certainty of daily humiliation in life,"[110] as one astute Wall Street journalist wrote. Shepherd even bristled when Studs Terkel posited that Jean covered the collective American past; Jean said: "You think it's the past...I'm writing about American *rituals*...the two-week vacation, the graduation, the Sunday afternoon dinner, the prom, the coffee break, which is as ritualistic from one end of the country to the other...in Tacoma they're sitting around with the same looks on their faces as everywhere else."[111]

Even at the height of his popularity, however, Jean never felt as accepted or valued as he would have liked, insisting that WOR would have enjoyed nothing better than for him to play the top twenty songs and otherwise shut up. His show was never listed by the *New York Times*, and he made a big deal of that on the air. Someone at the *Times* claimed it wasn't a music show, therefore there was no category for him.

In what can only be called a brilliant coup, Jean single-handedly created demand for a book that he'd not only not written, but that hadn't been published. He jump-started a mammoth buzz for a pseudo-book he called *I, Libertine* that wound up on the 1956 *New York Times* list of new books. *Publisher's Weekly* frantically called for information on this hot new release. Egged on by Shepherd, listeners flocked to bookstores to buy the nonexistent novel by the fictitious Frederick R. Ewing. Set in England during the 1700s, *I, Libertine* followed the bawdy adventures of Lance Courtnay—by day a respected citizen, by night loutish rake. In the final irony, Ballantine Books, who'd been after paperback rights, persuaded Jean to actually write the book together with author Theodore Sturgeon.

Many times I saw Shepherd perform Saturday-night gigs at the Limelight Café where he held sold-out crowds in the palm of his hand. He would pace around, work himself up, really get into character. A New York critic called him "often brilliant and sometimes a trifle mad."[112] He did hundreds of live shows at colleges across the country and played to throngs of excited fans at Carnegie Hall as well as Boston Symphony Hall. Excerpts from his radio shows turned into essays that showed up in everything from *Playboy* to the *Village Voice* to *Car and Driver*. His four books, *In God We Trust, All Others Pay Cash*, *Wanda Hickey's Night of Golden Memories and Other Disasters*, *The Ferrari in the Bedroom*, and *A Fistful of Fig Newtons*, all drew on his radio monologues.

In our interview, we discussed the roots of his 1983 classic movie, *A Christmas Story*, a sardonic look at a holiday he'd hoped to call *Satan's Revenge*: "(The script) was taken from the first chapter of my novel, *In God We Trust, All Others Pay Cash*. I wrote the script based on that one chapter, where Ralph is waiting in line to see Santa Claus. He had this fantastic hangup about owning a Red Rider Model 200 single shot BB gun. It originally ran in *Playboy* as an antiwar parable, that's how they read it in Europe. It's interesting that they got it in Europe, but not here. Everyone at MGM was shocked by how commercial it was; it led the box office for five weeks; this was when *Yentl* was out, *Tender Mercies*, *The Right Stuff*. Now I'm working on a sequel, called the *Revolt of the Mole People*, which eventually became *My Summer Story*.

"Ralph is dragooned into going to summer camp—he doesn't want to go—and he's put in a cabin called the Mole cabin, and all the other cabins have names like Leopard, Tiger, Wolf, Moose. He's in the Mole cabin with a bunch of little fat guys with thick glasses. The high point comes when Ralph finishes his leather wallet on which he's carved the face of Roy Acuff and presents it to his old man."

In the mid-seventies, problems began to surface at WOR. Just when Shep was about to be released from the show for not being commercial enough, he did an impassioned commercial for Sweetheart Soap, which happened not to be a sponsor.

When his listeners learned he'd been canned, they reacted in such numbers that not only was he reinstated, Sweetheart Soap signed up as a sponsor. In fact so many sponsors wanted in that once again Jean's show felt too packed with ads, and eventually WOR cancelled the show in 1977.

A year after WOR dropped him, he advertised in various interviews that he was glad to be free of radio, insisting that his best work was as a screenwriter and novelist, therefore undermining work that he was surely proud of. At a public radio convention he was asked to address, there was a question and answer period following one of his typically brilliant monologues. A young man stood up and said "Mr. Shepherd, you must be in love with radio," Shepherd bristled and said, "No, you got that backwards. Radio should love me." He continued to trash radio with his signature wit: "all the girls in school who wouldn't let you copy their algebra papers are now wearing granny glasses and running public-radio stations."[113] He was finally able to complete the sequel to *A Christmas Story*, which was called *My Summer Story*, released in 1994, as well as the PBS *American Playhouse* TV series, which also featuring the Parker family. Episodes included: *Ollie Hopnoodle's Haven of Bliss*, *The Great American Fourth of July and Other Disasters*, *The Star-Crossed Romance of Josephene Cosnowski*, and *The Phantom of the Open Hearth*, all produced by Fred Barzyk. Jean also did *Shepherd's Pie* for New Jersey Public Television.

He talked in detail about the series in our interview: "The first one, *Phantom of the Open Hearth*, covers the life of one family over one summer. I've always been a real believer in the seasons: I believe the seasons play a great role in our lives; you're not the same in December as you are in mid-August. This one takes place in June just as the school year is ending and Ralph is about to go to his junior prom. There's the whole business of the early summer; the old man is walking around outside having his first bout with crabgrass; there's a certain meanness in the house.

"The second was the *Great American Fourth of July and Other Disasters*. Now Ralph is involved in the big Fourth of July parade and all that business, and it's July and its hot and the meanness is worse, 'cause if you know anything about Midwestern heat in an industrial town in mid-July you know that heat is an entity, so you can hear the mosquitoes buzzing outside, fist fights breaking out...

"In the *Star-Crossed Romance of Josephene Cosnovsky*, Ralph's a senior, a very different kind of guy now that it's November; he's gripped by his first truly serious girl madness coupled with the underlying theme that the grass is always greener, that you're always excited by things foreign. For him it was Josephene, a Polish girl who moved in next door. She had mysterious overtones of Heddy Lamar, Ingrid Bergman; there was an aura of stuffed cabbage about her, which is very exciting since he comes from a meatloaf family, so when he kisses her for the very first time it's something cataclysmic, the music just rises enormously..." Cue the Warsaw Concerto...

Losing Shep in 1999 was a shock to the system. How unlikely that someone so full of life could be silenced. He had also become a dear friend to both myself and Joyce. In the end he was living with his fourth wife Leigh on Sanibel Island, where he told me, "You could hear nothing but the grunting of the alligators in the bayous...."

Uncertainty shrouded much of his personal life, including circumstances around earlier marriages and his refusal to acknowledge two children born to his first wife. Bitterness characterized his later years, possibly because he never felt appreciated until *A Christmas Story*, which had been a major turning point in his life.

In my mind, without Shepherd, it's hard to picture Spalding Gray, Garrison Keillor, or even David Sedaris coming onto the scene. Bill Griffith, creator of the comic *Zippy the Pinhead*, said he'd lie awake nights listening to Shep spin his tales, and credits Jean as one of his major inspirations. Media critic Marshall McLuhan called him "the first radio novelist."[114] Five thousand hours of radio in New York is an accomplishment not realized by many.

The truth is, I still get excited when I hear the show's opening bars: the trumpeting silliness of "The Bahn Frei Polka" by Eduard Strauss. Whenever I need a Shep fix these days, I visit www.flicklives.com where I can listen to such classics as: "Taxi! Hey You, Taxi!," "Super Fink Is Here," or "Salute to Slobs."

I like to think I "got" Shepherd, through all the walls he might throw up, despite his tendency to relentlessly be "on"; that I understood the chronic need, in this business, to be appreciated and heard. During one interview, Shep recalled our phone conversation after an American Playhouse episode had just aired. He said, "Leigh and I were drinking champagne, and she sighed and said, it's over now, which I knew because the PBS plant lady had come on; and the phone rang, and I thought, who could it be, my manager, my agent? Our director? I rushed to the phone and it's you, Ron, and you said, without any preamble, 'You've created an unrivaled American classic; it reminded me of the *verismo* quality in Italian opera.' There's only one other critic who picked that out, this guy in Los Angeles: he said, 'Shepherd is the only guy working in movies who uses Wagnerian operatic themes in an American setting.'"

After I told him I remembered our conversation very well, there was a pause. Shep said, "But did you laugh?"

I assured him we were laughing nonstop.

Just like all of us, I believe—in radio or not—Shep needed to know that he wasn't just a voice in the darkness, but that someone was out there, listening to every word he said.


## SUGGESTED READINGS

All Shepherd lovers should be familiar with the book, *Excelsior, You Fathead! The Art and Enigma of Jean Shepherd*, by Eugene B. Bergmann, Applause Books, 2005.

This is the definitive book on the artistry of one of America's greatest humorists. Bergmann includes interviews with people who knew his subject including yours truly. With verbatim transcripts of Shepherd's radio shows and rare photos, the author offers a fascinating insight into the life of a complex and creative genius. Bergmann dedicates his book: "To the memory of Jean Parker Shepherd, who gave so much of his real as well as imagined self to us all."

For more on this icon of American radio, visit: www.FlickLives.com: A Salute to Jean Shepherd. The site includes biographical information as well as a large collection of Shepherd's works, including hundred of hours of radio shows, images, and writings. Through this site, where you can order his recordings, it seems as though his work is almost bigger in death than in life, and I'm heartened to know that his work is being introduced to a whole new generation.

# If You're For It, I'm Against It

Author's collection.

Mort Sahl and me

Mort Sahl made his inauspicious debut on Christmas night in 1953 at a San Francisco club called the Hungry i. Dressed in a cardigan sweater, slacks, loafers, and open collar, and holding a rolled up newspaper, he walked onstage and launched his comic career using material inspired by politics of the day. Just by being Mort, he broke the mold: that of the slick nightclub comedian telling canned wife jokes and oozing show-biz smarm. His delivery was nervous and staccato; the content full of tangents, asides, and digressions, yet somehow he always circled back to make his brilliant, often stinging point. From that night on he was considered the premier satirist of our time, influencing, among others: Lenny Bruce, Dick Gregory, Woody Allen, Jay Leno, George Carlin, and Shelly Berman.

Born in Montreal in 1927, Mort and his family soon moved to Los Angeles. His mother said that he was talking at seven months and "spoke like a man of thirty" at age ten.[115] Hoping to go to West Point, Sahl was drafted instead and ended up in the Alaskan Air Force, where he got into trouble for editing the newspaper "Poop From the Group," and was given eighty-three consecutive days of KP for editorials about alleged military payoffs. After his service, he went to the University of Southern California "to please my father,"[116] majoring in city management and traffic engineering, and graduated in 1950.

After a few fruitless years trying to find writing gigs, Sahl was broke and starving. His girlfriend suggested he try the "Hungry i," which meant "hungry intellectual." She said, "The audiences are all intellects, which means if they understand you, great, and if they don't, they'll never admit it..."[117] Though the crowd threw pennies and peanuts at Mort, the owner saw something in him and kept him on. Instead of "how-fat-is-my-wife" jokes, Mort put it on the audience to connect a few more dots than they were used to; in other words: think. Satire forces the listener to consider the object being satirized in order to laugh along. He told the crowd that conservatives were now dressing in charcoal gray suits "because modern science was looking for a color more somber than black."[118]

The audience started to catch on, not only at the Hungry i, but at legions of hip clubs and with the liberal press. That said, Sahl was never someone to be pigeonholed politically. Behind his genially rumpled manner, he insulted everyone including Senator McCarthy, and presidents from Eisenhower to Nixon to Obama, even nabbing a *Time* cover story which was unheard of at the time. He fearlessly tossed out lines like, "I'm not so much interested in politics as I am in overthrowing the government,"[119] and "If you were the only person left on the planet, I would have to attack you. That's my job."[120]

Sahl became not only the first comedian to make a comedy album, but one after another they became best-sellers; this at a time when topical humor simply wasn't to be found on television. He also worked as a speechwriter for John F. Kennedy, Ronald Reagan, Ross Perot, George Bush, and Alexander Haig, and as a script doctor during the early eighties.

I met Mort back in the 1970s when he appeared at a local Boston club. Every year I'd do an annual birthday tribute to bandleader and musical innovator Stan Kenton. I'd read somewhere that Kenton had a major impact on Mort's early career and that Mort was a great lover of Stan's music. With that in mind, I invited him to be a guest on my show. It turned out to be a revelation! He knew the names of every member of every addition to the Kenton band from the early days in Balboa, California, to the final years up to Stan's death.

Since Mort pioneered his style of stand-up in night clubs that showcased primarily jazz music, it was only natural that he adopted some of that free-flowing jazz form as well as the dry, ironic wit of jazz musicians. He was fluent in jazz lingo I like to call "Sahl Speak": gasser, chick, drag, cool it, bugged, the most, dig it, weirdo, wild, shakin', and wigged. Jazz musicians loved him. Sahl cites jazz pianist and bandleader Stan Kenton as his most important performing influence: "Stan, of course, was a great artist, but he was a voice of defiance, and did it on his own terms."

During our interview, Mort recalled the first time he met Stan: "The band manager approached me and said, you always come to see the band, would you like to meet Stan? I was scared to death, totally intimidated." The fact was he'd routinely skipped classes at USC after staying up all night following Kenton's band. Soon after Mort appeared in San Francisco, Kenton invited him to go on the road with the band. Mort was incredulous; he asked Stan, "What if I bomb?" Stan replied, "Don't worry...I'll have the band play louder!"

Mort told me, "In the end, Stan became like a father to me...he had expectations but he always knew how to talk to you; he'd say to me: God endowed you with ideas: value them. Or if there was a problem at a gig and he was talking to a club owner, he'd say: we have a bit of discord; how are *we* going to solve this. He included you."

Stan Kenton's band had a wild, frenzied sound that roared onto the scene during the staid, Ozzie and Harriet, Eisenhower years in the 1950s. I asked Mort his theory of why the band took hold the way it did: "Because of Stan's conviction. He hired fourteen young arrangers for the Innovations band. When you heard Kenton, you knew it was him. He'd say again and again, the purpose of the band isn't to knock the audience out, it's to knock ourselves out. He was a man on a mission. But keep in mind, the East coast really pilloried the band, it was too much for them at first...June Christy and Stan, and they were like heretics...and though some people thought the band was frenzied or stentorian, when they did the ballads like 'Here's That Rainy Day' or 'Lush Life' they had one of the most romantic sounds."

From everything I'd heard about Kenton, he loved being on the road. Mort laughed and said, "A reporter interviewing him on his bus said Stan, you've been all over the world; what's your favorite place? He said 'You're sitting in it.' Back east, down south, out west and due north, we went everywhere and a lot of nowhere. The buses were also segregated in the sense that the intellectuals—guys who wanted to read—sat up front, while the partiers hung out in the back. Also, Stan made two seats for musicians because he said they were never treated right. And a Kleenex box in the ceiling. He said it would straighten out your head to get on the bus after a gig."

Mort continued: "Stan was a way of life; he changed everybody that came after him, and he did things for people. When Lee Konitz joined the band and it wasn't working out, Stan said I have a guy in Los Angeles you're going to study with; he's a classical player. Lee studied for six months, six hours a day, just blowing; then he took over the section and he was lead alto. One night Stan woke him on the bus and asked him: how do you feel now? How are your chops? Lee flexed his biceps and said, 'Stan, I feel like King Kong.'"

We also discussed Mort's relationship with Frank Sinatra during the Kennedy era. Sahl had become one of Kennedy's speechwriters, and Sinatra had called on him to assist his efforts in promoting JFK and his ascendancy into the White House. Mort said, "I'll never forget meeting him for the first time. We were in Beverly Hills at an Italian restaurant and he walks in with his entourage. He sat down next to me and said, 'You want to go on my label? I want you on my label. Money's no problem.' He could enthuse you; he was like a tidal wave. He said, 'My life is terrific; it's the envy of everybody.'"

Sinatra, always the loyalist, stuck by his friend and hired Mort to record a comedy album for his new label Reprise, called "New Frontiers." Their years-long friendship blew hot and cold, but Mort told me he had the opportunity to attend many historic sessions at Capitol Records. Mort said Frank liked to have friends around when he was making those classic albums.

I asked for Mort's take on Sinatra's acting career: "He thought Marlon Brando was pretentious. Frank knew every line of every script he was assigned, and he knew it before the first day of rehearsal. He thought the most energy was in the first take, which some saw as lazy, but that was his view. He did a diverse number of roles...he came out of a generation that thought chicks were the prize...you ran the race and if you won you got the princess. He believed that movies were our literature. That romance was the background music.

"But when Frank sang about lost love it was never whining, it was real; he thought love was essential. The man would give you a house or a car or pay your hospital bill...then the demons would move him and he'd go the other way. Nancy always protected him; she loved him. Frank had to be moved emotionally or he wouldn't get involved with you. If he thought you were a jivey artist he just wouldn't be there. But it was all him, don't forget, He was invested in the tunes...think of him as an artist giving credit to the composers right up till the end. When I worked with him during the Kennedy campaign I never introduced him; he'd come out while I was wrapping up and stand behind me and smile as if to say: isn't he incredible...and that would make me a hero. Then he'd come to the mic and just start singing."

Following the election, Mort got on the wrong side of the Kennedy Democrats by going after the new president. Mort said, "I only have a few months to tell these jokes before they become treason."[121] Peter Lawford told the comic, "You're going to get it, you'll see...you made lots of enemies. Nobody wants you!"[122] The assumption was that Sahl was the Democrat's boy, but as Gerald Nachman said in *Seriously Funny*, "Mort was nobody's boy,"[123] with a personality that over time cost him "friends, colleagues, club dates, managers, wives, and girlfriends."[124] His claim that the assassination had been a conspiracy nearly cost him his career: Ed Sullivan among others refused to have him on his show. Sahl's tirades against the Warren commission's findings on the JFK assassination combined with early anti-Kennedy jokes alienated him from much of his audience. It seemed he'd lost that critical distance between himself and his material.

In Sahl's 1976 memoir, *Heartland*, he vents about his dating past, veering into misogynistic rants: "I used to go out exclusively with actresses and other female impersonators."[125] Gerald Nachman cuts him a little more slack: "He's not a misogynist: he's a totally disappointed romanticist."[126] To me he said, "I agree with Paul Osborne, the screenwriter who said 'men marry women hoping they'll never change and women marry men hoping they will.' Each is understandably inevitably disappointed." *Newsweek*'s August 9 review said the book portrayed him as a "double casualty of the Kennedy years,"[127] outlining the alleged attacks on him by the Kennedy Mafia, followed by his inability to get work and an income that plunged from a million to $13,000.

Mort has appeared at such venues as Mr. Kelly's in Chicago, the Village Vanguard in New York City, and the Crescendo in Los Angeles. His fans ranged from James Jones, Saul Bellow, Leonard Bernstein, Marlena Dietrich, and Woody Allen.

Allen once said, "He was the best thing I ever saw…he was like Charlie Parker in Jazz. Mort was the one. He totally restructured comedy. A great genius who appeared and revolutionized the medium."[128] Allen was just coming up as a shy, nebbishy comedy writer when Sahl owned the crowds at the Hungry i. Before he saw Sahl, Allen hadn't committed to doing stand-up, but after he did, it was either commit or drop the idea entirely. About Sahl's approach he said, "It wasn't that he did political commentary—as everyone kept insisting. It's that he had genuine insights."[129]

Mort celebrated his eightieth birthday at a 2005 appearance at Jimmy Tingle's Comedy Club in Somerville, Massachusetts. Still shocking the system as he has for over half a century, he was as sharp and incisive as ever. He had the audience in the palm of his hand and received a standing ovation.

The next day over lunch at a cozy Italian restaurant in East Boston, Mort seemed relaxed and reflective. We talked mostly about jazz, Stan, Dizzy, Bird, and his early days on the road. Afterward, when I dropped him off at his hotel in Cambridge, he said, "Hang in there babe, and keep swingin'."

## Favorite Quotes

"I'm for capital punishment. You've got to execute people. How else are they ever going to learn?"[130]

"A Yuppie believes it's courageous to eat in a restaurant that hasn't been reviewed yet."[131]

Re: Larry King's book, "I read parts of your book all the way through."[132]

On Ed McMahon after Johnny Carson retired: "Ed is still laughing in case Johnny has said anything funny at home."[133]

"The Democrats don't want anyone to be born, while the Republicans don't mind if you're born, they just don't want you to live long enough to collect Social Security."[134]

Sahl recalled a meeting with George Bush, who said to him, "This is a dirty job, but that's what you elected me to do." Sahl's retort: "We didn't elect you that much."[135]

## RECOMMENDED READING
### Seriously Funny...The Rebel Comedians of the 1950s and 60s

By Gerald Nachman
Published by Pantheon Books, NY

One of the best books written about some of the most revolutionary comics of our time. Mort is on the cover and is the subject of the first chapter, "A Voice in the Wilderness." I had the pleasure of interviewing Gerald Nachman about this and his other book entitled *Raised on Radio*, a marvelous history of radio's golden age.

*Mort Sahl's America*, audio CD by Eugene McCarthy and Mort Sahl, June 1997

# The Importance of Being Ernest

Author's collection.

Ernest Borgnine with Joyce Della Chiesa

256

To my delight, there were only two degrees of separation between myself and Ernest Borgnine. My ex-wife Jackie's husband, Charlie Brown, who has since passed, met Ernie in Las Vegas about twenty-five years ago. Talk about separated at birth! Both down-to-earth outdoorsmen, they loved to hunt, fish, race cars, and drive RVs. Charlie even invited Ernie and his wife Tova to his lodge in Alaska, where he'd built a special room he named the Ernest Borgnine room, which was completely dedicated to the man, his movies, and their incredible friendship.

It was my great fortune to interview Ernie in 2008. I'd just finished reading his memoir, *Ernie*, in which he revels in his love of acting and movie-making. In it he summed up his personal philosophy, which was inspired by an old sign he saw dangling from a street vendor's chestnut cart on a wintry day in Manhattan: "I don't want to set the world on fire," the sign said, "I just want to keep my nuts warm."[136] Though I enjoyed quite a few laughs reading the book, I confessed to him that many passages moved me to tears. Ernie said, "That's okay, Ron, a man shouldn't be afraid to cry. I've read the book over a few times and some things in there leave me teary-eyed too. People cry—men, women—except maybe cowboys...but I'll bet when they get out there in the wilderness with their horses under a moonlit night, who knows, they probably bawl their eyes out."

Ernest was born Ermes Effron Borgnino in 1917 in Hamden, Connecticut, to Camillo and Anna, Italians who immigrated to America at the turn of the century. He joined the navy as an apprentice seaman right out of high school and spent a decade in the service. Back at home he found himself directionless, casting about for what to do next and applying for dead-end factory jobs he never really wanted. He told me, "One day I'd come back from looking for work feeling completely fed up. I remember thinking, forget it, I'll just go back to the service another ten years, get my pension, and call it a life. I said all this to my mother. She was quiet a minute, then out of nowhere she said, 'Have you ever thought of becoming an actor? You like to make a damned fool out of yourself in front of people; why don't you give it a try?' And it's like a light went off and I said, 'Mom, that's what I'm gonna be,' and by God, ten years later Grace Kelly was handing me an Oscar. And it was like falling off a log."

As his book attests, Ernie's mother Anna was no ordinary woman. He shared with me, "She was a countess in real life, and it was frowned upon for royalty to go on the stage, so maybe there was something satisfying for her to see me act. We were so close—she used to take me everywhere; we did everything you could imagine together. She showed me things I never even thought of exploring, took me to Italy, La Scala, Milano. When we came back she started getting this illness no one could diagnose. For years the poor soul suffered...there were times I had to move away from the house to my uncle's, but she had the idea she couldn't do enough for me...we learned later she had tuberculosis. But she was tough, too...the epitome of cleanliness. You could eat off the floors; I know because I used to scrub them. She'd show up with a white glove, and if it wasn't done right I didn't go to the movies that week."

Ernie attended the Randall School of Drama in Hartford, then moved to Los Angeles in 1951, landing supporting roles right away, most notably as a male nurse in the play *Harvey*. His husky voice, boxer's face with his signature gap-toothed grin, and stout build made him a natural to play the heavy. It was as the ruthless sergeant 'Fatso' Judson, Frank Sinatra's killer in 1953's *From Here to Eternity,* that Borgnine found real traction in his career, and he was signed to a seven-year contract.

Ernie said, "It was the first time I got to work with Frank Sinatra, never mind Burt Lancaster, Deborah Kerr, and Montgomery Cliff. I'll never forget the first hour of shooting...I'm up on a platform playing piano, badly, and they're all below me dancing. Sinatra looks up and says, Knock it off, Fatso, we're trying to dance down here. I got up from the stool *very* slowly. Frank looked up and said, 'Jesus Christ, he's ten feet tall!' Anyway, from then on, Frank was the dearest person I've ever known in my life.

"When you hear Frank sing you really feel he's telling you a story, and it's the most beautiful thing in the world. We don't have that any more. Unintelligible racket and screaming and suddenly it's over? Where's the song? When Frank sang, he sang to *you*."

I asked Ernie what it felt like to constantly play the heavy: "You think that was bad, have you seen *Emperor of the North Pole*? I used to say to my wife Tova, Am I really that bad? And she'd say, Of course not, you're acting. But still, it was frightening. The nasty things I had to pull out of me to become these people...and the next day I'm playing Mr. Nice Guy."

Though briefly typecast as a brooding villain in *Johnny Guitar* and the Western *Vera Cruz*, Ernie landed an against-type role as a lonely butcher looking for love in Paddy Chayefsky's 1955 film, *Marty*. His performance won him an Academy Award as well as top honors from the New York Film Critics' Circle as well as Cannes. Ernie reminisced about his experience and the relative simplicity of movie-making at the time: "It was shot in black and white in the Bronx. Remember, there were no computers then, and we made it on a shoestring. There was something so natural, just intuitive about the process then. I remember Fellini—what a wonderful filmmaker—he'd take anybody and put them in. He'd grab a guy on the street and say, 'Hey buddy, what are you doing? Come over here, viene qui, we're gonna put you in a movie, we're gonna put you on a bicycle and teach you how to do this.'"

Ernie continued appearing in movies (a jaw-dropping 201 films are listed on his IMDB page) until landing his signature role as a boat-skipper on the wacky TV series *McHale's Navy*, reprising the role in 1964 for the show's film. The bumbling crew of PT-73 became an instant hit, solidifying a lifelong friendship with Tim Conway who played Ensign Charles Parker. After the show's cancellation, Ernest quickly made his way back to the big screen, taking on the role of General Worden in 1967's *The Dirty Dozen*.

In the seventies and eighties, Borgnine appeared in a number of acclaimed television movies including *The Trackers, The Ghost of Flight 401,* and *The Poseidon Adventure*; as "Dutch" in Sam Peckinpah's *The Wild Bunch*; and as a helicopter pilot in the hit series *Airwolf*. He even reprised his *Dirty Dozen* character in a series of TV movies: *The Next Mission, The Deadly Mission*, and *The Fatal Mission*.

In 1995 he was introduced to a whole new generation as doorman "Manny Cordoba" on the NBC sitcom *The Single Guy*. During a relative slowdown in his career in 1996, he took the time to indulge his passion for travel, and took off in a customized motor home, finally seeing the country and meeting and spending time with friends, family, and fans. This tour became a 1997 documentary called *Ernest Borgnine on the Bus*.

His gruff, expressive voice made him a natural for voice-overs; he could be heard in *All Dogs Go to Heaven* and *Small Soldiers*. In 1999 he took on the recurring voice role of "Mermaid Man," a superhero admired by the absorbent phenomenon *Spongebob Squarepants*, the top rated cable cartoon (where he re-teamed with friend Tim Conway as Mermaid Man's sidekick, "Barnacle Boy.") Perhaps the hardest working nonagenarian in show business, Ernie shows no sign of stopping. As I write these words he's working on two movies: *The Lion of Judah* and *Night Club*, and has just won the Screen Actor's Guild Lifetime Achievement Award.

I asked Ernie how he felt about the mind-blowing evolution of movies and movie-making over the past sixty years: "What can I say? I tend to always go back to the Turner Classics, though every time I watch one of my movies I say: you dummy, you could have done better! Movies then had a moral; they were realistic in their own way. Today it's all explosions and sex; to me it's not entertainment. It seemed as though we said so much more with less—it's sexier not showing the sex. We did more with an ankle those days than you can imagine…Bette Davis, Kay Francis, Joan Crawford…to me they were everything.

"Mickey Rooney said when studios collapsed, we just lost it. A director used to really watch your expression—the look in your eyes, your stance, everything; do as many takes as needed, *then* he'd say good, print. Today it's hurry up and shoot and let's go; the clock and money are ticking away. We're overdoing computer graphics, the noise level is insane, and what's with the commercials?"

I asked who he admired in his younger days, who influenced his career, and if there was any newer talent that had caught his eye:

"Gary Cooper, hands down, and Wallace Beery, he was it for me. You know, I'm proud to have won an Academy Award as a character actor; behind the scenes they're the ones that make the picture, a lot of the time. They make the leading actors look good. You give a character actor a chance, and they'll steal it away from you.

"Walter Brennan was a character actor par excellence, he won three Oscars; he almost did away with Bogey in *To Have and Have Not*.

"Orson Welles—what a sweetheart—met him in a bar in New York. My eyes were as big as plates when he walked up and shook my hand. He said, 'Ernie, if you don't win this Academy Award, there is no justice. You deserve it for *Marty*.' I said, th-th-th-thank you Mr. Welles!

"Every time I see Clint Eastwood, I say, Clint, you still haven't used me—come on—I'm right here! I think he's marvelous, same with Ron Howard; but Clint calls his own shots, and he hits two out of three times, and that's not bad. His brilliance is that he's *paid attention*, he knows his stuff; he's become a great director and actor because of it. Nothing's left hanging in his work; he creates well-defined, complete movies with a beginning, middle, and an end.

"Johnny Carson was a man unto himself; few people will approach his standards; he was a gentleman, a man for all seasons, the world was his oyster and everybody loved that oyster.

"In terms of younger talent, I'm certainly not watching it like I should, but I like Gary Sinise—terrific actor—and Kevin Kline when he gets the right parts. These days you have to take what you can get. It's tough, the choices are pitiful, so this younger pool of talent is left stranded in a way."

Though happily married to his present wife, Tova, for close to forty years, Ernest does leave four other wives in his wake. He was previously married to Rhoda Kemins, who he met while in the navy, Mexican actress Katy Jurado ("beautiful, but a tiger"[137]), Ethel Merman, and Donna Rancourt; he has one daughter with Kemins and two children with Rancourt. All in all, Borgnine hasn't spent much more than a year since 1949 unmarried.

He spoke to me frankly about married life: "It's been a wonderful life, it's brought me some wealth...if I didn't have those five wives to worry about I'd have been a *very* wealthy man. I always wanted a family with kids I could do things with, just the normal things, and we did for a while, but my first wife...it seemed like everything was fine as long as we were poor. The minute I got the Academy Award she put on the dark glasses and became Mrs. Ernest Borgnine. She kept saying I gotta go have my hair straightened—she had this beautiful curly hair—it was one thing after another.

"Once during my first marriage I called my wife—I was so excited, and tired—and said I'm done with the picture, I'm coming home. She said you can't come home. I said, what do you mean, I can't come home? She said there's no room for you...I said no room for me in my own house? Well...my family's visiting, she said. I'd been making a picture with Allan Ladd—a great actor—so I took him home with me, probably for moral support, who knows. At home I got a very weak 'hello, how are you'; I felt like a stranger in my own home. One thing led to another and that was the end of that."

Ernie called his month-long marriage to Ethel Merman in 1964 "the worst mistake of my life. I thought I was marrying Rosemary Clooney."[138] Riding high on the success of *McHale's Navy* at the time, Ernie said things started coming unglued during the honeymoon when he received more fan adulation than she did. Merman was loaded for bear. "By the time we got home, it was hell on earth," Borgnine recalled in a 2001 interview. "And after thirty-two days I said, 'Madam, bye.'"[139] Merman remained single after her divorce, devoting a chapter in her 1978 autobiography to the marriage which consisted of one single, blank page.

Ernie said, "I've always had this image of the ideal family in my mind...and I pretty much have it now, but the kids are scattered around, and even though they visit occasionally it's not like I always envisioned it, you know? But Tova's been incredible, so I'm thankful for her."

Borgnine has worked with countless greats including Helen Hayes, Clark Gable, Joan Crawford, Spencer Tracy, Gary Cooper, Montgomery Cliff, Bette Davis, Jimmy Stewart, and Kirk Douglas. He's appeared in comedies, westerns, war dramas, horror films, Biblical epics, even a musical. He's been a good guy, cop, crook, murderer, mob boss, Western villain, an Amish farmer; he's been Jewish, Asian, Irish, Swedish, Mexican, whatever the part called for. Now he's the oldest living person to have won an Oscar, and told me that "acting is still my greatest passion."

I said, Ernie, you're ninety-three years old and bursting with energy and life, what's your secret? He laughed and confided, "Well, I got into trouble on Fox News answering this one...I leaned over and whispered my secret in this guy's ear, and it went around the world and back again, got something like six-hundred thousand hits, and so I'm not going to tell you now, but I think it's the idea of thinking young. You can't just sit in your chair and read the paper and nod off. Old man, my foot! Think and be young, and as long as you do that, you got it made. In terms of work, as long as they need me, I'm ready."

## FAVORITE MOVIES

*Marty* (1955)

Ernie won his Oscar for best actor that year.

*From Here to Eternity* (1953)

Two classic performances by Borgnine and Sinatra.

*The Wild Bunch* (1969)

Ernie teams up with an all star cast in this landmark Sam Peckinpah Western.

# What's Up, Doc?

Chuck Jones

Creator of more than three hundred animated films over a span of sixty years—three of which won Academy Awards—Chuck Jones helped shape the humor of three generations. During the golden age of animation from the mid-thirties to the early sixties, he helped breathe heart, action, soul, and life into some of Warner Brothers' most famous characters: Bugs Bunny, Daffy Duck, Elmer Fudd, and Porky Pig. Those he created himself included: Wile E. Coyote, Pepe lePew, Michigan J. Frog, and Marvin Martian. Many of these cartoons still enjoy world-wide recognition every day. He also produced, directed, and wrote the screenplay for *Dr. Seuss' How the Grinch Stole Christmas* as well as the feature length film *The Phantom Tollbooth*.

The minute I learned Chuck Jones would be in Boston for an exhibit at a Newbury Street gallery, I leapt at the chance for an interview. How could I miss this? If you can place a value on joy and laughter, then you can estimate the importance of Chuck Jones' life's work. My entire childhood was marked by those thrilling moments when the Warner Brothers logo appeared on the screen at the Adams Theater in Quincy. It was then that I knew, burrowing into my seat with popcorn and Raisinettes, that the bad B movie was over and I could finally see what Bugs was up to.

Jones looked like a man who made cartoons: with his wispy mustache and impish quality he could have been one of his own creations. He was a delight to interview. Born in Spokane, Washington, in 1912, he grew up in Hollywood and was shaped by its wonders. As he explained in our interview,

"We lived on Sunset Boulevard, right across from Hollywood Hills, and my father had an orange grove, which he sold eventually. Charlie Chaplin's studio was two blocks from us, so we'd walk over and watch him work right through a wire fence. These were silent movies, remember, so they didn't need the sound protection. We were friends with a very popular comedy team called Ham and Bud. Bud was a midget and I'd run around in his costumes when I was six years old, and he came to visit us. We saw Mary Pickford during the war—she came riding her white horse down Sunset at the head of the 160th infantry.

"We got to watch lots of stunt work done over the ocean. A guy doing wing-walking was much more likely to live falling sixty feet from these planes that flew at a very slow pace over the ocean; in any case it sure beat landing on the ground. Anybody who happened to be there could be in the crowd scene, and the pay was lunch. I was far too young to be an actor. On the other hand, they tried out my six-month old brother to play a baby, but he failed; he really was one odd-looking kid."

In his autobiography, *Chuck Amuck* (the follow up was *Chuck Reducks*), Jones credits his father, an unsuccessful businessman, for instilling in him his artistic tendencies. For every new enterprise his dad started, he'd order reams of stationary, envelopes, and business cards. As the failed endeavors piled up, so did the paper, which his dad insisted Chuck and his siblings make use of as soon as possible. They drew constantly. Later, in one of Jones's art school classes, a professor proclaimed that each student had one hundred thousand bad drawings in them that they must first get out of their systems before the good stuff could appear. Jones remarked with some relief that since he was well past the two hundred thousand mark, he hoped that he'd purged himself of the dreck.

After graduating from what is now called the California Institute of the Arts, Jones drew pencil portraits on the street for a dollar apiece. He "blundered into animation" in 1932, as a cel washer for former Disney animator Ub Iwerks. It was there he met Dorothy Webster, who he married the same year.

Chuck spoke in detail about the earliest days of animation: "In the twenties we lived on *Felix the Cat, Alfalfa,* the early *Terry Tunes.* William Randolph Hearst actually had a studio at the time called the Metropolitan; he put out *Mutt and Jeff* and *Maggie and Jigs.* Now, keep in mind: they didn't do much; just walked and ran and jumped up and down...but the big deal was that no one had seen characters move before, so it was a delight just to see a still picture move, just like when sound came out. You didn't have to do much, just make some kind of noise or speak, and people were in awe.

"The kind of animation we did then was inspired by Disney's *The Three Little Pigs*: their actions were very much alike with small differences that revealed their character. We may have been young but we weren't dumb: we saw something we could grab hold of. Disney went on to make *Snow White and The Seven Dwarves,* all who looked mostly alike except for Dopey, but again, they were characterized by how they moved and whether they sneezed, things like that. (Disney's) golden age in short subjects was from 1933–41 when their best people went on to make features like *Fantasia.* We didn't feel in competition exactly, we were just learning from what we saw. So by the middle of the war we'd begun to really understand how to animate characters.

"The personality of our characters became more irreverent, more—how can I say this—perky? Definitely insouciant. By post-war everyone was doing that; everything was breaking wide open."

By 1936 Jones had been hired by Fritz Freleng as an animator for the Leon Schlesinger Studio, working under Tex Avery. Along with animators Bob Clampett, Virgil Ross, and Sid Sutherland, space grew tight, so they moved into an adjacent building on the lot they christened the "Termite Terrace." Jones' very first cartoon was *The Night Watchmen* featuring a cute kitten later called Sniffles. His work was called overbearing and lacking in humor at the time; Jones admitted he finally "learned to be funny" with his cartoon *The Dover Boys* in 1942.

In the late forties and fifties he created much of his best-regarded work including Claude Cat, Marc Antony and Pussyfoot, Charlie Dog, Michigan J. Frog, and his three most beloved creations: Pepe lePew, the Road Runner, and Wile E. Coyote. The Road Runner cartoons in addition to *Duck Amuck, One Froggy Evening,* and *What's Opera, Doc?* are considered masterpieces.

Of Pepe lePew, Jones said, "Pepe is the individual I always wanted to be, so sure of his appeal to women that it never occurs to him that his attentions might be unwelcome, or even offensive. I tried to make Pepe's confidence a part of my own personality, hoping to share in his sexual success. On the screen it worked."[140]

Jones explained in our interview how he believed the quality of the cartoons went up along with a deepening understanding of each character. I tended to view Bugs as a tough character, with an urban quality almost like the Bowery Boys. Jones explained: "The Bowery Boys were a take-off on the Dead End Kids; they were definitely tough. Bugs is more...capable. He's the type of character I'd like to have been, personally...but he's not the way I am. He's an extension. I'd like to be a male Dorothy Parker. Hell, I'd like to be a male Burt Reynolds. Someone with that kind of charisma. Most of our characters are failures or prone to making a mistake, like Wile E. Coyote or Daffy Duck or Yosemite Sam. I think that's one reason people respond, because we're all much more familiar with mistakes than triumphs. Anybody's lucky to get one triumph a year...if I made ten pictures a year and one was outstanding in the public eye, then I'd be thrilled. Historically, more have become endearments to the public than I ever expected.

"When Bugs came out, took a bite of the carrot and said, 'What's up, Doc?'—well, that was a breakthrough in character, this attitude he took on. This was when Tex Avery did the *Wild Hare,* when Elmer Fudd was introduced. Mel Banks, (who was the voice of Bugs), said Bugs was a little stinker, but that's wrong; he's urbane, sophisticated. He didn't waste words.

"He could befuddle Daffy when trying to decide whether it was duck season or rabbit season. Daffy said: 'I know what the trouble is: pronoun trouble!' The point is this: we never made pictures for children—we never made pictures for adults! We made them to entertain ourselves. And there was no such thing as a Nielson rating. The only feedback we ever got was from the exhibitors."

As I thought back to all the wild, far-out situations Jones created for his characters, I wondered about the extent of his artistic freedom at the time of Jack Warner and Louis B. Mayer. Evidently, indifference fostered and unfettered their creativity. Jones said, "In the beginning, when we worked for Leon Shlesinger and sold (the cartoons) to Warner Brothers, our shorts were sold bundled with the features: they were sold before they went out, which is why we could experiment so much. Later that changed, but at the time, Jack Warner knew nothing about the animation department. I didn't even meet him until I'd been directing for fifteen years, in the business for twenty-five.

"One day they invited us for lunch in the so-called executive dining room where we'd never before set foot. By that time we worked under producer Eddie Selzer, who hated laughter; a great person to put in charge of an animation studio! Anyway, they knew money was coming from somewhere...Harry Warner said to me: 'I don't know a thing about our cartoons except we make Mickey Mouse.' That showed how little they thought about the short subject, till TV came along and then they realized, hey, there's money here..."

I asked Jones about critics and if they had an influence on what he created. He said, "We could make very public mistakes because there really wasn't anything to compare us to; we were lucky that way. No one knew how to critique animation; the first real critics came along in 1962."

What did curtail creativity were some strict editing requirements for short subjects. Jones said, "It's very specialized work. We were bound by a tough discipline; Spielberg, Scorsese, and Lucas were always surprised to hear that we edited our work to the second. The editor had to learn to time the picture, do the key drawings for an entire film, then send it to the animators, recording the dialogue first. It had to measure out at exactly five hundred and forty feet, which is six minutes, because the exhibitor needed exactly that to build out their program to two hours, which would be a feature, newsreel, and short subject, perhaps a coming attraction. We did thirty cartoons a year for fifteen years this way.

"I think it's true of all creative work that you have to operate within a discipline, and in the end, that becomes an advantage. It forced us to come up with an ending that worked; with a Pepe lePew or Bugs story, it took a lot of artifice to make it work. We always started Bugs out in a natural rabbit environment: him burrowing through the earth in a carrot patch or in the forest; he's provoked by someone and he fights back, which is something we'd all like to do, with the style he does it."

In the early sixties, Jones and his wife Dorothy wrote the screenplay for the animated feature *Gay-Purr-ee*, featuring the voices of Judy Garland, Robert Goulet, and Red Buttons as Parisian cats. The final cartoon Jones created before the cartoon studio closed in 1963 was *The Iceman Ducketh*.

With business partner Les Goldman, Jones wasted no time ramping up again. He began Sib Tower 12 Productions, pulling in most of his gang from Warner Brothers; soon MGM contracted with Jones and his staff to produce new Tom and Jerry cartoons. His short film *The Dot and the Line* won an Oscar in 1965 for best animated short.

Jones stepped into television as his Tom and Jerry series lost steam, producing and directing *How the Grinch Stole Christmas* and *Horton Hears a Who*, though his main focus was *The Phantom Tollbooth*. Though MGM closed their animation division in 1970, Chuck once again struck out on his own with Chuck Jones Productions, producing children's television series as well as the 1979 compilation film, *The Bugs Bunny/Road Runner Movie*. After the death of his first wife, Jones met and married Marian Dern.

In the eighties and nineties, Jones sold cartoon and parody art through his daughter's company, Linda Jones Enterprises. Though not much of a fan of contemporary animation, he never stopped evolving and adapting, even creating new cartoons for the Internet based on his character Thomas Timberwolf. In 1999 he established the Chuck Jones Foundation to recognize, support, and inspire excellence in the art of character animation.

The Wagnerian mini-epic *What's Opera, Doc?* was inducted into the National Film Registry for being "among the most culturally, historically, and aesthetically significant films of our time."[141] To me, *What's Opera, Doc?* explained Wagner in a way that had never been done before on film. Jones outlined some of what went into extracting a cartoon from an opus-length opera:

"We had a wonderful musician named Carl Stalling, who helped us lay out the story on score sheets; he wrote in all the musical accents we should be aware of. All of our pictures needed to be laid out in musical terms, so we had to be thinking musically all the time.

"We took the fourteen hours of Wagner's Ring Cycle and squashed it down to six minutes. A picture like that or the *Rabbit of Seville*, the music has to be just right. We had an eighty-piece orchestra playing Wagner. But I also added lots of timpanis; they gave a wonderful feeling, especially for that guttural roll just before a character is shot out of a cannon."

I told Jones that my best musical analogy for Bugs and his gang was a tight yet loose-knit swinging jazz ensemble, like what Benny Goodman did with Lionel Hampton. Chuck said, "You're the first one, Ron, who's noticed that similarity to swing music and early jazz. These cartoons are orchestrated, but you had to make room for the funny, you know?"

I asked Chuck what he was most proud of. Without a doubt, it was the characters he created, or helped to develop: "The fact that an ordinary person who's not just big muscles—Bugs—can go and handle matters with cleverness and intelligence is very important, and I think the coyote's a hero because he never stops trying. The sympathy is with him, not with Road Runner—you know he's invulnerable—but the coyote represents the multiplication of all the problems everybody has.

"I guess it's pretty simple: to find that people get joy out of something that you got joy doing is the greatest accolade that can happen."

As Bill Schaffer stated so brilliantly in his article for *Senses of Cinema*, "Whaddya Know, It Dithintegrated" published after Jones' death in 2002:

> "Like the rest of us, Chuck Jones, the man, was born at a definite date into a mortal body; one destined to die at an equally definite time and place. Unlike most of us, however, Chuck Jones, the animator, was able to isolate and revivify selected aspects of his personality, endowing them with a completely different, elasticised, effectively immortal kind of life."[142]

The mother of a four-year-old put it much more simply. Upon hearing that Chuck Jones had died, her child asked through her tears, "Does this mean the bunny won't be in the barber chair any more?" The answer was, "No, sweetie, the bunny will be in the barber chair forever."[143]

## Favorite Cartoons

### The Carl Stalling Project: Music from Warner Bros. Cartoons from 1936–1958

Stalling was the man behind the scenes at "Termite Terrace" who composed and arranged the music for the golden age of animation at Warner Bros.
Selections include:

"Dinner Music for a Pack of Hungry Cannibals"
"Beep Beep"
"Nutty News"

"Gorilla My Dreams"
"Hot Cross Bunny"
"Behind the Meatball"
"What's Cookin' Doc"
"I Got Plenty of Mutton"

Warner Bros. CD 926027-2

**Looney Tunes: Golden Collection**

A DVD that includes 60 of the finest and funniest cartoons from Warner Bros. Fully restored and uncut. As it says on the cover "A 24–Carrot Gem of a Collection. Anything less would be DETHPICABLE!"

Warner Bros. DVD 31284

# Dean of American Crime Fiction

Robert and Joan Parker

Photo courtesy of Joan Parker.

I'll never forget reading Robert B. Parker's novel *Ceremony* and quite literally stumbling upon a mention of—myself. Spenser was on a stakeout in his car listening to *MusicAmerica* on the radio to pass the time. What an odd and wonderful feeling to be part of a quintessential Boston scene. My show had suddenly become immortalized under the masterful pen of the dean of American crime fiction! I began to wonder about the role music played not only in Parker's life, but in his writing and creative life as well, and invited him for an interview on my show.

Parker was a prolific best-selling writer who churned out more than sixty books including Westerns, historical fiction, young adult novels—even a marriage memoir—but it was his detective fiction for which he'll be best remembered: specifically for the thirty-seven novels that star Spenser, Parker's signature creation: the soft-hearted bruiser with a taste for doughnuts and the Red Sox. In fact, Parker gave Spenser a number of his own traits: he was a great cook, boxer, weightlifter, jogger, and dog lover—both Parker and Spenser owned a series of short-haired pointers, all named Pearl.

Born in Springfield, Massachusetts, in 1932 to working-class parents, Parker went to Colby College in 1954, then served as a radio operator with the U.S. Army in Korea. I asked him about his earliest musical influences: "I grew up in New

Bedford listening to Bob and Ray. Also Fred B. Cole, Norm Nathan, Dave May-nard, Norm Prescott, Stan Richards, and Bill Marlowe. There was a brief flourish of Benny Goodman and swing from 1938 to the advent of Elvis in the fifties; it all seemed to pass so quickly! I think keeping this kind of music alive—swing, big band, jazz vocalists like Jimmy Ricks—has always been a struggle. I came back from Korea and thought: who's this guy with the funny haircut and the shoes? I'd never heard of Elvis. And he was the king.

"Both my sons are performers; my oldest, David, is a choreographer in New York, my younger, Dan, is an actor and singer in Los Angeles. They both dig this music. They know it well, in fact David has choreographed to it and Dan sings it in cabaret; it endures with people who like music. Duke Ellington said: 'there's two kinds of music, good and bad.' I'd say a lot is just plain bad."

Parker earned his M.A. in English from Boston University, then took work as a technical writer, copy writer, ad exec, and briefly as a state cop before returning to academia in 1962, finally earning his PhD in detective fiction from Boston University in 1971. His dissertation, called "The Violent Hero, Wilderness Heritage and Urban Reality," discussed the exploits of fictional private-eye heroes created by Dashiell Hammett, Raymond Chandler, and Ross Macdonald. Parker became a full professor at Northeastern but retired in 1976. By that time he'd already published five Spenser novels and it was clear that fiction was his true calling. That year Parker's *Promised Land* won the Edgar Allen Poe award from Mystery Writers of America.

Parker once observed that he created Spenser because Chandler was dead, and he missed Philip Marlowe. So much so that in the eighties Parker was chosen to complete *Poodle Springs*, a novel that Raymond Chandler had not finished at his death in 1959. He added thirty-seven chapters to Chandler's opening four, "(sounding) more like Chandler than Chandler himself,"[144] as the novelist Ed McBain once remarked. A year later Parker published *Perchance to Dream*, a sequel to *The Big Sleep*.

One of the best ways to understand Robert Parker the man is to look at the iconic characters he created. Spenser is named after Edmund Spenser, a contemporary of Shakespeare's who examined knightly virtues. He was never given a first name; the rumored reason being that Parker couldn't choose between the names of his two sons: David and Dan. Spenser's longtime love was psychologist Susan Silverman, who one could argue is a stand-in for Joan, Parker's wife of fifty-two years, also a therapist. Where Susan represents Spenser's rational side, Hawk, Spenser's asocial, violence-prone partner and hit man, is another extension of Spenser's personality. The narrative friction that resulted from reconciling Spenser's domestic, rational side with his penchant for brutality resulted in this constant stream of creation that was Parker's life work.

The first few pages of *The Godwulf Manuscript*, the book in which Spenser debuts, reveal the heart and soul of what readers came to love about him: his self-confidence, impatience with grandiosity, and sharp wit:

"'Look, Dr. Forbes,' Spenser says to the long-winded college president who is hiring him. 'I went to college once. I don't wear my hat indoors. And if a clue comes along and bites me on the ankle, I grab it. I am not, however, an Oxford don. I am a private detective. Is there something you'd like me to detect, or are you just polishing up your elocution for next year's commencement?'"[145]

Parker even dealt with his marital problems indirectly in his books. He separated briefly from Joan in 1982 just when Susan left Spenser. Later, in *A Catskill Eagle*, Spenser rescues Susan from captors and they reconcile their differences. Susan wants Spenser to move in with her, but Spenser had come to like having his own place. After a two-year separation from Joan, they reunited with a new arrangement—they bought a fourteen-room 1869 Victorian home with independent apartments. Robert took over the ground floor where he kept up his prodigious five-page-a-day, four-novel-a-year output. "It comes easily," he told the *Wall Street Journal* in 2009, "and I don't revise because I don't get better by writing a new draft."[146] Afterwards he would hand the manuscript over to Joan for her thoughts. Parker never planned how the story would turn out until he'd written it, making the writing process an adventure both for the characters and the author.

Another groundbreaking element of Parker's work was his inclusion of characters of varied races and sexual persuasions, perhaps influenced by his openly gay sons. Hawk and Chollo are African American and Mexican American; there are Chinese, Russians, even a gay cop and a gay mob boss, Gino Fish. This modern sensibility, in my mind, counts almost as much as the pleasure he gave to his millions of readers for the stories he told.

To help prepare for our interview, Parker ticked off a list of his favorite artists: Nat King Cole, Sinatra, Mel Tormé, Carol Sloane, Ella Fitzgerald, Carmen McCrae, Stan Kenton, Ben Webster, Dave McKenna, Bobby Hackett, and Dave Brubeck. He was especially enraptured with Ella: "She was the first lady of song. I'll never forget going to the movies when I was a little kid and watching her sing 'A-Tisket, A-Tasket.' She was probably seventeen, eighteen years old: tall, gangly, with the purest voice that's ever sung popular music; it was like a flute—amazing what she did with that voice. To me she seemed eternally young, the same at sixty as twenty, and she could sing anything, jazz, ballads, novelties."

I asked him whether he listened to music as he wrote his books. "Absolutely not. One of the things a writer does is listen to the way the words sound in his head. I listen before I write, or after. I drive across the country a couple of times a year with Pearl the wonder dog, because we have a house in Los Angeles and I won't fly her out, so I drive out with her every winter. I listen constantly on the way out, nonstop. You'd be amazed at all the music-of-your-life stations there are across this country! Still, there are fewer and fewer places to hear this kind of music any more. In any case, I like the simple, strong, melodic lines that resonate with the way I write. There is a noir music: the blues, lots of saxophones. Ben Webster *is* noir with that guttural sax that he plays."

To help him picture Spenser in different situations, Parker loved to listen to Lee Wiley among others. He said, "Some singers just call up Spenser's Boston for me: he's in black tie, she's in an evening gown...Lee Wiley has a smoky, sophisticated, urban café society sound to her. I remember getting a 45 one Christmas...she sang with Bobby Hackett...I thought that was one of the great albums of all time. Late thirties to fifties female singers had that quality, not now...now they all sound like Joni Mitchell with that soprano hoot, as opposed to Anita O'Day, June Christy, Chris Connor...these women just sounded to me like they'd lived more.

"But don't forget: Spenser also digs the big band stuff; also jazz greats like Stan Getz, J. J. Johnston, Stan Kenton. I prefer vocals, but that's what I grew up on.

"Hawk's musical tastes, believe it or not, are a bit more sentimental. He's a Ben Webster or Coleman Hawkins guy; he loved a smoke-filled jazz club, Charlie Parker, Miles Davis. But he also dug Olatunji, Afrocuban drums."

Parker also talked about the astounding popularity of his books in Japan: "For reasons which are a complete mystery to me, I'm a cult figure in Japan. There are Japanese couples who name their children Spenser. I was there in '89 with Joan and the kids; four hundred people lined up for my book signing. I felt like a rock star. My wife said it's because they think I'm a sumo wrestler. Thanks, Joan! Anyway, there's a great deal of entrepreneurship around Spenser in Japan; there's a cosmetic line called Spenser's Tactics. I've written endorsements for various products over there.

"There's even a Spenser jazz book out there—a collection of all of Spenser's jazz. They went through all the books, took note of every piece of music that's mentioned, and compiled it. They sent me five copies, but I can only read the title and my name."

I asked him how he felt about the *Spenser for Hire* TV series. "Mixed, I guess. I didn't like everything about the show but I thought it was well shot. Tourism in Boston really exploded because of it, any location tie-in will do that; look at *Cheers*. Japanese nationals who wanted to see Spenser's Boston really came in droves. We spent a million dollars a week, did twenty-two shows a year for three years plus the pilot, so sixty-seven shows in all. Four were shot in Toronto, where we pretended we were in Boston."

Parker claims to have met Joan at a party when she was three years old, remembering her when they met again at Colby College in Maine, and marrying her in 1956. I asked if they shared any musical history. He explained, "Before we were married—a long time ago, we've been married for thirty-nine years—we used to listen to Matt Dennis, and 'Eager Beaver' by Stan Kenton. Sinatra's 'Violets For Your Furs' still bring a lump to my throat.

"I also loved Jimmy Ricks' deep bass voice—he played with The Ravens. It's the best I've ever heard, and it seems to emanate so effortlessly from him. They sang 'Love is the Thing.' If Joan and I do have a song, that's it. It was playing the night I proposed. She said, oh, all right."

Robert Parker was just the man I'd imagined him to be: charming, unflappable, and of course an incredible storyteller. He liked beer and baseball but knew the key to his success and happiness was to sit at his desk and spin out his daily pages, which is what he was doing up to the moment of his death in January of 2010. I don't know of any other author who not only captured the essence and spirit of Boston, but imbued such a masculine character as Spenser with such class and heart.

"He was a master of the genre, as many have noted," said Helen Brann, who represented Mr. Parker for forty-two years. "And he was the most fun, the most real, highly intelligent, witty, down-to-earth, warm, endearing guy I've ever known. I adored him."[147]

Parker visited *MusicAmerica* toward the end of its eighteen-year tenure, and I cherished his support at what had been a bittersweet time for me. "This music formed me," he said, "my sensibilities, Spenser's...it colored the backdrop of these books. It's been fun talking about what my characters listen to; in fact, in some cases I didn't consciously think about it until today. Thanks for including me in celebrating this country's musical heritage and everything this show has been about."

## FAVORITE BOOKS

*Ceremony*

*Poodle Springs*

Also: the movie *Appaloosa*, a 2005 Western Parker wrote that stars Ed Harris.

# Fabulous People/ Simple Food

## INTRODUCTION

Through the years, Ron and I have been fortunate not only to witness countless performances, but also to actually spend time with celebrities just talking and relaxing together in our home over a good meal. That said, there were times I'd get butterflies in my stomach while I whipped up a sauce or frosted a cake in the kitchen. It was the role reversal: suddenly I became the entertainer, while they became my audience.

Such pressure! But as a chef, I felt pretty confident.

The fact is, we all have to eat, and Ron and I are passionate about what we eat and love to share that enthusiasm with others.

Making simple, delicious food for the likes of Tony Bennett, Ben Heppner, Stan Getz, Dizzy Gillespie, Dave McKenna, and Eileen Farrell was a thrill. When musicians are on the road and traveling, it's no surprise that they relish a home cooked meal in a quiet setting with no gawkers in the background. It was a joy to be able to set a hearty table for these performers and, in the process, call them our friends.

*—Joyce Scardina Della Chiesa*

## Dizzy Gillespie: Braised Short Ribs and Key Lime Pie

One night when Dizzy was working a late gig in Harvard Square and was afraid he wasn't going to have time to eat, I delivered ribs and pie to him. I slow cooked the ribs with a hot, tart sauce. They were at the "falling off the bone" stage and finger-licking good by the time they reached him. His lime pie had a flaky golden "blind baked" crust with a hint of ginger, a sweet–tart creamy custard, and fresh whipped cream. Delightful, he says!

# Dave McKenna: Chicken Provencal

Cooking for Dave was one of my great joys. He was the most enthusiastic audience I could imagine: he loved everything I made for him, but this dish was one of his favorites. I sautéed the chicken, then simmered it with a touch of garlic, anchovies, capers, beautiful tomatoes, fresh parsley and basil, a touch of dried oregano, salt, and pepper and served it with penne, rigatoni, or maybe even spaghetti and always freshly grated Parmesan. For wine: Always a hard decision...the red or the white?

# Eileen Farrell: Veal Scaloppini Marsala

Eileen would never start a meal without first enjoying a nice cocktail of top shelf vodka with one rock! Maybe followed by another...only then would we segue into veal scaloppini marsala with angel hair alfredo on the side and lots of grated cheese. Tiramisu and Sambuca would top it all off. Eileen had countless funny, provocative stories about her time at the Met. After she left the Met and the legendary Maria Callas had been fired by General Manager Rudolf Bing, Madame Callas was quoted as saying, "What is the Met anyway...they don't have Farrell!"

# Tony Bennett: Simple, Classic, Italian Food

A little pasta, usually penne or rigatoni. A simple sauce of chopped tomatoes sautéed with a touch of garlic. Fresh herbs, salt and pepper...not too spicy. Add the boiled pasta to the saucepan and toss thoroughly. Shave the beautiful Parmesan on top! Served with an arugula salad and hours of talking about the great artists, from Michelangelo, John Singer Sargent, Daniel Chester French, and David Hockney to his own work. Tony lives and breathes art. "Vissi d'arte!" For wine: a glass of red, perhaps Tuscan.

# Ben Heppner: Spaghetti all'Amatriciana

In our dining room with the horned old victrola, Ben and his agent Bill came for dinner on an evening when Ben had a night off from his Boston engagement. I served spaghetti with bacon, sweet onions, crushed red pepper, and chopped tomatoes and basil followed by a lightly dressed, lemony Caesar salad. When I brought out the second course: crispy chicken cutlets with sautéed spinach, Bill exclaimed, "Oh my God, I thought that was it!" For wine: Pinot Noir and a Sardinian white.

## Stan Getz: Surf & Turf

On our roof deck one evening in the South End of Boston, Stan and Charlie Lake, aka "The Whale," enjoyed the antics of our cat Radames as I served surf & turf: grilled sirloin steak and large marinated grilled shrimp with sautéed mushrooms, zucchini, onions, and herbs, with oven-roasted potatoes coated lightly with extra virgin olive oil, fresh rosemary, tomatoes, toasted pine nuts, and crumbled gorgonzola. After our meal, as if on cue, Radames jumped into his wicker basket and Stan picked him up. Standing there holding the cat in his basket, Stan looked at him and said, "I dig you, man." We all laughed as the two hip cats stared at each other, just grooving away. We have a little snapshot of that moment.

## Illinois Jacquet, aka Jean Baptist "Illinois" Jacquet: Fried Chicken with All the Fixings

Illinois danced up a storm when I served him crispy succulent fried chicken with mashed potatoes and brown gravy, butternut squash puree with a touch of cinnamon, and long green beans cooked with diced onions and tomatoes. His feet did this crazy dance imitating all the mashing and squashing the meal involved, all the while sharing musical history with the embellishments of a seasoned performer. Lemon meringue pie and coffee capped off the meal.

## Richard Cassilly and Patricia Craig: A Christmas Feast

We shared so many memorable evenings and holidays with this famed opera couple. Dick loved to cook, and while some meals took elaborate planning and preparation, others were spontaneous and whimsical. One Christmas we had fresh oyster stew with hand-made star- and snowman-shaped oyster crackers, whole poached salmon with lemon caper hollandaise sauce, Potatoes Anna, fresh green beans with toasted almonds, all washed down with lots of champagne. Desserts were pumpkin pie with whipped cream flavored with maple syrup and apple cranberry crisp with peppermint gelato.

## Robert Merrill: Lobster Savannah

Marion and Bob Merrill loved to come to Boston, mostly I think to have lobster Savannah with us. This dish is prepared by baking the lobster, removing the tail meat and adding to it sautéed confetti-diced sweet peppers and shallots, cooked down with brandy and finished with cream. Replace combined mixture in the shell, sprinkle with cheese, and bake until hot and bubbly. A person could just pass out with happiness after a meal like that!

## Ruby Braff: Dorchester Bouillabaisse

Turns out, the impudent man likes the impudent oyster! Ruby was an angel when he held a coronet to his lips, but look out if he had an ax to grind and you were there; few were spared! But the irascible Ruby showed his loveable side when it came to dining. He believed nothing could beat a certain Cape Cod fish house not too far from his cottage. We ended up going there with him and it wasn't bad for a local fish place, but of course I thought I could do better. When he visited us in Boston I served him up a pot of seafood I learned to make in Marseilles many years ago. Ruby flipped! He went on and on, stammering while he slurped everything down with crusty French bread with rouille (a fiery sauce of hot peppers, garlic, olive oil, and breadcrumbs). It wasn't exactly a Bouillabaisse, but it was my New England version of it.

## Ernest Borgnine: Fresh Baked (and Caught!) Halibut

Ron and I met Ernie in an odd way. Ernie's best pal was a man named Charlie Brown. Ron's former wife Jackie and Charlie were partners and had a lodge in Alaska where they spent their summers, mostly fishing for halibut and salmon. Jackie and I had become great friends; she was more like a sister-in-law and often invited us to spend time up there. How lucky we were! Not only did we learn how to fish, but we met this wonderful actor with roots in Connecticut and Italy, and now Hollywood. These people adored each other. There was no pretense; just fish, tall tales, and movie and race car stories. And nothing could beat freshly filleted halibut on a white-hot charcoal grill. I brushed the fish with olive oil, coarse salt and pepper, Tabasco, and lime, then placed it on the grill with corn in the husk with the silk removed and big baked potatoes. I served mushrooms sautéed in butter along with the fish, corn, and potatoes. And as the Northern Lights began to glow in the sky, it was time to serve the Baked Alaska!

## Norman Kelley: A Summer Table of Earthly Delights

Norman loved to entertain at his house in South Easton, Massachusetts. After a few of his meals of "Loaf Surprise," usually with crushed potato chips mixed in or garnished with canned onions rings, I told Norman I would help him with his parties. One Fourth of July there was a gathering at one of his friend's houses with Norman insisting we have whole salmon as was traditional in Maine. "I'll bring the salmon...having it flown down you know!" When we all came together at his friend's eighteenth-century farm house; the day was hot but promising, an elaborate table set in a beautiful low-ceilinged room for the celebration. People had brought various items and dropped them somewhere near the kitchen...his friend seemed oblivious to this. Nothing was happening to forward the dinner. Everyone was having a marvelous time with cocktails and talk. By the time I inquired as to how we were going to proceed, the friend was ossified! I look in the kitchen: disaster! I looked in Norman's cooler...a huge whole salmon, head on, not scaled and very slippery. Something had to be done and fast. With Patton in command, in less than two hours, the serving table groaned under the weight of the following: poached salmon with dill sauce; orzo with cucumber, tomatoes, parsley, and feta; bruschetta with baby grilled eggplant and zucchini; and fresh mozzarella with basil and the greenest of extra virgin olive oil; and fresh strawberries, shortcake biscuits, and whipped cream. Norman was in his glory. "Come on...there's nobody like us!" was his constant toast!

## Harry Ellis Dickson: Roofdeck Lazy Lasagna

We enjoyed several dinners with Harry, many of them at our house. From our roof deck overlooking the South End, we would sip red wine while Harry regaled us with stories of the classical music world, Arthur Fielder, and the Pops. A typical meal that delighted Harry was pasta, an assortment of antipasti including grilled calamari, baked ziti with fresh mozzarella and tiny meatballs...a kind of lazy man's lasagna, and a lemon tart with blueberries, which we ate with a delicious Auslese wine.

## Gunther Schuller: Pot Roast with Sauerbraten

When Gunther Schuller arrives at your house, he's usually armed with a small jar of pickled onions for the Gibson he hopes you will provide. He did each time he came to ours, though I tried to stock up on onions as soon as I caught wind of how much he loved a "Sapphire" Gibson. As Ron showed Gunther his studio, I busied myself with the pot roast with sauerbraten, which was finally baking in the oven after days of marinating in red wine. To go with the meat were dumplings and noodles, and red cabbage sautéed with toasted caraway seeds. It was beautiful to behold, and Gunther seemed to relish the meal and enjoy himself thoroughly. We topped it all off with snifters of brandy while we listened to our guest's wonderful adventures and insights.

# Jean Shepherd: Picnics at Blueberry Rock

We visited Jean and his wife Leigh Brown in Sanibel Island, Florida, as well as their summer home in Maine, where we would cook in as often as eat out. They also came to my restaurant, the Turtle Café, or our home, when Jean was shooting one of his movies in Boston. I soon learned that even though Jean was in one way an everyman, he had strong opinions about food. One evening at the Turtle, we not only closed the place but stayed after all the help had gone home while Jean finished the cheesecake and another pot of coffee. That night we dined on fresh made gnocchi with tomatoes, oregano, basil, and parmesan, roasted local cod with butterbeans and greens, sautéed chicken marsala with shitake mushrooms, and pork tenderloin with a ginger five-spice rub and sweet potatoes. I believe a bottle or two of Italian wine disappeared along with this feast. On a visit to Waterville, Maine, I compiled a picnic to take to Jean and Leigh's favorite spot, Blueberry Rock, a huge boulder we climbed to enjoy an incredible vista of mountains and ocean. I brought pate with toast points and cornichons, hummus with pita bread, mini-tuna and egg triangles with olive toothpicks, sliced salami and breadsticks, prosciutto wrapped around cantaloupe, and chunks of Parmesan followed by local strawberries and homemade biscotti. We toasted the great Blueberry Rock with a bottle of Italian Prosecco.

# Dick Johnson: The Man Loved Everything!

Dick Johnson was the most ebullient man. He could turn a room on just by walking in—upbeat, suave, impeccably dressed, smiling…always excited to do the gig. He might have come from another late night gig hours away, but he was always just happy to be there. He loved to play, he loved his work, and he loved to eat. His dear wife Rose is a legendary cook, and I could never go toe-to-toe with her meatballs! That said, I had the pleasure of cooking for Dick a few times, and it was a joy. The cat was always so appreciative, and we enjoyed every meal together. We shared a variety of foods: grilled sausages, mounds of seafood spaghetti, penne ala vodka, buttered mushrooms, and always a dry martini or two!

# Bobby Short: Chateaubriand with Béarnaise Sauce

A beautiful glow in a soft pink light. That is how Bobby appeared to me when I first saw him at the Carlyle in New York City. The most sophisticated and urbane man, he was a delight to watch and a joy to have as a friend. He often worked in Boston and would extend his stay to visit with many of his New England friends. Bobby loved taking his friends out on the town, but I finally got him to come to our house for dinner. Though he said he preferred the simple life, he also loved great food and wine. That evening I prepared gravlax and served it with iced Veuve Cliquot champagne. For our main course, we segued into Chateaubriand with classic Béarnaise sauce and broiled portobello mushrooms with baby leeks and individual potatoes gratinee. This course was served with a Chambolle-Musiney, which was as over the top as the meal. Bobby was glowing; he was so happy to have a home-cooked dinner and one that was so memorable. He was gracious and so grateful, it almost makes me cry when I think of it. But hold on, for dessert, we had a Grand Marnier soufflé with fresh raspberry sauce...and oh, was it grand!

## Luciano Pavarotti: A Post-Concert Feast

Several years ago, when the great tenor first came to Boston, there was an after concert reception at the Wang Center, which was at the time known as the Metropolitan Center. I was so excited not only to hear him sing, but to attend the reception afterwards. Knowing he loved to eat, I was bold enough to send some food to his dressing room earlier that day. I kept my fingers crossed, hoping he would get it. After the concert, I stood at the end of a long line of admirers waiting to meet the maestro. I could see that he was seated at a long table, signing programs and exchanging pleasantries with his adoring public. But I discovered, as I drew closer, that there was something else he was doing. Twirling spaghetti with one hand as he signed programs with the other! Amazing! When my turn finally came I asked if he'd received the goodies I send him. He immediately put down his fork, wiped his mouth, stood up, and exclaimed: "Amore! You are divine...and so delicious too!" as he hugged and kissed me on both cheeks. I totally swooned, and I believe Ron had to pull me away. Here's what I sent to Luciano: grilled shrimp marinated in fresh garlic, olive oil, oregano, and lime juice; tiny tomatoes stuffed with creamy smoked eggplant and olives; chicken wrapped in prosciutto drizzled with aged balsamic vinegar; and a healthy chunk of Parmigiano-Reggiano, sliced hard salami, and homemade breadsticks.

# acknowledgments

I am so fortunate in my over fifty years in broadcasting to have worked with some of the most talented and creative people in the industry. Wrapping up this book, I thought it only fitting that I mention many of the colleagues, friends, and relatives who have contributed in some way to my life.

I'll never forget my childhood teachers in Quincy: Miss Frances Mahoney who instilled in me a love of history; Arnold Rubin, my science teacher; and Steve Goodyear, who taught me the basics of the Italian language. To classmate Gerry Dempsey: thanks for sharing your copy of *Variety* with me—it changed my life! Many professors at Boston University's School of Public Communications were an inspiration: especially Victor Best, who instilled in me the importance of good diction and losing my regional accent.

I don't know where I would be if Arnie (Woo Woo) Ginsberg and General Manager Ralph Weinman of WBOS hadn't taken a chance and hired me for my first radio gig in 1959. I owe the same debt of gratitude to Don Otto for my stay in the '60s at WBCN, where I was privileged to work with Peter Wolf, Al Perry, Charles Laquidara, Jack Kearney, and my old BU friend, Bill Wayland, among others.

In the world of opera: I thank Milton Cross for bringing opera live from the Met to countless rapt listeners on Saturday afternoons. To Francis Robinson, Assistant Manager of the Metropolitan Opera, and John Tischio, President of the New England Opera Club: my heartfelt thanks and admiration. And of course, kudos to Arthur Puopolo at Court Square Styling, the opera barbershop where you could listen to Pavarotti and get a trim at the same time.

At WGBH, many thanks to David Ives, Henry Becton Jr., Jon Abbott, Marita Rivero, Emily Rooney, and Ben Rowe. To my on-air colleagues, present and past: Robert J. Lurtsema, Bill Cavness, Eric Jackson, Steve Schwartz, Laura Carlo, Kathy Fuller, and Greg Fitzgerald; and from the world of television: Fred Barzyk, who was instrumental in bringing Jean Shepherd to public television; Dick Robinson, founder of the Society for the Preservation of the American Songbook; and Dick Golden: thanks for sharing in this wonderful journey together.

I am grateful to the unmatched professionals at the Boston Symphony Orchestra including Mark Volpe, Managing Director; Tony Fogg, Artistic Administrator; Bernadette Horgan, Director of Public Relations; Gregory E. Bulger of the Gregory E. Bulger Foundation, underwriters for our BSO broadcasts; Brian Bell, producer of our BSO broadcasts; Jim Donahue, sound engineer; Judge Francis J. Larkin for securing underwriting for our early Tanglewood broadcasts from the International

Laborers Union of North America; Jamie and Ethan Berg at the Winthrop Estate in Lenox; and of course the men and women of the BSO.

It's not easy writing about music. Here are a few scribes who have made their words sing: Jack Thomas, Ernie Santosuosso, Ed Symkus, Bob Blumenthal, Richard Dyer, Lloyd Schwartz, and Joe Fitzgerald.

In the culinary world, there are so many friends to thank: From the Turtle Café days, Spider Landevin, his son Reuben and wife Gabriella; Leo Romero and his partner Iory Allison at the Casa Romero; Anita Baglaneas of Jules Catering in Somerville; Elio Richie and his son Frank of Raphael's South Shore Country Club in Hingham; David Colella, General Manager of the Colonnade Hotel; Lefty, Tony, Pat and Rockie at Santarpio's, one of the world's best pizzerias; Joe Baker's Sugar Bowl on Dot Ave; Kevin Tyo and his restaurant 224; Sal and Joe Sconamiglio of Patsy's restaurant in New York City—Frank Sinatra's favorite!—John Chan of Chan's restaurant in Woonsocket, RI; Gordon and Fiona Hammersley of Hammersley's in Boston; and Tommy and Pat Floramo of Floramo's where, of course, the meat falls off the bone!

More dear friends who have made my life richer include: Judge Francis J. Larkin and his wife Virginia; Liz Muir; Frankie Dee of GMMY radio; Dr. Michael Goldberg and his wife Fran; attorney Ron Itri and his wife Maria; movie producer and film scholar Dale Pollock and his wife Susan; singers Renee and Maria Rancourt; Chris Sarno and Janet; Jay and Kate Rooney; sculptor Lloyd Lilly; Arthur and Louise Gobbeo: thanks for the wonderful nights spent listening to opera at your home; Bob and Justine Graham; Carolina Tres Balsbaugh, Al Vega, Dr. Arthur Wills and wife Hannah, John Gillespie, Bob and Suzanne Lobel, Gary and Michelle Cohen; Bob Merwin; Terri Anthony and wife Karen Carbone: thank you for introducing me to Frank Sinatra, Jr.! opera lover Ruth Wells; John and Betsy Henning: our dear friends and traveling companions; Paul Nash and Mark McLaughlin; and Ed Pelletier who just turned 90 and doesn't look a day over 39!

For keeping me in Steve Reeves shape, I tip my hat to Chris and Greg at Gold's Gym on Broadway; thanks also to Dr. Harry Anastopoulas; Dr. Bert and Hannah Wills; and Dan and Norma Frank at the Price Rehabilitation Center.

I am so grateful for my friends at the magnificent Cathedral of the Holy Cross in Boston's South End where I've been privileged to be a lector for the past 25 years.

For their unflagging kindness and support: Mayor Tom Menino; Keith Lockhart; Joan Bennett Kennedy for her friendship and inspiration to write this book; Jonathan Soroff of *The Improper Bostonian*; opera diva Melanie Campbell; June and Dan Weiner of Galaxsea Travel, sponsors of our annual Sinatra cruise; Lawrence (Cha Chi) Loprete; Art Singer, founder of the Massachusetts Broadcasting Hall of Fame; Bill Russell and his partner Bruce Bossard; jazz promoters Sue Au Clair, Ted Belastock, Al Julien, and Fred Taylor; Mary Toropov, former coordinator of the WGBH Learning Tours and Peter Strauss of The Grand Tour; WPLM General Manager Alan Anderson and his staff; DeeDee Rose; Nick Muscato; Pam Donnaruma and staff at the Post Gazette; Vita Paladino, managing director of the Gottlieb

Archive Center at Boston University; my manager and partner Paul Schlosberg; Paul Kelly of Kelly Communications; David Mugar of Mugar Productions; Jordan Rich and Ken Carberry at Chart Productions; Joe Chinzi of *Ontray Magazine*; Chuck Sozio; attorney George Handran; and sports legends Gino Cappelletti of the New England Patriots and Sam Mele of the Boston Red Sox.

For early inspiration and help recording interviews for this book, many thanks to Sue Asci and Jeanne Horos Denizard. I thank attorney Lucy Lovrien for her legal wisdom and guidance, and publicist Jennifer Prost for getting the word out. At Pearson, I am indebted to Bill Barke, Ziki Dekel, Jeanne Zalesky, Kay Ueno, Karon Bowers, Daryl Fox, Sally Garland, Mary Dalton-Hoffman, Wendy Gordon, Phil Olvey, Bayani DeLeon, and Stephanie Chaisson, and Jessica Werley at Integra-Chicago, for their top notch professional guidance on this project.

To my son Aldo, Mary, and my dear grandchildren Tia, Gabby, Nico, Dominick, and Donovan, you have my love always. Thanks also to Robert Chiesa and his wife Madeleine; Joe Breken; and to my aunt Jane Wotton for taking me to one of my first operas at the Boston Opera house when I was 12 years old.

Love and gratitude to my extended family: Tony and Stephanie Treco; Dr. Gail Cave; Uncle Ken Treco and his daughter Yvonne; Aunt Dolores Wells who makes the best jam in the world; John and Lynn Scardina; Richie and Rosie Scardina; Joyce's sister, Donna; and the Lasorsa family of Staten Island.

Ditto for Giacomo, the world's greatest cat, who loves opera.

For the incomparable but not overly ostentatious Erica Ferencik who I discovered thanks to Jordan Rich. Collaborating with Erica has been one of the highlights of my life on this planet: I couldn't have done it without her. Thanks for the peanut butter sandwiches, and to her husband George for letting us sit there for hours in the living room reminiscing, laughing, and finally sorting it all out.

To my many fans and friends, if I have missed mentioning you, my deepest apologies.

This book is dedicated to the memory of my mother and father, Aldo and Florence Della Chiesa, for giving me life and inspiring in me my abiding love of art, music, and culture.

Finally, to the love of my life, my wife Joyce. You're all the world to me.

## Chapter 3

1. Nat Hentoff, *Jazz Is*, 54.
2. Bryan Marquard, *Boston Globe*, "Radio Legend Jess Cain Is Dead at 81," February 14, 2008.
3. Jack Thomas, *Boston Globe*, "The Gentle Voice of the Night," October 31, 1996.
4. Robert A. MacLean, *Boston Globe*, "Channel 7's Henning Reassigned," June 4, 1981.

## Chapter 4

5. www.ThinkExist.com
6. http://wbghalumni.org/profiles/c/cavness-bill/

## Chapter 5

7. Joe Fitzgerald, *Boston Herald*, September 13, 1995.
8. Globe Staff, *Boston Globe*, "The Day the Music Dies?" August 31, 1995.
9. Edward Grossman, *Harper's Magazine*, "Jean Shepherd, Radio's Noble Savage," January, 1966.
10. Joe Fitzgerald, *Boston Herald*, September 13, 1995.
11. Rob Hoerburger, *New York Times,* "Eartha Kitt, A Seducer of Audiences, Dies at 81," December 25, 2008.
12. Whitney Balliett, *The New Yorker*, 1973.

## Chapter 6

13. Globe Staff, Boston Globe. "The Day the Music Dies?" August 31, 1995.
14. Joe Fitzgerald, *Boston Herald*, September 13, 1995.

## Chapter 7

15. Joyce Curel, *Post-Gazette*, "Ron Della Chiesa, Mr. MusicAmerica," December 25, 1992.

## Chapter 8

16. Matheopoulos, *Domingo, My Operatic Roles*, Foreword.
17. Ibid.
18. Anthony Tommassini, *New York Times*, "Richard Cassilly, American Tenor, Dies at 70," February 4, 1998.

19. Donal Henahan, *New York Times*, April 4, 1986.

20. Institute of Jazz Studies, Rutgers University Libraries, http://newarkwww.rutgers
    .edu/ijs/bc/index.html, "Benny Carter: Eight Decades in American Music," 2009.

21. John Twomey, http://www.jazzsight.com/jazzsightprofiles.html, "The Troubled
    Genius of Stan Getz."

22. Ibid.

23. http://www.buddyrich.com/index.php

24. Whitney Balliett, *American Musicians II: Seventy-One Portraits in Jazz*, 263.

25. http://www.jazzprofessional.com/interviews/Ruby%20Braff_1.htm, "Ruby Braff:
    I've Always Hated the Trumpet."

26. *Associated Press*, George Shearing obituary, February 15, 2011.

27. Jake Coyle, *Chicago Sun-Times*, "Jazz Pianist George Shearing Dies at 91," February
    15, 2011.

28. Ellington, Edward Kennedy, *Music Is My Mistress*, 265.

29. Tony Bennett, Will Friedwald, *The Good Life*, 55.

30. John Lewis, *AARP Magazine*, "Tony Bennett," July–August, 2003.

31. John J. O'Connor, *New York Times*, "Tony Bennett and MTV: Talk About
    Bedfellows," June 1, 1994.

32. Lynn Elber, *Associated Press*, "Clint Eastwood tells Tony Bennett story for
    American Masters," September 5, 2007.

33. James Gavin, *Intimate Nights-The Golden Age of New York Cabaret*, 211.

34. Terry Gross, National Public Radio website, "A Conversation With Bobby Short,"
    March 21, 2005.

35. Short, *Black and White Baby*, 65.

36. Ibid.

37. Stephen Holden, *New York Times*, "Sounds Around Town," June 1, 1990.

38. Whitney Balliett, *The New Yorker*, "New York Voices: Fourteen Portraits By
    Whitney Balliett," 1970.

39. http://www.mckuen.com

40. Howard Reich, *Chicago Tribune*, "Magic in Music for Frank Sinatra, Jr.," February
    12, 2010.

41. Ibid.

42. Hat Nat Hentoff, *Wall Street Journal*, "The Other Frank Sinatra," September 1,
    2009.

43. Wil Haygood, *Washington Post*, "Come Fly With Me," July 9, 2006.

44. Nat Hentoff, *Wall Street Journal*, "The Other Frank Sinatra," September 1, 2009.

45. Wil Haygood, *Washington Post*, "Come Fly With Me," July 9, 2006.

46. Jerry Fink, *The Las Vegas Sun*, "Frank Sinatra, Jr. will play to a Vegas unlike his
    father's," September 2, 2009.

47. Wil Haygood, *Washington Post*, "He's Got a Big Heart and his Pop's Voice but Just
    a Shadow of His Success," July 9, 2006.

48. Nat Hentoff, *Wall Street Journal*, "The Other Frank Sinatra," September 1, 2009.

49. Ibid.

50. Ibid.

51. Howard Reich, *Chicago Tribune*, "Magic in Music for Frank Sinatra, Jr," February
    12, 2010.

52. Josh Getlin, *Los Angeles Times*, "Finally, a show with standards," July 18, 2006.

53. http://www.spaceagepop.com/torme.htm

54. NPR Biography, Nancy Wilson, http://www.npr.org/people/2101390/nancy-wilson

55. *Time Magazine*, "Singers, the Great Pretender," July 17, 1964.

56. "Footsteps of Civil Rights Leaders Placed in International Civil Rights Walk of Fame at Martin Luther King, Jr. Historic Site," http://www.prweb.com/releases/2005/9/prweb280109.htm

57. Richard Harrington, *Washington Post*, "Singer Rosemary Clooney, finishing on a high note," July 1, 2002.

58. Rosemary Clooney, *Girl Singer*, 86.

59. Richard Harrington, *Washington Post*, "Singer Rosemary Clooney, finishing on a high note," July 1, 2002.

60. http://www.rosemaryclooney.com/biography.html

61. Stephen Holding, *New York Times*, "Recalling Rosemary Clooney, Fondly and Lyrically," January 9, 2007.

62. http://www.rosemaryclooney.com/tribute/index.htm, Rosemary's Friends Remember.

63. Joan Merrill, *National Public Radio*, Jazz Profiles: Rosemary Clooney, http://www.npr.org/programs/jazzprofiles/archive/clooney.html

64. Sammy Cahn, *I Should Care*, 23.

65. Ibid.

66. Ibid.

67. *Sammy Cahn Songbook*, Warner Brothers Publications, Inc., 1986.

68. Robin Armstrong, "Sammy Cahn," www.enotes.com

69. Conati et al., Marcello, *Encounters with Verdi*, 303.

70. Sachs, Harvey, *Reflections on Toscanini*, 154.

71. Kimmelman, Michael, *New York Review of Books*, "Music, Maestro, Please!" review of *The Letters of Arturo Toscanini*, November 7, 2002.

72. Ibid.

73. Ibid.

74. Ibid.

75. Sachs, Harvey, *Reflections on Toscanini*, 9.

76. Ibid.

77. Lebrecht, Norman, *The Book of Musical Anecdotes*, 245.

78. Joan Anderman, *Boston Globe*, "'Where the word ends,' he keeps on," February 1, 2009.

79. Ibid.

80. Ibid.

81. *Los Angeles Times*, obituary, "Harry Ellis Dickson, 94; Boston Violinist and Pops Conductor," March 31, 2003.

82. Anne Midgette, *New York Times*, obituary, "Harry Ellis Dickson, 94, Violinist and Conductor in Boston," April 2, 2003.

83. *Los Angeles Times*, obituary, "Harry Ellis Dickson, 94; Boston Violinist and Pops Conductor," March 31, 2003.

84. Dickson, Harry Ellis, *Gentlemen, More Dolce Please: An Irreverent Memoir of Thirty-five Years in the Boston Symphony Orchestra*, 87.

85. Ibid.

86. Ibid.

87. *The Rest is Noise*, blog by Alex Ross, the music critic of the New Yorker, "David Raksin, 1912–2004," http://www.therestisnoise.com

88. Ibid.

89. Paul Zollo, *American Songwriter*, "American Icons: David Raksin," January 1, 2008.

90. *The Telegraph*, Obituaries, David Raksin, August 12, 2004. http://www.telegraph .co.uk/news/obituaries/1469196/David-Raksin.html#

91. *The Film Music Society*, http://www.filmmusicsociety.org/resources_links/ composers/raksin.html

92. Andrew Druckenbrod, *Pittsburgh Post Gazette*, "Music Preview: PSO, Van Hoesen to premiere Previn's Harp Concerto," March 6, 2008.

93. Bernard Holland, *New York Times*, "One Couple Fulfilling Three Musical Roles," March 12, 2005.

94. *Publisher's Weekly*, November, 1991.

95. Ibid.

96. http://www.andre-previn.com

97. David Mermelstein, *New York Times*, "Recordings; Emulating the Romantics in a 'Love Song,'" March 28, 2004.

98. Gerald Nachman, *Seriously Funny*, 266.

99. Ibid, 271.

100. Ibid, 277.

101. Ibid, 267.

102. Ibid.

103. Ibid.

104. Donald Fagen, *Slate*, "The Man Who told A Christmas Story: What I Learned from Jean Shepherd," December 22, 2008.

105. Jean Shepherd, CD: "Will Failure Spoil Jean Shepherd?" Collector's Choice, 2002.

106. Gerald Nachman, *Seriously Funny*, 269.

107. Ibid.

108. Ibid, 270.

109. Ibid, 273.

110. Ibid, 275.

111. Ibid, 268.

112. Ibid, 279.

113. Ibid, 281.

114. Larry McShane, *Seattle Times*, "'First Radio Novelist' Jean Shepherd Dies—Indiana Native was Multimedia Performer," October 17, 1999.

115. Gerald Nachman, *Seriously Funny*, 51.

116. Ibid.

117. http://www.mortsahl.com

118. Ibid.

119. Gerald Nachman, *Seriously Funny*, 53.

120. Ibid.

121. Ibid, 81.

122. Ibid.

123. Ibid, 80.

124. Ibid.

125. Mort Sahl, *Heartland*, Houghton Mifflin Harcourt, 167.

126. Gerald Nachman, *Seriously Funny*, 78.

127. *Newsweek*, review of *Heartland*, August 9, 1976.

128. Gerald Nachman, *Seriously Funny*, 95.

129. Ibid.

130. http://www.mortsahl.com

131. Ibid.

132. Steven Winn, *San Francisco Chronicle*, review: Mort Sahl's America, October 9, 1996.

133. Gerald Nachman, *Seriously Funny*, 93.

134. Ibid.

135. Amy Kaufman's Interview with Mort Sahl, *Los Angeles Times*, October 17, 2008.

136. Ernest Borgnine, *Ernie*, 35.

137. IMDB biography, http://www.imdb.com/name/nm0000308/bio

138. Ibid.

139. Ibid.

140. Chuck Jones, *Chuck Amuck: The Life and Times of an Animated Cartoonist*, 153.

141. http://www.animazing.com/gallery/pages/bio_chuckjones.html

142. Bill Schaffer, *Senses of Cinema*, "Chuck Jones," http://www.sensesofcinema .com/2002/great-directors/jones/

143. http://www.chuckjones.com

144. Ed McBain, *New York Times*, "Philip Marlowe is Back, and in Trouble," October 15, 1989.

145. Robert B. Parker, *The Godwulf Manuscript*, 45.

146. Sarah Weinman, *Los Angeles Times*, "Robert B. Parker left a mark on the detective novel," January 20, 2010.

147. Bryan Marquard, *Boston Globe*, "Mystery Novelist Dies at 77," January 19, 2010.

# text credits

Excerpt from *Jazz Is* by Nat Hentoff, p. 20: From *Jazz Is* by Nat Hentoff. Copyright © 1976 by Nat Hentoff. Published by Limelight Editions. Reproduced by permission of Hal Leonard Corporation.

From Globe Staff, "The Day the Music Dies?" *Boston Globe*, pp. 63–64: From "The Day the Music Dies?" from *The Boston Globe*, August 31, 1995. Copyright © 1995 Globe Newspaper Co. Reproduced by permission of PARS INTERNATIONAL.

Letter by Frank Haigh, Brockton Enterprise, p. 65: From letter by Frank Haigh to *The Enterprise*. Copyright © 1995 by Frank N. Haigh. Reproduced by permission of the author.

Excerpt from Whitney Balliett, *American Musicians II*, p. 148: From *American Musicians II: Seventy-Two Portraits in Jazz* by Whitney Balliett. Copyright © 1986, 1996 by Whitney Balliett. Published by Oxford University Press, Inc., New York. Reproduced by permission of the author's agent, The Jennifer Lyons Literary Agency, LLC.

Excerpt re Joe Williams (Quote by Cassandra Wilson); listed as from www.wicn.org but is in the same NYT article as the following excerpt, pp. 173–174: From "Jazz Singer of Soulful Tone and Timing is Dead at 80," by Jon Pareles from *The New York Times*, March 31, 1999. Copyright © 1999 by The New York Times Company. Reproduced by permission of PARS International.

Excerpt from Jon Pareles, "Jazz Singer of Soulful Tone and Timing is Dead at 80," p. 174: From "Jazz Singer of Soulful Tone and Timing is Dead at 80," by Jon Pareles from *The New York Times*, March 31, 1999. Copyright © 1999 by The New York Times Company. Reproduced by permission of PARS International.

Three excerpts from Sammy Cahn, *I Should Care*, pp. 212–216: From *I Should Care: The Sammy Cahn Story* by Sammy Cahn. Copyright © 1974 by Sammy Cahn. Published by Arbor House Publishing Co. Reproduced by permission of Sammy Cahn Music Company.

Excerpts from Gerald Nachman, *Seriously Funny*, p. 253: From *Seriously Funny: The Rebel Comedians of the 1950s and 1960s* by Gerald Nachman. Copyright © 2004 by Gerald Nachman. Reproduced with permission of the author.

Excerpts from Paul Krassner, "Review of Mort Sahl Tribute," pp. 254–255: From "Review of Mort Sahl Tribute" by Paul Krassner from *Huffington Post,* July 6, 2007. Copyright © by Mort Sahl. Reproduced by permission of Mort Sahl.

Quotations from WGBH interviews conducted by Ron Della Chiesa from 1972–2009 on WGBH Radio (pp. 89–273). Copyright © 1972–2009 WGBH Educational Foundation. The WGBH Interviews: Luciano Pavarotti, James McCracken, Ben Heppner, Marcello Giordani, Jose Van Dam, Jerome Hines, Eleanor

Steber, Aprile Millo, Dizzy Gillespie, Benny Carter, Illinois Jacquet, Stan Getz, Buddy Rich, Joe Venuti, Lionel Hampton, Ruby Braff, George Shearing, Dick Johnson, Joe Williams, Tony Bennett, Bobby Short, John Pizzarelli, Mel Tormé, Nancy Wilson, Rosemary Clooney, Sammy Cahn, Arturo Toscanini, Gunther Schuller, Harry Dickson, David Raskin, André Previn, Jean Shepherd, Mort Sahl, and Carlo Bergonzi.